COSPAR COLLOQUIA SERIES

VOLUME 10

MICROSATELLITES AS RESEARCH TOOLS

PERGAMON

Sponsored by

Committee on Space Research
International Academy of Astronautics
Swedish National Space Board
Academia Sinica
Ministry of Education
National Science Council
Aeronautical and Astronautical Society of ROC
Center for Aerospace Science and Technology, ITRI
Tainan Hydraulics Laboratory
National Central University
National Cheng Kung University

Organized by

COSPAR Committee in Taipei
Institute of Aeronautics and Astronautics
National Cheng Kung University, Tainan, Taiwan

Scientific Organizing Committee

S.D. Choi (KAIST, Taejon)
G. Hall (BNSC, UK)
J.C. Hong (NCU, Chung-Li)
F.B. Hsiao (NCKU, Tainan)
W.H. Ip (MPAe, Lindau) (Vice Chair)
D.H. Lee (Kyunghee, Seoul)
R. Lundin (SISP, Kiruna) (Chair)
S. McKenna-Lawlor (St. Patrick, Ireland)
A. Nakajima (NAL, Tokyo)
H. Saito (ISAS, Kanagawa)
R. Twiggs (Stanford Univ., Palo Alto)

Local Organizing Committee

A.J. Chen (NCU, Chung-Li)
F.B. Hsiao (NCKU, Tainan) (Chair)
L.H. Lee (NCKU, Tainan) (Vice Chair)
C.I. Lin (NCKU, Tainan)
C.T. Liu (NTU, Taipei)
J.J. Miau (NCKU, Tainan)
Q.C. Sung (NCKU, Tainan)
F.C. Tang (NCKU, Tainan)
C.L. Tseng (NCKU, Tainan)
K.C. Yen (NSYSU, Kaohsiung)

MICROSATELLITES AS RESEARCH TOOLS

Proceedings of COSPAR Colloquium on Microsatellites as Research Tools held in Tainan, Taiwan, 14-17 December 1997

Edited by

Fei-Bin Hsiao

Institute of Aeronautics and Astronautics,
National Cheng Kung University,
Tainan, Taiwan

PERGAMON
An imprint of Elsevier Science

ELSEVIER SCIENCE B.V.
Sara Burgerhartstraat 25
P.O. Box 211, 1000 AE Amsterdam, The Netherlands

First edition 1999

Library of Congress Cataloging in Publication Data
A catalog record from the Library of Congress has been applied for.

ISBN: 0 444 50196 7

∞The paper used in this publication meets the requirements of ANSI/NISO Z39.48-1992 (Permanence of Paper).
Printed in The Netherlands.

CONTENTS

Session IV: Lesson Learned (II)

Session VI: Innovations (II)

Session VII: Scientific Applications (I)

Session VIII: Scientific Applications (II)

Session IX: Scientific Applications (III)

Poster Session II: General

Session X: Education

PREFACE

The COSPAR Colloquium on Scientific Microsatellites was held between December 14-17 1997 in the Institute of Aeronautics and Astronautics (IAA), National Cheng Kung University (NCKU), Tainan, Taiwan, gathering about 150 participants from 18 countries in 5 continents. In order to reflect the increasing importance and interest of the microsatellites in high technology and scientific applications in space, the Colloquium on Microsatellites as Research Tools was organized to promote its usage and technology development and to foster the international cooperation, especially in the area of Asia pacific region. The Colloquium started in the first day's morning with the Welcome Address provided by Professor Gerhard Haerendel, President of COSPAR, but unexpectedly read by Dr. Wing H. Ip to express his wishes and regards to the success of this Colloquium. The Opening Session was then followed by three distinguished guesses speaking. It was truly honored to have the participation of Dr. Jin Wu, Minister of Education and former President of NCKU who clearly addressed the current and prospective of space education and research activities in Taiwan. Prof. Chao-Han Liu, President of National Central University and also the space mentor and navigator of space research in Taiwan, then gave a very up-to-dated presentation about the space development in Taiwan. The final Opening Session was concluded by Professor Cheng-I Weng, President of NCKU and host of the Colloquium, showing his ambitious willingness of leadership in space research in NCKU in addition to his warm welcome to all honored guests and participants from abroad and domestic. The Keynote Presentation was given by the honorable guest Dr. Jin-Fu Chang, Vice Chairman of National Science Council, to speak on "Striving toward a technologically-advanced nation - vision for the 21st Century's science and technology."

Both oral and poster presentations of the contributed papers were organized in the colloquium in two days. Five major themes arranged in serial were addressed: regional development, lessons learned, innovations, scientific applications, and education. A special session was organized as well by the Organizing Committee and supported by the National Space Program Office to present its development of the Taiwan's satellite program and the current status of ROCSAT-1, which is scheduled to be launched at the end of 1998 or beginning of 1999. Dr. David F.H. Chu, the Principal Investigator of ROCSAT-1, gave the speech and answered all questions as related to this satellite. The Round Table Discussion Session was arranged by Dr. Rickard Lundin and Dr. Wing Ip at the second day's afternoon to wrap up all comments and ideas in response to the paper and poster presentations. At least two conclusions were remarked: Microsatellite in general is a very good means for doing space research and technology development and a suitable vehicle to promote international collaborations. In addition, COSPAR Headquarters should consider in the near future to host the second microsatellite colloquium such as the current one in realizing and enhancing the microsatellite and related activities especially for the developing countries in the Asia-pacific region.

Apart from the indoor paper presentations, two technical field trips outdoors were arranged after all paper sessions, which included the visits to the Aerospace Science and Technology Research Center of NCKU in Tainan County and the Hsinchu Science-based Park in Hsinchu City. The Hsinchu's trip by luxurious bus contained the visits to Science Park Administration, National Radiation Synchrotron Laboratory, Microelectronic Technology, Inc., and National Space Program Office, where to witness the real hardware of the ROCSAT-1.

It should be grateful for the technical support by Dr. R. Lundin and Dr. Wing Ip who chaired the Scientific Organizing Committee, and the local support by Prof. Fei-Bin Hsiao and Lou C. Lee who chaired the Local Organizing Committee. The excellent logistic support by the staffs and students in the Institute of Aeronautics and Astronautics of NCKU were highly acknowledged. Special thanks extend to such agencies who provided the financial supports as COSPAR, Academia Sinica, Ministry of Education, National Science Council, Aeronautical and Astronautical Society (Taiwan), Center for Aerospace Science and Technology, Tainan Hydraulics Laboratory. National Cheng Kung University serving as the main sponsor university sincerely thanks to the COSPAR Headquarters for providing the opportunity with honor to host this international event.

Finally, I would like to express my sincere thanks to Miss Linda Lin, who help type and edit all submitted manuscripts, for her excellent performance and extreme patience, and to Elsevier Science Co. for publishing this volume as the COSPAR Colloquia Series.

Fei-Bin Hsiao, Ph. D.

Editor

KEYNOTE ADDRESS

Gerhard Haerendel, COSPAR President

Dear Distinguished Delegates, Ladies and Gentlemen,

I against my original intention I had to cancel my participation in this symposium on short notice, because my presence was urgently required in Germany, in order to be present during a critical phase of operations of our EQUATOR-S satellite, which was launched on December 2. Therefore, this message has to be read to you, I apologize and hope for your understanding. The topic of this Colloquium, Microsatellites as Research Tools, is particularly close to my heart since I started promoting research using small satellites about a decade ago. I am very pleased to see that the concept is becoming accepted in the scientific community, that many meetings now address the topic from one angle to another, and that results are being obtained from research out with such tools.

However, let me also express a warning that we not be carried away by the enthusiasm arising from opportunities created by small and mini-satellites. In my mind these tools have a role complementary to that of big observatories and space probes, adding flexibility, short lead-times, and alternate means of procurement and of sharing responsibility. For many tasks in astronomy, solar system research, and Earth observations big missions will be indispensable and, after careful analysis, may prove to be more cost effective than a number of small missions accomplishing the same work.

As we approach the end of the year, it seems fitting to take a moment to acknowledge certain recent successes that will certainly have to a greater or lesser degree an impact on our work and in some cases perhaps even on your discussions this week. I cannot but admire the elegance and technical prowess of the Pathfinder mission to Mars, the success of which this past summer so justifiably enthralled tens or perhaps even hundreds the potential of robotic exploration in space, especially encouraging in this era of limited funds.

In the past year we also witnessed successful launches of important space projects like the NASA/ESA Cassini/Huygens mission to Saturn and its satellite Titan and the deployment by ISAS of the Haruka space radio-telescope, the space segment of the MLBI Observatory. Let me also mention the spectacular success of the Italian-Dutch Beppo-SAX satellite for X-ray astronomy. Its detection of the X-ray afterglow of gamma-ray bursts, combined with optical and radio data, seems to have positively proved, after three decades of debate, that these enigmatic bursts originate at cosmological distances.

I should like also to warmly salute the obvious scientific and technological maturity of our Spanish colleagues who successfully launches their first satellite (Minisat-01) and express my satisfaction with the positive decision by ESA to rebull the Cluster mission.

To return to matters of more immediate concern, I have no doubt that this Colloquium will be very fruitful given the dynamism that currently characterizes the Taiwanese space community. We have all been impressed by the great strides made in recent years by our colleagues here. More generally, COSPAR looks with great expectation and hope to the invigorating drive of many of the scientists whose countries are represented at this meeting to master the techniques and technologies enabling man to exploit successfully the potential of mini and microsatellites.

I would be most happy if your discussions these next few days also resulted in reinforced international cooperative efforts, for I am convinced that it is only through greater cooperation among space researchers on a wide scale that we can overcome limitations imposed by ever tighter budgets. COSPAR shall, of course, continue its efforts to promote international collaboration, and as the President of the premier organization working to this end specifically on behalf of space researchers, I am pleased to inform you that as a participant in this meeting you will be conferred the status of Associate in our committee.

As you may know, COSPAR's next big event after this Colloquium will be its 32nd Scientific Assembly which will be held in July 1998 in Nagoya. Those of you who have seen the Call for Papers know that this Assembly will be a memorable event because of the strength of the science that shall be presented. The Nagoya Assembly will also be a special event because it marks the 40th anniversary of our committee's service to the international space research community. I hope many of you will attend the Assembly next summer in order to help us celebrate this important anniversary and because I am sure that you will all find in the program scientific events that suit your individual interests.

I would particularly like to draw your attention to the fact that COSTAR'98 will provide an opportunity to increase interaction between the scientific and engineering communities through a joint event organized with the International Astronautical Federation (IAF) entitled "Science and Engineering Aspects of Solar System Exploration". As mutual understanding and smooth cooperation between the two communities is a springboard for success in all our fields. I that exploration and utilization of space is dependent on collaboration between researchers and engineers and on the close coordination of their efforts. Even the greatest degree of interaction between the research and engineering communities cannot be deemed excessive. Fortunately, this situation is well-understood and appreciated by the majority of those concerned which includes many of us here today.

On this note, distinguished delegates, ladies and gentlemen, I would like to wish you all a most agreeable and worthwhile Colloquium.

ROCSAT PROGRAM AND SOME RELATED RESEARCH TOPICS

Jia-Ming Shyu

Director, National Space Program Office, 8th Floor, No. 9 Prosperity First Road, Hsin-Chu Science-Based Industrial Park, Hsin-Chu, Taiwan 30077, ROC

ABSTRACT

While the ROCSAT-1 project proceeds to the integration and test stage, we expect to launch the first satellite of the Republic of China with three payload instruments on board in December 1998. The following ROCSAT series are now in planning. Among them a supporting program for "microsatellite as research tool" with international cooperation is also under consideration. Besides introducing ROCSAT-1's recent development, this presentation will suggest some areas of interest for further discussion. The results of the discussion may be included into the future NSPO's mission oriented research topics of the following ROCSAT projects.

PREFACE

The ROCSAT program started from October 1990. The first runner of this program, the ROCSAT-1 project, granted it's spacecraft contract on June 1994. Subsequently, the ROCSAT Ground Segment (RGS), the Ocean Color Imager (OCI) payload, the Ionosphere Plasma and Electrodynamic Instrument (IPEI) payload, the Experimental Communication Payload (ECP) and the integration and test facilities were contracted to foreign and domestic companies. Under the close cooperation between NSPO and contractors, this project is going on smoothly and has finished integration of spacecraft and payloads. Now the satellite is under testing. We expect to launch ROCSAT-1 some time between December 1998 and March 1999 with the launch vehicle LMLV-1 from Cape Canaveral, Florida, USA.

Since all three payloads of the ROCSAT-1 have been delivered and integrated to satellite, and science teams of OCI, IPEI and ECP have been established and are preparing for the experiments quite a few time, it is now impossible to add on any extra payload to the satellite. Any researcher, who has interests about the experiment using these payloads, may contact directly with following primary investigators (PI) of the science teams:

OCI --- Prof. Sien-Wen Li, National Taiwan Ocean University (NTOU)
e-mail: lihw@sun4.oce.tours.ntou.edu.tw
IPEI --- Prof. Shin-Yi Su, National Central University (NCU)
e-mail: 2700146@ncu865.ncu.edu.tw
ECP --- Prof. Yen-Hsyang Chu, National Central University (NCU)
e-mail: 1272665@ncu865.ncu.edu.tw
Prof. Szu-Lin Su, National Cheng Kung University (NCKU)

e-mail: ssl@eembox.ncku.edu.tw
Prof. Yun-Chang Chen, National Tsing Hua University (NTHU)
e-mail: jjong@alumni.nctu.edu.tw

After the launch of ROCSAT-1, the downlinked data of ROCSAT-1 will be open to science community worldwide through Science Data Distribution Centers (SDDC) located in NTOU (for OCI), NCU (for IPEI) and NSPO (for ECP).

The mission life of ROCSAT-1 is two years and design life is four years. The extended mission in the third and fourth years after launch has not been decided yet. We will consider any meaningful suggestions when the time is near.

THE TREND OF SMALL/MICRO SATELLITE AND CONSTELLATION

As we all well know, the recent satellite development trend is "smaller, faster and cheaper" as advocated by NASA. Besides, we can add "constellation" as one of the new trends. Smaller is possible through functional simplification, faster is conceivable through reduction of documents and tests, and cheaper can be achieved through deduction of documents and tests, and cheaper can be achieved through deduction of redundant parts and modulation. Constellations to upgrade the function from static to dynamic or from non-real-time to real-time through mission repeat and it will bring far more benefits than of a single satellite.

The selection of subsequent ROCSAT missions is concentrated on small satellites and based on the requirements survey, supply sources investigation, and the trends of space technology development. It shows that small satellites and microsatellite constellation are suitable for domestic demand on remote sensing, meteorology, and science research.

RESEARCH TOPICAS OF FOLLOWING ROCSAT SERIES

While ROCSAT-2 is defined as a remote sensing satellite and ROCSAT-3 is planning as a microsatellite constellation for meteorology, some traditional major research topics for the main missions are under planning. Besides, there are some research topics needed to discuss.

Detection of Groundwater Over Utilization

The southwest coast of Taiwan has suffered from flood in recent years. The flood was induced from landsinking, which according to the report [1], cost us in average 11 Billion NT dollars every year. The over suction of groundwater by aquaculture farmer is the major cause of landsinking. The Environment Protection Bureau warned this situation, but was not capable to identified anyone who excessively pumped the groundwater out to feed his fish pond.

If we can use suitable instrument, such as infrared sensor from LEO satellite to get the water temperature distribution map with isothermal line, then we can see the cold/warm water sources. These cold/warm water sources are identical to the place where the groundwater, with different temperature as nearby areas, pours in. In addition, the density of isothermal lines will show the volumetric flow of the groundwater. It is therefore possible to identify the offender of the environment protection.

Rain may disturb the water temperature distribution. However, on raining days the fishfarmer will not use groundwater to dilute their fish pond. So, there is no need to remote sense water temperature in raining days.

Microclimate

The resolution of the cloud maps provided by the GEO meteorological satellite is about 3 km nowadays. This resolution is obviously too rough for small, mountainous and densely populated area like Taiwan. We need more accurate microclimate information such as cloud maps with resolution about 200 m and delivered every hour. Through integration of GEO cloud maps with lower resolutions and LEO cloud maps with higher resolution we expect to have near continuos pictures of cloud motions. From these pictures and other weather informations we may differentiate and derive local wind speed, vapor, air pressure, air temperature, etc. more precisely.

Qualifying of Space Use Components

To secure the reliability of the satellite, it is important to test the components of the spacecraft as well as payload instruments before launch. We use ground test facilities to simulate space environment for verifying the survivability of the satellite. Nevertheless, it is still needed to prove its reliability in the real space environment.

Since constellation has many small satellite with the same conditions, we have more test opportunities to test components or instruments. The construction and circuitry of small satellite is mostly rather simple, and the down data of the satellite is far less than the big satellite. Therefore, it is easier to add on some ground-qualified components or instruments for space proving.

Space Gravity Field of Earth

The gravity of the earth is not homogeneous around the earth at the same altitude. And the changes of the gravity at the same moment in different places are not thoughtfully investigated so far. To better understand our planet earth it is desirable to measure the gravity of the earth using constellations of equally distributed satellites. The high accurate position determination with the GPS receiver on board of the micro-satellite constellation may precisely determine the gravity field.

Sprites

On the terrestrial upper atmosphere there are many newly discovered phenomena associated with thunderstorms, such as red sprites, blue jets, elves, lightening-induced electron precipitation, pairs of VLF pulses and gamma-ray flashes of atmospheric origin. Observations and theories of these lightning associated physics have been the subject of special topics sessions on several international scientific conferences. Direct observations of sprites to date, mainly from ground and airborne, have been limited to their visible or near-infrared (400-900 nm) emissions. However, numerous micro-physical processes deriving from these phenomena are expected on general theoretical grounds to released energy over a broad band of wavelengths extending from the ultraviolet into the infrared. It is difficult to observe the ultraviolet emission from ground and a systematic observation from satellite becomes important. Use of the sprites imaging science payload on the ROCSAT-2 program will provide a first hand systematic global coverage observation data of these upper atmospheric lighting related phenomena and will have certain impacts and contributions to the international science community.

CONCLUSION

Although the small and microsatellite as space research tool is well known since the beginning of the space age in 1950s, its application is limited to amateur area since then. The function and performance of microsatellite improved quite a lot in recent years in conjunction with the development of ASIC, VLSI based microprocessor, lightweight antenna, lightweight spacecraft structure, and MMIC, etc. The miniaturization of satellite and payload instrument enables microsatellite get breakthrough in many science and practical application fields.

The lower cost of microsatellite allows more satellites to perform designated missions. The smaller size and lighter weight of microsatellite makes the launch of microsatellite/constellation either dedicate or piggyback possible. Moreover, since the concept of backup and reliability of microsatellite is different from big satellite's, cost reduction from elimination of redundant components and direct utilization of commercial/military components make the micro-satellite more competitive. The future of microsatellite as research tools is therefore bright. The usage of microsatellite constellation in ROCSAT-3 is hence feasible.

AN INTRODUCTION TO THE KITSAT PROGRAM AND THE ACTIVITIES AT THE SATREC IN KOREA

Soon-Dal Choi, Byung Jin Kim, and Ee-Eul Kim

SaTReC, KAIST, 373-1 Kusung-Dong, Yusung-Gu, Taejon 305-701, Korea

ABSTRACT

The Satellite Technology Research Center (SaTReC) in the Korea Advanced Institute of Science and Technology (KAIST) is an institute for education and research in the field of satellite engineering, space science, and remote sensing. SaTReC was selected in 1990 as an Engineering Research Center' (ERC) by the Korea Science and Engineering Foundation in order to make it the center of excellence in satellite engineering. It has successfully produced and operated the first and second Korean microsatellites, KITSAT-1 and 2. A brief description of the history of SaTReC is presented. KITSAT-1 and 2 that produced some valuable scientific and satellite engineering results are described. The current mission in the KITSAT series, KITSAT-3 which is an engineering test satellite, is introduced. The activities of the SaTReC in remote sensing and its participation in international collaboration are also summarized. The future plan of the development of the new generation satellite, KITSAT-4, is finally envisioned.

INTRODUCTION

The Satellite Technology Research Center (SaTReC) in the Korea Advanced Institute of Science and Technology (KAIST) is an institute dedicated to the education and research in satellite engineering, space science, and remote sensing. In 1990, the SaTReC was selected as an Engineering Research Center (ERC) by the Korea Science and Engineering Foundation (KOSEF), which marked the beginning of the space activities in Korea. Since then, the SaTReC has acquired the capability to manufacture micro and small satellites through the international cooperation and the interdisciplinary collaboration in the KAIST. By sending a number of KAIST graduates abroad since 1989, the SaTReC has obtained the trained personnel in various fields related to the space.

The SaTReC has successfully developed and operated two micro satellites, KITSAT- I and 2, the first two satellites of Korea. Both of them are equipped with several payloads to perform experiments in satellite engineering, space science, and earth observation. In addition to the scientific results, they produced the valuable information on the satellite engineering and gave the indispensable experience to engineers. The SaTReC is presently developing KITSAT-3, a small engineering test satellite, based upon the experience and knowledge acquired from the two previous missions. It has several improved payloads such as the Space Environment Scientific Experiment (SENSE) and the multi-spectral optical camera. The SaTReC has also established the KAIST Remote Sensing Center (KRSC) in 1995 to perform the research in remote sensing and to support the acquisition, preprocessing,, and distribution of the remote sensing data in Korea.

The SaTReC has actively participated in various international activities and is looking forward to the international cooperation in its future missions. The SaTReC will play a key role in promoting Korea as one of active countries in space science, satellite engineering, and remote sensing.

OVERVIEW OF KITSAT-1 & 2 MISSIONS

The KITSAT-1 was developed through the international collaboration with the University of Surrey in the UK. Ten KAIST graduates from the SaTReC were sent to the University of Surrey as postgraduate students and were deeply involved in the development of the UOSAT-F. With the invaluable experience they gained from this program, they played key roles in the development of KITSAT-1. Returning back to Korea, they independently developed the next micro-satellite, KITSAT-2. Figure 1 shows the configuration of KITSAT-2.

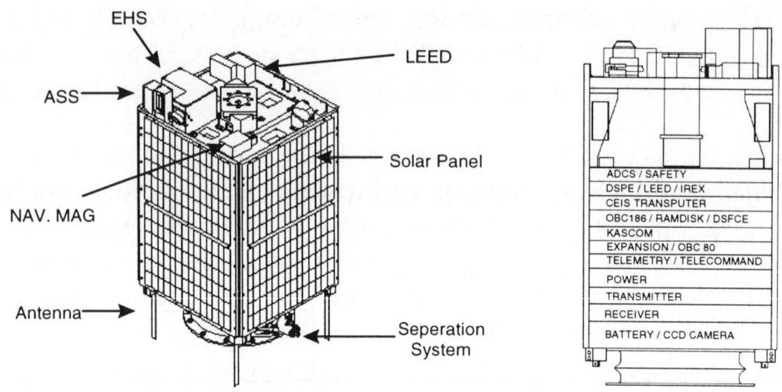

Figure 1. Configuration of KITSAT-2.

These two satellites, which are very similar in structure but different in their orbits, provide a unique opportunity to study the effects of the radiation environment characterized by their orbits. Both KITSAT-1 and 2 carry simple space science experimental modules to measure the radiation particles. Each of them also has two Earth observation CCD cameras, one for wide angle and the other for narrow angle observations. In cooperation with a Korean industry, the Samsung Electronics, a color CCD camera was developed and used as the wide angle camera sensor in KITSAT-2. The general features of two satellites are shown in Table 1.

Table 1. General features of KITSAT-1 and 2.

	KITSAT-1	KITSAT-2
Altitude	1330 Km	800 Km
Inclination	660	98.7 0
Size(mm)	352 × 356 × 670	352 × 356 × 670
Weight	48.6 Kg	47.5 Kg
Power	3OW max	3OW max
Attitude	Gravity Gradient Boom	Gravity Gradient Boom
Control	Magnetorquer : Pointing 5°	Magnetorquer : Pointing 5°

OBC	OBC 1 86, OBC80	OBC186, KASCOM
Payloads	- Digital Store and Forward Comm. Experiment (DSFCE) - CCD Earth Imaging System (CEIS) - DSP Signal Processing Experiment (DSPE) - Cosmic Ray Experiment (CRE)	- Digital Store and Forward Comm. Experiment (DSFCE) - CCD Earth Imaging, System (CEIS) - DSP Signal Processing Experiment (DSPE) -KAIST Sat. Computer (KASCOM) -Low Energy Electron Detector (LEED) - IR Sensor Experiment (IREX)

The Cosmic Ray Experiment (CRE), one of the main payloads of KITSAT-1, has the primary objective to study the space radiation environment encountered by earth-orbiting satellites in the Van Allen belt[1]. KITSAT-1 also provides the facility to monitor SEUs occurring in the program memory and 12 MByte RAM disk memory of the primary onboard mission control computer (MCC). On KITSAT-2, there is the secondary MCC, named KASCOM, to replace OBC186 in case of its failure as redundancy. KASCOM is a 32-bit INTEL 80960 based MCC, another objective of which is to test its capability as the primary MCC for future small satellites.

The Digital Store and Forward Communication Experiment (DSFCE) provides an open access store-and-forward communications for the amateur HAM throughout the world. KISAT-1 and 2 have identical communication systems which use standard PACSAT protocol with the data transmission rate of 9600bps and have onboard 12Mbytes of CMOS SRAM for the communication experiment.

In August 1992 , the first Korean satellite, KITSAT-1 was successfully launched from Kourou, the French Guiana, by the Ariane 42P rocket as an auxiliary payload to TOPEX/POSEIDON. Following the successful operation of KITSAT-1, the second Korean satellite, KITSAT-2, was launched from Kourou, the French Guiana, as an auxiliary payload to SPOT-3 in August 1993. Since then, both satellites have produced valuable data for the space science and satellite engineering and images of the Earth's surface. Figure 2 shows the global map of SEUs in the program memory of KITSAT-1 on the left-hand side and that of KITSAT-2. They exhibit the different distribution characterized by their orbits and indicate the South Atlantic Anomaly(SAA). Figure 3 is an image of the Myo-Hyang mountains obtained with the narrow angle camera on KISAT-2. It clearly shows terrestrial features such as coast lines, rivers, fields, and mountains[2].

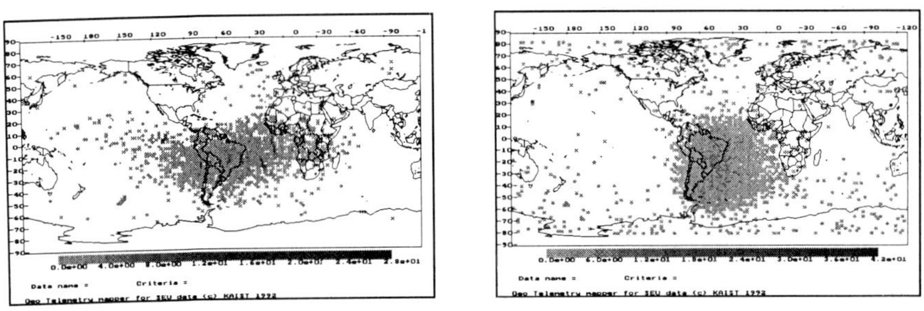

Figure 2. Global map of SEUs in the program memory of KITSAT- 1 (left) and KITSAT-2 (right)

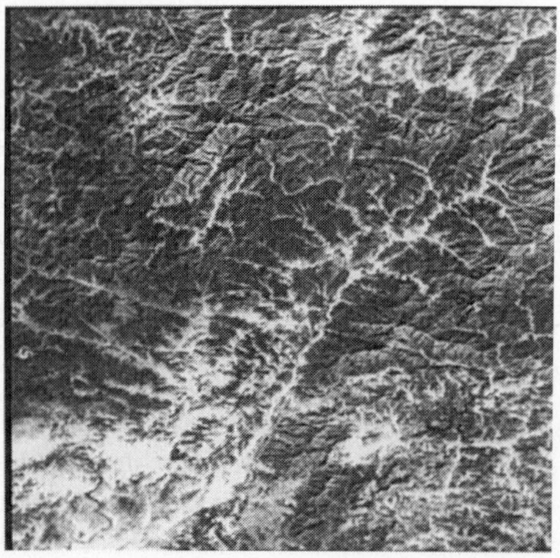

Figure 3. Image of the Myo-Hyang Mountains obtained with KISAT-2 narrow angle camera.

KITSAT-3 MISSION

At the present, the SaTReC is developing the next micro satellite in the KITSAT series, KITSAT-3. It is an engineering- test satellite scheduled to be launched by the Indian PLSV rocket in mid 98. Its primary objective is to provide an opportunity to elaborate various technologies for high performance micro satellites and to qualify them in the space environment. In addition, it has payloads for the remote sensing and space science. The satellite system was designed to satisfy the requirements of low cost, low power and small size. The successful operation of KITSAT-3 will verify new technologies adopted to the design and will lead to the development of the next generation satellites in the KITSAT series. The objectives of KITSAT-3 are summarized as follows.

Primary, objectives

- Design of a unique satellite bus system
- Development of the high accuracy 3-axis attitude control system
- Realization of the high data rate transmission system
- Implementation of the modular command and data handling network
- Design and evaluation of the solar panel deployment mechanism

Secondary objectives

- Design of a pushbroom remote sensing camera: Multi-spectral Earth Imaging System (MEIS)
- Collection of the scientific data on the space environment

The system specifications of KITSAT-3 are as follows.

- Launch Date: mid 1998
- Launcher: Indian PSLV
- Orbit : 720 km, Sun-synchronous
- Dimension: 495 × 604 × 852 mm
- Weight < 110 Kg

- Power: maximum 150 W
- Attitude Control: 3-axis stabilized (0.5° Pointing Accuracy)
- Payloads: Remote Sensing Payload (MEIS)
  - 13.8m GSD, Three-channel (Red, Green, Near IR)
  - Space Science Payloads (SENSE)
    - High Energy Particle Telescope (HEPT)
    - Radiation Effects on Micro-Electronics (REME)
    - Scientific Magnetometer (SMAG)
    - Electron Temperature Probe (ETP)

Figure 4 shows the exploded view of KITSAT-3. The Multi-spectral Earth Imaging System (MEIS) for remote sensing, the Space Environment Scientific Experiment (SENSE) for space science, and other major subsystems are indicated in this figure.

The optical camera in MEIS is a pushbroom camera with three linear CCDS. It produces multispectral images of various terrestrial features in three observation wavebands that are similar to those of SPOT. The incoming light is split by three prisms and one CCD is assigned to each waveband. The optical layout of the camera and its operational concept are depicted in Figure 5. The images acquired from the camera are stored in the solid-state recorder that provides the memory capacity of 10 Gbits in total. The solid-state recorder includes flash and SRAM memory blocks developed by the Samsung Electronics. The SRAM memory blocks are specially manufactured to reduce the physical size by using the three-dimensional packaging technology.

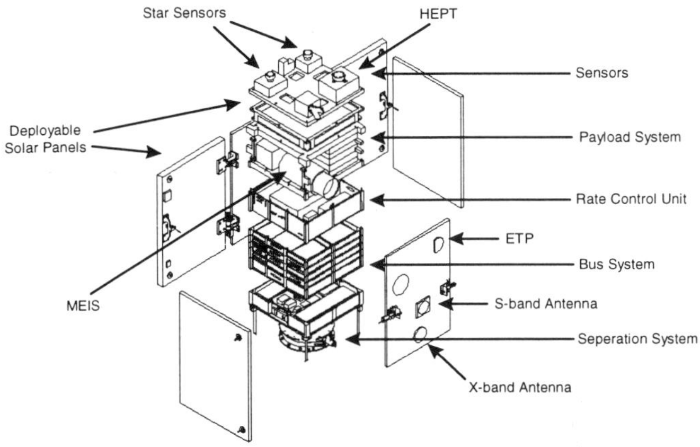

Figure 4. Exploded view of KITSAT-3.

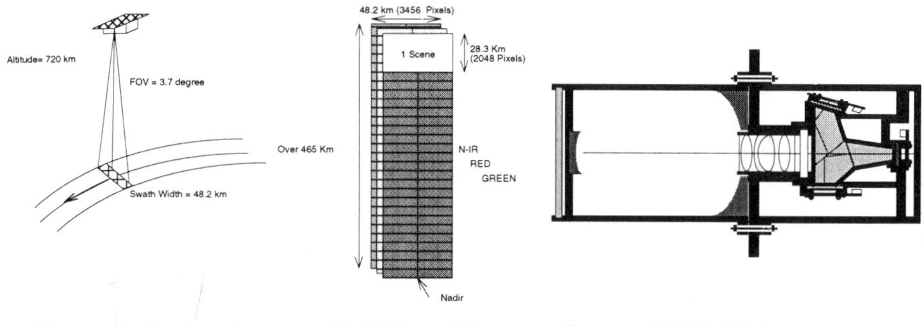

Figure 5. Optical layout of MEIS and its operation on KITSAT-3.

SENSE consists of four smaller payload modules as described above. They are all designed based on the past experience acquired from the space science payloads of KITSAT-1 and 2. The scientific objectives of SENSE are as follows[3].

- Understanding of the physical mechanisms in the radiation belt
- Measurement of the pitch angle distribution
- Study of the third radiation belt
- Measurement of the electron temperature in the ionosphere
- Study of the anomalous electron heating in SAA
- Measurement of the geomagnetic field disturbances
- Research of the radiation effects on the micro-electronics

REMOTE SENSING ACTIVITIES

The remote sensing plays a crucial role in the resource management and environmental monitoring of the Earth. The KAIST Remote Sensing- Center (KRSC) was established in 1995 as a branch of the SaTReC to perform the dedicated research and development in the field of satellite remote sensing and to support the acquisition, preprocessing and distribution of satellite remote sensing data. It aims to acquire and preprocess the data in near real-time through its facility and provide the efficient user data network services.

The image data from remote sensing satellites are acquired through a 13-m dish antenna at the KRSC and a 10-m dish at the Radio Research Laboratory. The raw image data acquired from MOS-1b, JERS-1, and SPOT-1/2 satellites are preprocessed to generate the image products that are ready for analysis by the appropriate users. The KRSC preprocessing system produces the image products ranging from transcription products to precision geocoded products. The processed data are distributed on media such as CCT, Exabyte, and CD-ROM. The on-line browsing is also supported through the WWW. Figure 6 shows the block diagram of the KRSC remote sensing system. Two sample images from JERS-1 and SPOT-1 are also shown in Figure 7.

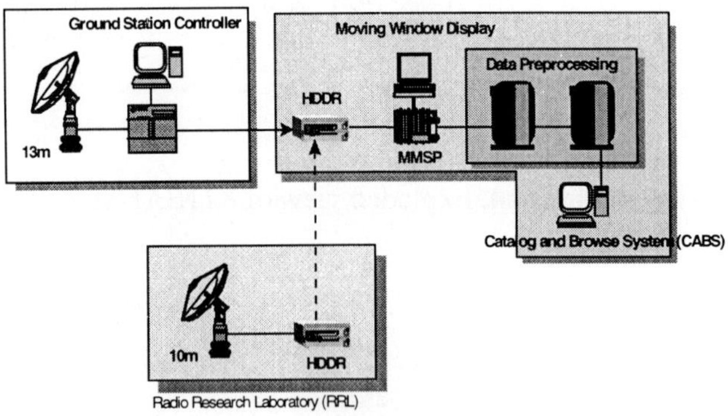

Figure 6. Block diagram of the KRSC remote sensing system.

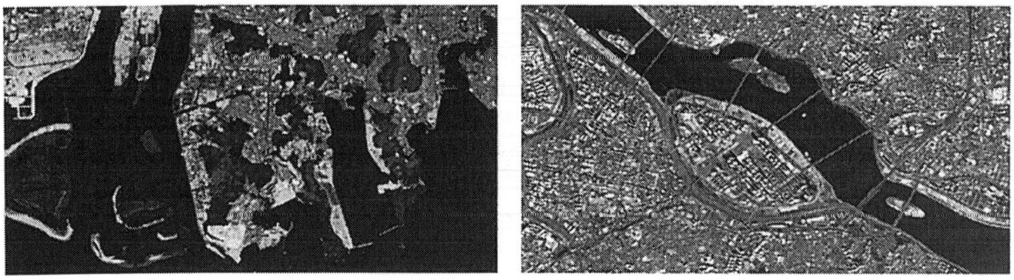

Figure 7. Two sample images from KRSC remote sensing system; a part of Pusan from JERS-1 (left) and a part of Seoul from SPOT-1 (right).

INTERNATIONAL COOPERATION

Many recent developments from the international joint projects for space technology are providing various opportunities for us to leap forward into space. In this context, recognizing the need for the globalization of space technology, the SaTReC has been sending a number of KAIST graduates to the USA, Japan, the UK, and France since 1989 with the mission to learn the cutting-edge space technology. Some of them have already returned and are playing key roles in the SaTReC.

Along with the manpower training program, an overseas joint research program was set up at the University College London in the UK to carry out the international cooperation projects in the field of Synthetic Aperture Radar and optical system technologies. In addition, the SaTReC is actively participating, in the international cooperative space activities such as APRSAF, AP-MCSTA, ESCAP, OOSA, and COPUOS and has an active role in building a mutually beneficial international community.

FUTURE PLAN - KITSAT-4 MISSION

The SaTReC's next generation satellite in the KITSAT series, KITSAT-4 is in the stage of proposal. It is expected to have increased size and weight compared with KITSAT-3 and to have new features for the satellite engineering. Four space science payloads have already been proposed for KITSAT-4; far ultra-violet spectrometer, low energy plasma detector, Langmuir probe, and scientific magnetometer. Other payloads Such as the W-CDMA mobile communication experiment, the data collection system, and the remote sensing, optical camera with the 5-m IFOV are also under consideration. The details of its specifications and payloads will be defined and selected during the mission analysis period in the near future. In addition, the international cooperation with the UK, the USA, Singapore, and Australia are under discussion. Figure 8 depicts the proposed configuration of KITSAT-4 at the present.

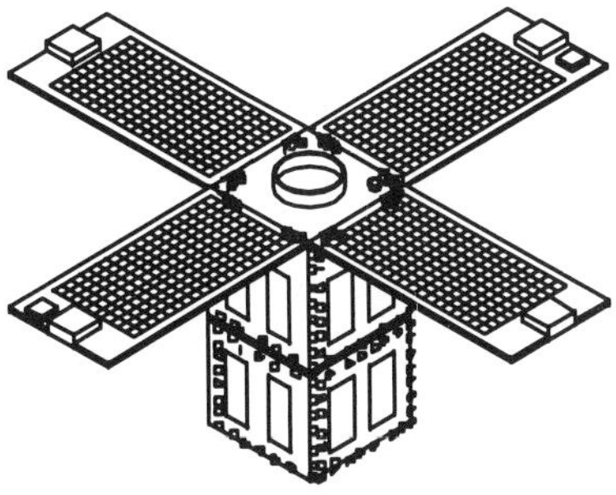

Figure 8. Proposed configuration of KITSAT-4

CONCLUSION

With the successful development and operation of KITSAT-1 and 2 micro satellite on the basis of the international collaboration, the SaTReC has established the solid foundation for the space activities in Korea. In addition to the valuable results from their space science and Earth observation payloads, they produced the experience personnel in space science, satellite engineering, and remote sensing. With the aim to develop cost-effective and efficient micro and small satellites, the SaTReC is developing an engineering test satellite, KITSAT-3. The success of this mission will verify the various new space technologies adopted and lead to the development of new generation satellites such as KITSAT-4. The KRSC in the SaTReC performs the research and development in the satellite remote sensing and supports the acquisition, preprocessing and distribution of the satellite remote sensing data in Korea. The international cooperation has been the fundamental philosophy of the SaTReC throughout its history. The SaTReC will pursue its goals and perform tasks to promote Korea as one of the advanced countries in the space and eventually to improve the quality of life of the mankind.

REFERENCES

C.I. Underwood, D.J. Brock, P.S. Williams, S. Kim, R. Dialo, P.R. Santo, M.C. Brito, C.S. Dyer, and A.J. Sims, Radiation Environments with Cosmic Ray Experiments On-Board the KITSAT-1 and PoSAT-1 Micro-Satellites, IEEE Trans. Nuc. Sci., Vol.41, No.6, pp.1353, 1994.

I.P. Lee, D.K. Sung, and S.D. Choi, Experimental Multimission Microsatellites -KITSAT Series, Proc. of the 7th AIAAFUSU Conference on Small Satellites, Utah State Univ., 1993

Y.-H. Shin, D.-H. Lee, S.-M. Park, J.-J. Lee, and K.-W. Min, Space ENvironment Scientific Experiment (SENSE) on KITSAT-3, Proc. of the 9th AIAA/USU Conference on Small Satellites, Utah State Univ., 1995

MICROSATELLITE'S R&D ACTIVITY IN JAPAN

Atsushi Nakajima

National Aerospace Laboratory, Space Technology Research Group, 7-44-1 Jindaiji-Higashi-machi, Chofu-shi, Tokyo 182, Japan

ABSTRACT

Recent years, the utilization of microsatellites under 100 kg has become of major interest. In the past, five piggyback microsatellites have already launched by H-I and H-II rockets in Japan. The outlines of these satellites are described in this paper. The launch opportunities will be increased after the operation of H-IIA launcher and opened for various countries. Since 1990, Small Payload Workshop (SPWS) has studied on small satellites, whose weight range under a few hundred kgs. The SPWS has also started the study of 50kg-class microsatellites by organizing research subgroup; Piggy-back Microsatellite Working Group (PMSWG) in 1995. The objectives of this working group are to discuss about the concrete missions of piggyback satellites of H-IIA launcher, proposed from the potential users belongs to national institutes, universities and private companies. According to the first questionnaires, 61 mission ideas are proposed. The PMSWG has discussed about these proposed ideas and some realizable missions were summarized and also made clear the requirements from user side.

INTRODUCTION

Recently, research and development of small and microsatellites are attracting interest of many researchers in various fields of science and engineering. From the first launch of the Japanese large launcher H-I, the opportunities of piggyback satellites are given when the excess payload margins are available. Five piggyback satellites were already launched by Japanese H-I and H-II rockets in the past. The first Japanese piggyback payloads are MABES (Magnetic Bearing Flywheel Experimental System) developed by National Aerospace Laboratory (NAL) and JAS-1 (Japan Amateur Satellite-one) developed by Japan Amateur Radio League, Inc. (JARL), which were launched by H-I rocket test flight #1 in 1986. The second opportunity was provided for DEBUT (Deployable Boom and Umbrella Test Satellite) developed by NAL and JAS-1b developed by JARL in 1990. The third one is for JAS-2, which was launched for the first time by the second generation large launcher H-II in 1996.

In 1990, the study group targeting on small satellites and their related technology, whose name is Small Payload Workshop (SPWS), was established. The objectives of the SPWS are to review the development of small and microsatellites and their launching system, to study possible missions, to exchange information, to pick up and study the issues to be solved and to contribute to the good and efficient development of the future plan of small and microsatellite utilization. More than 40 organizations are participated in the SPWS now. Three to four meetings and a symposium are held every year and these activities are documented. In 1995, the SPWS has started the study of microsatellites, the weight range under 100kg, by organizing research subgroup, PMSWG. This subgroup focused on the 50kg-class microsatellites which will be launched by H-IIA rocket as piggyback payloads in the near future. The launch opportunities as piggyback payloads will also be opened for various countries.

PIGGYBACK SATELLITES LAUNCHED IN JAPAN

Five piggyback satellites were launched in Japan, four of them were by H-I and one by H-II launchers. Two satellites are technology demonstration satellites developed by NAL and three are amateur radio satellites developed by JARL. On August 13, 1986, EGS (Experimental Geodetic Satellite) as main satellite was launched by H-I TF#1 with two piggyback payloads from Tanegashima space center in the polar orbit. After the separation of main satellite, one piggyback microsatellite, JAS-1, was separated from the 2nd stage of H-I rocket and the another payload, MABES, was remained at the top of the 2nd stage as planned and experimented during the mission life. On February 7, 1990, the sixth H-I rocket has launched two microsatellites, DEBUT and JAS-1b, with main satellite MOS-1b. Two satellites were separated in order of DEBUT and JAS-1b with time interval about 4 seconds after the separation of MOS-1b and the orbital change of the rocket with residual fuel. On August 17, 1996, the third Japan amateur radio satellite JAS-2 was launched as auxiliary payload of the Advanced Earth Observation Satellite (ADEOS) by H-II rocket. The next flight opportunity of 50kg microsatellite will be provided in 2000 as a secondary payload of ADEOS-II satellite and open for all over the world

Technology Demonstration Satellites Developed by NAL

NAL has been researched and developed on new technologies for the future space applications (A.Nakajima *et.al.*,1988). Some technologies originated by NAL were already practically used in the operational satellites. The on-orbit verification tests of newly developed hardware and software are very important step in the development phase. For the technology demonstration, on-orbit experiments using small and microsatellites were performed two times in the past, MABES and DEBUT projects.

MABES. Magnetic bearing experimental satellite, MABES (the nickname is JINDAI;site of NAL) was launched into a 1500km nearly circular orbit of 50 deg. inclination angle. The total weight is 294kg including structure to connect main satellite and 2nd stage of rocket. The desired mission life is only 3 days which is enough time to evaluate the fundamental characteristics of magnetic bearings under zero-g condition. For the minimum cost achievement, only the unchargeable primary Ni-Cd battery with 35 Ah was used as power source and refurbished TT&C subsystem, manufactured as PFM for the ETS-IV satellite, was used and functioned well for the telemetry and command transmission. The ground facilities were also reused and supported by NASDA (National Development Agency of Japan).

Since 1978, NAL has been developed magnetic bearings for space and industrial applications. The advantages offered by contactless magnetic bearings over more conventional bearings are; very high reliability through electronic redundancy, unlimited life time due to the absence of abrasion and wear, high vacuum operating compatibility, no stiction and low rotational drag, very low torque noise and transmissibility of vibrations, stiffness properties adjustable electronically, virtually gimballing capability without any additional mechanisms, no contamination problems due to lubricants. Several types of magnetic bearings, from one degree-of-freedom (DOF) active magnetic bearings to five DOF active type, have designed/manufactured and evaluated. 2-axis actively controlled magnetic bearings with permanent magnets and electromangets have the characteristics of simple arrangement and easy construction. The most preferable feature of this type is its virtually flat-shaped form, which enables achieving a high moment of inertia/mass ratio flywheel. 2-axis active magnetic bearings is a candidate of practical use as reaction and momentum wheels for attitude control actuator of spacecraft. Figure 1 is the photograph and Figure 2 shows the cross sectional view of the magnetic bearing flywheel for space experiment.

Figure 1. M.B. Flywheel.

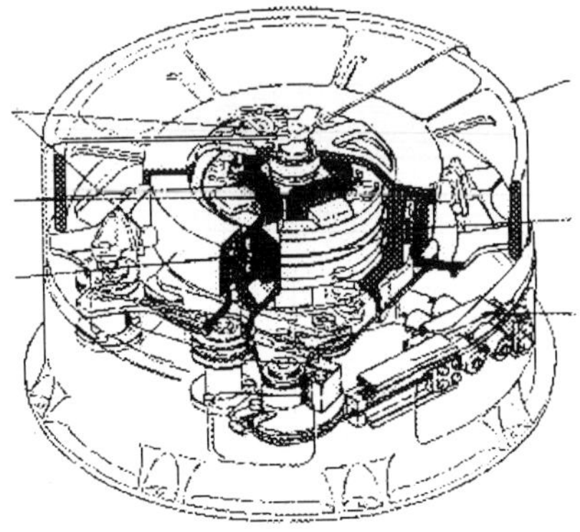

Figure 2. Cross Sectional View of M.B.

DEBUT. Technological demonstrations of deployable flexible boom and umbrella-like structures were performed on orbit using DEBUT microsatellite (the nickname is ORIZURU) (A.Nakajima *et.al.*, 1991). The experimental objectives of DEBUT were the in-orbit verification of the mechanical systems of aerodynamic braking system (ADB) and flexible boom and also the development of small and low cost bus system. ADB and flexible boom are possible candidates of future microgravity realization systems such as boomerang and tether satellites.

Boomerang satellite is a free flying payload operated in the proximity of the large mother spaceship such as space station. Flying freely around the mother spaceship, the boomerang satellite realizes the microgravity (10^{-6}-10^{-7}g). After the high quality microgravity experiments are finished in a free flying condition, an ADB system is opened to compensate the difference in ballistic coefficients of the satellite and the mother spaceship. The very small amount of aerodynamic force in Low Earth Orbit (LEO) would slightly change the relative position and velocity between the both spacecraft. Then the satellite comes back to near the mother spaceship with very small relative velocity just like boomerang. It is easily captured into the mother spaceship by means of a manipulator. Utilizing the boomerang satellite, repeated microgravity experiments can be executed by controlling the ADB system without any fuel. Tether satellite connected with the space station with thin tether will also provide the microgravity environment (10^{-4}-10^{-5}g) by isolating the g-jitters generated mainly by crew activities and life support systems. Microgravity experimental payloads are easily changed by deployment/retrieval of the tether. The boom is one of the possible candidates for the initial release/final recovery mechanism of the tether satellite.

The experimental items of DEBUT are repeated operations of ADB open/close, boom deployment/ retraction and ADB panel vibration measurement. ADB is composed of thin CFRP panels, which are connected with each other by flexible hinges. Open/close motions were repeated 52 times under day and night environments and it was proved that the ADB system had no problems with respect to mechanical and electrical points. The flexible boom is one of the coilable lattice masts without any hinges between each member, called hingeless mast. It was deployed and retracted by unreeling or reeling the center strap driven by a d.c. motor. The experiments were 34 times repeated and the appropriateness of the boom system under space environment were also experimentally proved. ADB panel vibration excited by open/close of ADB itself and also by deployment/retrieval of boom were also measured and evaluated the data. The bus system of DEBUT was developed within 1.5 years. The satellite cost is significantly depend on the development time, DEBUT has proved the realization of "faster" and "cheaper" microsatellite.

Figure 3 shows the DEBUT configuration at full deployment mode. The main features of two satellites developed by NAL are summarized in Table 1.

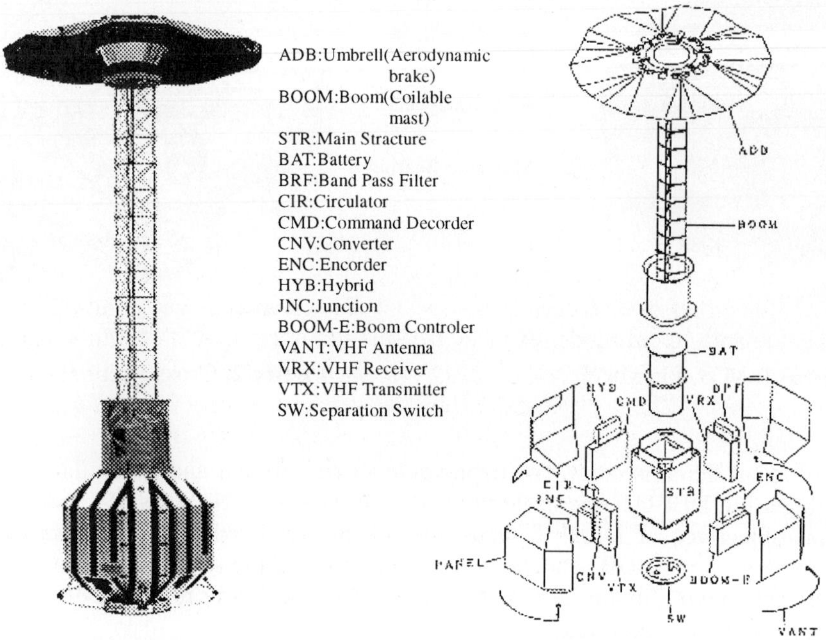

Figure 3. DEBUT configuration at full deployment.

Table1. Features of NAL's Technology Demonstration Satellites "JINDAI" and "ORIZURU"

	MABES ("JINDAI" 1986-61C)	DEBUT ("ORIZURU" 1990-13B)
Mission	To evaluate the characteristics of the Magnetic Bearing Flywheel under zero-g condition, which will be used as attitude control component of satellites	To prove functions of major structure elements of tether/boomerang satellites which will be used as future space microgravity environment utilization
Development Organizations	NAL, Mitsubishi Electric Corp., Toshiba Corp.	NAL, NEC Corp., Nissan Motor Co., Japan Aircraft Manufacturing Co.
Launch and Orbit		
Time	05:45:00, Aug.13, 1986 (JST)	10:33:00, Feb.7, 1990 (JST)
Launch Vehicle	H-I TF#1 (H15F)	H-I F#6 (H21F)
Launch Site	Tanegashima Space Center	←
Orbit	1500~1600km, Circular	900km (Perigee)~1700km (Apogee)
Period	116 min.	106 min.
Inclination	50 deg.	99 deg.
Satellite Dimension		
Configuration	Rectangle	Polyhedron of 26 faces
Size	1m (W)×1m (D)×1.5m (H)	φ0.44m×0.79m (H) (Launch mode) φ0.90m×1.94m (H) (Deploy mode)
Weight	294 kg	50.3 kg
Attitude Control	No control	No control
Power Supply	35 Ah NiCd (discharge only)	95 Ah Lithium (Li (CF) n) (discharge only)
TT&C	Uplink 148.27 MHz Downlink 136.11 MHz, 256 bps, 1.2 W	Uplink 148.27 MHz Downlink 136.89 MHz, 2048 bps, 0.2 W
Life	3 days	10 days

Radio Amateur Satellites Developed by JARL

Since 1961, only four years later after the first artificial satellite Sputnik 1, many amateur radio satellites were launched in the world. A lot of radio amateurs, cooperate internationally, have made effort to construct and operate amateur satellites by themselves. JARL, the most amount of wireless stations are participated in the world, desired his own radio amateur satellite in the 1980's. After negotiations with national organizations, the first private, but public-opened amateur satellite, JAS-1 was launched in 1986. The nickname of JAS is "Fuji", then the JAS-1 is internationally called as Fuji-OSCAR 12 (FO-12). JARL has developed three radio amateur satellites, FO-12, FO-20 and FO-29. JARL is now operating FO-29 and FO-20, which is still partially alive 8 years after launch (expected life time is 3 years).

JAS-1 and JAS-1b. The first amateur radio satellite in Japan, JAS-1 (FO-12) was launched in 1986 and the second JAS-1b (FO-20), which is a back up satellite of JAS-1, was launched in 1990 after some modifications. The objectives of JAS-series are world-wide communications between radio amateurs via JAS, both in analogue and digital mode, exercise the satellite tracking and operation and the certification of the in-orbit performance of transponder developed by radio amateurs. The modifications of FO-20 are satellite dimension, solar cells and antennas. The average electric power of FO-12 generated by $2\times 2cm^2$-979 silicon solar cells was about 8 watts, which were not enough for the continuous operation of the digital communication. The width of FO-20 was enlarged 10% and the solar cells were changed to gallium arsenide (GaAs) with higher conversion efficiency. The average electric power generation was improved to 11 watts with $2\times 2cm^2$-620 and $1\times 2cm^2$-900 cells, resulted in the comfortable satellite operation. By applying a ring-shaped turnstile antenna in FO-20, more stable uplink from the ground stations and digital/analogue transmissions with the same antenna were realized. Figures 4 and 5 are the photographs of FO-12 and FO-20.

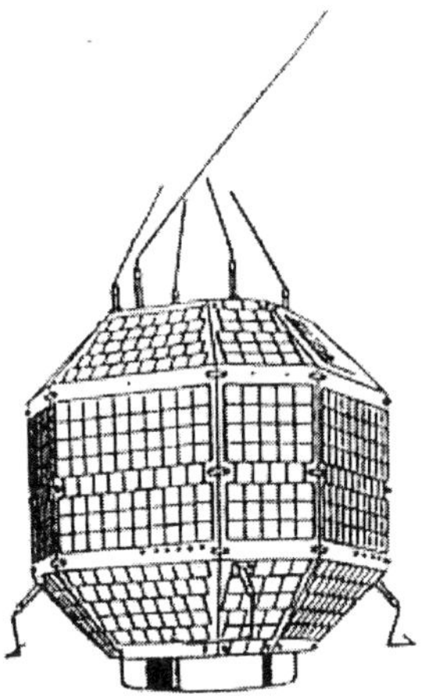

Figure 4. JAS-1 (FO-12).

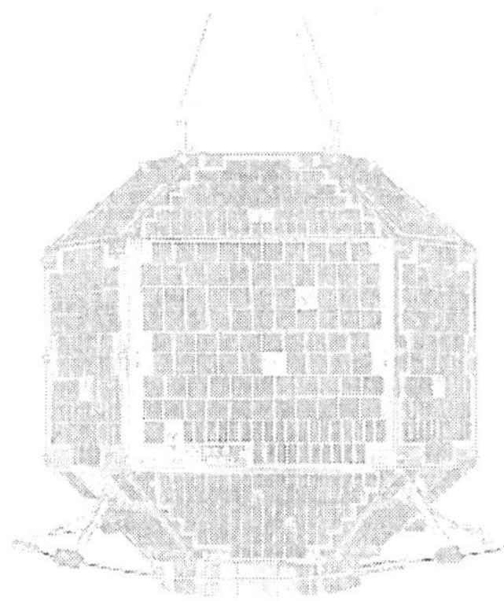

Figure 5. JAS-1b (FO-20).

JAS-2. The third amateur radio satellite, JAS-2 (FO-29), was launched by H-II rocket in 1996 (F.Yamashita *et.al.*,1996). FO-29 is almost the same configuration as FO-20, but the maximum power generation is improved to over 20 watts by applying the attitude control system. Sun sensor, geomagnetic sensor and magnetic torquers are used for the orientation control, spin control and nutation damping. The attitude of FO-29 is carefully monitored and controlled to increase the electric power and to keep the good thermal condition. Some detailed characteristics are as follows.

Analogue transponder: Linear transponder for phone and CW. The uplink frequency is 145.90 to 146.00 MHz with 100 kHz bandwidth and the downlink is 435.80 to 435.90 MHz with 100 kHz bandwidth. Beacon signal frequency is 435.795 MHz and modulated by CW. TLM total transmitting power is 1 watt including 100 mW beacon.

Digital transponder: Store-and-forward packet processor. The four-channel uplink frequencies are 145.85, 145.87, 145.89 and 145.91 MHz and one-channel downlink is 435.91 MHz. The bit rate is 1200 or 9600 bps; bi-phase Manchester code on FM signal with 1200 bps for uplink and PSK 9600 bps for downlink. Total transmitting power is 1 watt.

Figure 6 is the photograph of FO-29 and the main features of three satellites are shown in Table 2.

Figure 6. JAS-2 (FO-29).

Table 2 Features of JAS-series

	JAS-1 (Fuji-OSCAR12)	JAS-1b (FO-20)	JAS-2 (FO-29)
Launch and Orbit			
Time	Aug.13,1986	Feb.7,1990	Aug.17,1996
Launch Vehicle	H- I (2-stage) TF#1	H- I (2-stage) F#6	H-II F#4
Launch Site	Tanegashima Space Center	←	←
Orbit	1500 km	900×1700 km	800×1300 km
Period	116 min.	106 min.	106 min.
Inclination	50 deg.	99 deg.	98.6 deg.
Satellite Dimension			
Configuration	Polyhedron of 26 faces	←	←
Size	400mm (W)×470mm (H)	440mm (W)×470mm (H)	←
Weight	50kg	←	←
Attitude Control	Permanent Magnets	←	Permanent Magnets, Magnetic and Sun Sensors
Power Supply	Silicon, 8.5W mean	GaAs , 11W mean	GaAs, 22W max
	NiCd 6 Ah	←	←
Antennas			
Reception	Mono-pole		
Transmission	Turnstile for JA and JD	Ring-shaped Turnstile	←
Communications			
Frequency	Uplink 144 MHz	←	←
	Downlink 430MHz	←	←
Transponder	Analogue and Digital	←	←
	Store-and Forward Packet Communication (1200bps)	←	←
	1W Output	←	←
Life	3Years (Phased Out, Nov.1989)	3Years (Partially Alive)	3Years (In Operational)

LAUNCHING SYSTEM

Various type of rockets will be available for microsatellites launch system as piggyback payloads in the world. Three types of rockets are operational in Japan, they are H-II, J-1 and M-V launchers. H-II has the launching capability of 2-ton class geostationary satellite and will be replaced to more powerful H-IIA launcher at the beginning of the 21st century. The nearest piggyback microsatellite will be launched by H-II in February, 2000. After this launch, launch opportunities of piggyback microsatellites using H-IIA will be provided almost every year and opened for all countries. J-1 solid rocket is composed of a SRB of H-II for the first stage and the upper stages of M-3SII for the second/third stages. A few hundred-kg small satellites can be launched in LEO. M-V, newly developed solid boosters, has the launching capability of 2-ton class large satellites in LEO and is planned to launch planetary missions. The both solid launchers will not be expected to launch piggyback microsatellites at the moment.

H-II and H-IIA Launchers

The first launch of H-II rocket was successfully done on 4 February, 1994 and the fourth launch of H-II gave the first launch opportunity of piggyback microsatellite, JAS-2, described in the former chapter. The launch configuration is shown in Figure 7 (The actual installation is shown in Figure 6). The same configuration will be applied for the second piggyback microsatellite (launching in 2000). For the multiple launching of piggyback satellites, an example configuration is shown in Figure 8. Maximum 8 microsatellites can be launched simultaneously and injected into each orbit and non-separated payloads will also be launched at the same time. For non-separated missions, some experimental components and bus system are installed in the structure to support the main satellite. The launch opportunities are strongly depend on the mass margins of the launcher and adequate mission proposals.

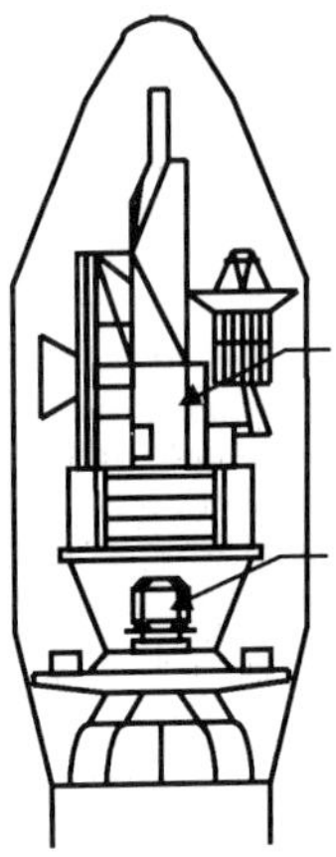

Figure 7. Single Piggyback Launch Configuration (JAS-2, 1996 launch).

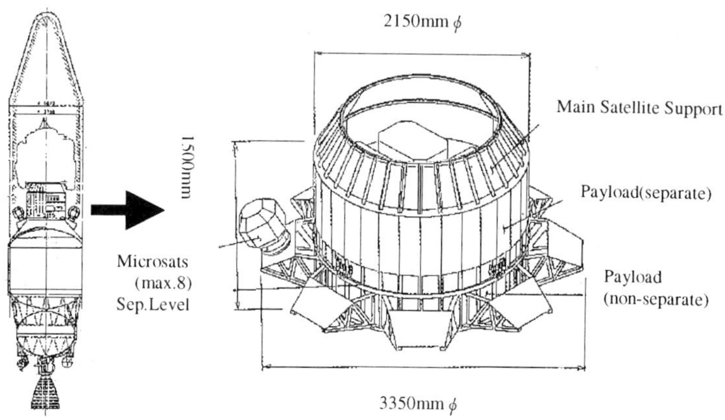

Figure 8. Multi-piggyback Launch Configuration (Proposal Configuration).

SMALL PAYLOAD WORKSHOP (SPWS)

The PMSWG, research subgroup of the SPWS, has surveyed the potential users about microsatellites utilization and 61 mission ideas were proposed in 1996. They are summarized as follows.

Mission Requirements

For the mission success, some demands for satellite attitude control system and separation system of the launcher are requested.

Attitude Control. The most microsatellites launched in the past have adopted the gravity gradient attitude stabilization or slowly tumbling (forced spinning or no active attitude control) because of the limited budget and requested attitude accuracy. The investigation results of the proposed missions show that more than 50 % missions require 3-axis attitude stabilization in order to meet the future extensive demands such as high bit-rate data transmission, high accuracy pointing, fixing to the inertial frame, high quality micro-gravity environment and so on. Low-cost, simplified 3-axis attitude control package must be developed as soon as possible. As a 3-axis attitude control actuator, small magnetic bearing reaction wheel is now developing in NAL (A.Nakajima *et.al.*,1997). From the viewpoint of the existing technology, gravity gradient stabilization is easily adopted for some missions. In our case, 16% missions require this attitude control system. Spin stabilization is not a major attitude control system required from our missions. Figure 9 shows the distribution of attitude control system for microsatellites.

Mass and Volume. Standard piggyback microsatellite is limited with mass and volume under 50 kg/0.125 m^3. For more spread of microsatellite utilization, 75 kg/0.6m (dia.)×0.8m mass/volume are desirable. The spring ejection system is adopted for microsatellite separation in a usual case.

Separation. In order to adopt a 3-axis attitude stabilization, the initial angular momentum given to the satellite from the spring mechanism must be restricted as small as possible. The maximum angular momentum accumulation of reaction wheel is designed an order of 0.2 – 0.5 Nms at most. Magnetic torquers or gravity gradient torquers will be used for the initial angular momentum decrease of the satellite and desaturation of reaction wheels.

Orbit. H-IIA rocket has the capability of re-start burn, the piggyback satellites will be injected into another orbits from that of the main satellite. Some missions require their own desired orbit.

Proposed Missions

The proposed missions are divided into 4 categories, which are technology, observation, communication and biology. The distribution is shown in Figure 10.

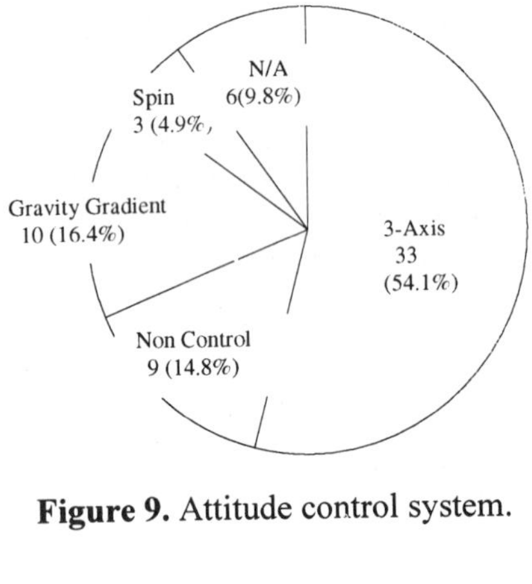

Figure 9. Attitude control system.

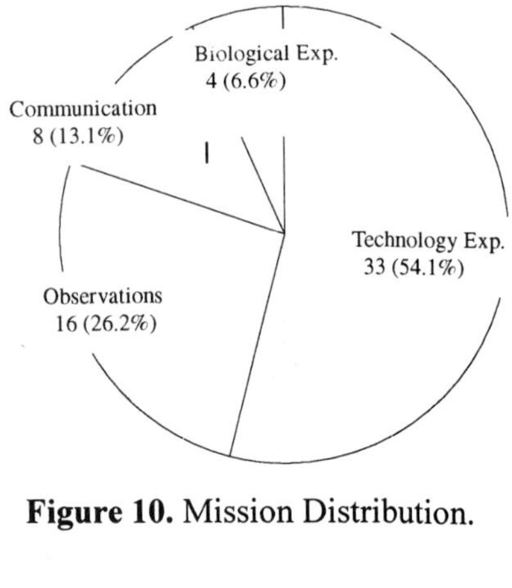

Figure 10. Mission Distribution.

More than 50 % are concentrated on the science and technology missions. They are technology verifications of satellite bus system and onboard components utilizing commercial-base products, aero-brake and recovery technologies, tether technology, planetary investigations, on-orbit maneuvers, robotics, microwave transmission, refrigeration, LH_2 storage, mechanical pump, liquid experiments and so on. Missions related to the Earth and astronomical observations and survey of debris distributions take the second place, about 30 % of total. In the communication field, safety and emergency communications and positioning are also very important missions. Small animals and plant growth experiments are proposed as biological experiments and they are necessary for the future manned space activities. The PMSWG has discussed and proposed some missions for the near future realization.

Tether Satellite. Space tether system has variety of applications such as microgravity realization, electrodynamic orbit boosting or de-boosting, artificial gravity generation by rotating two bodies connected with tether, scientific measurements of upper atmosphere. The objectives of microsatellite tether system are to verify the mechanical system such as reel, ejection, and retrieval mechanisms, control system such as position keeping, pendulum motion damping and stable retrieval/capture of the tethered subsatellite and the measurements of dynamic behaviors. Figure 11 is a concept of Tether Experimental Microsatellite proposed by NAL (S.Yoshimura *et.al.*,1997). The masses of mother satellite and subsatellite are 50 kg and 20 kg, respectively. The tether length is about 1 km. In this experiment, the reel mechanism for deployment and retrieval is mainly tested under zero-g condition.

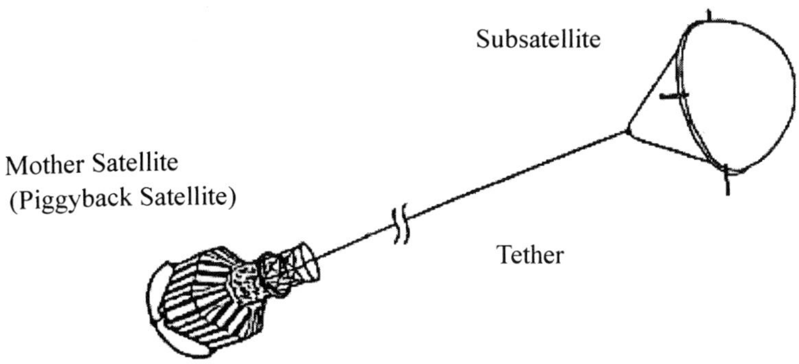

Figure 11. Tether Experimental Microsatellite.

Whale Ecology Observation Satellite (WEOS). In the field of communications, microsatellite can be used as a data relay satellite. WEOS is an example mission. This satellite is designed to monitor the ecological data of whales in the Antarctic ocean. Polar orbit microsatellites gather ecological data from the probes attached to whales and store them before data transmission to the ground station. WEOS and probes, proposed by Chiba Institute of Technology, are already designed/manufactured and wait for flight opportunity (T.Hayashi, 1997). The same missions targeting on various animals and birds act important roles for the knowledge of their global activities.

Telemedicine Satellite. Another example of communication satellite is a telemedicine small satellite. The medical data of human beings living in the area of Africa and South America, for example, where ground-based communication network system was not yet enough serviced, are transmitted to the medical facilities and adequate medical assistance will be received via telemedicine small satellites. Instead of microsatellite, a few hundred-kg class small satellite will be necessary for the telemedicine, because of high power transponder and stable communication links. Microsatellite will be used as precursor mission.

Tokai University School of Medicine has proposed this mission (I.Nakajima *et.al.*, 1997).

Onboard Components Verification Satellite. By applying new technologies in time, the satellite bus system could be miniaturized with high operating performance. Micro-electronics technology has changed the situation in the past dynamically and the microsatellite plays an important role in technology demonstration program in space. Commercial-based parts and components will be applied for onboard after on-orbit verification tests using microsatellites. Many companies list up electric parts to be tested in space.

Optical Image Transmission Satellite. Space Telescope is now operational for gathering very high quality optical images without atmospheric influences. Amateur astronomers and also those who are interested in space want to take photographs of stars, planets, the Earth and something like that from satellite. The image data are necessary for the monitoring of manned space activities or unmanned robotics movements, for example, and furthermore they will be used for the evaluation of the malfunctioned artificial satellites ("Eye ball nanosatellite" was already tested in STS). By applying the image data compression technology, JPEG or MPEG, microsatellite space telescope can be realized and the obtained image data can be transmitted to the ground under the limitation of microsatellite power supply. "Planetarium satellite" is an idea from students.

CONCLUDING REMARKS

Recent activities on microsatellites in Japan are described. The framework on launch opportunity of piggyback microsatellites using Japanese H-IIA launchers is now preparing for the organizations all over the world who intend to develop microsatellite for the public domain applications. The continuous activities of the SPWS and other societies have played an important role in the realization of a non-commercial launch of microsatellites. From the beginning of the 21st century, launch opportunity will be given almost every year.

REFERENCES

A. Nakajima and C. Murakami, Research and Development of Magnetic Bearings, *Proceedings of the 16th International Symposium on Space Technology and Science (ISTS)*, pp.1059-1064, Sapporo (1988).

A. Nakajima, M. Abe, Y.Nishio and H.Yahagi, Space Experiments of Deplyoyable Boom and Umbrella Test Satellite (DEBUT), *Acta Astronautica Vol.25, No.12*, pp.765-773 (1991).

A.Nakajima, N.Kakuta, T.Soeda G.Shibata and T.Sekine, 2-Axis Actively Controlled magnetic Bearings, 4th International Symposium on Magnetic Suspension Technology, *NASA CP* (to be published), Gifu (1997).

F. Yamashita, T.Kondo, K. Kitade, Y. Masumoto and Y. Abe, Development of Amateur Radio Satellite "JAS-2", 96-e-33V, ISTS, Gifu (1996).

I. Nakajima, Y. Sawada, T. Maeda, D. L. Martin, S. Nagano and N. Hamano, Telemedicine Using Small Satellite MEO/HICAT for International Medical Communications, COSPAR Colloquium on Scientific Microsatellites, Tainan (1997)

S. Yoshimura, A. Nakajima and R. Yasumitsu, Numerical Simulation of the Tether System Dynamics in the Space Verification Test using a Small Piggyback (Mother) Satellite, 48th International Astronautical Congress (IAF), IAF-97-A. 3.07, Turin (1997).

T. Hayashi, Outline of the Whale Ecology Observation Satellite System, COSPAR Colloquium on Scientific Microsatellites, Tainan (1997).

RUSSIAN SMALL SATELLITES AND MEANS TO LAUNCH*

Michael Yu. Ovchinnikov

Keldysh Institute of Applied Mathematics, Russian Academy of Science, 4 Musskaya Sq., Moscow 125047 Russia

ABSTRACT

The brief review of Russia small satellite projects that have been realized or are under consideration to be realized and means available to launch them into orbits are presented. Some of projects represent idea only without concrete scheme of execution.

INTRODUCTION

The changes occurring in economic and political life of Russia are the real basis for integration of Russian space science and industry in international market of space service. Just the high technologies, accumulated for long years of resistance in "Cold War" plus the new economic conditions, though rather "rigorous" and sometimes non-friendly in favour to manufacturer, are capable to push the effective integration processing in global market on the basis of initiative and competition. Let's consider it on the example of the small satellites that are intensively developed high-technology subject over the world.

REALIZED AND PERSPECTIVE PROJECTS OF SMALL SATELLITE IN RUSSIA

Due to vast territory of Russia the suppressing majority of projects of small satellite developed now in Russia is directed on creation of regional and interregional communication and remote satellite networks.

The existing deficit of means of communication is huge indeed and despite obvious liquidity of this segment of market, it remains practically unsaturated. The poorness of the long-term investments directed into the development of space facilities forces the designers of communication means to search the approaches for inexpensive and not requiring a long-time period design and manufacture of equipment and space facilities. From the cost and duration of development, and launching up into the orbits points of view the small satellites satisfy these conditions in a good way. To be can due to unsufficient capacity of power sources, small allowable volume and weight of on-board equipment these satellites are used predominary for creation low-orbit satellite communication networks. To cover all needed areas on the Earth's surface the satellites are to be placed on the near-polar orbits. The low-orbit satellite network permits to be satisfied both by present opportunities for arrangement of board equipment but also by low-powerful portable transmitters. The communication between correspondents located outside the zone of

* Presented at COSPAR Colloquium on Small Scientific Microsatellites: Microsatellites as Research Tools, 14-17 December, 1997, National Cheng Kung University, Tainan, Taiwan.

visibility of one satellite, is executed either by transportation of message memorized in on-board computer to zone of visibility of correspondent as an electronic mail means, or by organization of intersatellite communication network. This network permits to realize the communication in real scale-time between any points of the Earth's surface. Utilization of the low-orbit satellites on near-polar orbits permits in many cases to refuse the geostatinary satellites with their demerits.

In Russia during the whole period of space industry development the small satellites were regularly orbited. They served mainly experimental purposes for testing instruments and structure elements for large spacecrafts. However, later they became more independent of large satellites. Partially, a few well-known projects of russian small satellite were considered in [1]. Basically, they belong to communication satellites. Other intensive developed direction is satellites for remote sounding. They will be here considered too. The division of satellites into communication, remote sounding, technological and other groups is performed later where it is possible.

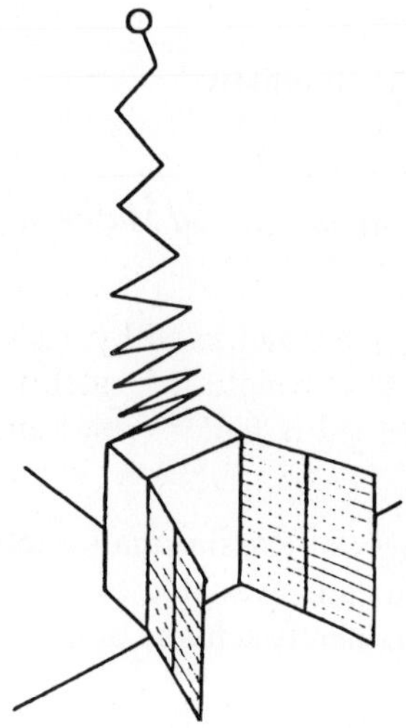

Figure 1. Gravity-gradient oriented satellite ISKRA-5.

RADIO-AMATEUR AND PROFESSIONAL COMMUNICATION SATELLITES

A series of small satellites for scientific research and non-professional radio-communication purposes were designed at Moscow Aviation Institute. One of them, ISKRA-5 satellite (Fig.1) equipped with passive GGACS and eddy current damper (ECD) was piggyback launched with METEOR-PRIRODA satellite (1981). GGACS with boom of original design was manufactured by All-Russia Scientific Research Institute of Electromechanics (VNIIEM) located in Moscow.

The Moscow Institute of Thermo-Facilities has developed and launched the experimental satellite START into the orbit (1993) by own rocket START that is modernized conversional missile SS-25. Its weight is about 250 kg. The satellite is provided by passive GGACS and actually was the dynamical sample with communication intention.

Satellite Personal Communication System (SPS-Sputnik) was developed by Enterprise "Makeyev Design Bureau" located in Miass City, South Ural area, for a low-orbit communication satellite network creation [2]. Satellite mass is 280 kg; it had passive GGACS with ECD designed by VNIIEM [3]. Launching by own conversional SHTIL-3N missile was planned. It was first experience of this Enterprise in development of satellite and its payload.

The further development of communication systems on the basis of SPS-Sputnik above has resulted in development of low-orbit data-collecting communication system for tracking of mobile and stationary objects intended for wide used application [4]. It is assumed the network on 700 km circular orbit will consist of 36 small satellites. The quantity of serviced users supplied by terminals of tracking by weight not more than 2 kg is estimated near a few hundred thousand. The location of the object is determined with accuracy of one hundred meters. Duration of autonomous activity of terminal is about one year. The duration of this system creation including the time of satellite arrangement in orbit is about 4 years.

Federal Space Program of Russia includes two multisatellite low-orbit systems of communication based on small satellites. They are GONETS and SIGNAL systems. The Goverment via Russian Space Agency takes part in funding of project GONETS [5]. The small satellite GONETS is developed by NPO of Applied Mechanics in Krasnoyarsk City, Siberia and launched as conversional one in frameworks of the project under the same name for communication in electronic mail mode. The lifetime of satellite is 4 years. Two primary satellites GONETS-D (1993) are launched for demonstration goal. Installation of six satellites constellation by TSICLON rocket is possible. The weight of satellite is 250 kg, weight of payload is 60 kg. The transmission rate of data is 2.4, 4.8, 9.6, 64 kbps. Satellite is supplied by GGACS with ECD and a system of preliminary magnetic orientation [6]. The draft of satellite is presented on Fig.2 [7].

On first stage (the system GONETS-DL) the space segment consists of six satellites placed on near circular orbits by height about 1500 km and inclination 83' in two mutually perpendicular planes with three satellites in each plane. In hereinafter the development of space segment consisting 45 satellites is planned.

Low-orbit communicational system SIGNAL is developed by joint-stock company International Concern of Space Communications (KOSS). The space segment with the same-named satellites is designed by Enterprise "NPO Energy" placed in Koroljev City (former Kaliningrad) near Moscow. It is assumed the satellites to be launched into the circular orbit with height 1500 km and inclination 74'. The total number of satellites in system is equal to 48 (12 satellites placed in 4 orbital planes with 90' difference in their ascending node longitudes). Lifetime is 6 years. The satellite mass is 300 kg, payload mass is 70 kg. The satellite is intended to provide the data relay and duplex telephone communication. The data rate is 2.4 kbps. The sketch of a satellite is presented on Fig.3. The launch of first satellite SIGNAL was planned in 1997. They may be orbited in pair by COSMOS or in six by TSICLON launchers [8], [9].

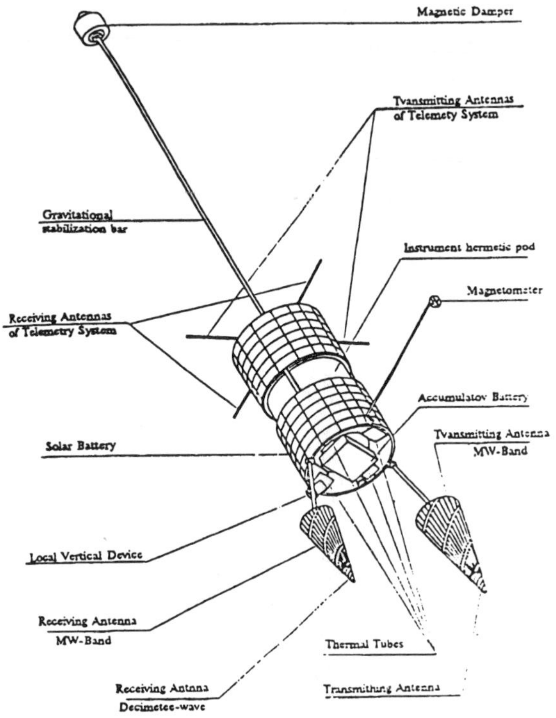

Figure 2. Draft of GONETS satellite.

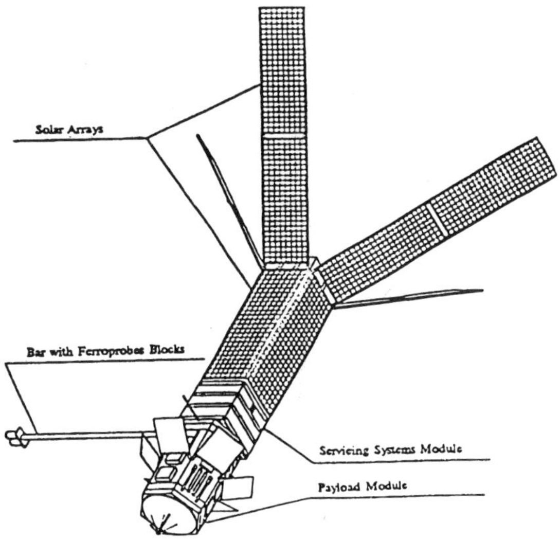

Figure 3. The satellite SIGNAL draft.

Low-orbit communicational system GLOBSAT was developed by Enterprise "Khrunichev State Rocket Center" located in Moscow [10]. From 30 till 66 satellites are suggested to be included into this system. It is intended to provide low cost communication in phone, fax, and digital exchange information modes. The structure of satellite should be the simplest one. Passive GGACS is used. Launching by conversional modernized ROKOT rocket is planned. The successful experimental launch of radio-amateur satellite RADIO-M with mass about 120 kg was carried out on 26th of December, 1994 by this rocket.

THE EXPERIMENTAL AND TECHNOLOGICAL SMALL SATELLITES

VNIIEM has designed and launched in sixtyth small satellite OMEGA for testing the three-axes flywheel ACS. Its ACS was considered as a basis for ACS currently used on large satellites METEOR, RESOURS-O, GOMS.

MAK-A satellite is developed [11] in Scientific Research Institute of Applied Mechenics and Electrodynamics (NIIPME) at Moscow Aviation Institute (MAI) on the basis of a PION satellite [12] and is supplied by original passive aerodynamic ACS with soft-magnetic damping rods (Fig.4) [13]. Rods are installed into non-magnetized tubes mounted on the sides of two truncated cones. Rods are collected into triplets in that three rods make mutially perpendicular structure. This satellite dedicated for autonomous measuring the atmosphere density in current point of satellite location on orbit by on-board accelerometer. The aerodynamic restoring torque was provided by displacement of center of mass of satellite relatively its center of pressure coinciding actually with geometrical center of sphere. The orientation of satellite was to be determined by measurements of magnetometer and accelerometer.

Figure 4. Aerodynamic oriented satellite MAK-A (half of cover is placed on the left).

In NIIPME also the series of non-oriented microsatellites MAK-1, MAK-2, and MAK-3 with universal service module and various payload modules intended for geophysical measurements was developed [14]. The weight of satellites does not exceed 25 kg, weight of payload - 7.7 kg. The lifetime on orbit is not more than month at orbiting through the lock of space station MIR. The work performed in MAI and later in NIIPME is typical for educational institute, that is, the students take part in development, design, manufacturing of a satellite, and further in process of data received from satellite during whole period of their staying in an institute. The same approach to educational process is used in Samara Avionic-Space University with Enterprise "Central Special Design Bureau" located in Samara City. Six passive non-oriented PION satellites with mass 50 kg launched as piggyback (1989-1992) with RESOURS-F satellites [12] are a result of this activity.

At present, the system based on PION satellites is named ROSS (Radar-Optical Satellite System). It is intended to clarify the characteristics of density of upper atmosphere model and to improve the methods of external tracking for small space objects. The outline of accommodation of small satellites ROSS (actually PION) on main satellite RESOURS-F board as a piggyback for orbiting into the orbit of hight $250 \div 300$km with lifetime from one to two months is shown on Fig.5 [15].

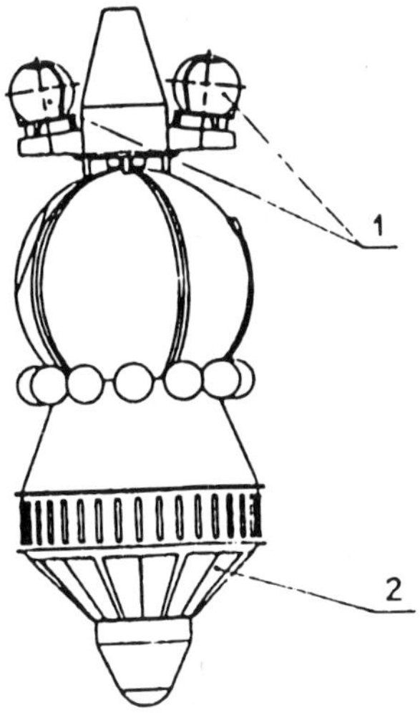

Figure 5. Outline of two PION-type satellites installation (1) on RESOURS-F satellite (2)

Figure 6. The satellite MOZHAETS.

It is interesting the elaboration of the Mozhayskiy Military Space-Engineering Academy. The students of the Academy developed on the basis of serial satellite manufactured by NPO of Applied Mechanics the educational non-oriented small satellite MOZHAETS [16]. The satellite is intended for student training in methods of satellite guidance and control, analysis of telemetry information by regular means of educational command - measuring complex. Besides, performance of educational purposes of this satellite is intended for research of influence of radiation flows on capacity for work of on-board electronic devices at height of $700 \div 800$ km. Students are developing other small satellite for supervision of geomagnetic field to predict the earthquake. Perhaps, it is the first attempt of a military educational establishment to use small satellite in educational process. The ideas to use small satellite for remote training of staff including military one is considered in [17].

Two projects for investigations with orbital station together are known. According to one of them [18] through station's lock the microsatellite for detailed review of external surface of station, environmental its atmosphere, interception and recovery to the station of small objects in its vicinity is ejected from station. By two-impulse maneuver the satellite is transferred into elliptical orbit with revolution period equal to station's one and for safety not intersecting its orbit.

In another project [19] the subsatellite of weight about 100 kg is kept near to station on distance upto 12 km using the variable ballistic factor. The change of this factor may be performed by attitude re-orientation of non-symmetrical satellite relative to the vector of center mass velocity or by deploying aerobraking sail.

Institute of Earth's Magnetism (IZMIRAN) placed in Trotsk City near Moscow with other organizations develops [20] experimental electromagnetically clean satellite COMPAS of weight about 70 kg (Fig.7). It is assumed to apply combination of attitude control systems using interaction of satellite with gravitational field of the Earth, aerodynamic drag and gyroscopic effects of spinning body. The accuracy of orientation is about 10', accuracy of attitude determination – 1°. The planned height of orbit is 400 km, inclination $-65° \div 78°$. The satellite is supplied by low-power thruster to keep the given height of orbit during seven months. The satellite is intended to study the electromagnetic and plasma precursor of earthquakes. It is assumed, that as well as SPS-Sputnik both satellites will be launched by "Makeev Design Bureau" using conversional submarine missiles like SHTIL. At present, on the basis of COMPAS satellite the universal small-sized space platform COLIBRY [21] intended for accommodation of various payloads and launching as a Piggyback is being developed under supervising by private Company Vneshnauchpribor. The height of its orbit depends on the requirements of customer and is defined by launcher used. It is considered rockets TSICLON, PROTON, COSMOS. The capacity of power supply available for payload is 50 W with 27 V. Accuracy of center mass position determination is about 100 m. There are a system of telemetry and radiochannel for transmission of information to on-ground station. Board memory capacity is 64 Mbytes. The duration of manufacturing is about one year.

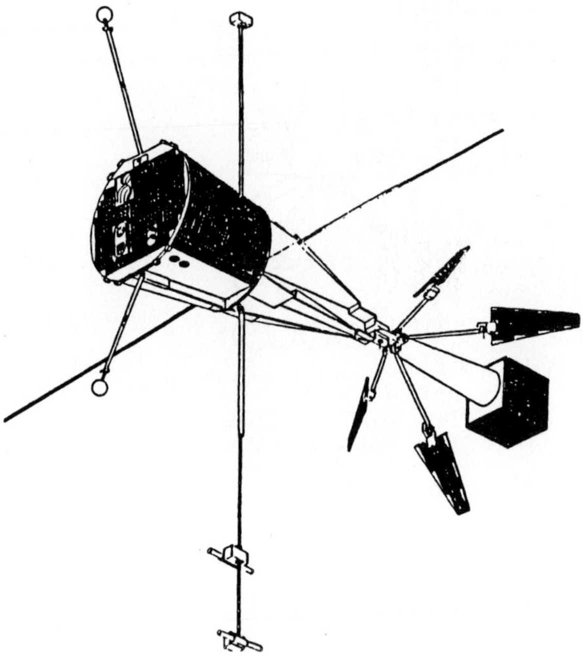

Figure 7. Draft of COMPAS satellite.

The series of small satellites is developed by VNIIEM. It was carried out the Phase-A stage of project for Universal Small Space Platform (USSP-1), which was provided by passive GGACS with ECD [22]. Satellite's mass is near 60 kg. Piggyback launch with own RESOURS-0 satellite was planned. It was not manufactured but VNIIEM has got experience in small satellite designing. Actually, it was the pilot-project in this Enterprise, which nudged its leaders to consider the small satellites as a serious and perspective production. Later the Technical Proposals for Universal Small Space Platforms USSP-2 and USSP-3 were developed. The both platforms are provided by active flywheel ACS. The USSP-2 mass is about 400 kg, attitude control precision—$6 \div 10'$. The basical payload of USSP-3 is radio equipment of COSPAS/SARSAT system intended to help for the disaster swrecked people[23]. The total mass is about 140 kg. The attitude control precision to the Earth is $6 \div 10'$, to the Sun—1°. USSP-3 is to be launched into the circular polar 900 km orbit performed by conversional rocket START-1. Increasing the total mass of USSP-3 upto 250 kg reduces the orbit height till 600 km.

The project of microsatellite intended for monitoring of the atmosphere, the Earth's ground and water surface is considered [24]. Its payload realization is based on the idea to use the tunable acousto-optical filter in ultra-violet and visible bands of frequency enabling to ensure the programmed varying of bands of passing with $\triangle\lambda$ < 1nm in area $0.2 \div 0.9$ μm and using the counter of photons with high spatial resolution as a recording sensitive sensor. Thus, electrically controlled polarizer of input radiation is also used. The sensitivity of system permits to film under over-night conditions. The height of the satellite orbit is 500 km and more. At monitoring in narrow angle of sight the spatial resolution is about 100 in, instant field of sight is 10×0 km2. The given approach permits to ensure the small orbital platform, not requiring precise three-axes orientation and stabilization of image on the sensor display, which simultaneously performs the role of astrosensor for attitude determination.

Settling on tunable acousto-optical filter approach the project of microsatellite for optical hyperspectral monitoring with spectral resolution better than 1 nm is proposed in [25].

As to satellites intended for monitoring of environment and remote sounding of the Earth, at present, the significant quantity of projects are being developed.

SMALL SATELLITE ORBITING BY RUSSIAN LAUNCH MEANS

At present, there are three real possibilities to launch a small satellites. They are as follows:

- piggyback style when a small satellite is put into the orbit on the main satellite, orbital station or space shuttle boards and further separated at the moment required;
- by missile or light launchers when one launcher is able to orbit one or a pair of small satellites;
- heavy launcher used to put into orbit a constellation of small satellites alone or with main satellite simultaneously.

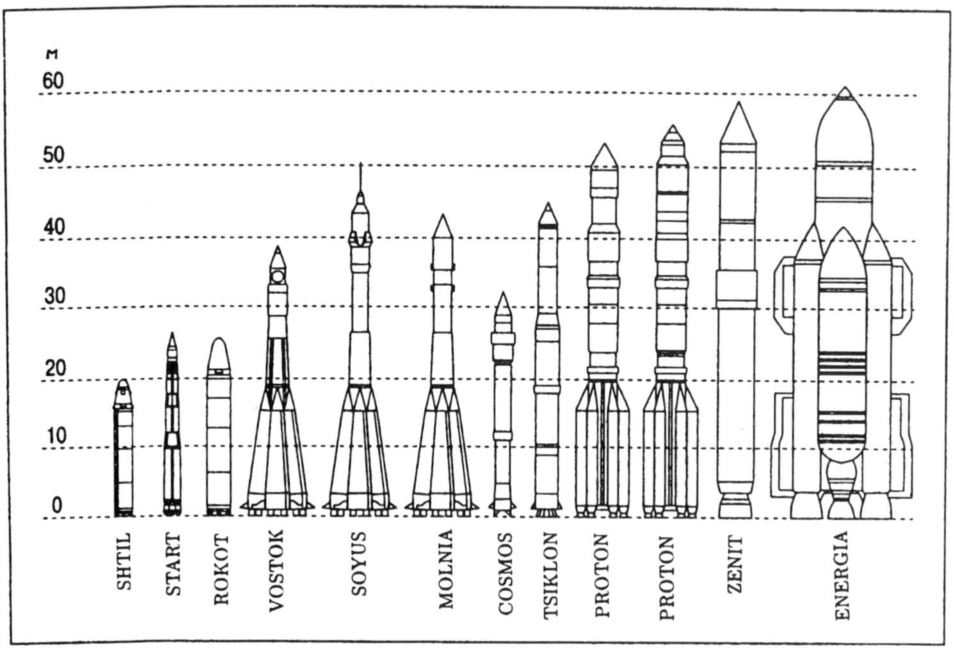

Figure 8. Main russian launchers.

At least, four principal programs of commercial launching are being realized now in the world. The most large-scale one is Arianespace's program which serves half of world space market. The next is NASA's program using Space Shuttle and expendable launchers. Last time Chinese space program based on the Long March launcher is made more active. Lastly, Russian commercial space enterprises started their attacks on the international space market. Under wide spectrum of launchers available on international market let us consider the means including conversional launchers offered by Russia for putting small satellites into orbits and from the author's point of view are the most appropriate for this purpose (Fig.8). This picture includes the base rockets. Some of them have modifications. Means to launch a small satellite by russian carriers and main spacecrafts are adduced in [26].

PIGGY-BACK LAUNCHING

Perhaps, the cheapest way to put a satellite into orbit is a piggyback launching. This opportunity arises owing to reserve of weight available at the launch of main comparatively heavy satellite. In this case, the small satellite is ejected from the main satellite. Follows to note, that such way of launching has the irrefutable advantages. The term "rider" for its designation even exists. The fact is that before the small satellite separation the main satellite already moves in predictable and determined motion of center of mass and around it, which, as a rule, is characterized by slow angular velocity. Moreover, if necessary the main satellite before small satellite separation may be reoriented in proper angular position and, by this, initial conditions of small satellite motion may be maintained in enough limits.

In Russia the piggyback launch may be realized on remote and meteorological satellites RESOURS-O, METEOR-3, RESOURS-F, NADEZHDA, and some others. We shell quote their general characteristics, which may be useful for potential customers.

RESOURS-O satellite is developed and manufactured for the ecological monitoring of the Globe. The launch of this satellite is executed into sun-synchronous orbit with inclination of 98' and altitude of 670 ÷ 690 km. The maximal deviation of the altitudes relatively to the Earth's ellipsoid surface after orbit achieved is equal to ±30 km. The problem of increasing the altitude of orbit upto 800 km is considered. The weight of satellites RESOURS-O1 is 1900 kg, RESOURS-O2—2400 kg, including payload 500 kg and 900 kg respectively. The accuracy of orientation of satellite RESOURS-O2 relatively the orbital reference system's axes is equel to 6'. It means, that separation of the small satellite will be executed with given initial conditions, certainly, with accuracy agreed with the separation system effect. After separation owing to unequal resistance of the atmosphere and initial separation pulse the RESOURS-O and small satellite will move over the different altitude but close to each other circular orbits distinguished in altitude of 0.1 ÷ 10 km. The total weight of piggyback payload is upto 200 kg. The installation of 4 small satellites is admitted. The

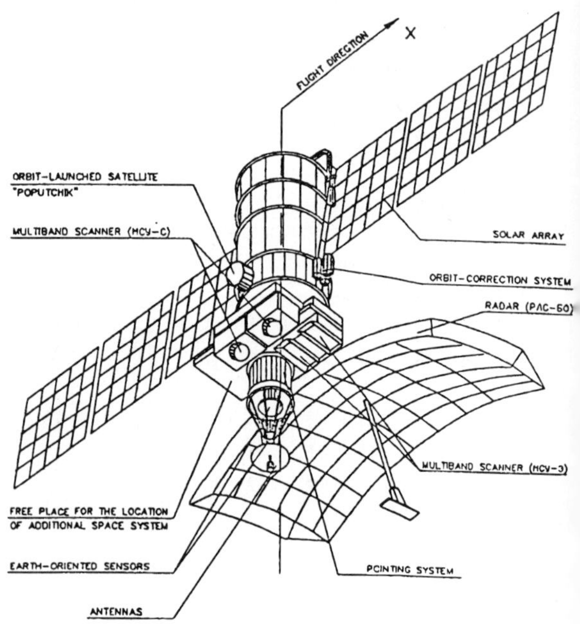

Figure 9. The draft of RESOURS-O spacecraft with small satellite installed.

view of satellite RESOURS-0 with one small satellite installed is shown on Fig.9. It is assumed small satellites SAFIR-2 (Germany), TECHSAT (Israel) and FIZO (Australia) will be orbited on RESOURS-O4 board.

METEOR-3 satellite is also developed and manufactured by VNIIEM. The satellite is intended for the hydro-meteorological observing the Earth's surface. The launch is executed into near-polar orbit with inclination of 82.5° and altitude of 1200÷1250 km (METEOR-3) and 900÷950 km (METEOR-3M). The weight of satellite METEOR-3 is 2200 kg, METEOR-3M—2500 kg including payload 500 kg and 900 kg accordingly. The accuracy of orientation of satellite METEOR-3M relatively the orbital reference system axes is equal to 10 arc minutes. The weight of piggyback payload is upto 200 kg. There is a positive experience to orbit small satellites by METEOR main satellite. This activity is planned to be gone on.

RESOURS-F satellite is developed and manufactured by Enterprise "Central Special Design Bureau". The satellite is intended for remote sounding of the Earth. The satellite is launched into the near-circular orbit with inclination of 82° and altitude of 250÷270 km. Notice, that these orbits are preferable to launch the small satellite provided by aerodynamic attitude control system due to appreciable the atmosphere resistance. On RESOURS-F satellite two places for installation and following ejection of small spherical satellites of PION and MAK-A satellites' type are available. The weight of each small satellite is 50 kg, diameter of its case is 0.33 m. Six such satellites were launched yet. Outline of two small satellites installed is shown on Fig.5.

NADEZHDA and PARUS satellites. Since 1969 Enterprise "NPO Poljot" placed in Omsk City has started serial light COSMOS-rocket. Upto the present moment more than 700 launches of this rocket with 20 failures only were realized. It is considered as the most reliable launcher over the world. This is two-stages liquid propellant rocket. Its launch is proceeded from on-ground launch-table-site. The COSMOS-rocket can launch spacecraft of weight upto 1500 kg on circular orbits of hight to 1700 km, as well as on sun-sinchronous orbits. The main satellites are NADEZHDA and PARUS, on which the small satellites FAISAT and ASTRID (on NADEZHDA) and UNAMSAT-B (on PARUS) were launched. Thus, the small satellites were installed on the top of main satellite. Besides, the variant of four nanosatellites installation of 5÷7 kg weight each on the separate structure is developed that permits not to be restricted by particular structure of given main satellite. The outline of FAISAT and ASTRID installation on main satellite is shown on Fig.10 [27]. The various ways of small satellite injection — in autonomous regime and by on-ground commands—are possible.

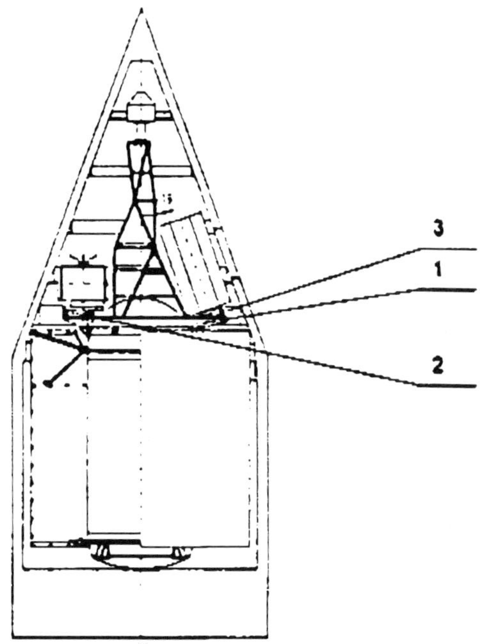

Figure 10. The main satellite NADEZHDA with installed small satellites; 1—ring-adapter, 2—injection system of ASTRID, 3—injection system of FAISAT.

LIGHT LAUNCHERS

Pursuant to Russian-American Agreements about the medium-range missile liquidation, and also by expiration of missile life service, which are standing in fighting trim, their obsolescence, at suitable updating instead of their destruction, it would be necessary to consider the problem of their utilize as launchers for small satellites. For realization of this opportunity in Russia the joint-stock companies NTTs "Complex" (START and START-1 launchers), "Ural-Cosmos" (VOLNA and SHTIL launchers), and others were established.

START family launchers for low Earth orbit reach are built on the base of multi-stage steering solid propellant on-ground based mobile missiles. The Prime-developer of them is the Moscow Institute for Thermo-Facilities. Their features consist in that the rockets may be launched from any point of the Globe without preliminary preparation of launch position. The main characteristics of these carriers are adduced in Table 1. The general orbiting characteristics of the START family launchers are presented in Table 2. The joint installation under fairing and group launching into the orbit of several small satellites, conformably with their dimensions and weights are possible.

Table 1. The main characteristics of the START family.

characteristics	launchers	
	START-1	START
stage quantity	4	5
initial mass, ton	47	60
length, m	22.7	28.8
max diameter, m	1.8	1.8

Table 2. The payload (kg) orbited by START family launchers.

launcher	orbit altitude, km			
	300	500	700	1000
TART-1	410	310	220	110
START	640	505	390	240

There are three variants of their injection scheme: "direct" injection into circular orbit; injection into a transfer orbit using a suborbital trajectory with its subsequent correction by means of satellite devices-, putting the payload into the assigned orbit's point with high accuracy. The third variant is especially important for the satellite orbital constellation building.

About 300 successful starts-up of SS-25 missile were executed, which is used as a basis for START family launchers development. The main technical decisions and characteristics of START-1 rocket were confirmed by first start-up on March 25, 1993 from Plesetsk Cosmodrom. The first successful working start-up was executed successfully on March 1997, when by order of Space Military Forces the small satellite ZEYA manufactured by NPO of Applied Mechanics was launched into the orbit. Thus, the accuracy of orbit altitude was about 1 km, revolution period—0.16 sec, orbit inclination—0.6'. The international contracts for launch of satellites by START from Svobodniy Cosmodrom at Russian Far East are signed.

ROKOT launcher. Also the intercontinental ballistic missile SS-19 may be attributed to the conversional use of military missiles. The "Khrunichev State Rocket Center" updated SS-19 missile into the 'launcher ROKOT of light type, which includes itself the accelerate block of SS-19 missile and speed-up block BRIZ. The speed-up block BRIZ is the third stage of launcher and is intended for arrangment of the payload and put it into the low orbit. Three successful test launches were executed. In particular, the radio-amateur satellite was launched in 1994. The launch is possible to be executed from two under-ground mine-typc starting installations into the low orbit with inclination of 65° and 98°. The opportunities to build up the starting points in Plesetsk to launch into the near-polar orbits with inclination of 82° and Svobodniy in Russian Far East are considered. For the nomenclature expansion on dimensions and configurations of "launchable" payloads the new fairing is developed. It will allow to put the payload into the orbits in accordance with the Table 3.

Table 3. Payload weight to be launched by ROKOT launcher.

orbit altitude, km	orbit inclination, deg	payload weight, kg
300	65	1600
1000	98	900
1000	82	1400
1500	82	1200

THE FAMILY OF LAUNCHERS BASED ON SUBMARINE MISSILES

The State Rocket Centre "Makeev Special Design Bureau" is unique russian enterprise for creation missiles for Navy. At present, this Enterprise offers the carriers for launching the various payloads. The main characteristics of missiles, on the basis of which the launchers are being developed are listed in Table 4 [29]. The power capacities of launchers created on missile basis are submitted in Table 5. Here the number over line (under line) corresponds the height of orbit of 200 km (700 km).

The launcher VOLNA is dedicated to delivery the recoverable apparatus of weight upto 650 ks including payload of 110 kg with length 0.8 m and diameter 0.5 m into suborbital trajectory. The duration of weightlessness of scientific instruments (overload—14-4g) is equel to 22 minutes, the highest point of tragectory is 2300÷3000 km. A number of successful launches with civil payloads was carried out.

Table 4. The main characteristics of submarine missiles.

Missile	VOLNA	SHTIL-1	SHTIL-2	SHTIL-3	RIF-MA	PRIBOY
Index	SS-18	SS-23	SS-23	SS-23	SS-20	SS-20
initial mass, ton	34.4	40	40	45	81	103
dimensions, m:						
length	14.4	14.8	17.3	18.7	20	30.5
diameter	1.8	1.9	1.9	1.9	2.4	2.4
stage quantity	2	3	3	4	4	5
payload volume, m^3	1.3	1.5	3.0	3.6	10	8

Table 5. The power capacities of launchers (kg)

launcher	VOLNA	SHTIL-1	SHTIL-2	SHTIL-3	RIF-MA	PRIBOY
orbit inclination:						
0°	430/185	265/90	410/220	950/730	1500/1200	200/1700
70° ÷90°			620/430	950/650	1600/1200	

On SHTIL-1 the zone of payload is located between second and third stages and has the ring shape. The zone of payload on SHTIL-2 is located in the top part of last stage. SHTIL-3 requires more deep modernization of a third stage and installation of additional stage with apogee thruster. Launches of these rockets are assumed to be executed from on-ground starting complex located in point with latitude about 770 North or from submarine board in the free global ocean. From listed launchers, very likely, SHTIL-2 is the most suitable under requirements of customer (30]. The experience in creation of autonomous marine-missile complexes permits to develop the space-rocket complex of air basing. This complex is created on basis of SHTIL-3A being the updated SHTIL-3 rocket and provides the launch from heavy plane AN-124.

MOLNIA LAUNCHER

There is an opportunity to launch a small satellite as a piggyback by MOLNIA launcher executed by Enterprise "Lavochkin Association" [31] located in Khimki City near Moscow, This is a heavy four-stages launcher able to put 1.9 ton into the high elliptical orbit from Plesetsk and Baykonur. Since 1966 till end of 1996 more than 262 launches with 14 failures only were realized. At present MOLNIA rocket executes the following launches:

- main MOLNIA satellite on orbit with $h_\pi \approx 500 \div 600$ km, $h_\alpha \approx 40000$ km and inclination 62.2° ,
- main PROGNOZ satellite on orbit with $h_\pi \approx 750$ km, $h_\alpha \approx 200000$ km and inclination of 62.8° .

In 1995 the launch of Indian satellite IRS-LC from Baykonur Cosmodrom on sun-synchronous orbit of height about 820 km was successfully carried out.

"Lavochkin Association" has executed 5 launches of small satellites by MOLNIA rocket. They are French -2 (1974, with main satellite MOLNIA), Chech MAGION-4 (1995, with main satellite INTERBALL-1), American-Russian SKIPPER (1995, with main satellite IRS-IC), Argentinean μ SAT and Chech NAGION-5 (1996, with main satellite INTERBALL-2).

In more details the information about possible variants of piggyback launches by MOLNIA rocket is listed in Table 6. The variant A when payload is installed on the fourth step or on the main satellite, β — payload is installed between third and fourth stages.

Table 6. Payload to be launched by MOLNIA and orbit parameters achievable.

Parameter	variant number			
	1	2	3	4
Payload weight, kg	$50 \div 100$	$100 \div 200$	350	500
Payload diameter and height, m	0.8×0.8	TBD	0.8×0.8	TBD
Perigee altitude, km	600	200	600	200
Apogee altitude, km	40,000	600	40,000	600
Orbital period, hour	12	2	12	2
Launchers with payload per year	1-2		1	
place of payload installation	A	B	A	B

EPILOGUE AND ACKNOWLEDGEMENTS

The present state of technology available at the russian space industry enterprises permits them to be involved into international space market. It mostly concernes market of launchers and heavy spacecrafts where technologies utilized are more conservative. Quite more flexible market of small satellites requires fast adaptation of companies involving in small satellite area of activity to very speed changes in the market.

The considerable factor affecting the cost of satellite are the overhead charges of enterprises where the satellite is designed and manufactured. It is expediently to fix the attention on this question in more detail. Traditionally the development and manufacturing of spacecrafts in Russia and other countries engaged the enterprises—monsters with quantity of employees in several tens of thousand. On the stage of space industry constitution and development of large projects, especially during the intensive race of armament it was quite justified. Only under the concentration of efforts of huge teams it was possible that so difficult problems could be solved in utterly compressed period.

At present, when the powerful financial support by goverment orders is away, and commercial market of space service in Russia is in rudimentary state, the attempts to organize the development and manufacturing of real cheap small satellites is coming to grief. It is possible to explain it by the various reasons. I see the following main reason. It lies not in scientific, engineering and even not in technological areas. The reason consists in functionals, which are necessary to be optimized during development and manufacturing of small satellites. To support the large enterprises in proper conditions the rich fundings are required, that is, participating in gross projects, processing the large financial and material means are quite needed. To achieve it the appropriate "publicity" is required alongside with other factors. Clearly, that the participation in small satellite projects hardly increases the rating of large enterprise to receive a powerful order. Moreover, the participation in such projects can have a negative effect, certainly, if we don't say about large projects, including the small satellites as a component.

The enterprise developing the small satellite project should confess the opposite principle, that is, this satellite is to be the most cheap one. In this case only it will be competitive on the market of small satellites, where the victory will fall to the most cheap project.

For this reason, the appearance of small private companies followed by private investments along with the market of cheap satellite units and elements will contribute to russian space-industry companies to be involved successfully into very competitive market of small satellites.

The situation on the market of investments and changes in minds of people interested in successful integration of Russia into the small satellite manufacturers' community provide the positive tendencies in this direction.

I'm very grateful for all my russian and foreign colleagues who discussed with me the problems involved in this paper. These discussions assisted me to recognize the present situation on international small satellite market. It permited to present my own points of view on problems formulated in this paper. Thanks again.

I wish to thank Professor Fei-Bin Hsiao from National Cheng Kung University, Tainan, Taiwan and his colleagues for useful help that made possible participation in the Colloquium. This work is performed under the particular financial support by Russian Basic Sciences Foundition (project # 97-01-00910).

REFERENCES

M. Ovchinnikov, Small Satellite Projects in Russia. Proceedings of International Conference on Small Satellites: Missions and Technology, 9-13 September, 1996, pp.121-127.

A.Burlaka, P.Petrov, V.Sudarev, Satellite Personal Communication System "SPS-Sputnik". Proceedings of International Conference on Satellite Communications, 18-21 Oct., 1994, Moscow, Russia. ICSTI, Moscow, 1994, v.2, pp.107-113.

M.Yu.Ovchinnikov, Gravity-Gradient Attitude Control System for Small Communication Satellite. Preprint AAS/GSFC International Symposium on Space Fight Dynamics, Greenbelt, Maryland, USA, 26-30 Apr., 1993, v.1, N AAS-93-282, 15p.

I.I.Velichko, S.M.Efendiev, A.I.Burlaka, The Satellite Information and Communication System. Proceedings of the 2nd International Conference on Satellite Communications, 23-27 Sept., 1996, Moscow, Russia. ICSTI, Moscow, 1996, v.1, pp.115-117.

Yu.G.Milov, The State and Prospects of Development of Space Complexes for Communication Broadcast in Russian Federation. Proceedings of the 2 nd International Conference on Satellite Communications, 23-27 Sept., 1996, Moscow, Russia. ICSTI, Moscow, 1996, v.1, pp.25-29.

V.Arbuzov, E.Korchagin, A.Deev, G.Phaleev, The Efficiency Enhancement of the Space Low- Orbit System GONETS. Proceedings of International Conference on Satellite Communications, October 18-21, 1994, Moscow, Russia. ICSTI, Moscow, 1994, v.2, pp.113-118.

M.V.Tarasenko, The Military Aspects of Soviet Cosmonautics. Moscow, Nikol, 1992, 162p.

I.B.Dunaev, V.I.Ivanov, V.N.Kozlov, Low-orbit System SIGNAL for Communication via Satellites. The Concept of Development and Prospects of Communication. Proceedings of the 2nd International Conference on Satellite Communications, 23-27 Sept., 1996, Moscow, Russia. ICSTI, Moscow, 1996, v.1, pp.201-207.

I.Dunaev, V.Ivanov, B.Chertok, Low Earth Orbit Satellite Communication System SIGNAL. Update Status and Perspectives of Development of the System. Small Satellites for Earth Observation, Digest of the International Symposium of IAA, 4-8 Nov., 1996, Berlin, Germany, p.35-38.

R.Alimov, V.Bondarick, U.Burlakov, V.Kukhtevich, O.Shipulya, S.Yurin, GLOBSAT Low-Orbit Communication System. Proceedings of International Conference on Satellite Communications, October 18-21, 1994, Moscow, Russia. ICSTI, Moscow, 1994, v.2, pp.105-106.

G.Malyshev, V.Kulkov, S.Pritis, S.Voevodin, Yu.Egorov, I.Bakhmetjev, Small-sized Satellite with Aerodynamic Stabilization for Research of Upper Atmosphere MAK-A. Paper on XVI Scientific Readings on Cosmonautics, Section: Fly Vehicles. Designing and Structure, Moscow, 3-6 February, 1992.

V.Bass, Yu.Tarasov, The PIONs in Flight. Aviation & Cosmonautcs, 1990, N 5, pp.40-41.

M.Yu.Ovchinnikov, V.I.Pen'kov, Aerodynamic Stabilization System of Small Scientific Satellite. Proceedings of the ESA International Symposium on Space Fight Dynamics, Darmshtadt, Germany, 30 Sept.-4 Oct., 1991, ESA SP-326, pp.347-353.

G.Malyshev, V.Kulkov, V.Kukin, N.Shmelev, R.Bychkov, The Small-Sized Diagnostic Satellite for Researclie of the Upper Atmosphere and Ionosphere and Execution of Active Experiments MAK-1, MAK-2, MAK-3. The Paper on XVI Scientific Readings on Cosmonautics, Section: Fly Vehicles. Designing and Structure, Moscow, 3-6 February, 1992.

ROSS—Space Complex for Research of the Upper Atmosphere of the Earth. Leaflet. SAKU, Samara, 4p.

M.Miroshnikov, The Mozhayskii Academy—Cradle of Russian Military Space Science. The Tele-Satellite, March, 1997, pp.26-28.

V.Merkulov, A.Zaitzev, M.Ovchinnikov, V.Fateev, The Concept of Small Satellite for Education and Scientific Purposes. Abstracts of the 2nd International Aerospace Congress, 31 Aug.-5 Sept., 1997, Moscow, Russia, p.372.

L.Kerstein, V.Kulkov, G.Malyshev, V.Sergjuk, Satellite for Servicing and Interaction with Orbital Station "Inspector-Pilot". Small Satellites for Earth Observation, Digest of the International Symposium of

IAA, 4-8 Nov., 1996, Berlin, Germany, pp.451-452.
N.Eismont, M.Pivovarov, V.Khrapchenkov, Use of the Aerodynamic Drag for Microsatellite Relative Motion Control in the Vicinity of the Earth Orbital Station. Small Satellites for Earth Observation, Digest of the International Symposium of IAA, 4-8 Nov., 1996, Berlin, Germany, pp.451-452.
V.N.Oraevsky, V.M.Chmyrev, V.I.Kostenko, The Microsatellite For Geophysical Research and Monitoring Emergencies and Technological Disasters from Space. Small Satellites for Earth Observation, Digest of the International Symposium of IAA, 4-8 Nov., 1996, Berlin, Germany, p.84.
The COLIBRY Project. The Platform of Service Systems for Microsatellite for Scientific and Applied Purposes. The fair brochure. Moscow, Vneshnauchpribor, 1997, 3p.
M.Yu.Ovchinnikov, A.I.Dyachenko, Universal Small Space Platform: Orientation Aspects. Papers of International Symposium on Spacecraft Ground Control and Fight Dynamics, 7-11 February, 1994, Sao Jose dos Campos, Brazil, 11p.
Yu.V.Trifonov, Yu.S.Solomonov, Small Multipurpose UNISAT Launched into Orbit by START-1 Launcher. Small Satellites for Earth Observation, Digest of the International Symposium of IAA, 4-8 Nov., 1996, Berlin, Germany, p.463.
I.Rodionov, M.Ovchinnikov, The Microsatellite for Remote Sounding of Environment. Small Satellites for Earth Observation, Digest of the International Symposium of IAA, 4-8 Nov., 1996, Berlin, Germany, p.225-229.
I.Rodionov, M.Ovchinnikov, The Optical Hyperspectral Monitoring System Based on Microsatellite. S& RC "Reagent", M., 1997, 25p.
M.Ovchinnikov, Russian Launch Opportunities for Small Satellite. Proceedings of IAA International Workshop "Small Satellites for European Countries Emerging in the Space Technology", 7-10 May, 1996, Dublin, Ireland, 1996, v.2, 19p.
N.Ivanov, B.Nesterov, Features of Piggy-Back Launch of a Satellite with Spacecraft NADEZHDA Together. Paper on International Competition-Seminar of Small Satellite Projects "The Possibility of Launching the Small Space Devices Including as Constituting Payload, Using Russian Rockets", 15-17 July , 1997, Moscow, International Institute East-West, 4p.
Opportunity to Launch Small-Sized Satellites by Rocket START and START-1. Paper on International Competition-Seminar of Small Satellite Projects "The Possibility of Launching the Small Space Devices Including as Constituting Payload, Using Russian Rockets", 15-17 July, 1997, Moscow, International Institute East-West, 12p.
I.Velichko, N.Obukhov, G.Sitiy, Yu.Nikulin, Yu.Usolkin, The Launchers Based on Technologies of Submarine Missiles, Space Bulletin, 1995, v.2, N 1, pp.24-26.
I.Velichko, N.Obukhov, The Commercial Rocket Complexes for Small Satellite Launching. Proceedings of 2iid International Conference on Satellite Communications, September 23-27, 1996, Moscow, Russia. ICSTI, Moscow, 1994, v.2, pp.101-114
V.Asjushkin, S.Ishin, A.Smirnov, The Integration and Launch of Small Satellites by MOLNIA-rocket. Paper on International Competition-Seminar of Small Satellite Projects "The Possibility of Launching the Small Space Devices Including as Constituting Payload, Using Russian Rockets", 15-17 July , 1997, Moscow, International Institute East-West, 11p.

THE IRISH AND OTHER EUROPEAN SPACE INITIATIVES

Susan M.P. McKenna-Lawlor,

Space Technology Ireland, National University of Ireland, Maynooth, Co. Kildare, Ireland.

ABSTRACT

In many small countries within Europe specialists (e.g. scientists, engineers, and policy makers) have for many years already been, and are presently, actively engaged in areas of space science, space applications and the scientific evaluation of space related data. These various national groups form a wider community in that they experience many common problems and difficulties which have precluded their being able, hitherto, despite their experience in space related activities, to proceed to develop their own small satellites. An account is provided of Irish lead pan-European initiatives, under the aegis of the International Academy of Astronautics, to provide support structures to promote the concepts and processes required by such countries 'Emerging in Space Technology' to gain access to space using small economical satellites. The consequent setting up within the Academy of a new Sub-Committee of the already existing Committee for 'Small Satellite Missions' entitled 'Small Satellites for European Countries Emerging in Space Technology' is described. This Sub-Committee has a philosophy of 'learning by doing' and it is hoped that the group will ultimately succeed in launching a microsatellite bearing an innovative scientific experiment. A wish has been expressed by participants in this Sub-Committee that its remit be extended to include not only European, but representatives of all countries 'Emerging in Space Technology.

SMALL SATELLITES FOR DEVELOPING COUNTRIES AND FOR COUNTRIES 'EMERGING IN SPACE TECHNOLOGY'.

The International Academy of Astronautics (IAA) which was founded in August 1960 in Stockholm by Thedore Von Karman, includes among its objectives that of producing effective results in the field of international cooperation in the space sciences. The motivation underlying this intention is that a spirit of cooperation would resultingly be fostered that can transcend national boundaries and cultures.

In 1988 Gerhard Haerendel of Germany initiated a special activity within the Academy on the topic of 'Small Satellites' when he organized a dedicated session on 'Inexpensive Scientific Satellites' at the International Astronautical Congress in Bangalore. This initiative resulted, over time, in the establishment of a full Committee within the Academy on ‘Small Satellite Missions’. The purpose of this Committee is to develop and promote the concepts and processes required by various user communities to gain access to space using small economical satellites, and associated launches. In this regard, with the object of promoting the participation of Developing Nations (as defined in political terms by the United Nations) in future space activities for the overall benefit of mankind, a Subcommittee of this Committee entitled 'Small Satellites for Developing Nations' was set up in 1992 and this body has, since that time, held a number of successful international meetings and workshops aimed at focusing the effort of Developing

Nations to gain access to space using small satellite technology.

In 1993 I was myself invited to join this newly formed IAA Sub-Committee and, thereafter, particularly at a Workshop held by the group in Brazil during the Summer of 1994, 1 became aware that the problems and objectives of 'Developing Nations' in gaining access to space using small satellites are fundamentally different from the problems and objectives of small European nations, such as my own country Ireland, which already have an existing background in space technology and can thus, now, realistically aspire to fly a small satellite mission (see Table 1 which summarizes the scientific and associated technological experience of Ireland as a representative small European country in participating, in various roles, in scientific missions launched by the European, American and Russian Space Agencies). I highlighted the difference between the problems of 'Developing' and 'Space Emerging' European countries at a plenary session on Academy Day in Jerusalem in 1994 and, following various meetings and discussions thereafter, it was decided by the Academy to mount a Workshop at the National University of Ireland, Maynooth on the topic 'Small Satellites for European Countries Emerging in Space Technology I (May 7-10, 1996), with a view to determining if a community of 'Space Emerging Nations' indeed exists within Europe with an interest in this subject.

The need within 'Space Emerging' countries of the European community for this kind of Workshop was confirmed by the arrival at Maynooth to participate in the Workshop of representatives of twelve countries (namely Austria, Denmark, England, Germany, Holland, Hungary, Ireland, Norway, Portugal Russia, Sweden and Switzerland). Expressions of interest in and support for the topic were also in parallel received from Bulgaria, The Czech Republic, Finland, Spain and Turkey. In addition to the formal presentations of some twenty two scientific and technical papers, an important product of the Workshop was the deliberations and recommendations of its members which, in view of their interest to the present COSPAR Colloquium on 'Scientific Microsatellites as Research Tools'. are summarized below in Section 2.

SELECTED DELIBERATIONS OF DELEGATES AT THE IAA WORKSHOP HELD IN IRELAND IN MAY, 1997 ON'SMALL SATELLITES FOR EUROPEAN COUNTRIES EMERGING IN SPACE TECHNOLOGY.

DAY 1

The participants were divided into three groups to discuss problems associated with the initiation/implementation of a small satellite program. Each Group Chairperson collated inputs from his/her Group Members and these were then formally presented in plenary session and generally discussed in the context of trying to achieve a consensus. While there were some differences, the majority view on the following topics emerged over the following four days.

Definition of Countries Emergin2 in Space Technology and Their Proposed Mission Philosophy

'Countries Emerging in Space Technology' were defined to be countries with a technical knowledge base and some space experience which are striving for small satellite missions to exploit the new, cost effective, possibilities they offer.

It was the view of the meeting that any small satellite program undertaken should:

- Have strongly focused goals
- Produce world class science
- Be affordable for 'emerging countries' (as well as for 'emerging' institutes and universities)

- Have an educational component

The program could include innovative dedicated 'Earth Observation' missions providing monitoring of particular phenomena over, extended periods. Telecommunications and other commercially oriented missions were deemed, in contrast, to be the province of specialist personnel within individual countries.

It was recognized that experimenters require to know the functionality and characteristics of their small satellite bus at a level comparable with their knowledge of the behavior of their individual instruments.

It was debated that the name 'Small Satellite Mission' has an undesirable connotation since 'small' might be construed as, somehow, 'second rate'. A 'Small Satellite Mission' was considered by the Group to be 'small' in terms of what it costs. The physical mass is not relevant.

Mission Support and Required Mission Elements

'Small Missions' are appropriate to individual countries, or to multinational groups co-operating together. As such, they constitute what can be defined as 'part of the national space program within a country'. Major space agencies are not seen to constitute leading partners in such an activity.

Although the primary source of support for such missions will probably come from Government, sponsorship should be sought (for example from industry - perhaps on a matching funds basis). Advantages for industry can arise through (a) strengthening its technology base, thereby rendering a particular company more competitive in applying later for commercial contracts; (b) through associated advertising. Other strategies for securing financial support should be explored if this is required.

The following steps were identified as necessary to the mounting of a 'Small Space Mission':

- Preparation of a Feasibility Study (interdisciplinary)
- Secure funding/lobby for the Project using the Feasibility Study as an explanatory document
- Establish Management Structures
- Detailed design and development
- Manufacture and procurement
- Assembly/Integration/Test/Calibration
- Launch campaign/launch
- Mission operations
- Data dissemination/analysis

Items that should be procured/decided include:

- The Feasibility Study/Mission Analysis
- Platform
- Payload
- Ground Station
- Management/support staff
- Launch services
- Mission operations
- Insurance
- Orbit Choice

It was recognized that not all orbits can be reached with the same frequency of secondary launch opportunity. Bearing this in mind, it was recommended that the scientific objectives of 'Small Missions' should be chosen having regard to the availability of access to those orbits required to implement the desired measurements - having regard to the fact that secondary payloads are lofted into Polar, Geostationary Transfer and orbits of high inclination relatively frequently. Sunsynchronous and Geostationary Transfer Orbits were identified as particularly favored choices among the experimenters present.

Getting the Work Done (Facilities Required)

It was recommended that a catalogue listing upcoming launch opportunities should be compiled and kept updated within the Group. A need was also felt for a catalogue specifying where affordable test/calibration facilities outside of the main space agencies and prime contractors can be found within Europe. Group members undertook to collect this information for collation, and later distribution, by the Chairman.

The unfavorable aspects of waiting for a 'free launch' were noted and it was agreed that there is 'no such thing as a free launch'. Experience indicates that to pay for a secondary payload launch is a better alternative.

Insurance

It was generally agreed that 'Small Mission' hardware should be insured to cover ground transportation and handling. An alternative to launch/post launch insurance could, however, be the construction of a redundant satellite. The decision in a particular case to carry insurance to cover launch/post launch failure depends, in the last analysis, on the level of available funding.

Ground Support Activities

It was agreed that small, relatively cheap, local automatic ground stations would be sufficient for 'operations' activities. On-board GPS capability obviates the need for ground station hardware to provide in-orbit determinations. A ground tracking station can provide a strong focus for public interest.

DAY 2

Getting Technical Expertise to Build a Satellite.

As a general philosophy it was considered that it is not necessary for experimenters to learn how to construct all required satellite subsystems. These items are available commercially and do not need to be developed 'in-house'. A participative role for local industry in the activity of mounting a 'Small Space Mission' is, however, desirable and can take many forms (e.g. production of attitude control and data management systems, building of flight model scientific instruments etc.). A structured approach to the gaining of necessary expertise involves training of personnel and the nurturing of national skills.

Training

- Through Summer Schools mounted under IAA patronage (the Academy should be approached to mount such courses)
- A limited number of those who have attended such Summer schools can then go on for post graduate training abroad (involving hands on experience) extending over 1-2 years (brain drain

issues Should he carefully considered in individual cases)

Sources of Training

Experienced Research Institutes/Universities might host, for training purposes, staff/Ph.D. students from countries 'Emerging in Space Technology', particularly within the context of jointly mounted missions. The EC Human Mobility Program could provide support for such visits.

ESA stipends and job opportunities provided within ESTEC at the Space Science Department already enable young people to learn how to participate in space projects. Courses at the International Space University also provide useful experience.

Technology transfer to personnel involved in a particular Project can, in addition, be gained through 'hands on' experience at other organizations with relevant expertise. For example, spacecraft manufacturers typically provide support to their customers in learning to successfully exploit the hardware/software (including 'turn-key systems) they offer commercially.

Nurturing National Skills

Nurturing of national skills requires two kinds of support, namely:

- Acquisition of Systems competence
- Structuring of the national substrate

Acquisition of systems competence can be achieved by:

- A national returning to his/her country with the required knowledge (reverse brain drain) where technical support is provided by personnel trained through one or more of the processes described in the previous section
- Time-limited hiring of carefully selected Consultants
- Outright acquisition of core foreign personnel

Structuring of the industrial substrate requires:

- Identification of local potential capability to participate in space activity
- Encouragement to such groups to participate
- Injection of suitable procedural standards

Facilities required to Get the Work Done

Access should be available to all, or to a subset, of the following, depending on individual circumstances

Workshop
Clean Room
Environmental testing, facilities
Spin Balance Test/deployment test /separation mechanism test
Thermal/solar simulation
Antenna range

Appropriate procedures and technical standards should be adopted in construction/testing/calibration.

DAY 3

Framework for Mounting a Mission

A 'groupe de reflexion', including industrial and scientific elements, could produce a mission theme involving scientific and technological development.

A mission framework (defining participating countries, a budget ceiling and selection procedures for experiments) should be agreed at Governmental level.

An Announcement of Opportunity (AO) should be issued simultaneously in all participating countries which would allow (within a predetermined budget) scientific, technical and cultural inputs to the mission and generate a feeling of competition in these different areas.

A successful scientific proposal should have the following characteristics:

- Capability to provide innovative world class science
- A dedicated instrument package focused on a primary scientific goal
- Maximization of the possible scientific gain (e.g. through integrated on board campaigns)
- A time scale relevant to the primary goal but within the constraints set by a small mission
- Use of technological innovation if this is required to support the mission objectives.
- A high public profile
- Small mission compatibility
- Suitability for co-operative implementation/exploitation

All of the Scientific, Technological, Educational and Cultural proposals submitted should undergo (a) peer review rating, (b) selection by a Steering Committee; (c) final review and acceptance at political level.

The selected mission should then be implemented.

The view was strongly held that such an initial mission should be mounted in the context of a longer term program. In this context. a thorough multidisciplinary evaluation of its success should be carried out as an input to the implementation of the next step in the program.

DAY 4

Selected Workshop Conclusions/Recommendations

The experiences of' the Workshop confirmed the existence of a community of European countries 'Emerging in Space Technology'.

The creation of a dedicated Sub-Committee within the framework of the International Academy of Astronautics was recommended to foster the interests and needs of this community. It was anticipated that (among other useful functions) the new Sub-Committee would provide a forum where world class science programs could be developed, and partners identified to co-operatively implement them.

I was myself, in my capacity of Chairperson of the Maynooth Workshop, authorized to present the conclusions and recommendations of the Workshop at the next meeting of the Academy in Beijing, China in October 1996.

FORMATION OF A NEW SUBCOMMITTEE OF THE IAA ENTITLED 'SMALL SATELLITES FOR EUROPEAN COUNTRIES EMERGING IN SPACE TECHNOLOGY'

The recommendations of the Maynooth Workshop were unanimously supported by the IAA Committee on 'Small Satellite Missions' at Beijing and I was immediately invited to draw up and submit to the Academy a document describing the objectives and goals of the proposed new Sub-Committee on 'Small Satellites for European Countries Emerging in Space Technology'. I was also required to provide, in parallel, 'Five Year Plan' for the activities of the new Sub-Committee. These documents were duly prepared and submitted to the Academy and, in due course, the foundation of the new Sub-Committee was authorized and this Sub-Committee had its first formal meeting at the IAA Meeting held in Turin, Italy in October 1997.

Objectives and Goals

The Objectives and Goals of the new Sub-Committee, which are based on a philosophy of 'learning by doing', give to this body an identity and profile different from that of other Committees/Sub-Committees presently within the Academy. These objectives and goals comprise:

- Selection of an innovative space experiment to fly on a microsatellite
- Exploitation of this experiment to provide the focus for a learning experience in determining how such a representative payload could be successfully flown
- Identification of partners to participate on a practical level and/or as a learning experience only in the experiment
- Advancement through all stages of preparation for and flying the mission, putting in place along the way opportunities for participants to acquire necessary technical and managerial skills to carry out the work.
- Publication of scientific and technical papers at all stages of the Project

Terms of Reference

Participation in the new Sub-Committee is presently by representatives of European countries with a technical knowledge base and some space experience which are striving for small satellites to exploit the new, cost effective, possibilities they offer.

A structured approach (having regard to the deliberations of the Maynooth Workshop) will be adopted over time by the Sub-Committee to nurture national expertise within particular countries through, for example:

- Mounting Summer Schools (ideally under IAA patronage) on scientific and technical subjects
- Facilitating processes whereby experienced Institutes/Universities would host, for training purposes, staff/Ph.D. students on a trans-national level
- Opportunities will be provided to bring together representatives of countries / universities / research institutes already successful in building and launching small satellite for scientific / educational / commercial purposes, so that they can share their experiences, at meetings of the Sub-Committee with those striving to mount the Mission. Representative of companies providing small satellites,

satellite subsystems and launch opportunities will also be given the possibility, at appropriate times, to describe what is commercially available to support small satellite missions.

- Papers produced through the scientific and technical activities of members of the Sub-Committee will be contributed to the IAA (Acta Astronautica), the IAF and to COSPAR.

Five Year Plan

Year One

- Send out a call for Proposals for an innovative scientific mission to be flown on a microsatellite, within a given budget envelope
- Establish a panel of experts within the Academy to review the merit of the Proposals received
- Set up a meeting of the Subcommittee at which all the Proposals are presented and one is formally chosen
- Publication of the winning Proposal

Year Two

- Hold a Workshop to explain how to prepare a Feasibility Study for the Project
- Decide which countries will be practical partners in the Project and which will participate at a 'learning level' only
- Establish Management Structures to take care of the scientific, technical, educational. financial, and cultural aspects of the program. Involve political decision makers in these structures from the participating countries
- Set up a formal agreement re data rights and other legal issues
- Start the work of investigating how to implement the project
- Lobby for necessary funding

Year Three

- Hold a Workshop to discuss problems encountered in educational, financial, technical, and other areas
- Put in place Workshops/Summer Schools under Academy patronage to provide information on how to overcome the perceived difficulties
- Investigate launch opportunities
- Encourage experienced Institutes to host, for training purposes, personnel working on the Project Ensure that the acquisition of necessary systems competence is equally gained by all participants
- Formulate a Design and Development Plan for the Project
- Produce a Project Overview Report for publication

Year Four

- Proceed to manufacture and procurement
- Organize assembly/integration/test/calibration activities
- Put in place Workshops/Summer Schools under Academy patronage to ensure that all participants are at an equal level of competence in the above mentioned activities
- Produce a Project Overview Report for publication

Year Five

- Proceed to establish a ground segment
- Put in place Workshops / Summer Schools to train participants in operational skills / data handling / archiving etc.
- Proceed to launch; mission operations; data dissemination and analysis
- Produce a Final Report for publication on the Project

It is particularly noted that there is no obligation within the Academy to make this Project happen. Success will only come about through the efforts of the participants themselves. The program is designed to constitute a valuable learning experience which could, in the best case, be crowned by the triumph of flying an IAA microsatellite bearing an innovative scientific experiment.

The first meeting of the new Sub-Committee at Turin, Italy in July 1987 attracted a large international attendance and the Terms of Reference of the Sub-Committee and its Five year Plan were, at that time, generally adopted by Sub-Committee Members. Plans are presently in train within the Group to prepare and issue an Announcement of Opportunity for an innovative scientific experiment to fly aboard a microsatellite (within a given budget envelope) as per the program of Year 1. A second Workshop is planned to be held at Maynooth to advance this activity.

An important outcome of the first meeting was that participants expressed the strong view that the remit of the Sub-Committee should be extended to include interested participants from all countries of the world and this view will be conveyed to the Academy at its next meeting in September/October 1988 in Melbourne, Australia.

CONCLUSION

It is recognized that, in many small European countries, specialists (including scientists, engineers, and policy makers) have been already, and are presently, actively engaged in areas of space science, space applications and the scientific evaluation of space related data. Despite this activity, these national groups form a community in that they have many common problems and difficulties which have precluded their, hitherto, being able to develop their own small satellites.

A new Sub-committee has been set up under the aegis of the IAA to provide support structures to develop and promote the concepts and processes required by 'European Countries Emerging in Space Technology' to gain access to space using small economical satellites. This Sub-Committee has a philosophy of 'learning by doing' and it is hoped that the group will ultimately succeed in launching a microsatellite bearing an innovative scientific experiment.

A wish has been expressed by participants in the Sub-Committee that its remit be extended to include interested representatives of, not only European, but of all countries in the world.

Table 1. Space Missions in which Ireland, as a representative small European country has already participated, or is presently in process of participating, through the design/construction/testing of space mission hardware and software and/or engaging in space data reduction/interpretation. Participation typically occurs either through a local scientist holding the position of Principal Investigator (PI); Team Leader (TL), Co-Investigator (Co-1); Working Group Member (GM), Official Manufacturer (OM) or Co-Opted Scientist.

Agency	Mission	Instrument	Experimenter's Status
ESA	GIOTTO/Halley GEM/Earth Flyby GEM/Grigg - Skjellerup	Energetic Particle Detector (EPONA)	PI
	SOHO	Low Energy Detector (LION)	Team Leader for LION
	CLUSTER(4 s/c)	Mass Spectrometer (RAPID)	Co-I
	ISO	Imaging Photo-Polarimeter (ISOPHOT)	Co-I
	ROSETTA	The COSAC Experiment	Co-I
	STEP	Payload Working Group	GM
	INTEGRAL	Study Team	Co-I
NASA	Apollo 11,16,17	Cosmic Ray Experiments	Co-I
	SKYLAB	White light coronagraph/soft X-ray spectrometer. (Data handling/ reduction/interpretation).	Co-opted Scientist
	LDEF	Ultra-heavy Cosmic Rays	Co-I
	Solar Maximum Mission	Ultraviolet Spectrometer (Data handling/interpretation)	Guest Investigator
	WIND	Radio Spectrometer (WAVES)	Co-I
	Relativity Mission (Gravity Probe B)	Energetic Particle Detector (EPD)	OM
Russian	PHOBOS (2 s/c)	Low Energy Particle Detectors (SLED 1, SLED 2)	Co-PI
	MARS-96	Low Energy Particle Detector (SLED- I 1)	Co-PI+OM
	MARS-96	Ionospheric Monitor (MARIPROBE)	CO-I+OM
	MAIZS-96	Magnetospheric Monitor (FONEMA)	Co-I
	Mars-96	Magnetic Field/Plasma Electron Investigation (MAREMF)	Co-I

OUTLINE OF THE WHALE ECOLOGY OBSERVATION SATELLITE SYSTEM

Tomonao Hayashi, Shigeru Hosokawa, Hideho Tomita

Chiba Institute of Technology, 2-17-1,Tsudanuma, Narashino, Chiba 275,Japan

ABSTRACT

The mission of the “Whale Ecology Observation Satellite System” has started from the opportunity of the first Satellite Design Contest held in Japan in 1993. Since then effort towards realizing the system has been made at the Chiba Institute of Technology.

This system consists of a small dedicated satellite in a polar orbit, probes attached to whales, and a ground station. By relaying the probe data at the satellite, whale ecology data, which have been unknown will be retrieved at the ground station. This system will also effectively be applied to many other fields.

INTRODUCTION

Conservation of whales has been repeatedly discussed for many years, but ecology of whales is not yet clear because of the enormously wide scope of their active regions. To solve this difficulty and to extend our knowledge of whales, we are developing a Whale Ecology Observation Satellite(WEOS) System. This system consists of three segments:(1) a small dedicated satellite (WEOS) of 50kg in weight on a polar orbit of around 800km high, (2)probes attached to plural whales in various oceanic regions, in which several sensors are installed, and (3) a ground station for satellite tracking, operation and data acquisition.

GENERAL SCHEME

Three segments stated above are shown schematically in Fig.1. Ecological data obtained by probes are sent with their identification code through UHF telemetry to the WEOS, which are stored in a memory onboard the satellite. When the WEOS passes over the ground station, it sends a command signal in UHF band to the WEOS, and the stored data are sent back to the ground station together with the house keeping data of the WEOS.

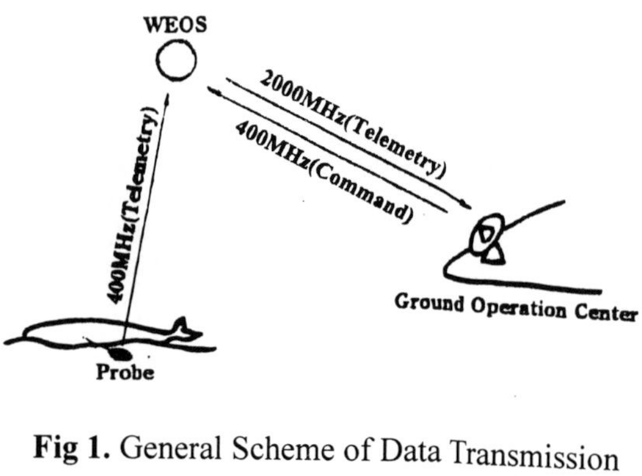

Fig 1. General Scheme of Data Transmission

SATELLITE(WEOS)

Flight configuration of the WEOS is shown in Fig.2. An extendible mast with hingeless mechanism is deployed after putting satellite into its orbit, which serves for gravity gradient stabilization of the attitude. Mission overview of the WEOS is shown schematically in Fig.3. Antennas of the WEOS on the orbit will thus be kept always to face the earth's surface. Some of the remained nutation foreseen will be actively damped by using a magnetic torquer. Communications link is set up in the scheme as shown in Table 1.

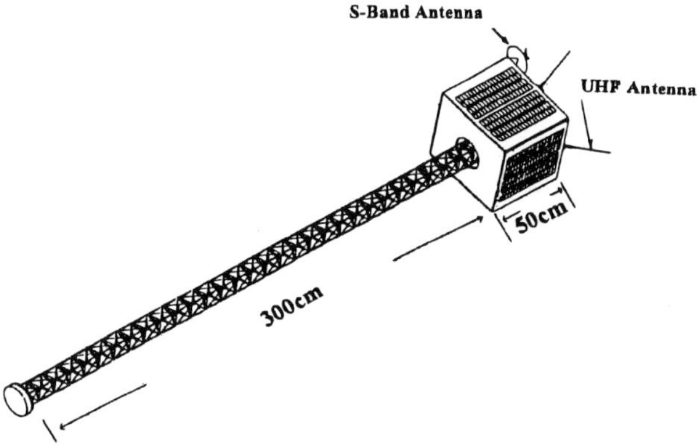

Fig 2. External View of the WEOS (mast expanded)

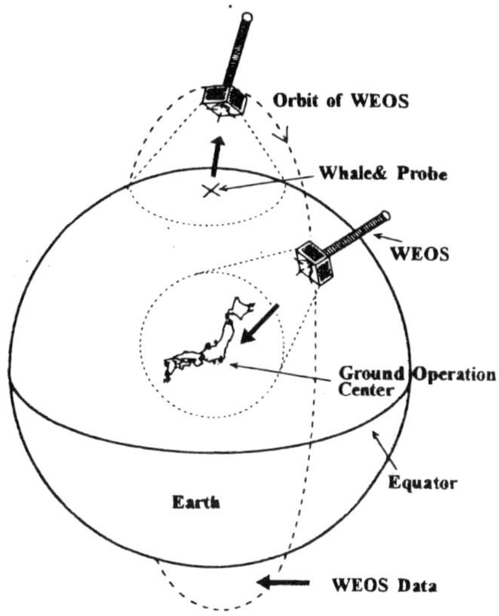

Fig 3. WEOS Mission Overview

Table1. Communications and Frequencies

Communication	Signal	Frequency	Code & Bit Rate	Note
Whale→Satellite Probe	Telemetry	401…402	NRZ-S…CONV. …BPSK Bit rate; 300bps Symbol rate; 600bps	From Earth to Space
Satellite→Ground Station	Telemetry	2000	NRZ-S…BPSK 1200bps	From Space to Earth
Ground→Satellite Station	Command	400	NRZ-S… Manchester…FSK 1200bps	From Earth to Space

Since the data transmission from the probe to the WEOS would be in a crucial circumstance, namely: (1)both transmission and reception are conducted with low gain antennas without man intervention for each side, (2)power available for the probe is not so ample, and the radiated power is limited to be around 1 watt, (3)the frequency allocated for this link is in UHF band, and fairly amount of interference and

Doppler shift are unavoidable, and (4)the WEOS has to make quick acquisition of signal from the probe since the time period for being surfacing of the whale occasionally would not be so long. To overcome these difficulties and to achieve reliable reception of the uplink signal at the satellite, a sensitive receiver is being developed, in which by avoiding the interference signal by an FFT processor, and also narrowing a phase-lock loop band for the carrier frequency, quick and reliable acquisition of the signal is established. The block diagram of the onboard system is shown in Fig.4. Electric power for this system is around 15W and is supplies by four solar cell panels attached to the side of the WEOS. The mission life is designed to be more than 2 years. The WEOS is now nominated as the candidates of the piggy-back launch of H-2 rocket by NASDA at the opportunity of the launching of the ADEOS-2 (Advanced Earth Observation Satellite 2) in the fiscal year of 1999.

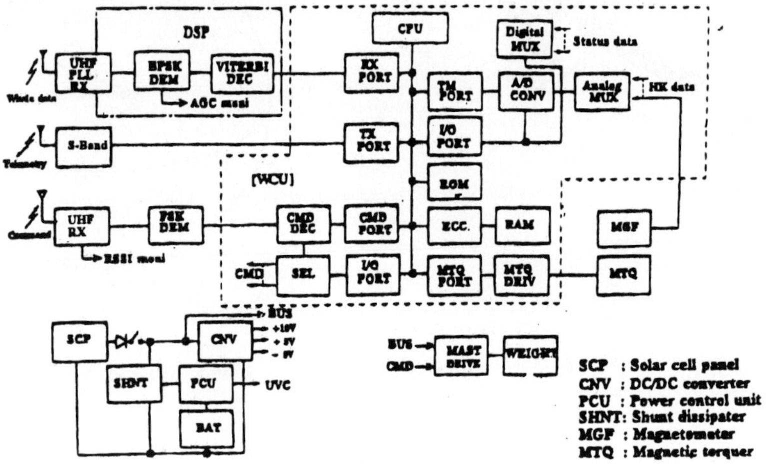

Fig 4. Block Diagram of WEOS

PROBE

We would like to attach a probe to each of plural whales in various oceans. In the probe following sensors are installed, namely, position (GPS receiver), pressure, temperature, geomagnetic field, and acoustic signal. The data obtained by the last four sensors in the sea are stored in a memory device, and when the whale surfaces for breathing, those data are combined with the position data from the GPS receiver, and are sent to the WEOS through an UHF transmitter. The probe floating on the sea surface is schematically shown in Fig.5. Electric power needed for the instruments in the probe will be generated by a kinematic motion of a pendulum inside the probe, which is excited by the swing motion of the probe through the translational motion of the whale. This idea is an application of the technology, which is practically utilized in a wristwatch "Kinematic" developed by the Seiko-Epson Co. To produce a proper motion of the probe, it is important to form the outer shape of the probe so as to exite Karman vortex behind the probe being towed. For designing the form of the probe we are cooperating with a study team of lures at the Tokyo University of Fisheries. Each probe is attached to a whale by a tiny pin, which is clud by Ti to avoid any hazard to living organization. Corresponding to the species of the whale, the length of the pin is chosen to be nearly equal to the thickness of its blubber. To make the attachment as gentle and firm as possible, several barb plates made of Ti-Ni shape memory alloy are fixed at the top of the pin. The barb is so designed as to be a quarter of a circle in the environmental temperature higher than 30℃ as shown in Fig.6. If the top of the pin were inserted into the muscle and pulled back somcthing, the barbs will be set on the fascia, and will be anchored tightly. The pulling force will then be distributed among the plates, and the pressure will be reduced to the level for avoiding necrosis. Encapsulation of medicine in the pin is also considered to protect the whale from infection.

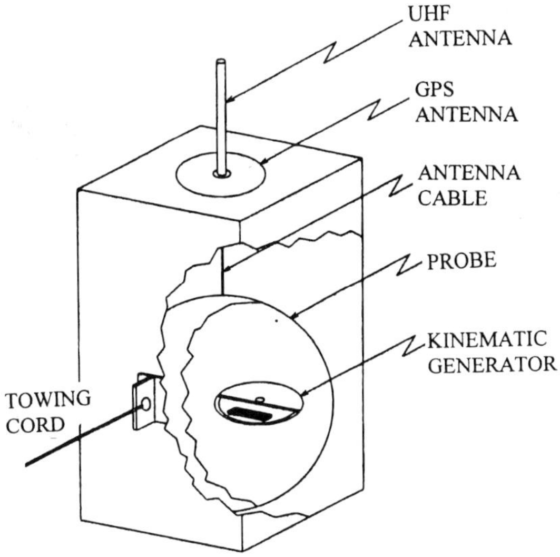

Fig 5. Configuration of Probe

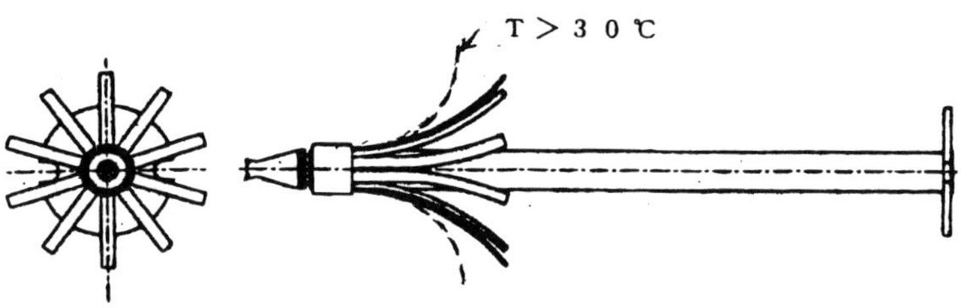

Fig 6. Conceptual View of Pin (the size is different corresponding to the specy of the whale)

GROUND STATION

In the design of the ground station matured technologies developed in the radio amateur community are adopted, which is very effective to reduce the cost. A ground station is being built at the Chiba Institute of Technology for tracking, control, data acquisition, and its analysis.

COST REDUCTION

To realize this system in low cost, we adopted commercial electronics parts for the WEOS with adequate tests. Measurement and test of parts and subsystems are conducted by students with educational considerations, which is also quite effective for reducing cost.

CONCLUSION

We are intending to provide a means for clearing the ecology of whales by using a small dedicated satellite on a polar orbit, probes attached to many whales, and a ground station. The data obtained by this system will provide valuable data for the conservation of whales. By accumulating migration patterns thus obtained, it would be possible to get the information of ocean environment by making use of whales as biosensors. This system would also be applicable to studying ecology of land animals.

ASTRID-2: AN ADVANCED AURORAL MICROPROBE

Lars G. Blomberg, G. T. Marklund, Per-Arne Lindqvist, and Lars. Bylander

Alfvén Laboratory, Royal Institute of Technology, Stockholm, Sweden

ABSTRACT

Astrid-2 is an advanced auroral microprobe with dual primary mission objectives; to do high-quality *in situ* measurements of the physical processes behind the aurora, and to demonstrate the usefulness of microspacecraft as advanced research tools. Mission success will open up entirely new possibilities to carry out low-budget multipoint measurements in near-Earth space. This long-desired kind of *in situ* measurements are the next major step forward in experimental space physics. Astrid-2 has platform dimensions of 45 x 45 x 30 cm, a total mass of just below 30 kg, and carries scientific instruments for measuring local electric and magnetic fields, plasma density and density fluctuations, ions and electrons, as well as photometers for remote imaging of auroral emissions. Attitude determination is provided by a high-precision star imager. Some 250 Mbytes' worth of scientific data will be received each day at the two ground stations. Astrid-2 will be launched as a piggy-back on a Russian Kosmos-3M launcher into an 83 deg inclination circular orbit at 1000 km altitude. Nodal regression will give complete coverage of all local time sectors every 3.5 months.

BACKGROUND AND GENERAL INTRODUCTION

As a continuation of the successful Swedish minisatellite programme that produced the auroral research spacecraft Viking, 1986 (Hultqvist, 1990) and Freja, 1992 (Lundin *et al.*, 1994); a microsatellite programme was embarked upon in 1993. One of the ultimate goals of the microsatellite programme is to facilitate inexpensive multipoint auroral *in situ* measurements. The first microsatellite in the Astrid series, Astrid-1, was launched in January 1995, carrying an energetic neutral atom imager as its main payload. The flight was successful, and served as a proof of concept for a scientific microsatellite. Inspired by the success of Astrid-1, the second spacecraft in the series, Astrid-2, was given a project go-ahead in the spring of 1995. Astrid-2 carries a much larger payload than Astrid-1, and is focussed on diagnosing auroral processes in the Earth's upper atmosphere and ionosphere. Astrid-2 is intended as a proof of concept for auroral microsatellites, and includes innovative mechanisms that are further detailed below.

Astrid-2 will be launched as a piggy-back on a Kosmos-3M launcher from Plesetsk, Russia (N62.8, E40.3). The nominal orbit is circular at 1000 km altitude with an inclination of 83 deg. The initial right ascension of the ascending node will not be known until shortly before launch. However, relative to the Sun, the nodal regression rate is –1.7 deg/day, so in about 3.5 months all local times will be covered, regardless of the initial orientation.

The overall technical responsibility of the Astrid-2 mission is with the Swedish Space Corporation while the overall scientific responsibility is with the Alfvén Laboratory of the Royal Institute of Technology.

THE SPACECRAFT

The spacecraft platform is built by the Swedish Space Corporation, and is a box of dimensions 45 x 45 x 30 cm. With the solar panels deployted the dimensions are 170 x 110 x 30 cm. In addition, 7 booms will be deployed for the scientific instruments. The overall mass is 29 kg, of which almost 10 kg is payload. The spacecraft is spin-stabilized with a nominal spin rate of 10 rpm. There are 6 solar panels, two fixed ones and four deployable ones, which lie along the sides of the platform in their stowed position. Together they deliver a total of 90 W of power when sunlit. Since the solar panels all face in the direction of the spin axis, the spacecraft should be kept Sun-pointing to within 30 deg. The spacecraft surface including the solar panels is electrically conducting, to avoid differential charging and also to supply a sufficient area for the return current from spacecraft to plasma caused by current biasing of the electric field sensors, and by sweeps of the electric field and plasma density instruments. A picture of the Astrid-2 platform with deployed solar panels is shown in Figure 1.

Figure 1. The Astrid-2 platform with deployed solar panels.

MISSION OBJECTIVES - SCIENTIFIC AND TECHNOLOGICAL

Astrid-2 has dual primary mission objectives. On one hand it is a fairly sophisticated research satellite, with scientific objectives related to the physics of the aurora. On the other hand it is a technological pioneering mission where a number of innovative techniques and mechanisms will be flight tested for the first time. If both mission objectives are fulfilled satisfactorily, the mission will serve as a proof of concept for small inexpensive auroral research probes. This will pave the way for affordable multi-spacecraft missions in the future.

Scientific Objectives

The advanced payload of Astrid-2 allows for a number of scientific topics to be addressed. Below we give a non-exhaustive list of examples of such topics.

Electrodynamics of Aurora and Black Aurora. The auroral acceleration processes are complex. A fairly recent discovery by the Freja spacecraft (Marklund *et al.*, 1994; 1997), subsequently studied also by the FAST team, was an anti-symmetry between the electric field structures in the upward field-aligned current region associated with auroral particles and those in the downward current region. Previously the downward current region was thought of as devoid of acceleration processes. Freja has demonstrated the existence of upward acceleration of electrons in these regions, with particularly strong transverse electric fields as a result. The strong transverse electric fields arise when current closes through ionospheric regions of low conductivity. The upward acceleration region is typically found at lower altitudes than the auroral acceleration regions, and we expect Astrid-2 to be able to contribute significant new observational data to shed additional light on the processes operating.

Physics of Transverse Ion Heating. Transverse ion heating is thought to be an important mechanism for ion outflow from the upper atmosphere. The heating often takes place at fairly low altitude, so, Astrid-2 is in a good position to measure *in situ* the fields, waves, and particle properties associated with the heating process.

Sources of the Cross-Polar Potential Drop. The cross-polar potential drop arises from the solar wind's interaction with Earth's magnetic field. Freja data have been used to study the relative importance of high and low latitude dynamos (Blomberg *et al.*, submitted manuscript, 1998). Astrid-2 will provide new valuable data that may deepen our understanding of the interaction mechanisms.

E-B Correlation and its Scale Size Dependence. Electric and magnetic fields associated with static structures in the ionosphere are often correlated. The electric field maps, at least partially, upwards along the geomagnetic field, and the field-aligned current produces a transverse magnetic disturbance field. Assuming complete mapping of the electric field the height-integrated ionospheric conductitvity may be inferred from the ratio of the magnetic to the electric field. At least for large scale sizes the electric field normally maps well between different altitudes. Studying the degree of correlation for smaller scale sizes may yield additional information about the ionosphere-magnetosphere interaction processes.

Global Mapping of $\mathbf{E}$, $j_{//}$. To understand the fundamental relationship between the ionospheric electric field and the field-aligned currents on the global scale, statistical pictures based on data from the same platform are needed (e.g., Blomberg *et al.*, 1992). Such pictures are normally not found in the literature. Astrid-2 is well instrumented and is in an ideal orbit to produce this kind of pictures.

Global Mapping of B. With a precision magnetometer and a star imager for attitude determination mounted back-to-back, Astrid-2 will be used as a complement to the Danish Oersted satellite, whose primary objective is to map the geomagnetic field.

Sub-auroral and Equatorial Electric Field Structures. Strong poleward directed electric fields are commonly observed at sub-auroral latitudes in the pre-midnight sector. They are believed to be associated with closure of field-aligned currents through an ionospheric region with depleted plasma density and, thus, low conductivity. For a recent overview of observational results see Karlsson *et al.* (1998).

Waves and Pulsations up to 1 kHz. With a maximum sampling rate of the field instruments of 2048 s^{-1} wave frequencies up to the local proton gyro frequency are covered. Thus, a multitude of pure wave as well as wave-particle interaction phenomena may be studied.

Technological Objectives

A number of new designs and mechanisms are to be flight tested on Astrid-2. Below are given a few examples of newly developed sub-systems.

Wire Boom Deployment Mechanism.

The wire booms used by the EMMA instrument (see also next section) are of a newly developed design (Hellman, 1996), where the booms in the stowed position are wound around the spacecraft body (Figure 2). Each of the four booms is wound two full turns, resulting in a boom length of 3.3 m, including a 0.1 m stub outside of the probe. The booms and probes are held by mechanisms at each corner of the platform and released by pyro firing. The deployment is driven by a step motor, located in one of the four legs of the platform structure. The total mass of the boom deployment system is less than 2 kg, to be compared with 3.5 kg per unit for conventional mechanisms.

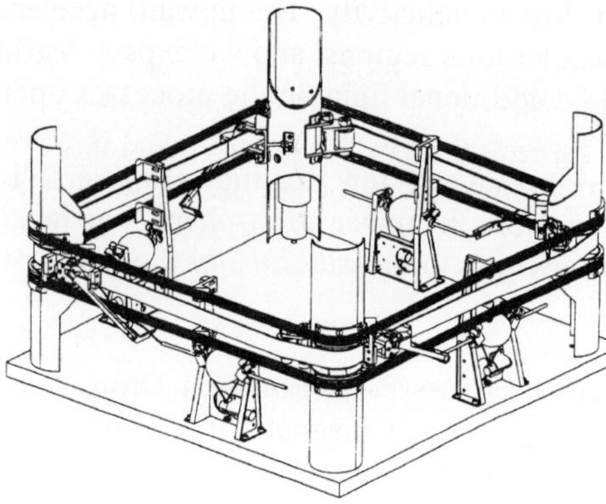

Figure 2. Schematic of the wire boom mechanism.

Star Imager for Attitude Determination.

Attitude determination will normally be performed by the on-board star imager which is a derivative of the Oersted star imager. The Astrid-2 design has been modified for use on a spinning platform. An accuracy of 1 arc sec for the spin axis and 20 arc sec for the spinning axes is foreseen. At times when the Earth is in the field of view of the star imager a new algorithm developed at Ålborg University for determining attitude based on magnetometer data will be employed.

Autonomous Ground Stations.

The ground stations to be used (further detailed below) are newly designed for the Astrid-2 mission. A high degree of autonomity has been a design driver. The ground stations will operate without manual intervention for extended periods of time. The controlling computer is remotely accessible and incoming data are time stamped using GPS time.

INSTRUMENTATION

There are 5 scientific instruments on-board Astrid-2. Together they form a fairly complete auroral physics payload. The instruments are:

- EMMA – Electric and Magnetic Monitoring of the Aurora.
- LINDA – Langmuir INterferomery and Density experiment for Astrid-2.
- MEDUSA – Miniaturized Electrostatic DUal-tophat Spherical Analyzer.
- PIA – Photometer for Imaging the Aurora.

• ASC – Advanced Stellar Compass.

All instruments are equipped with on-board memory, so that the sampled data can be stored temporarily for later transmission to ground. The location of the instruments on-board is shown in Figure 3.

EMMA

EMMA is an integrated electric field and magnetic field instrument performing simultaneous sampling of the potential of four electric field probes mounted on the wire booms in the spin plane (see above) and a tri-axial flux-gate magnetometer sensor mounted on the axial boom facing away from the Sun. Thus, the two spin plane components of **E** and the full **B** vector are measured simultaneously. In addition, since the electric probes are sampled individually, the spacecraft potential is continuously monitored giving information about the plasma density and also providing important information for the measurement of low-energy particles (which are attracted to or repelled from the detector if the spacecraft is at a different electrical potential from that of the plasma). The magnetometer sensor is mounted back-to-back with the star imager to maximize the precision of the spacecraft attitude knowledge for the scientific interpretation of the magnetometer data.

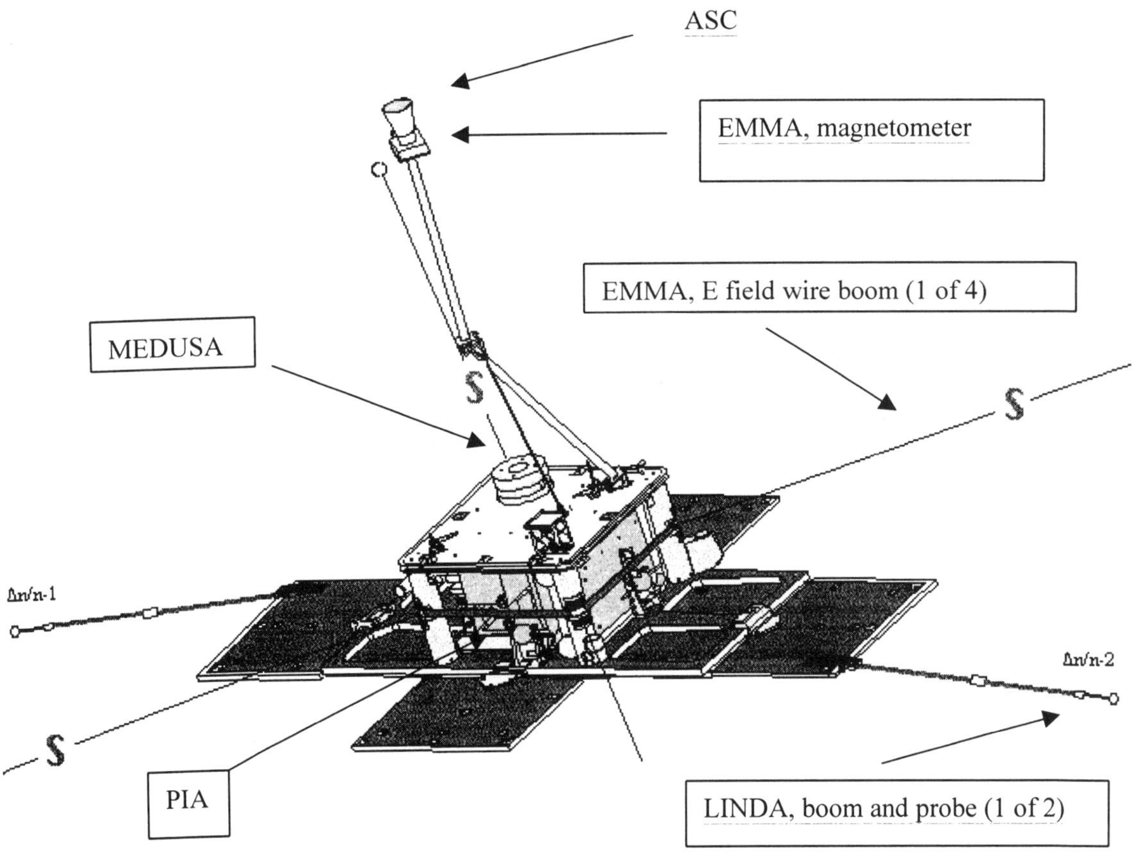

Figure 3. The Astrid-2 platform and instruments.

Sampling may be done at 16, 256, or 2048 samples/s. EMMA is equipped with a 12 MB semi-conductor memory for temporary storage of data outside times of contact with either of the ground stations. This memory is shared with the LINDA instrument (see below). In addition to passively monitoring the electric field, the probes may be actively swept in current or voltage to produce a current-voltage characteristic from which (at least in certain parameter regimes) information about plasma density, temperature, photo electron temperature, etc. may be derived.

The probes used for the electric field measurement are mounted on the 3.3 m long wire booms, have a diameter of 40 mm, and a titanium nitride (TiN) surface. TiN is used as surface material for the Langmuir probe on-board Cassini and is reported to work excellently. Because of the relatively short boom length the pre-amplifiers are located on the platform rather than inside the probes as is often the case when the booms are long.

The range of the measured electric field is ± 5 V/m, and the electric potentials are sampled with 16 bits resolution giving a bit resolution of 0.2 mV/m. The magnetic field is sampled with 20 bits resoultion. The range is ± 65,536 nT and the bit resolution 0.125 nT.

EMMA is built by the Alfvén Laboratory of the Royal Institute of Technology, Stockholm together with the Technical University of Denmark, Lyngby.

More details about the EMMA instrument can be found in Marklund *et al.* (1997) and in Blomberg *et al.* (1997).

LINDA

LINDA measures plasma density at up to 16 samples/s and density fluctuations at up to 8192 samples/s. The latter corresponds to a spatial resolution of about 1 m. With two Langmuir probes mounted on 0.8 m booms, which in the stowed position lie along the edges of the solar panels, interferometric measurements of moving plasma structures are possible. Voltage sweeps of the probes are also possible which yield information on plasma parameters as discussed above (EMMA subsection). LINDA is controlled by the same CPU as EMMA and the two instruments share on-board memory. Because of the high sampling rate a snapshot scheme will be used for data collection, where segments of the full wave form are captured intermittently. Also the LINDA probes have a TiN surface.

LINDA is built by the Swedish Institute of Space Physics, Uppsala.

MEDUSA

MEDUSA measures ions and electrons simultaneously in the spacecraft spin plane. There are 16 sectors in the plane of acceptance, and the maximum particle energy that can be detected is 18 keV per unit charge. For electrons up to 16 energy sweeps/s can be made. The corresponding figure for ions is 8 sweeeps/s. A variety of modes with lower resolution may be used to acquire overview data during prolonged periods of time.

MEDUSA is built by the Swedish Institute of Space Physics, Kiruna together with Southwest Research Institute, San Antonio, Texas.

PIA

PIA consists of three photometers. Each photometer measures four pixels and has a focal length of 250 mm. Two of them are mounted in the spin plane for spin-scanning imaging of auroral emissions. The sensors are sensitive to emissions in the wavelength band 125-150 nm, thus, in the case of aurora mainly oxygen emissions will be recorded. The third photometer is looking along the spin axis, hence, approximately in the Sun direction, is sensitive to Lyman α emissions (121 nm), and will mainly be used for atmospheric absorption studies. The sampling rate of the photometers is 256 s^{-1}.

PIA is built by the Max-Planck-Institut für Aeronomie, Lindau together with the Swedish Institute of Space Physics, Kiruna.

ASC

The ASC is a derivative of the star imager developed for the Oersted mission. The major difference from the Oersted mission is that the Astrid-2 platform is spinning, and thus, each star will show up as an arc segment rather than as a point on the image. The ASC uses CCD detectors for the imaging, and an on-board star catalogue, so that absolute values of the attitude are put directly into the telemetry. No post-processing on the ground will be needed. The attitude will be determined with an accuracy of 1 arc sec for the spin axis and 20 arc sec for the spinning axes.

The ASC is built by the Technical University of Denmark, Lyngby.

GROUND STATIONS AND TELEMETRY

Two ground stations will be employed for the Astrid-2 project (Figure 4). The primary ground station is located at the roof of the Swedish Space Corporation building in Solna, Sweden (N59.3, E18.0). On average the spacecraft will be above the Solna horizon about 2 hours per day. All commanding of the spacecraft will be done from this ground station. The second ground station is located on the South African base SANAE IV on the Antarctic continent (S71.1, W02.8), see Figure 5. Because of the higher latitude the contact time is longer at SANAE IV than in Solna, about 2.5 hours per day on average. The SANAE IV station is downlink only, no commanding is possible from there.

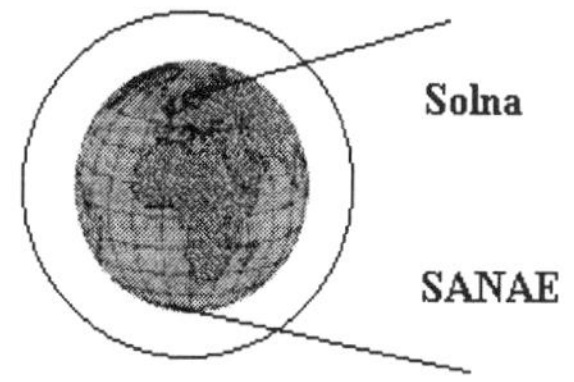

Figure 4. Location of the Astrid-2 ground stations.

The telemetry uses S-band, BPSK, and Viterbi-encoding, and has a net data rate of 128 kbits/s for the downlink, and 4.8 kbits/s for the uplink. With 4.5 hrs of contact time a data volume of approximately 250 MB can be transferred each day from spacecraft to ground. From the Solna station data are accessible to the instrument groups in real-time through Internet, and real-time commanding of the instruments is also possible. From the SANAE IV station a small subset of the data received may be retrieved via modem, but the primary means of data transfer is by ship, during the summer months when SANAE IV is accessible.

The ground station at SANAE IV has been set up in collaboration with the Physics Department of the University of Natal, Durban, and will be operated by their resident expedition members. Joint studies of Astrid-2 data and ground-based observations from SANAE IV's scientific instrumentation are planned.

Figure 5. The telemetry station at SANAE IV (left, during construction; right, completed).

IN-FLIGHT OPERATIONS

All instruments are designed to allow continuous operation. Both the instruments and the platform have also been designed to run without operator intervention for prolonged periods of time. Instrument commands may be uploaded in bulk to the Astrid-2 System Unit (ASU) for later execution. Power, attitude and spin control are all autonomous.

Two basic modes are planned; one with low resolution of the measured quantities, normally to be used at low latitudes and at times when consecutive ground station contacts are well separated in time; the other with medium resolution, normally to be used at auroral and higher latitudes. Nominally, in the low resolution mode EMMA will take 16 samples/s of **E** and **B**, LINDA will sample the absolute density but not density fluctuations, MEDUSA will sample particle fluxes in three directions, parallel, perpendicular, and anti-parallel to **B**, and PIA will not operate. Nominally, in the medium resolution mode, EMMA will sample at 256 s^{-1}, LINDA will sample density at 16 s^{-1}, and take snapshots of dn/n at 8192 s^{-1}, MEDUSA will sample all 16 sectors in the spin plane, and PIA will take 256 photometer samples/s. The ASC is foreseen to operate at all times when it has an unobstructed view of the sky.

In addition to the two basic modes, the instruments will at times be operated in a high resolution mode. In this mode EMMA samples at 2048 s^{-1}, LINDA samples dn/n at 8192 s^{-1} but with either more frequent or longer snapshots than in the medium resolution mode, while MEDUSA and PIA operate in the same way as in the medium resolution mode.

Data dumping will be done interchangingly at the two ground stations, so that northern as well as southern hemisphere data are dumped at both stations. This is required for operational reasons since the ground station contacts alternatingly occur before and after the high-latitude pass where the on-board memory fills up because the instruments are run at higher resolution. The long delay time in receiving the full Antarctic data set must also be taken into consideration. For example, data from campaigns run jointly with other instrument sets will normally be dumped at Solna.

DATA HANDLING AND ANALYSIS

The data received at the Solna station will be available to the instrument groups in real time. They will also be transferred electronically to the data archive at the Alfvén Laboratory directly after each pass. Data received at SANAE will be stored on DAT tapes. The DATs will be transported out of SANAE by ship when the ice conditions permit it.

Astrid-2 Summary Plots (ASP) will be made available on the World Wide Web within a day or two of receiving the data. A fairly sophisticated IDL based analysis software package has been developed for the scientific analysis. To facilitate easy data exchange intermediate data files are in ASCII format.

SUMMARY

Astrid-2 is the first complete microspacecraft for auroral research, featuring new developments for the platform as well as for the scientific instruments. Primary mission objectives are to do high-quality auroral research and to demonstrate the feasibility of using microspacecraft for this kind of science. If these objectives can be met successfully, it will open up brand new possibilities to perform inexpensive multipoint measurements of auroral processes. The spacecraft may also be operated at a low cost because of the many autonomous systems on-board as well as on the ground. Since operations are not a cost driver we expect to be able to operate the spacecraft beyond its nominal one year life time, if it still functions properly.

ACKNOWLEDGEMENTS

The Astrid-2 project is funded by the Swedish National Space Board and by corresponding agencies in the other participating countries. The Antarctic ground station is mainly funded through donations from the Wallenberg Foundation and from the Royal Institute of Technology We wish to acknowledge the many scientists and engineers who have contributed to this project in different ways. The pictures of the Antarctic ground station are by courtesy of the SANAE IV team.

REFERENCES

Blomberg, L. G., G. T. Marklund, and P.-A. Lindqvist, Viking Observations of Electric Fields, *ESA-SP-346*, 269 (1992)

Blomberg, L. G., G. T. Marklund, P.-A. Lindqvist, and L. Bylander, EMMA - The Electric and Magnetic Field Instrument on Astrid-2, *Royal Institute of Technology Report* TRITA-ALP-1997-04 (1997)

Hellman, H., Design of Wire Boom System for a Satellite, *Royal Institute of Technology, Division of Plasma Physics Internal Report,* ALP-1996-101, (1996)

Hultqvist, B., The Swedish Satellite Project Viking, *J. Geophys. Res*., 95, 5749 (1990)

Karlsson, T., G. T. Marklund, L. G. Blomberg, and A. Mälkki, Subauroral Electric Fields Observed by the Freja Satellite: A Statistical Study, *J. Geophys. Res*., 103, 4327 (1998)

Lundin, R., G. Haerendel, and S. Grahn, The Freja Science Mission, *Space Sci. Rev.,* 70, 405 (1994)

Marklund, G. T., L.G. Blomberg, C.-G. Fälthammar, and P.-A. Lindqvist, On Intense Diverging Electric Fields Associated with Black Aurora, *Geophys. Res. Lett*., 21, 1859 (1994)

Marklund, G., T. Karlsson, and J. Clemmons, On Low-Altitude Particle Acceleration and Intense Electric Fields and their Relationship to Black Aurora, *J. Geophys. Res*., 102, 17509 (1997)

Marklund, G., L. Blomberg, L. Bylander, and P.-A. Lindqvist, Astrid-2: A Low-Budget Microsatellite Mission for Auroral Research, ESA SP-397, 387 (1997)

AN OVERVIEW OF THE KOREA MULTI-PURPOSE SATELLITE (KOMPSAT)

Byung Kyo Kim

Korea Aerospace Research Institute, P.O.Box 113 Yusong, Taejon, Korea

ABSTRACT

Korea has initiated its own space program since 1990, successfully launching its first microsatellite into the earth orbit in 1992, and scientific sounding rockets in 1993. Broadcast and communications satellites were launched in geostationary orbit in 1995. A multipurpose satellite, KOMPSAT is under the joint-development with an American company with its launch target in mid 1999. Recently, the national space development plan has been approved in the Commission headed by the Prime Minister.

Korea Aerospace Research Institute(KARI) has been developing KOMPSAT, a small 510-kg earth observation satellite with an orbital altitude of 685 kilometers, jointly with a US-based firm, TRW. Once developed, KOMPSAT is expected to help Korea accumulate space technology in short time to eventually become one of the world's top ten nations in the space technology sector by the year 2010. KOMPSAT's payload will include high-resolution electro-optical camera, ocean observation camera, ionosphere measurement sensor, and high-energy particle detector.

This paper addresses an Overview of the Korea Mult-Purpose Satellite (KOMPSAT) system : its background, specifications and performance, design concepts, joint development description, development schedule, launch vehicle, AIT facility, ground station, etc.

KOMPSAT PROGRAM

The purpose of the KOMPSAT program is to satisfy the demand of the domestic satellite and strengthen national space development capability. Specific goals are to develop KOMPSAT for Korean Peninsula Observation, Ocean Observation, and Scientific Experiment as well as Koreanize main components in the spacecraft and payload. Development period is from Nov.,1994 to April, 1999, which is 54 months. Total budget is 193.4 billion won (about 0.2 billion \$). TRW (U.S.A.) is our overseas partner.

This program is getting joint support from several of Korean government organizations : MOST as secretariat, and MOIC, MOTIE, MOFE. Korea Aerospace Research Institute (KARI) is the general supervising organization of the program, which is performed by participation of Korean Industries : Korean Air & Doowon for structure and mechanisms, Samsung Aerospace for telemetry, command and ranging, Daewoo H.I. for attitude and orbit control, Hyundai Aerospace for electrical power, Halla H.I. & Hanwha Ltd. for propulsion. SaTReC, ETRI, Seoul National Univ., KAIST, Inha Univ., Yonsei Univ.,etc. are participating research institutes and universities.

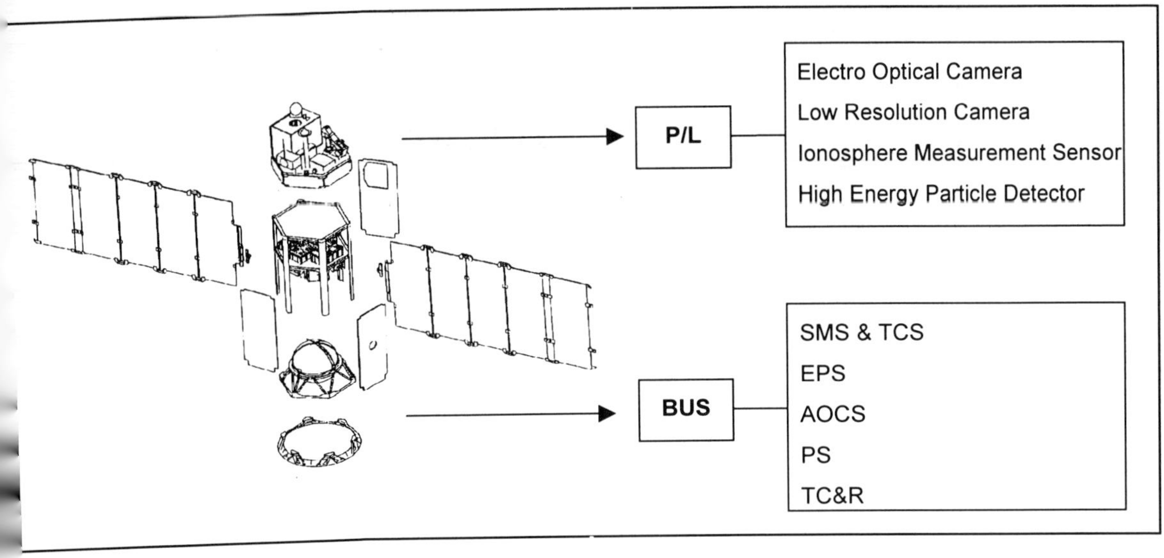

Figure 1. KOMPSAT Satellite.

SPECIFICATIONS AND DESIGN CONCEPTS

KOMPSAT, Korea Multi-purpose Satellite means a satellite which can load various payloads (observation and test equipment) by using developed bus system which is the basic structure of satellite. As shown in Figure 1 current payloads are an electro optical camera (EOC), a low resolution camera (LRC), an ionosphere measurement sensor (IMS), and a high energy particle detector (HEPD). Other payloads can be loaded on the same bus structure depending on the further needs.

KOMPSAT specifications and performance are as shown in Table 1, and operational concept is shown in Figure 2. KOMPSAT has a design life of more than 3 years, and a sunsynchronous orbit of 685Km altitude. Bus structure is made of honeycomb structure, and has 3 axis stabilization system of 0.1deg (2σ) pointing accuracy. Four payloads are loaded on the bus : an electro optical camera (EOC) for cartography, a low resolution camera (LRC) for ocean color monitoring, an ionosphere measurement sensor, and a high energy particle detector. KOMPSAT uses X-band down link for EOC and LRC data, and S-band up and down links for telemetry and command.

Table 1. KOMPSAT Specifications and Performance

Life	More than 3 years	
Orbit	Sunsynchronous, Crossing Time 10:50, Altitude 685 Km	
KGS Contact	2, 3 contact times per every 10 hours, One contact time 10~13 minutes	
Satellite Weight	510 Kg	
Power (EOL)	636 W	
BUS	Honeycomb Structure, Heat pipe, Heater 3 Axis Stabilization, 0.1 degree Pointing Accuracy (2σ) Hydrazine Thruster Solar Array	
Payload	EOC	1:25,000 Cartographic Image Acquisition Resolution 10m, Panchromatic Band 0.51 μ~0.73 μ, Swath 15 Km
	LRC	Resolution 1 Km, 6 Band 0.4 μ ~ 0.9 μ, Swath 800 Km
	Space Physics Sensor	IMS, HEPD

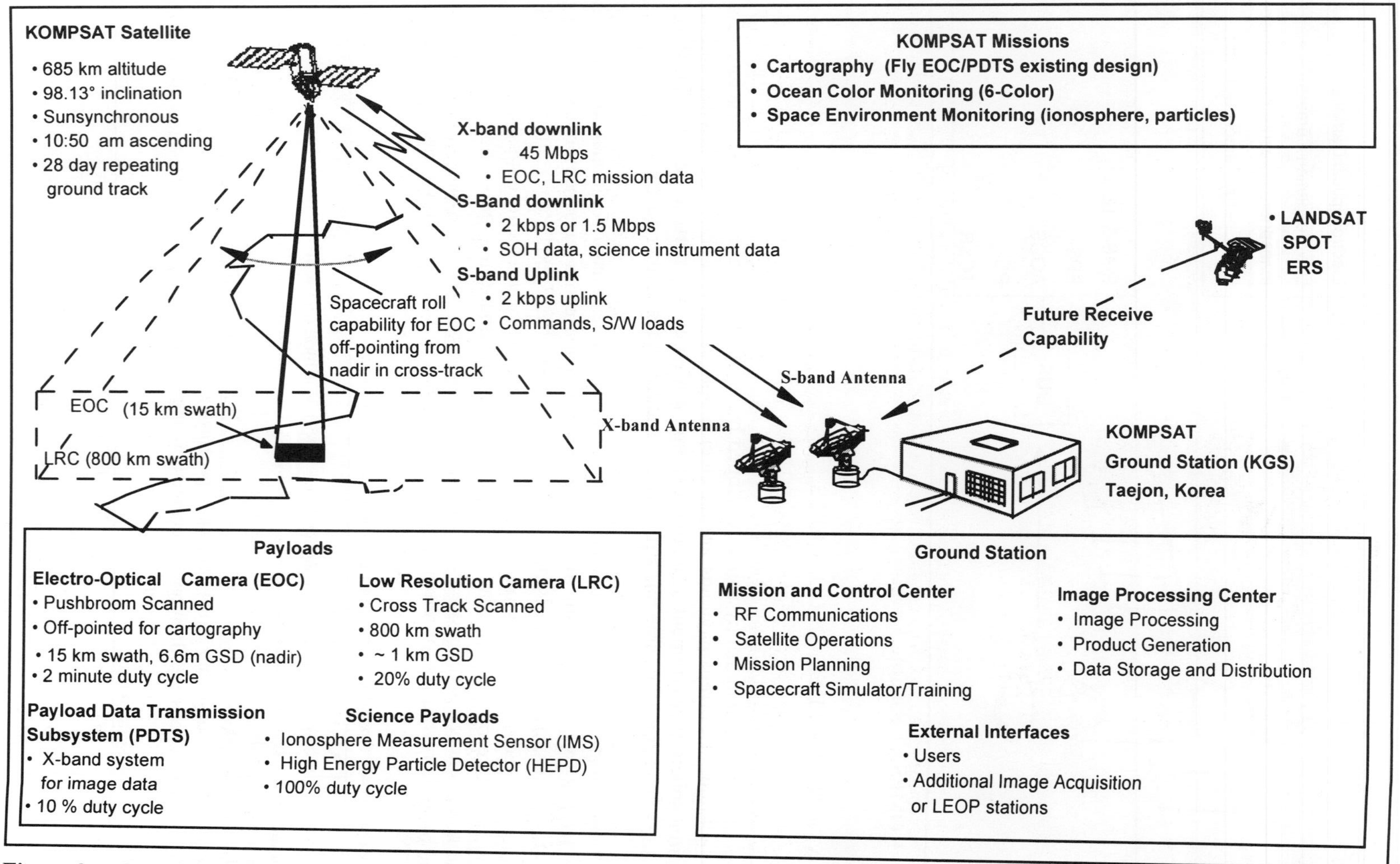

Figure 2. Operational Concept.

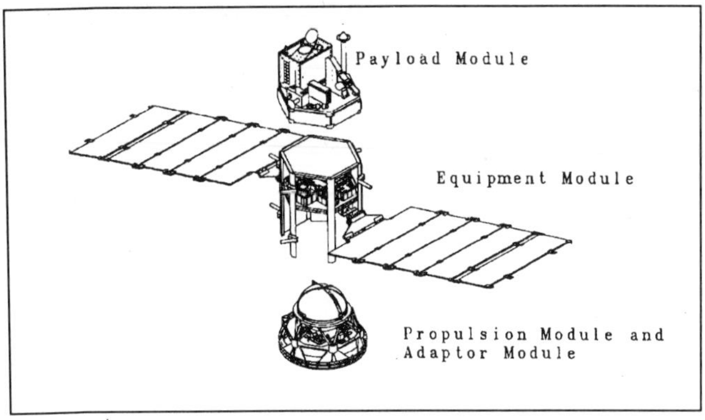

Figure 3. Structure and Mechanisms Subsystem.

The KOMPSAT bus is composed of 6 subsystems : the structure and mechanisms subsystem (SMS), the thermal control subsystem (TCS), the attitude and orbit control subsystem (AOCS), the electrical power subsystem (EPS), the propulsion subsystem (PS), and the telemetry, command & ranging subsystem (TC&R).

The structure and mechanisms subsystem is composed of the payload module, the equipment module, and the propulsion and adaptor module as shown in Figure 3. It supports all of the spacecraft components during induced and natural environment, including ground handling, transportation, launch and in-orbit operation. It is of the hexagonal modular type with the dimension of 235cm x 134cm x 690cm when solar arrays are deployed. Mass of the bus subsystem is 300kg. The thermal control subsystem keeps the proper operating temperature for the components of the satellite using second surface mirrors, multilayer insulators, and heat pipes. The attitude and orbit control subsystem performs attitude determination, attitude control, momentum unloading, solar array pointing and orbit adjustment. Electronics and processors, attitude sensors (the fine sun sensor, the coarse sun sensor, the conical earth sensor, the gyro reference assembly, the three axis magnetometer), and actuators (the reaction wheel assembly, the magnetic torquer assembly), constitute feed back loops. Its pointing accuracy is 0.1deg (2σ) and the knowledge stability is 2.0 arc-sec during 0.737 sec which corresponds to 5km of satellite movement. The electrical power subsystem provides the electrical power during mission life. It is composed of the solar arrays for electrical power generation, the battery for energy storage, the PCU (power control unit) for power distribution and the SAR (solar array regulator) for power regulation. The solar arrays are as shown in Figure 4, have two wings with 5 panels. Each wing has 2140 solar cells of 2.84cm x 4.638cm size. They generate 640w at the end of life. The propulsion subsystem shown in Figure 5 provides required thrust for attitude and orbit control. It has propellant tank containing 70kg of hydrazine propellant, the dual thruster module, the piping system, the fill and drain valves, the filters, and the pressure transducer.The telemetry, command and ranging subsystem receives telecommands from the ground station, distributes and executes commands, collects and downlink the state of health data to the ground station, and provides time reference and satellite position information using GPS data. It is composed of RF parts such as S-band antenna, transponder, RF assembly, on-board computer with MIL-STD-1553B data bus, and GPS antenna and receiver. Its data rate is 2Kbps for the real-time data and 1.5Mbps for the playback data. Transmitting power is 5 watts.

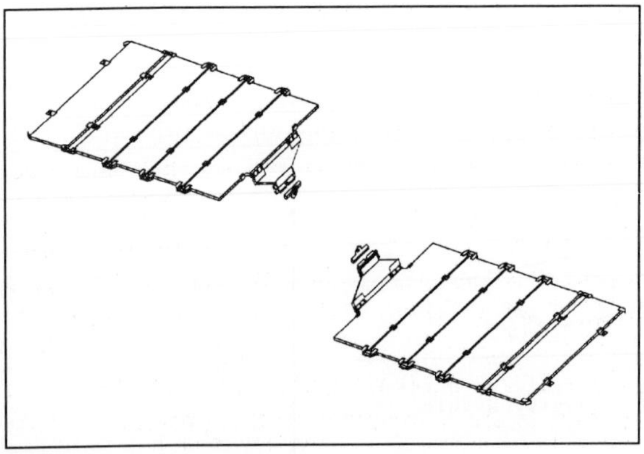

Figure 4. Solar Arrays.

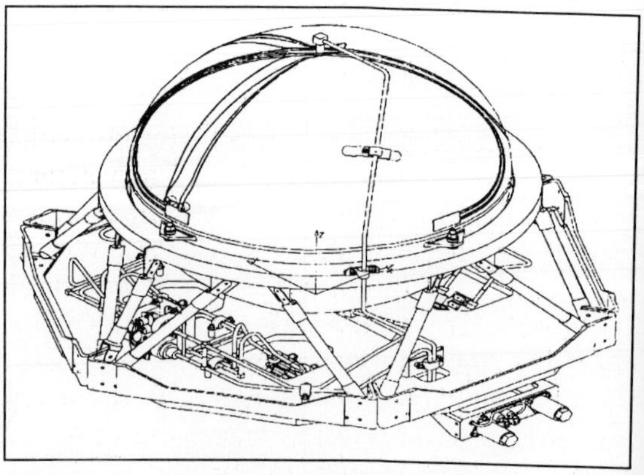

Figure 5. Propulsion Subsystem.

The electro-optical camera, Figure 6, is the main payload of the KOMPSAT. Its main mission is Korea cartography. It has 10m resolution, 15km of swath width. It scans 800km per orbit, using pushbroom type scanning. It uses panchromatic band of 510nm - 730nm. It obtains stereo image by up to ± 45deg. roll, and transmits data at 25Mbps of speed to the payload data transmission subsystem, which has the capability of storing data up to 2.5 Gbits at EOL and transmitting data at 45Mbps of speed to ground. The next payload is the low resolution camera, Figure 7, mission of which is ocean color monitoring. It provides 1km resolution image with 800km swath width. It scans 8,000km per orbit, uses 6 bands in 400nm - 900nm range. Data rate is 1Mbps. The ionosphere measurement sensor is the third payload which provides global ionosphere measurement for data base establishment. It is composed of electron temperature sensor, electron density sensor, and related electronic circuit. Measurement ranges are 10 - 10^6 electron/cm^3 for electron density and 0 -1 eV for electron temperature. The last payload, high energy particle detector will provide informations on radio-active rays and their effects on semi conductor components. It is composed of proton and electron spectrometer, linear energy transfer spectrometer, total dose monitor and single event monitor. Measurement ranges are 0.25MeV - 2.5MeV for electrons, 25MeV for α - particles, and 1MeV - 30MeV for protons.

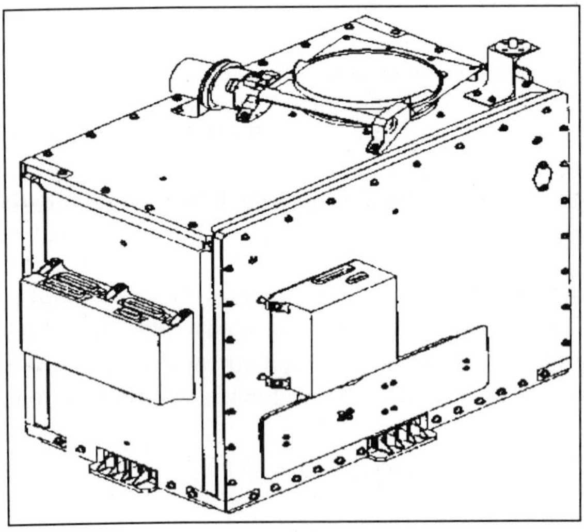

Figure 6. Eletro-Optical Camera (EOC).

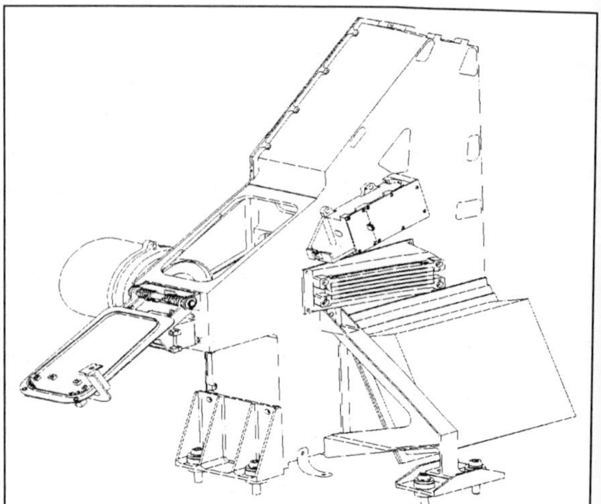

Figure 7. Low Resolution Camera (LRC).

JOINT DEVELOPMENT

KARI is developing KOMPSAT jointly with TRW of the U.S.A. Under the contract with TRW, a flight model, a proto flight model, and a structural development model is made with a test bed and a set of ground support equipment. TRW transfers technology for design, manufacturing, assembly and test of the spacecraft to KARI and participating Korean industries. KARI and Korean industry engineers participate in design phase at TRW. TRW dispatches manufacturing support engineers to Korean industries in manufacturing phase. TRW conducts AIT of PFM at TRW with KARI and Korean industry engineers. KARI conducts AIT of FM at KARI facility with the skill obtained from PFM AIT at TRW and the support of TRW engineers. TRW leads and supports launch preparation and early operation. The 30 components out of 48 major components are Koreanized in this program. Table 2 shows major Koreanized components.

Table 2. Major Koreanization Components of KOMPSAT

Subsystem	**Components**		
SMS & TCS	○SMS		
	- Payload module	- Electronics module	- Propulsion module
	- Solar panel support	- Adapter	
	○TCS		
	- Battery heat pipe	- MLI	- Heater
	- Thermostat	- Thermistor	- SSM
	- RTV, paint, etc.		
AOCS	- RDU	- Valve drive electronics	- Fine sun sensor
	- Coarse sun sensor	- Flight Software(AOCS)	
EPS	- Power control unit	- Solar array regulator electronics	
	- Solar array panel	- Panel-to-panel hinge	- Harness
	- Flight software(EPS)	- Deployment device controller	
PS	- Thruster	- Piping system	
TC&R	- On-board computer	- S-band antenna	
	- S-band transponder	- Flight software(TC&R)	

LAUNCH VEHICLE AND GROUND FACILITY

Taurus launch vehicle of Orbital Sciences Corporation in the U.S.A. has been chosen for KOMPSAT launch. KOMPSAT will be launched at Vandenberg Air Force Base, CAL, in August 1999. Taurus is a 4 stage solid propellant rocket, 27m long and capable of delivering 790kg of payload to 685km altitude KOMPSAT orbit.

KARI has constructed an assembly, integration and test facility as shown in Figure 8 for the FM AIT. It can handle 1000kg, 2.5m dia. satellites, has 1/360°, 1μ measurement accuracy and 10,000 class cleanliness. Environmental tests and EMI/EMC tests can be performed in this facility. KARI is constructing KOMPSAT ground station for mission control and data receiving and processing as shown in Figure 9. It will be ready by January 1999. Mission control center will have a 6-m S-band antenna and the mission control and operation subsystem. Data receiving and processing center will have a 13m X-band antenna, data acquisition and processing subsystem.

PROGRAM SCHEDULE AND STATUS

Figure 10 shows the master schedule of the KOMPSAT program. KOMPSAT program is now at PFM AIT stage. Although KARI, TRW and Korean industries have experienced some minor difficulties, KOMPSAT program is now on schedule. Modern technology of system and subsystem design, integration and test, subsystem manufacturing is being accumulated through on site experience. Korea is now preparing not only for a successful launch of its first satellite of practical purpose, but for its own development of the next satellite.

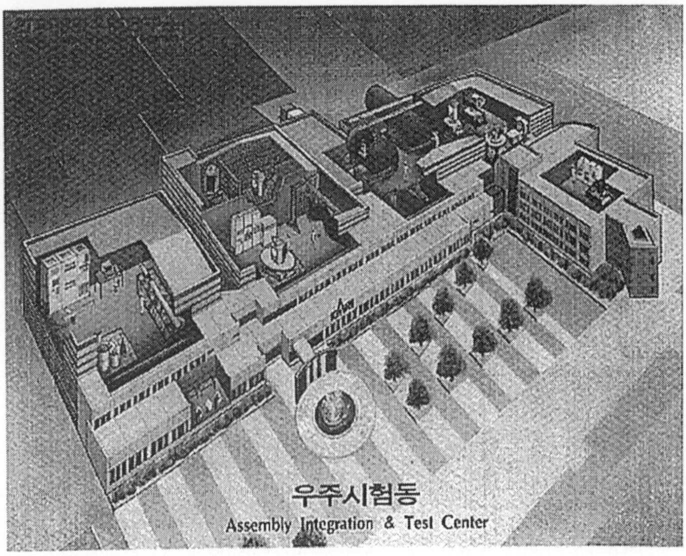

○ S/C ASSEMBLY FACILITY

- 1000kg, Φ2.5m Class
- 1/360 °, μm Accuracy
- 10,000 Class Cleanliness

○ S/C TEST FACILITY

- 1000kg, Φ2.5m Class
- 10,000 Class Cleanliness
- Component Test
- Launch Environment Test
- Orbit Environment Test
- EMI/EMC Test

○ Mission

Control, Operation, Data Acquistion and Processing of KOMPSAT

○ Constitution

Mission Control Center
- 6m S-band Antenna
- Mission Control & Operation Subsytem
- Simulator Subsystem

Data Receiving Center
- 13m X-band Antenna
- Data Acquisition Subsystem
- Data Processing Subsystem
- Value-Added Product Subsystem

Figure 9. KOMPSAT Ground Station.

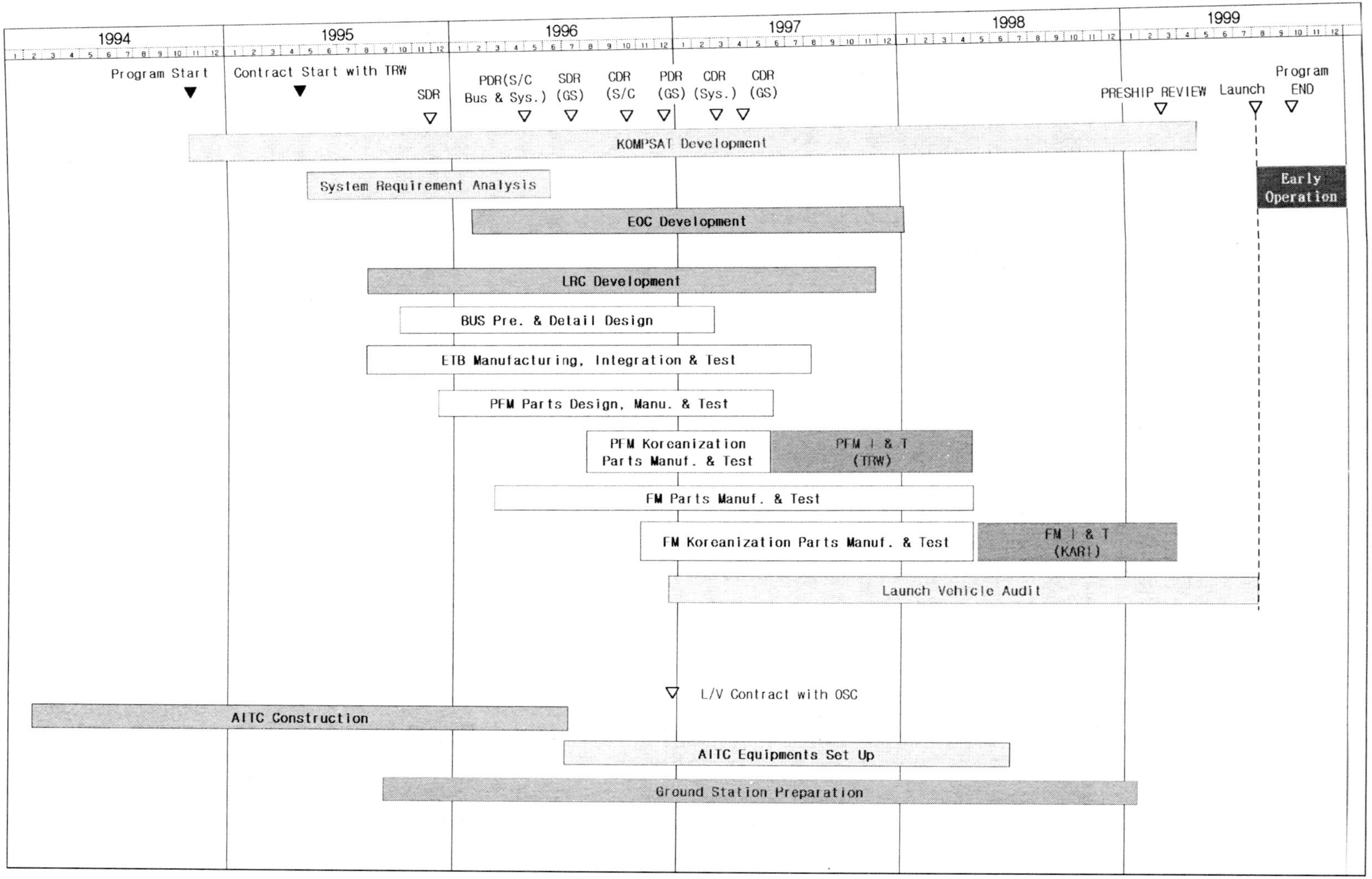

Figure 10. KOMPSAT Program Schedule.

SPACE RADIATION EXPERIMENTS ON KITSAT-1 AND KITSAT-2

Kyoung Wook Min, D. H. Lee, S. M. Park, and Y. H. Shin

Satellite Technology Research Center, Korea Advanced Institute of Science and Technology, Taejon 305-701, Korea

ABSTRACT

The results of the space radiation experiments carried out on board the first two Korean technology demonstration microsatellites are presented in this paper. The first satellite: KITSAT-1, launched in August 1992, carries a radiation monitoring payload Cosmic Ray Experiment (CRE) for characterizing the low-Earth orbit radiation environment. CRE measures the linear energy transfer of the high energy particle flux, total radiation dose, and the single event upsets (SEU). The second satellite: KITSAT-2 , launched in September 1993, carries various types of memories which have been monitored to test the possibility of the future use of these memories in space. These two satellites, which are similar in structures but different in orbits, provide a unique opportunity to study the effects of the radiation environment characterized by the orbit. The result shows that the SEU characteristics observed in the two satellites are different: single bit errors are dominant in the low inclination KITSAT-1 orbit, while multiple bit errors are prevalent on KITSAT-2 in the high inclination orbit. It is also found that monolithic memories are more vulnerable to the ionizing radiation than the hybrid memories.

INTRODUCTION

In August, 1992, the first Korean satellite (KITSAT-1) was launched from Kourou, French Guiana , by the Ariane 42P rocket as an auxiliary payload to TOPEX/POSEIDON. The spacecraft body is constructed as a stack of rectangular modules (Figure 1) and its total size is 35.2cm x 35.6cm x 67.0cm. Navigation Magnetometers, Earth Horizon Sensor, and the Analog Sun Sensor reside on top of the satellite, while two CCD cameras and antennas are attached to the bottom of it. Four side walls are surrounded by the solar cell panels. The satellite is stabilized mainly by the gravity gradient boom, although a magnetic torque can be employed as an active attitude control. The whole spacecraft weighs about 48.6kg.

The KITSAT-1 orbit with an altitude 1330 km is dominated by the inner Van Allen radiation belt. This region has a large population of high energy protons which contribute significantly to both long-term and transient radiation effects on micro-electronics (Gussenhoven *et al.* 1987). The purpose of the Cosmic Ray Experiment (CRE), one of the modules flown on board the KITSAT-1, is to study the space radiation environment as encountered by the satellite. CRE consists of two sub-systems: the Cosmic Particle Experiment (CPE) and the Total Dose Experiment (TDE) (Underwood *et al.* 1994). CPE measures the linear energy transfer (LET) spectrum (in a silicon detector) of ionizing radiation particles passing through the spacecraft over short intervals of time, while TDE provides a mechanism for measuring the long-term total accumulated ionizing radiation dose (in SiO_2) for various locations in and around the spacecraft. The TDE sensors are solid state RADFET/MOSFET dosimeters, which are monitored for changes in the threshold voltage. The primary on-board mission control computer (MCC) for KITSAT-1 is called OBC186, which uses an INTEL 80C186 as its CPU. Single event upsets (SEUs) in the program memory and the 13MByte RAM disk memory have been monitored since the beginning of the satellite operation.

In September, 1993, the second Korean satellite (KITSAT-2) was launched from Kourou, French Guiana, as an auxiliary payload to SPOT-3. KITSAT-2 is also composed of a stack of rectangular modules. It has roughly the same size and weight as the KITSAT-1. Its orbit is 800 km altitude with an inclination angle 98.7°, while the inclination angle of KITSAT-1 is 66°. In addition to OBC186, a 32-bit INTEL 80960 based computer, KASCOM, was boarded to test its capability as a primary MCC for future small satellite missions. KASCOM was also designed to serve as a backup MCC in case OBC186 fails. SEUs in the KASCOM, as well as in the program memory and the RAM disk memory, have been monitored and compared to each other to study the response of these different types of RAMs to the radiation environment. The results, together with those of KITSAT-1 SEU experiments, provide an opportunity to study the effects of the radiation environment characterized by the orbit.

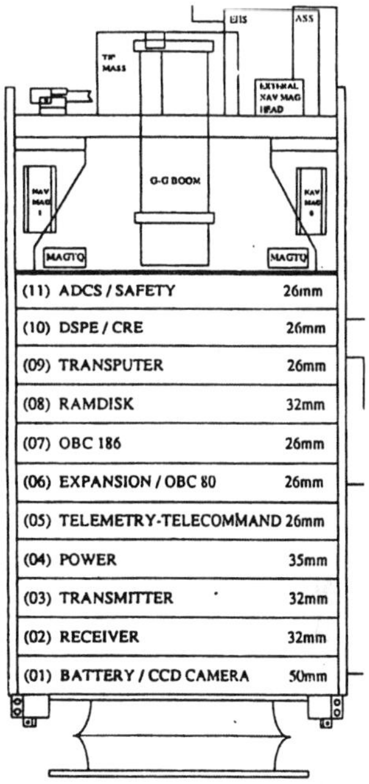

Figure 1. KITSAT-1 module.

EXPERIMENT DESCRIPTION

CPE monitors the LET spectrum in a single, fully depleted, 300μm thick, 30mm×30mm PIN diode which is housed in a separate screened aluminum box within the CRE module (Figure 2). The shielding by 10mm thick aluminum walls prevent most primary electrons and low-energy ions from contributing to the LET spectrum. This leaves the high energy proton, alpha and heavy ion content of the cosmic ray flux to be detected. However, as the system is only capable of measuring the deposited energy (in terms of the electronic charge produced in the diode), the particle-types and energies cannot be identified directly. The face of the detector is shielded by the rest of the spacecraft structure. These blocking materials limit the count-rate to less than 10,000 counts per second in the most sensitive channel when KITSAT-1 is passing through the heart of the

Figure 2. CPE analog board.

South Atlantic Anomaly (SAA). The silicon detector is connected to a charge amplifier and a pulse-shaping circuit which is then connected to a pulse-height multi-channel analyzer. The experiment is controlled autonomously by a 87C51 microcontroller with its own data storage RAM. The multi-channel analyzer has 512 channels in energy, each of equal width equivalent approximately to 0.05pC of charge deposited in the detector (i.e., an equivalent normal incidence LET of 16 MeV cm^2g^{-1}). However, the first four channels are sub-threshold, and the remaining 508 channels are used for the charge range approximately 0.2pC to 26.0 pC, equivalent to a normal-incidence LET range of 64 MeV cm^2g^{-1} to 8360 MeV cm^2g^{-1}. The detector deadtime determined by the hardware used is less than 10μs, so that the count rate up to 10^5 particles per second can be handled. These counts are integrated over a fixed period of 600 seconds, sub-divided internally into four periods of 150 seconds to improve the orbit-position/time resolution.

TDE measures the long-term accumulated ionizing radiation dose (in SiO_2) at various locations on-board the spacecraft. The sensing elements are RADFET dosimeters (modified power MOSFETs) which have a thick (> 0.1μm) gate-oxide to make them especially sensitive to the ionizing radiation. The TDE is based on the AEA Technology design originally flown on board UoSAT-3 as part of the Cosmic-Ray Effects and Dosimetry (CREDO) payload (Dyer *et al.* 1991). Exposure of the RADFET to radiation causes the formation of trapped holes in the gate oxide, which in turn causes a gradual shift in the threshold voltage according to the accumulated dose (Adams and Holmes-Siedle 1978; Dyer *et al.* 1993). Every five minutes the TDE is put briefly into the read mode, during which a constant current (6μA) is switched to each RADFET and the threshold voltage is measured.

The OBC186 memory system consists of a bootstrap ROM, a program memory, and a RAMDISK for data storage. A hardware Error Detection and Correction (EDAC) circuit is implemented to correct one bit errors and detect two bit errors in the program memory using Hamming code. Instead of using hardware EDAC circuit, RAMDISK uses a block coding similar to the Reed-Solomon code to correct one byte error from 252 byte data block by adding three byte codes to the data block. The file system task washes 1 KByte (KB) cluster of the RAMDISK area every second for SEU correction. KASCOM of KITSAT-2 is configured with the hardware EDAC protected 2 MByte (MB) program memory and 8 MByte data memory. Two different types of RAMs are employed to construct these memory systems: monolithic RAMs are used for the OBC186 program memories of KITSAT-1 and KITSAT-2 and for the data memory of KASCOM on KITSAT-2, while hybrid RAMs are used for the data memories of KITSAT-1 and KITSAT-2 and for the program memory of KASCOM on KITSAT-2. Hybrid RAMs are constructed by packaging four monolithic chips together and thus, have an over-all density lower than the monolithic RAMs.

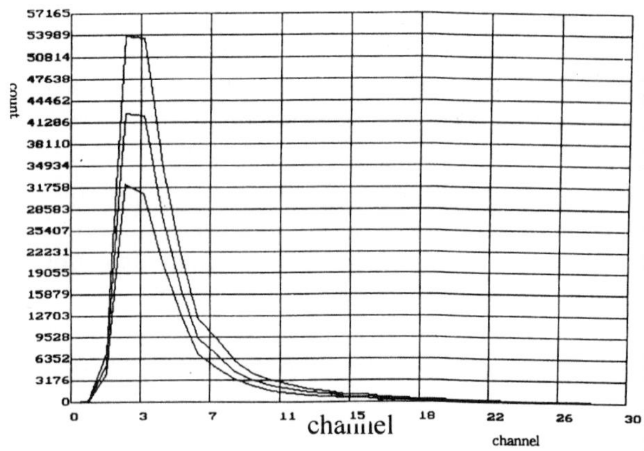

Figure 3. Changes of the LET counts after the solar flare. The three curves represent the counts/150sec at the South Atlantic Anomaly on March 15, 17, and 24, 1993, from the top, respectively. Three solar flare events were recorded during this period: M2 flare on March 12 and two M1 flares on March 16 and 24, 1993. It should be noted that the channels below 4 are sub-threshold.

RESULTS

Since the results of the CPE experiment are reported in detail elsewhere (Shin *et al.* 1994), we would like to discuss briefly one particular example here. Figure 3 shows the changes of the LET counts at the South Atlantic Anomaly after the major solar flare on March 12, 1993. The three curves are for March 15, March 17, and March 24, 1993, respectively. It is seen in the figure that the LET counts are affected by this event and the count rate decreases as its effect subsides, though the present measurement is done deep inside the spacecraft. However, the shape of the spectrum (hardness) does not seem to change much, probably because the heavy blocking materials prevent the real shape of the energy distribution of particles in the radiation belt from showing up in this measurement.

Figure 4 shows the result of TDE: changes in the threshold voltages for the two dosimeters located in the CRE module box during the period September, 1992 through October, 1996. It is clear from the figure that the RADFET-3, residing near the center of the module box and having complex structures on top of it, receives less ionizing radiation and shows smaller changes in the threshold voltage than the RADFET-1 which is located just inside the wall of the spacecraft. Another interesting feature seen in the same figure is the oscillatory behavior of the curves, with the amplitude growing in time. These oscillatory behaviors are identified as coming from the thermal annealing effects, as the TDE temperature and the threshold voltages shown together in Figure 5 are well correlated. The temperature change itself is due to the changes in the eclipse percentage of the spacecraft. The effect of the thermal annealing is verified in the laboratory experiment using the Co^{60} γ–ray source.

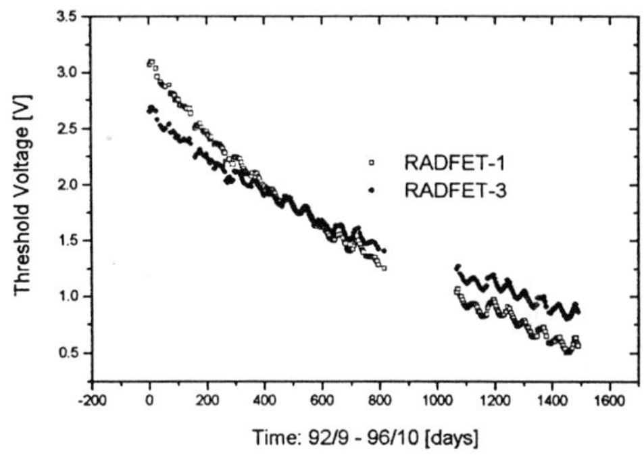

Figure 4. Changes in the threshold voltages for the two dosimeters during the period September, 1992 through October, 1996.

(a)

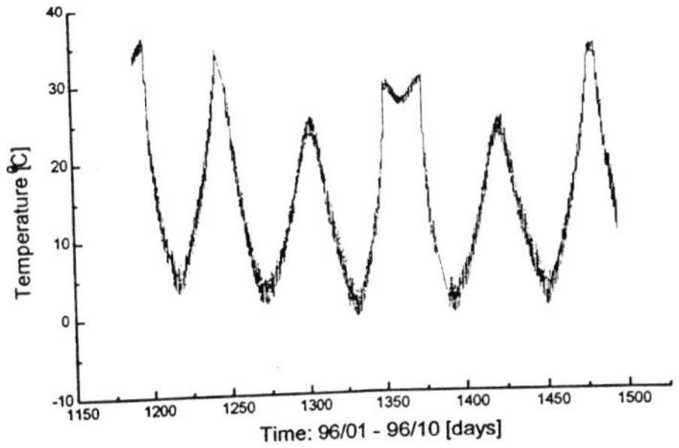

(b)

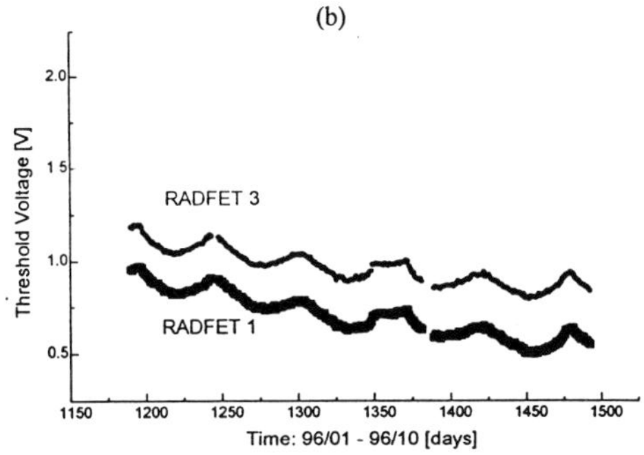

Figure 5. TDE temperature (a) and the threshold voltage (b) measured in 1996.

Figure 6 shows two cycles of the temperature change from 18°C to 72°C, while the RADFET sensor is irradiated at a constant dose rate with the total dose 0.35 Mrad in 4 hours. It is seen that the threshold voltage shift is affected as the temperature increases and the effect is larger during the second temperature peak than during the first one. It is conceivable that the thermal annealing effect is larger when the sensor has accumulated more radiation dose since it contains more holes which can escape from the traps as the temperature increases. This annealing effect is not small, as can be seen in Figure 4, and hence, it is necessary to calibrate the TDE sensors with this temperature effects taken into account for the better estimation of the total dose received. Nevertheless, crude estimation is possible by taking into account only the threshold voltage shift during the low temperature state with an extrapolation to the high temperature state, assuming that the dose rate received from the space radiation is same regardless of the TDE temperature. The results are 191 krad for RADFET-1 and 114 krad for RADFET-3 during this period.

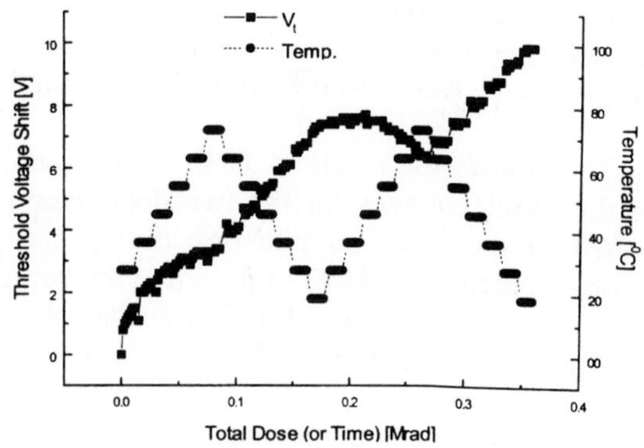

Figure 6. Change in the threshold voltage when the RADFET sensor is irradiated by Co^{60} γ–ray source as the temperature varies from 18°C to 72°C for two cycles.

Previous RAM tests have shown that many devices are susceptible to proton-induced SEUs and therefore, upsets in Low-Earth polar orbit tend to be concentrated in the SAA (Harboe-Sorensen *et al.* 1993). Figure 7 is the SEUs map of the KITSAT-1 OBC program memory in 1994 which confirms that SEUs are indeed concentrated in the SAA region. KITSAT-1 experiences more upsets in the SAA than KITSAT-2 because KITSAT-1's orbit is closer to the inner Van Allen belt in SAA. KITSAT-2 result also shows many SEUs in the polar region, north America, and south of Australia (Figure 8). Table 1, the SEU log of KITSAT-1 and KITSAT-2, shows that the characteristics of RAMDISK SEUs are different in these two satellites. On KITSAT-1 we observed single-bit errors (Sbit event) most of the time (34 events/30 days compared to 53 events/1 day for Mbit errors), while SEUs observed are usually the multiple-bit errors in a single byte (Mbit event) in the case of KITSAT-2 (148 events/98 days compared to 2 events/2 days for Sbit errors). This result seems to be associated with the orbital difference of the two satellites, since KITSAT-2, being in the high inclination orbit, are more susceptible to energetic cosmic particles which have an easy access to the polar region (Adams *et al.* 1991 ; Vampola *et al.* 1994). Another interesting feature is that the rate of SEU is different according to the type of the RAMs. Monolithic RAMs used for the program memories of KITSAT-1 and KITSAT-2 show 31.9 SEU rate per day for 1.536 MBytes in total (21 SEU rate/MByte), while hybrid RAMs used for the data memory of KITSAT-1 and KITSAT-2 show 179.9 SEU rate per day for 38 MBytes (4.7 SEU rate/Mbyte), indicating monolithic RAMs are more affected by the radiation than the hybrid RAMs.

Table 1. SEU Log Analysis of KITSAT-1 and KITSAT-2

	Program memory SEU rate/day (RAM size)	Data Memory SEU rate/day (RAM size)	RAMDISK Mbit Event (Events/Days)	RAMDISK Sbit Event (Events/Days)
1993. KITSAT-1	16.6 (512 KB)	61.6 (13 MB)		

1994. KITSAT-1	10.2 (512 KB)	59.9 (13 MB)	53/1	34/30
1994. KITSAT-2	5.1 (512 KB)	58.4 (12 MB)	148/98	2/2
1994. KASCOM	4.0 (2 MB)	29.5 (8 MB)		

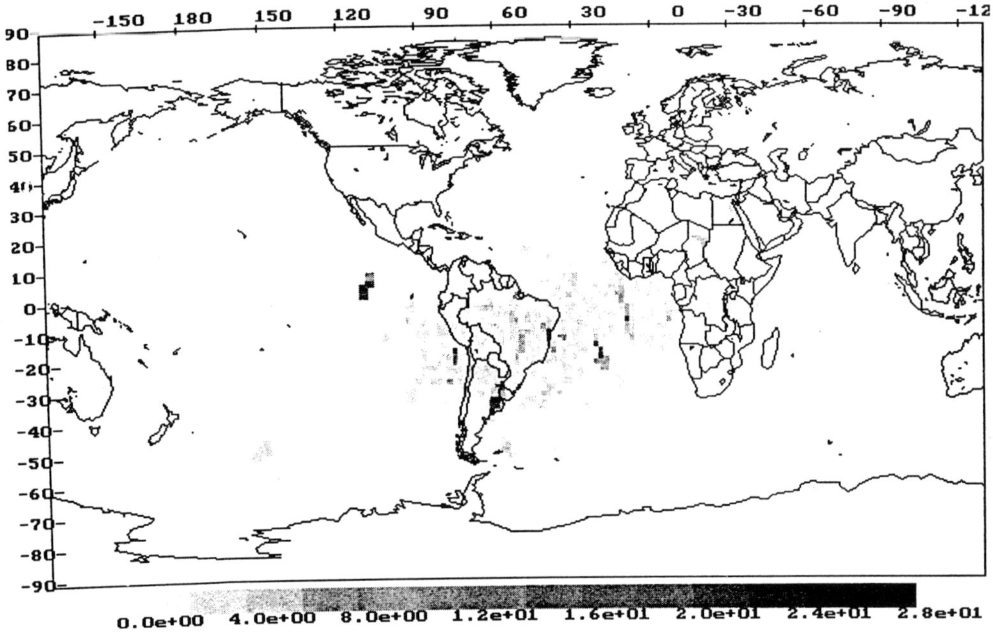

Figure 7. Global map of the KITSAT-1 program memory SEUs measured in 1994.

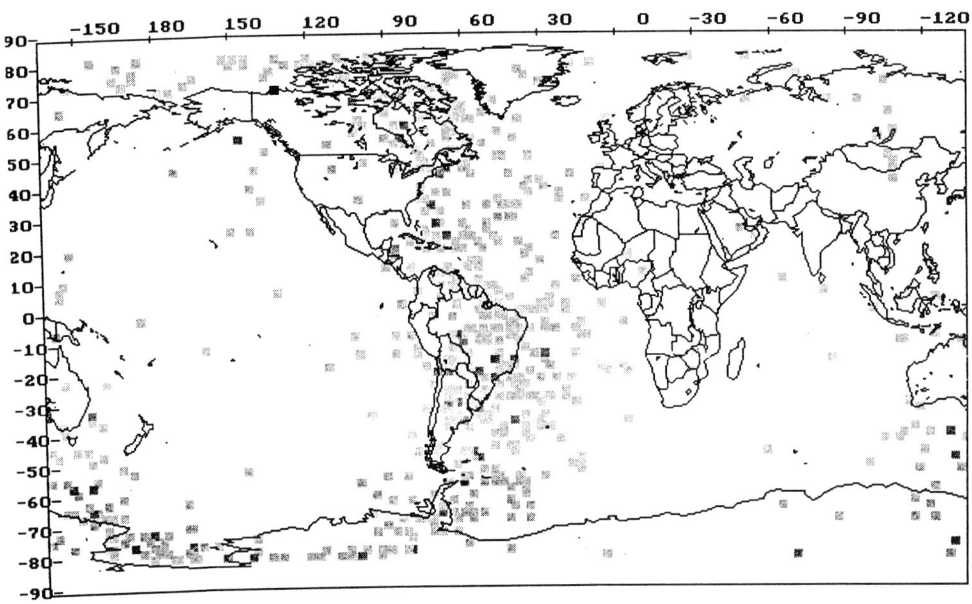

Figure 8. Global map of the KITSAT-2 KASCOM program memory SEUs from January through June in 1994.

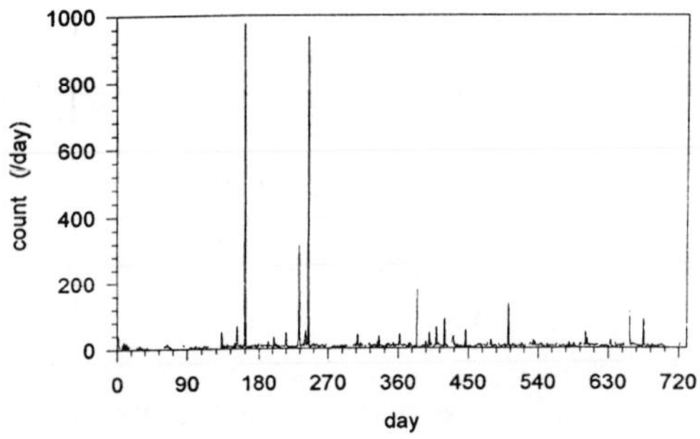

Figure 9. SEU rates occurred in the KITSAT-1 OBC program memory in 1993 and 1994.

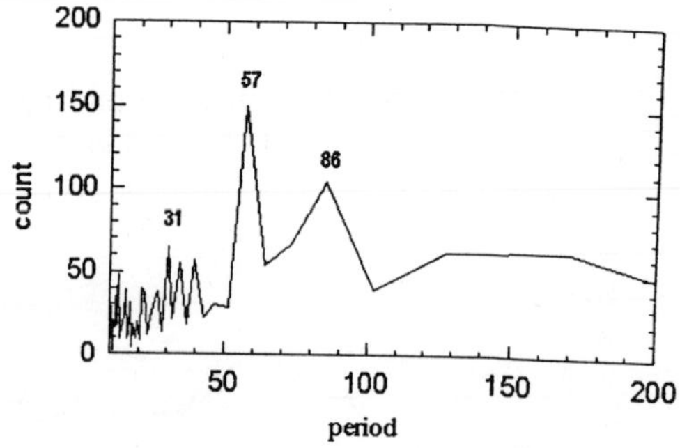

Figure 10. Period of the KITSAT-1 RAMDISK SEUs.

The KASCOM results also show the same trend: hybrid RAMs of the program memory shows 4 SEU rate per day for 2 MBytes (2 SEU rate/MByte), smaller than that of the monolithic RAMs of the data memory (29.5 SEU rate per day for 8 MBytes: 3.7 SEU rate/MByte).

Figure 9 shows the SEUs occurred in the KITSAT-1 OBC program memory in 1993 and 1994. Some of the peaks, such as those on days 162 (June 11, 1993), 385 (January 20, 1994), 420 (February 24, 1994), 657 (October 19, 1994), and on day 675 (November 6, 1994), match the proton flux enhancements observed by the GOES-7 satellite. However, some peaks do not apparently correspond to the changes of the space environment. In addition, the SEU rates are not linearly proportional to the proton fluxes.

Fourier transformation is performed to check the periodicity of the SEU rates obtained by the KITSAT-1 RAMDISK data. As Figure 10 shows the result, peaks correspond to the solar rotation period and its harmonics (Williams 1966). Instead of the 29 day solar rotation period, small peaks spread around this primary period with a largest one at the 31 day period. This obscurity comes from the fact that the satellite does not pass through the SAA in every orbit, rather it takes several days to resume its orbit close to the original one. Hence, the result we observe is the modulation of the solar rotation period with such orbital characteristics.

SUMMARY

The methods and results of the space radiation experiments carried out on board the KITSAT-1 and KITSAT-2 are discussed in this paper. These two satellites, which are similar in structures but different in orbits, provide a unique opportunity to study the effects of the radiation environment characterized by the orbit. The RADFETs on board the KITSAT-1 show the different changes in the threshold voltage corresponding to their shielding effects. The oscillatory behaviors seen in the curves of the threshold voltage are identified to come from the thermal annealing effects which, in turn, vary according to the changes of the eclipse percentage of the spacecraft. SEUs caused by SAA protons are observed in both satellites as they make regular passes through the SAA. The SEU characteristics observed in the two satellites are different: single bit errors are dominant in the low inclination KITSAT-1 orbit, while multiple bit errors are prevalent on KITSAT-2 in the high inclination orbit. Monolithic RAMs are found to be more susceptible to the ionizing radiation than the hybrid RAMs of lower packaging density. The SEU rates generally correlate with the proton fluxes, but there are cases in which the causes of SEUs are not clear. Also, the present analysis shows SEU rates are modulated by the solar rotation period and its harmonics.

REFERENCES

Adams, L., E. J. Daly, R. Harboe-Sorensen, A. G. Holmes-Siedle, A. K. Ward, and R. A. Bull, Measurements of SEU and Total Dose in Geostationary Orbit under Normal and Solar Flare Conditions, *IEEE Trans. Nuc. Sci.,* 38, 1686 (1991).

Adams, L., and A. G. Holmes-Siedle, The Development of an MOS Dosimetry Unit for Use in Space, *IEEE Trans. Nucl. Sci.,* 25, 1657 (1978).

Dyer, C. S., A. J. Sims, J. Farren, J. Stephen, and C. Underwood, Measurements of the Single Event Upset and Total Dose Environments in Polar Orbit, *IEEE Trans. Nucl. Sci.,* 38, 1700 (1991).

Dyer, C. S., A. J. Sims, P. R. Truscott, J. Farren, and C. Underwood, The Low Earth Orbit Radiation Environment and Its Evolution from Measurements Using the CREAM and CREDO Experiments, *IEEE Trans. Nucl. Sci.,* 40, 1471 (1993).

Gussenhoven, M. S., E. G. Mullen, R. C. Filz, D. H. Brautigam, and F. A. Hauser, New Low-Altitude Dose Measurements, *IEEE Trans. Nucl. Sci.,* 34, 676 (1987).

Harboe-Sorensen, R., E. J. Daly, L. Adams, C. I. Underwood, and R. Mueller, Observation and Prediction of SEU in Hitachi SRAMs in Low Altitude Polar Orbits, *IEEE Trans. Nucl. Sci.,* 40, 1498 (1993).

Shin, Y. H., K. W. Min, Y. W. Choi, and S. H. Kim, CRE Experiment on KITSAT-1, *J. Astro. Space Sci.,* 11, 131 (1994).

Underwood, C. I., D. J. Brock, P. S. Williams, S. Kim, R. Dialo, P. R. Santo, M. C. Brito, C. S. Dyer, and A.J. Sims, Radiation Environments with Cosmic Ray Experiments On-Board the KITSAT-1 and PoSAT-1 Micro-Satellites, *IEEE Trans. Nucl. Sci.,* **41**, 1353 (1994).

Vampola, A. L., M. Lauriente, D. C. Wilkinson, J. Allen, and F. Albin, Single Event Upsets Correlated with Environment, *IEEE Trans. Nuc. Sci.,* 41, 2383 (1994).

Williams, D. J., A 27-Day Periodicity in Outer Zone Trapped Electron Intensities, *J. Geophys. Res.,* 71,.1815 (1966).

SPACE AT SURREY: MICROSATELLITES & MINISATELLITES FOR AFFORDABLE ACCESS TO SPACE

MN Sweeting OBE FEng FIEE

Surrey Space Centre, University of Surrey, Guildford, Surrey GU2 5XH, UK
Tel: (44) 1483 259278, Fax: (44) 1483 259503
internet: m.sweeting@ee.surrey.ac.uk, WWW: http://www.ee.surrey.ac.uk/EE/CSER/UOSAT/

ABSTRACT

Changing world politics and military emphasis has brought considerable pressure on space agency financial budgets. Budgetary pressure, coupled with developments in micro-electronics electronics, has catalysed the use smaller satellites as a 'faster, cheaper, better' means of realising space missions - complementary to conventional large satellite systems. Affordable small satellites, however, require a very different approach to established space engineering techniques. Since 1979, the University of Surrey (UK) has pioneered cost-effective satellite engineering techniques for small satellites, and has developed a series of highly sophisticated, yet inexpensive, 'microsatellites' - built for around US$3M each. Twelve such microsatellites have been launched by Surrey into low Earth orbit carrying a wide range of satellite communications, space science, remote sensing and in-orbit technology demonstration payloads - both civil and military. Six microsatellite missions have also provided the focus and means for a highly successful satellite technology transfer and training programme between Surrey and emerging space nations such as Korea, Portugal, Pakistan, South Africa, Chile, Singapore, Malaysia and Thailand - leading to the launch of their first national satellites (eg. KITSAT-1&2; PoSAT-1, BADR-1, TMSAT & FASat-B) - providing rapid response, low risk and affordable access to space. The Surrey Space Centre offers a unique environment, combining both academic research and postgraduate teaching with commercial development & manufacturing of satellites through its spin-off company, Surrey Satellite Technology Ltd. A new 350kg minisatellite (UoSAT-12) is being built at Surrey for launch in April 1999 which carries more demanding payloads within the same cost-effective satellite engineering philosophy and will provide the basis for Surrey's first low-cost interplanetary mission to the Moon for launch in 2001.

INTRODUCTION

Following the first space launch in 1957, satellites rapidly became increasingly large and enormously expensive. Initially, the 'Space race' was an effective catalyst for the development of advanced technology as the super-powers strove to out-do each other and gain the advantageous 'Sigh ground' of space - irrespective of budget. However, as costs escalated and timescales lengthened with satellites generally taking many years to mature from concept to useful orbital operation, this process limited access to space to only a relatively few nations or international agencies. Changing world politics and military emphasis in the last decade has brought about a quiet revolution in space. Pressure on space agency financial budgets has increasingly meant that fewer (and bigger) satellites have been commissioned and new ideas, technologies and scientific experiments have found it difficult to gain timely access to space. The staggering developments in micro-electronics, stimulated by the consumer market as much as the military requirement, and the dramatic pace of consumer product development, caused space technology often to lag considerably behind that now taken for granted on Earth. The combination of reducing budgets for space and increasing capability of low-power micro-electronics have enabled a new breed of highly capable 'Smaller, faster, cheaper' satellites to realise many space missions - complementing the conventional large satellite systems still necessary for large-scale space science and communications services to small terminals. Indeed, in the field of commercial satellite communications, numerous constellations of small satellites have been proposed and are being built to provide a range of global services to hand-held terminals either for real-time voice or non-real-time data.

However, whether a particular satellite is 'large' or 'small' depends somewhat upon viewpoint. For instance the 51mall' satellites proposed for the Iridium mobile communications system weigh in at around 600 kg each whereas those for HealthNet email network are a mere 50kg! In view of this potential for confusion, the following classification has become widely adopted:

Class	***Mass (kg)***	***Cost (K)***
Large satellite	> 1000	>100
Small satellite	500 -1000	30 - 100
Mini-satellite	100 - 500	5 - 20
Micro-satellite	10 - 100	2 - 4
Nano-satellite	1 - 10	0.2 - 1
Pico-satellite	>1	>0.2

Although there have been many examples of large, small and even mini-satellites, however it is only relatively recently that capable *microsatellites* have shown that it is possible to execute both civil and military missions very effectively, rapidly, and at low cost and risk, for the following applications:

- Specialised Communications Services & Research
- Earth Observation & Remote Sensing
- Small-Scale Space Science
- Technology Demonstration/Verification
- Education & Training

However, a sustained, commercial, low-cost small satellite programme must also be matched by correspondingly inexpensive and regular access to orbit through formal launch service contracts - as it makes little sense to construct sophisticated yet inexpensive microsatellites if the launch costs remain prohibitively high. Early microsatellites were launched virtually for free on a 'favour' basis by the US & USSR, but these launch opportunities were infrequent and unpredictable. The breakthrough came in 1988 when Arianespace developed the **A**riane **S**tructure for **A**uxiliary **P**ayloads (ASAP) specifically to provide, for the first time, regular and affordable launch opportunities for 50kg microsatellites into both LEO and GTO on a commercial basis.

To date, some 18 microsatellites have been launched via the ASAP but, whilst it has been key to catalysing microsatellites world-wide, Ariane alone cannot now provide the number of launch opportunities into LEO needed to meet the burgeoning growth of small satellites. and so alternative, inexpensive launch options from the CIS (on Tsyklon, Zenit, & Cosmos) are now increasingly being used for micro/minisatellites. Within the last few years, the large stockpiles of ICBMs in the CIS have become available for use as small launchers through the demilitarilisation programme (e.g. SS-18/Dnepr; SS-19/Rockot; SS-25/START).

SURREY MICROSATELLITES

The University of Surrey has pioneered microsatellite technologies since beginning its *UoSAT* programme in 1978. From very modest experimental beginnings, its space-related research, post-graduate teaching and international commercial activities are now housed within a purpose-built *Surrey Space Centre* - with over 130 staff and postgraduate research students. The objectives of the Centre's programmes are:

- to research cost-effective small satellite techniques
- to demonstrate the capabilities of micro/minisatellites
- to catalyse commercial use of micro/minisatellites
- to promote space education and training

Figure 1. The Surrey Space Centre and SSTL facilities

Over the last decade, Surrey has established an international reputation as pioneers of innovative small satellites in a uniquely combined academic research teaching and commercial environment.

Surreys' first experimental microsatellites (UoSAT-1 & 2) were launched free-of-charge as 'piggy-back' payloads through a collaborative arrangement with NASA on DELTA rockets in 1981 & 1984 respectively [1,2]. Since then, a further ten low cost yet sophisticated microsatellites have been placed in low Earth orbit using Ariane and Tsyklon launchers for a variety of international customers and carrying a wide range of payloads.

Two new (50kg) microsatellites for Chile and Thailand are in Moscow awaiting launch on Zenit from the Baikonur Cosmodrome, and three new microsatellites for Malaysia, France and the US-DoD and an experimental (350kg) minisatellite are currently being prepared for launch in 1999.

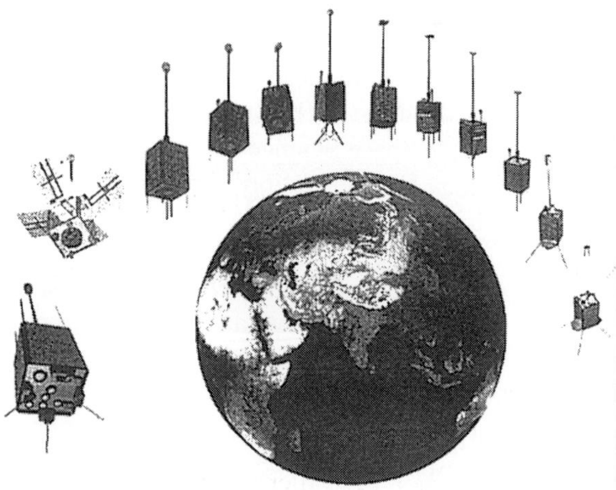

Figure 2. Surrey microsatellite missions in LEO

Table 1. University of Surrey Microsatellite Missions

Microsatellite	Launch	Orbit	Customer	Payloads
UoSAT-1	1984-D	560 km	UoS	Research
UoSAT-2	1984-D	700 km	UoS	S&F, EO, rad
UoSAT-3	1990-A	900 km	UoS	S&F
UoSAT-4	1990-A	900 km	UoS/ESA	Technology
UoSAT-5	1991-A	900 km	SatelLife	S&F,EO, rad
S80/T	1992-A	1330 km	CNES	LEO comms
KitSat-1	1992-A	1330 km	Korea	S&F,EO, rad
KitSat-2*	1993-A	900 km	Korea	S&F,EO, rad
PoSAT-1	1993-A	900 km	Portugal	S&F,EO, rad
HealthSat-2	1993-A	900 km	SatelLife	S&F
Cerise	1995-A	735 km	CNES	Military
FASat-Alfa	1995-T	873 km	Chile	S&F,EO
FASat-Bravo	1997-Z	835 km	Chile	S&F,EO
TMSAT	1997-Z	835 km	Thailand	S&F,EO
TiungSAT-1	1998-?	650 km	Malaysia	EO, Comms
UoSAT-12	1999-S	650 km	SSTL & Singapore	EO, Comms
PicoSAT	1998-S	650 km	USAF	Military
Clementine	1999-A	735 km	CNES	Military

UoSAT-1 & 2 both used a rather conventional structure - a framework 'skeleton' onto which were mounted module boxes containing the various electronic subsystems and payloads with a complex 3-dimensional inter-connecting wiring harness.

Following UoSAT-1 and UoSAT-2 (in 1984), the need to accommodate a variety of payload customers within a standard (ASAP) launcher envelope (400 × 400 × 600mm and 50kg), coupled with increased demands on packing density, electro-magnetic compatibility, economy of manufacture and ease of integration, catalysed the development at Surrey during 1986 of a novel modular design of multi-mission microsatellite platform. This modular microsatellite platform has been used successfully on *seventeen* missions, each with different payload requirements, and allowing the spacecraft to proceed from order-to-orbit in typically 10-12 months!.

The SSTL modular microsatellite has no 'skeleton' but rather a series of identical outline machined module boxes stacked one on top of the other to form a body onto which solar panels and instruments may be mounted.

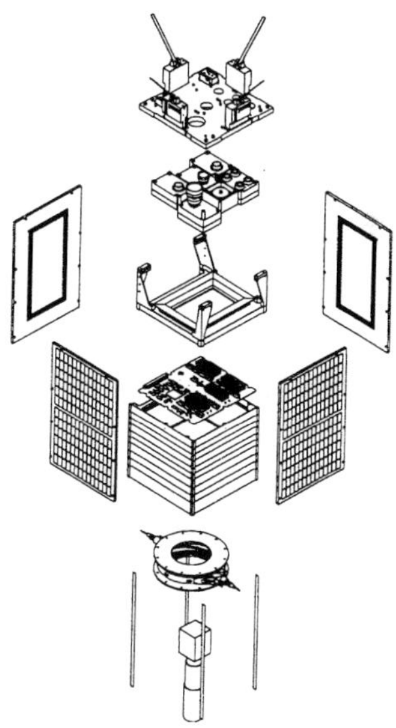

Figure 3. Exploded' view of SSTL microsatellite structure

Each module box, houses the various microsatellite subsystems - eg. batteries, power conditioning, on-board data handling, communications and attitude control. Payloads are housed either in similar modules or on top of the platform alongside antennas and attitude sensors as appropriate.

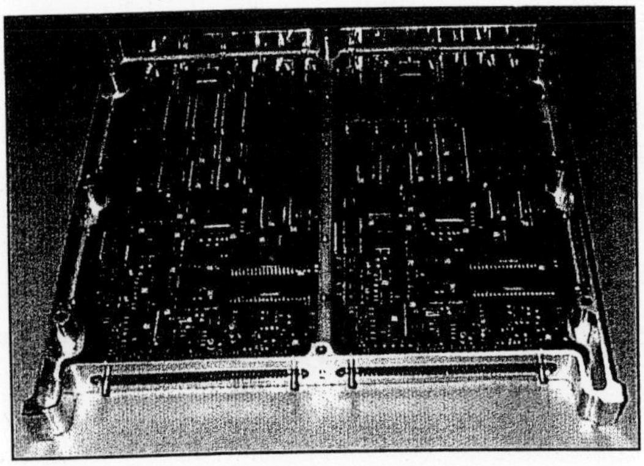

Figure 4. Typical SSTL Microsatellite Module

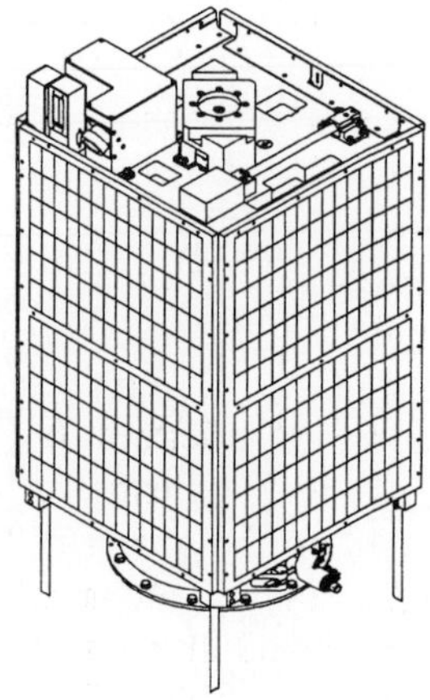

Figure 5. Integrated microsatellite structure

Electronically, the microsatellite uses modern, sophisticated, but not necessarily space-proven, electronic circuits to provide a high degree of capability. These are underpinned by space-proven subsystems - resulting in a layered architecture that achieves high performance with operational redundancy by using alternative technologies rather than by simple duplication.

Communications and Earth observation payloads require an Earth-pointing platform and so the microsatellite is maintained to within 1° of nadir by employing a combination of gravity-gradient stabilisation (using a 6-metre boom) and closed-loop active damping using electromagnets operated by the on-board computer. Attitude determination is provided by Sun, geomagnetic field sensors, and star field cameras, whilst orbital position is determined autonomously to with ±50 metres by an on-board GPS receiver. Electrical power is generated by four body-mounted GaAs solar array panels, each generating ~35W, and is stored in a 7Ah NiCd rechargeable battery.

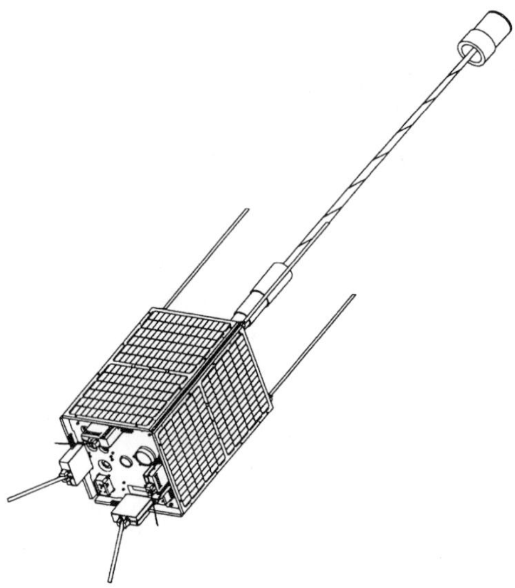

Figure 6. Microsatellite orbital configuration

Communications are supported by VHF uplinks and UHF downlinks, using fully error-protected AX.25 packet link protocols operating at 9.6 to 76.8 kbps, and are capable of transferring several hundred kBytes of data to brief-case sized communications terminals.

It is the On-Board Data Handling (OBDH) system that is the key to the sophisticated capability of the microsatellite. At the heart of the OBDH system is a 80C386 On-Board Computer, which runs a 500 kByte real-time multi-tasking operating system with a 128 MByte solid-state CMOS RAMDISK. In addition, there is a secondary 80C186 OBC with 16 MByte SRAM, two 20 MHz T805 Transputers with 4 MBytes SRAM, and a dozen other microcontrollers. A primary feature of the OBDH philosophy is that all the software on-board the microsatellite is *loaded after launch* and can be upgraded and reloaded by the Control Ground station at will thereafter. Normally, the satellite is operated via the primary OBC-386 and the real-time multi-tasking operating system. All telecommand instructions are formulated into a 'diary' at the ground station and then transferred to the satellite OBC for execution either immediately or, more usually, at some future time. Telemetry from on-board platform systems and payloads is similarly gathered by the OBC-386 and either transmitted immediately and/or stored in the RAMDISK whilst the satellite is out of range of the Control Station. The OBC's also operate the attitude control systems according to control algorithms that take input from the various attitude sensors and then act accordingly. Thus it is this OBDH environment that allows such a tiny microsatellite to operate in a highly complex, flexible and sophisticated manner, enabling fully automatic and autonomous control of the satellites systems and payloads.

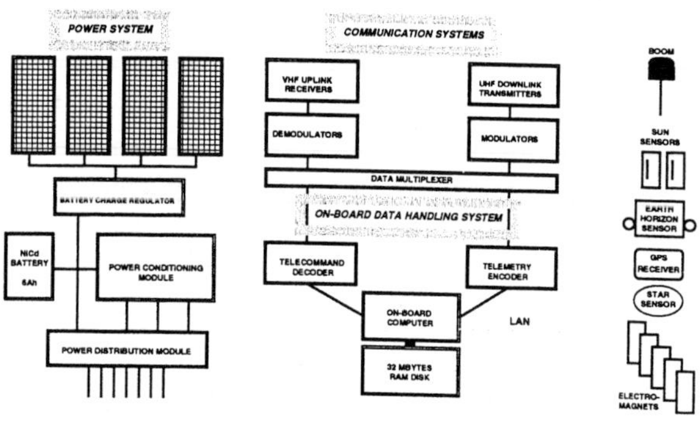

Figure 7. SSTL Microsatellite Platform Systems

The latest SSTL microsatellite platforms have enhanced sub-systems supporting the following major new features for greatly increased performance and payload capacity. The first version of the enhanced microsatellite platform was used on the FASat-A mission for Chile launched into LEO by a Ukrainian TSYKLON launcher in 1995 and has since been further developed and used on the new FASat-Bravo, TMSAT, TiungSat-1 & PICOSat missions. The very latest microsatellite platforms can provide:

- distributed telemetry/ telecommand for easy expansion
- 5 MIPS (386) OBC + 256 MB RAMDISK
- 1 Mbps on-board LAN
- autonomous GPS navigation (50m)
- attitude determination (0.001^{o})
- 0.25^{o} rms nadir pointing pitch/roll axes, 2° roll axis
- 128 kbps BPSK downlinks

APPLICATIONS OF MICRO/MINISATELLITES

The early UoSAT missions demonstrated the potential capabilities of microsatellites and generated considerable interest in applications such as digital store-&-forward file transfer and in-orbit technology verification, however wider applications such as space science and Earth observation were slow to develop. Emerging space nations, however, were quick to recognise the benefits of entering the space-faring community with an affordable, low-risk 'first step' via an extremely inexpensive yet realistic microsatellite programme.

The need to handle this growing interest, to catalyse wider industrial & commercial applications, and to generate regular income to sustain a research activity in satellite engineering at the University of Surrey without dependence on government funding, stimulated the formation in 1985 of a University company - *Surrey Satellite Technology Ltd (SSTL)*. SSTL provides a formal mechanism to handle the transfer of small satellite technologies from the University's academic research laboratories into industry in a professional manner via commercial contracts.

Since UoSAT-5, all the Surrey microsatellites have been design and built for individual commercial customers. The income generated by SSTL is then re-invested to support the academic activities of the Surrey Space Centre, which has now become the largest European 'centre of excellence' in satellite engineering research, teaching and applications.

The range of applications of microsatellites can be demonstrated by reviewing recent examples of payloads carried by the UoSAT/SSTL microsatellite platform

LEO Communications

Various constellations of small satellites in LEO have been proposed to provide world-wide communications using only hand-held portable terminals; these broadly fall into two main categories:

- real-time voice/data services (eg. Iridium, Globalstar)
- non-real-time data transfer (eg. Orbcom, HealthNet)

The close proximity of the satellites in LEO to the user and the consequent reduction in transmission loss and delay time appear attractive when compared to traditional communications satellites in a distant geostationary orbit - holding out the promise of less expensive ground terminals and regional frequency reuse. The communications characteristics associated with a LEO constellation pose, however, quite different and demanding problems, such as varying communications path & links, high Doppler shifts, and hand-over from satellite to satellite.

Currently, there is only one small LEO satellite service in full operation (HealthNet) which employs a 35 onstellation' of just two microsatellites, HealthSat-1 & 2, both built by SSTL for the network operator, SatelLife (USA). HealthNet, and the services waiting to be implemented (e.g. TemiSat, OrbComm, VITASat, GEMStar) use narrow-band VHF/UHF frequencies recently allocated to the 'little-LEO' services to provide digital data store-&-forward "Email" capabilities for use with small, low-power user ground terminals that can be located in remote regions where existing the telecommunications infrastructure is inadequate or non-existent [3].

These VHF/UHF frequency allocations exhibit such deleterious effects as multi-path propagation and, in particular, man-made co-channel interference which can very significantly reduce the performance that can be achieved in practice by the satellite communications system compared to that expected from a simple theoretical model. A thorough understanding of the real LEO communications environment at VHF & UHF is therefore necessary in order to select optimum modulation and coding schemes.

Figure 8. LEO VHF Interference Environment

The KITSAT-1 & PoSAT-1 microsatellites carry a Digital Signal Processing Experiment (DSPE) which has been designed to provide a sophisticated in-orbit test bed for research into optimising communications links with satellites in LEO. The DSPE comprises a TMS320C25 and a TMS320C30 with supporting PROM, RAM and data interfaces to the spacecraft communications sub-systems - enabling it to replace the hardware onboard modems with an in-orbit, re-programmable modem. The DSPE is being used in a research programme to evaluate adaptive communications links - continuously optimising modulation/demodulation techniques, data rates and coding schemes in response to traffic characteristics during the microsatellite's transit of the ground station.

The interference characteristics of the VHF LEO frequency allocations have been measured using an experimental communications payloads on the S80/T and HealthSat-2 microsatellite missions. In conjunction with a mobile ground station, S80/T has measured the VHF spectrum 'noise' and interfering signals to evaluate the use of VHF frequencies for a full-scale LEO communications service (S80). S80/T was completed by SSTL for the Centre National d'Etudes Spatiales (France), from proposal to launch, within 12 months!

Space Science

Microsatellites can offer a very quick turn-around and inexpensive means of exploring well-focused, small-scale science objectives (e.g. monitoring the space radiation environment, updating the international geo-magnetic reference field, etc.) or providing an early proof-of-concept prior to the development of large-scale instrumentation in a fully complementary manner to expensive, long-gestation, large-scale space science missions. SSTL missions have demonstrated that it is possible to progress from concept through to launch and orbital operation within 12 months and within a budget of £2-3M. This not only yields early scientific data but also provides opportunities for young scientists and engineers to gain 'real-life' experience of satellite and payload engineering (an invaluable experience for later large-scale missions) and to be able to initiate a programme of research, propose and build an instrument, and retrieve orbital data for analysis and presentation for a thesis within a normal period of post-graduate study.

UoSAT-3 & 5, KITSAT-1 & 2 & PoSAT-1 provide examples of the use of a microsatellite platform for collaborative space science research between the University of Surrey, the UK Defence Research Agency, UK AEA, the UK Science & Engineering Research Council, KAIST and Portugal.

A 'Cosmic Ray Effects & Dosimetry' (CREDO) payload monitors the near-Earth radiation environment and provides an important opportunity to validate ground-based numerical models with flight data yielding simultaneous measurements of the radiation environment and its effects upon on-board systems (especially SEUs in VLSI devices) [4].

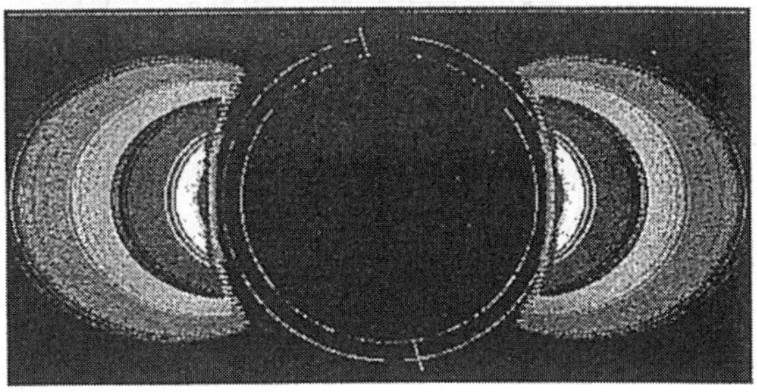

Figure 9. *LEO Radiation Environment from PoSAT-1*

Earth Observation

Conventional Earth observation and remote sensing satellite missions are extremely costly - typically in excess of £150M each - and thus there are relatively few such missions and the resulting data, whilst impressive, is correspondingly expensive. The development of high-density two-dimensional semi-conductor Charge-Coupled Device optical detectors, coupled with low-power consumption yet computationally powerful microprocessors, presents a new opportunity for remote sensing using inexpensive small satellites.

UoSAT-1 & 2 both carried the first experimental 2D-CCD Earth imaging cameras which lead to the development of the CCD Earth Imaging System on-board UoSAT-5, intended to demonstrate the potential of inexpensive, rapid-response microsatellite missions to support remote sensing applications. Clearly, the limited mass, volume, attitude stability and optics that can be achieved with a tiny microsatellite means that a different approach must be taken to produce worthwhile Earth observation. For these reasons, SSTL employs electronic cameras with 2-dimensional CCD arrays to gather imagery from its microsatellites. Because cameras capture whole images in a single snap-shot, they preserve scene geometry and are therefore immune to the residual attitude drift experienced on microsatellites.

The Earth Imaging System (EIS) on-board the UoSAT-5, KITSat & PoSAT microsatellites comprise an EEV (UK) 576×578 pixel area CCD digitised to 256 levels of grey. The digitised data is stored within a 2 Mbyte CMOS RAM which is accessed by two Transputers to allow the image data to be processed to enhance quality and compressed to reduce storage and transmission requirements.

The EIS data is transferred via the microsatellite's local area network to the 80C186/386 on-board computers and stored as files within the 32-128 Mbyte RAM DISK for later transmission to ground - some 60 images can be stored within the RAM DISK at any one time. The EIS is commanded to collect an image of a particular area of the Earth's surface by the on-board computer which operates a multi-tasking, real-time operating system responsible for the automatic (and in some cases, autonomous) operation of the microsatellite mission.

Ground controllers select a sequence of images of areas of interest anywhere on the Earth's surface and, checking the time and position of the microsatellite using an on-board GPS receiver and orbital model, instruct the on-board computer to collect the images according to a 'diary' that is loaded periodically in advance to the microsatellite. The PoSAT-1 microsatellite carries two independent cameras providing a wide-field ground resolutions of 2km for meteorological imaging and a narrow-field ground resolution of 200 metres for environmental monitoring, with 650 nm ($\pm$40nm) optical filters providing good separation of arid/vegetation and land/sea boundaries [5].

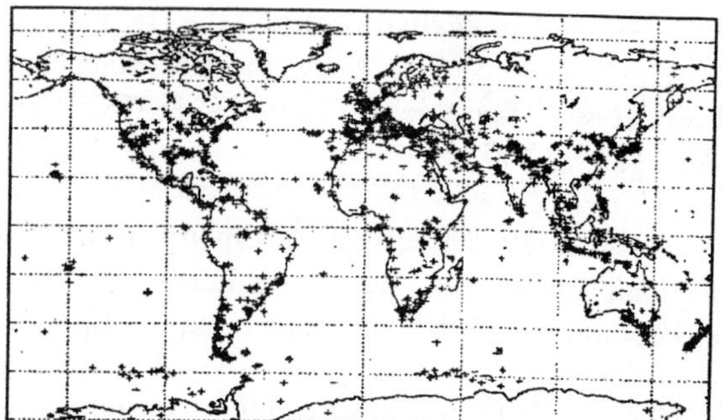

Figure 10. Daily, world-wide imaging by SSTL microsatellites

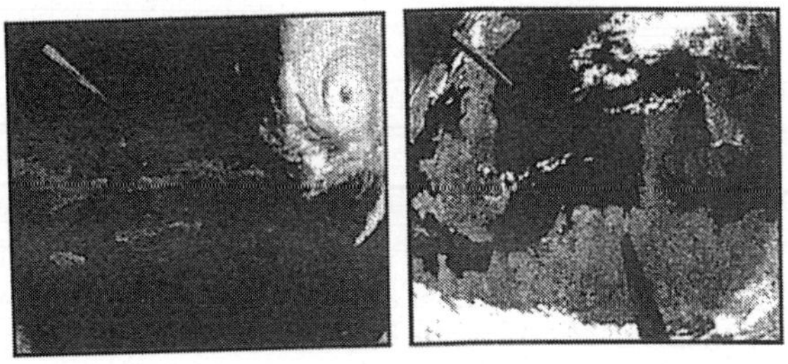

Figure 11. Meteorological images from PoSAT-1

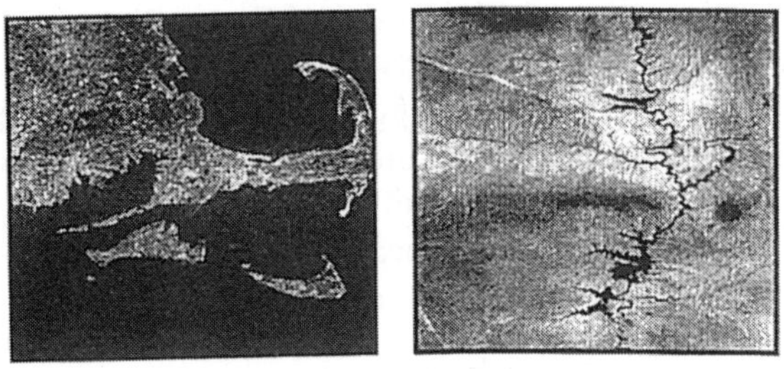

Figure 12. Narrow angle images from PoSAT-1

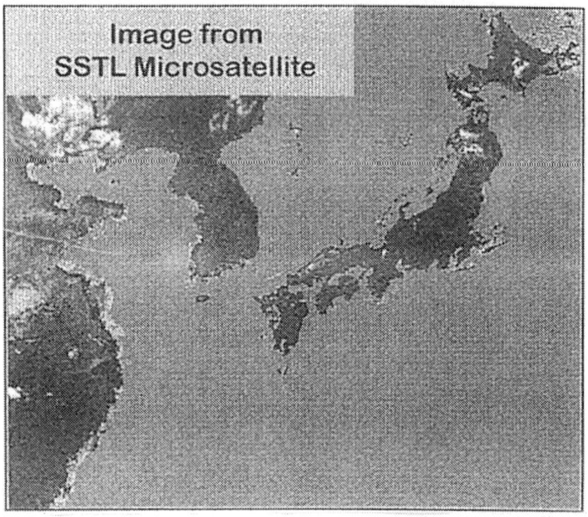

Figure 13. Meteorological images from PoSAT-1

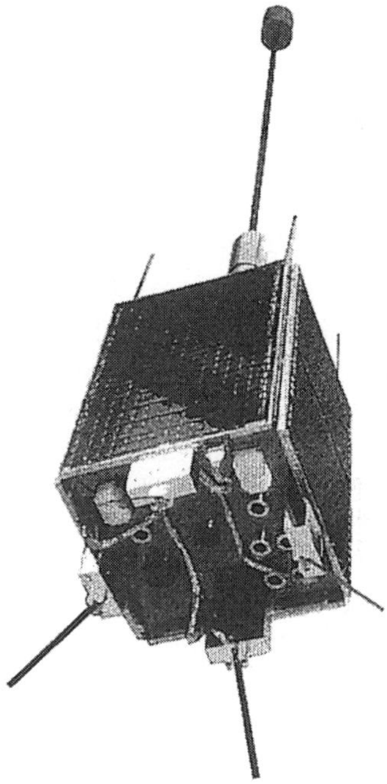

Figure 14. PoSAT-1 image compared to

The results from current UoSAT EIS in orbit have been remarkable however the next generation of small Earth observation satellites using the SSTL platform, such as FASat-B & TMSAT will support EIS cameras yielding better than 100-metre resolution with 3 spectral bands.

The TiungSAT-1 microsatellite will provide 75-metres resolution in 4 spectral bands and the latest SSTL microsatellites for launch in 1999 will provide 50m 4-band multi-spectral imaging with the capability of 1° (200km) off-nadir imaging coverage upon demand.

In-Orbit Technology Verification

Microsatellites also provide an attractive and low-cost means demonstrating, verifying and evaluating new technologies or services rapidly in a realistic orbital environment and within acceptable risks prior to a commitment to a full-scale, expensive mission.

UoSAT-based microsatellites have supported a wide range of such in-orbit technology demonstrations, covering:

· new solar cell technologies
· modern VLSI devices in space radiation
· demonstration of advanced communications
· 'pilot' demonstrations of new communications

For example, satellites depend upon the performance of solar cell arrays for the production of primary power to support on-board housekeeping systems and payloads. Knowledge of the long-term behaviour of different types of cells in the radiation environment experienced in orbit is, therefore, essential. The continuing development of solar cell technology, based upon a variety of materials and different process techniques, yields a range of candidate cells potentially suitable for satellite missions.

Unfortunately, ground-based, short-term radiation susceptance testing does not necessarily yield accurate data on the eventual in-orbit performance of the different cells and hence there is a real need for evaluation in an extended realistic orbital environment. UoSAT-5 carries a Solar Cell Technology Experiment (SCTE) designed to evaluate the performance of a range of 27 samples of GaAs, Si and InP solar cells and from a variety of manufacturers. When the sun passes directly overhead of the panel mounted on the body of the microsatellite, the monitoring electronics are triggered automatically and measure typically 100 current/voltage points for each cell sample. These data are then sent in a burst to the microsatellite's on-board computer, together with associated temperature and radiation dose data, for storage prior to transmission later to ground. SCTE measurements are taken repeatedly immediately after launch, when the radiation degradation is most rapid, and then at increasing intervals thereafter.

Military Applications

The demands of a military-style satellite procurement and the cost-effective approach to microsatellite engineering might, at first sight, appear incompatible! However, whilst retaining the essential characteristics of low cost and rapid response, a military version of the SSTL microsatellite platform with deployable solar panels has been developed to support various military payloads. The main differences between the 'commercial' and 'military' versions of the platform is in the specification of components and, particularly, in the amount of paperwork that traces hardware and procedures. An optimum trade-off between the constraints of a military programme and economy has been sought which results in an increase factor for cost and timescale of approximately 1.5 when compared to the ‘commercial’ procurement of the platform. The first use of the SSTL military microsatellite platform was on the CERISE mission designed and built for the French MoD and launched a 700km low Earth orbit by Ariane in July 1995. After a year of perfect operations, Cerise made history as the first operational satellite to be struck by a piece of space debris (rocket fragment) which severed its stabilisation boom. However, due to the flexibility of the microsatellite systems, SSTL engineers were able to re-stabilise Cerise by uploading new attitude control algorithms and return it to operations.

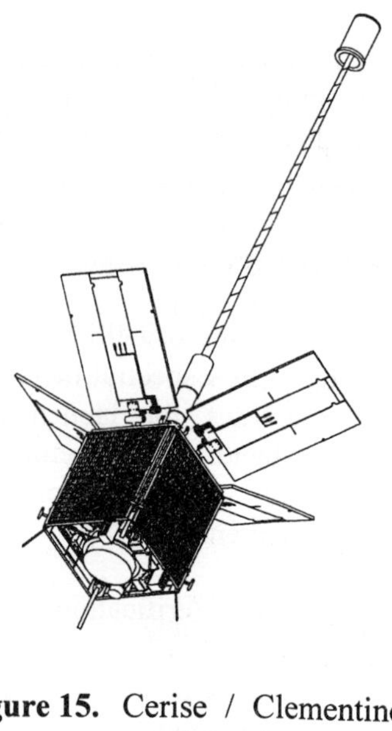

Figure 15. Cerise / Clementine orbital configuration

A second microsatellite for the French MoD is now being completed (Clementine) for launch into LEO in 1999 and a microsatellite (PICOsat) is being built for the USAF FCT programme.

The Geostationary Transfer Orbit (GTO) provides a good opportunity to study the effects of a severe radiation environment on satellite components - especially solar cells and VLSI components. Surrey has provided platform sub-systems and payloads to the UK Defence Research Agency (DRA - Farnborough) for their two Space Technology Research Vehicle (STRV-1) microsatellites which were launched into GTO in early 1994. STRV-1a & b carry a range of in-orbit technology demonstration payloads - particularly to study the effects of the space radiation environment on military satellite components

Education & Training Using Microsatellites

Although microsatellites are physically very small, they are nevertheless complex and exhibit virtually all the characteristics of a large satellite - but in a microcosm. This makes them particularly suitable as a focus for the education and training of scientists and engineers by providing a means for direct, hands-on experience of all stages and aspects (both technical and managerial) of a real satellite mission - from design, construction, test and launch through to orbital operation.

The very low cost, rapid timescale and manageable proportions makes this approach very attractive to emerging space nations who wish to develop and establish a national expertise in space technology through an affordable small satellite programme.

Figure 16. Technology Transfer to Chile with FASat-Alpha

Each technology transfer and training (TTT) programme is carefully structured according to the specific requirements or circumstances of the country or organisation concerned, but the first phase typically comprises:

- Academic Education (MSc, PhD degrees)
- Technology Training (seconded to SSTL)
- Groundstation (installed in country)
- Microsatellites (1st at SSTL, 2nd in country)
- Technology Transfer (satellite design licence)

Five highly successful international technology transfer programmes have been completed by Surrey & SSTL and three new programmes are now under way:

Table 2. SSTL technology transfer & training programmes

Country	***Dates***	***Satellites***
Pakistan	1985-89	BADR-1
South Africa	1989-91	UoSAT-3/4/5
S.Korea	1990-94	KITSat-1/2
Portugal	1993-94	PoSAT-1
Chile	1995-97	FASat-Alfa/Bravo
Thailand	1995-98	TMSAT-1
Singapore	*1995-99*	*Merlion payload*
Malaysia	*1996-98*	*TiungSAT-1*

A total of 70 engineers have been trained through these in-depth TTT programmes - a further 320 students from countries world wide have graduated from the MSc course in Satellite Communications Engineering unrelated to these TTT programmes.

MICROSATELLITE GROUNDSTATIONS

Compact and low-cost mission control ground stations have been developed by SSTL to operate the microsatellites once in orbit. These ground stations are based on PCs and are highly automated - interacting autonomously with the microsatellite in orbit -to reduce manpower requirements and to increase reliability.

Figure 17. SSTL Groundstation in Portugal for PoSAT-1

The SSTL Mission Control Centre at Surrey operates eleven microsatellites in LEO with just a single operator.

Figure 18. New SSTL Mission Control Centre at Surrey

SSTL MINISATELLITES

In response to growing payload demands for power, volume & mass - but still within a small-scale financial budget - UoSAT & SSTL are developing an enhanced, modular, multi-mission minisatellite platform capable of supporting missions up to 400 kg.

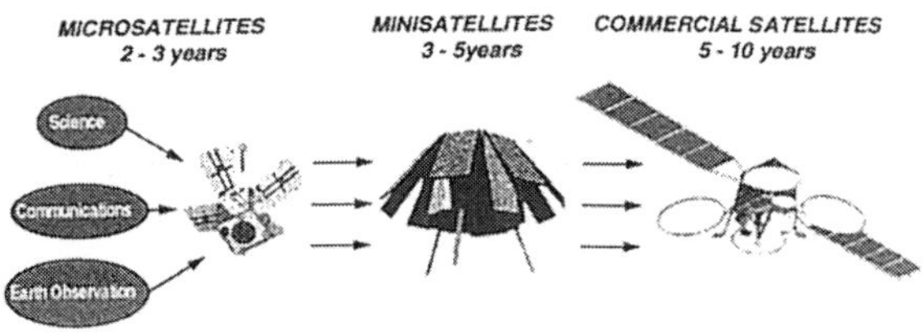

Once developing space nations have mastered microsatellite technology, the minisatellite provides a logical next step in the development of an increasingly capable national space infrastructure.

The minisatellite platform has been designed to meet a variety of mission objectives and capable of operating in different orbits, its primary features are:

- 400 kg total mass
- 150 kg payload capacity
- 1.2m diameter, 1m height
- 3-axis, 0.1 degree, attitude control
- GPS autonomous orbit & attitude determination
- 1 Mbps L/S-band communications links
- On-board propulsion for orbit manoeuvres
- cold gas thrusters for attitude control
- 300 watts orbit average power, 1 kW peak power

The SSTL minisatellite platform has been designed according to similar cost-effective principles that have proved so successful on the UoSAT/SSTL microsatellites - resulting in a basic platform cost of £5-8M. Considerable effort has been expended to ensure compatibility with a range of affordable launch options - on Ariane, CIS, Chinese & US (Pegasus) launchers.

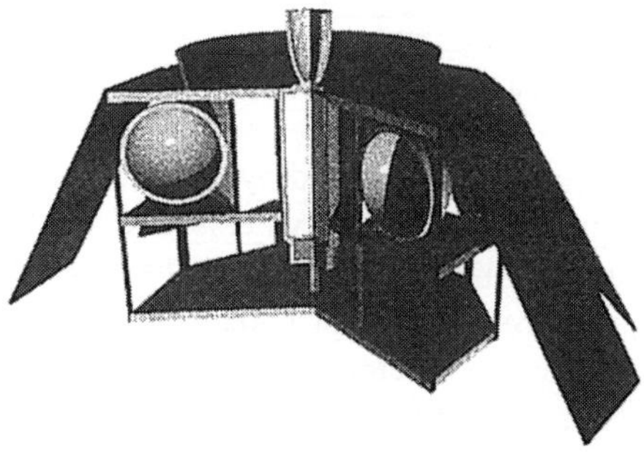

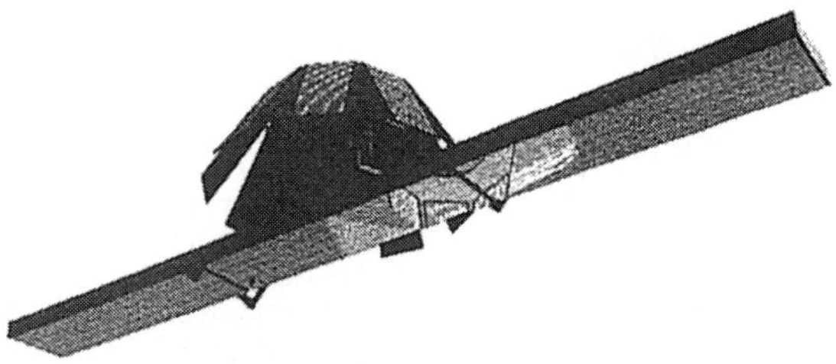

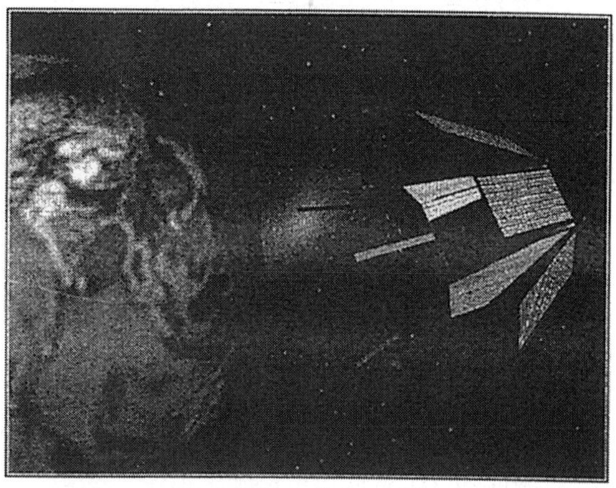

Figure 20. SSTL Minisatellite in LEO

UoSAT-12 Minisatellite

The first SSTL minisatellite (UoSAT-12) will carry experimental Earth observation & communications payloads and be launched by a converted SS18 ICBM (Dnepr) launcher into LEO in April 1999.

UoSAT-12 will carry 35-metre resolution multispectral and a 8-metre panchromatic CCD Earth cameras with powerful on-board image processing capabilities.

UoSAT-12 Minisatellite Multispectral (35m) and Panchromatic (8m) Earth Observation Capabilities

Sophisticated frequency-agile VHF/UHF and L/S-band DSP regenerative transponders will provide both real - time and store - & - forward communications to small terminals.

Three-axis control is provided by a combination of magnetorquers, momentum wheels and cold gas N_2) thrusters - whilst an experimental electric H_2O 'resisto-jet' thruster will provide orbit trimming and maintenance demonstrations for future network constellations.

NANOSATELLITES !

A tiny, highly integrated, 2kg 'nanosatellite' SNAP-1 is being built as a research project at Surrey for launch alongside UoSAT-12 in April'99 and intended to image the 45 other' minisatellite and launch vehicle. Future applications for the nanosatellite are for remote inspection of satellites and monitoring of deployments systems in orbit, and carrying small space science instruments requiring measurements with spatial diversity.

PROJECT MANAGEMENT

Solving the technical challenges associated with the design, construction, test and operation of a microsatellite is less than half of the story - in parallel with the technical considerations of the mission, effective project management is crucial to the realisation of a successful low-cost, sophisticated small satellite project.

Figure 21. Surrey SNAP-1 Nanosatellite

Affordable small satellites require a very different approach to management as well as technology if cost, performance and delivery targets are to be met. Several attempts at taking a traditional aerospace organisation to produce such satellites have failed because of the rigidity of management structure and 'mind-set'. Small teams (25 persons), working in close proximity with good communications, with well-informed and responsive management, are essential.

These characteristics are best found in small companies or research teams rather than large aerospace organisations, who may find it difficult to adopt or modify procedures necessary to produce affordable small satellites using staff and structures that are designed for conventional aerospace projects. The main ingredients for a successful small satellite project can be summarised as:

- highly innovative technical staff
- small, motivated teams
- personal responsibility for work rigour and quality
- good team communications, close physical proximity
- well-defined mission objectives and constraints
- knowledgeable use of volume, modern components
- layered, failure-resilient system architecture
- thorough burn-in
- technically-competent project management
- short time scale (prevents escalation of objectives!)

SUMMARY & CONCLUSIONS

The University of Surrey embarked upon the design of its first experimental microsatellite in 1978 and UoSAT-1 was launched by NASA in 1981 - since then a further eleven low cost yet highly sophisticated microsatellites have been built & launched into low Earth orbit. The UoSAT missions have demonstrated that microsatellites can play a useful role in supporting specialised communications, Earth observation, small-scale space science, and in-orbit technology verification missions.

Whilst obviously limited in payload mass, volume and power, but with very real and attractive advantages in terms of cost and response time, microsatellites offer a complementary role to traditional 'large' satellites by providing an alternative 'gap-filler' for affordable quick-response or exploratory missions for both civil and military objectives. Developing space nations have used rapid and inexpensive microsatellite projects to act as the focus for effective technology transfer and an affordable first step into orbit. A further 2 new microsatellite missions will be launched by SSTL in 1998 and 3 in 1999.

Building upon its success with microsatellites, Surrey is now building an affordable, modular (400 kg) minisatellite capable supporting more demanding payloads - but still within a cost of ~£5-8M - in order to stimulate further use of small satellites to complement large space programmes.

Surrey has established itself firmly as the international centre of excellence in academic research, teaching and commercial applications of small satellites.

ACKNOWLEDGEMENTS

The author would like to recognise the contributions of colleagues at and support from Surrey Satellite Technology Ltd, UK Defence Research Agency, Science & Engineering Research Council, Arianespace (France), Alcatel Espace (France), KAIST (Korea), PoSAT Concorcium (Portugal), Matra-Marconi Space (France), SatelLife (USA) and the University of Surrey. FASat-Alfa and FASat-Bravo is owned by the Chilean Air Force (FACH) and has been developed, constructed and will be operated under a Technical Collaboration Program between FACH and Surrey Satellite Technology Ltd (SSTL) U.K. TMSAT is a collaborative programme between Mahanakorn University of Technology (Thailand), the Thai Microsatellite Company (TMSC) and SSTL. The L/S-band communications payload on UoSAT-12 is supported by Nanyang Technological university, Singapore. TiungSAT-1 is supported by ASTB, Malaysia. PICOSat is supported by the USAF Foreign Comparative Test program.

REFERENCES

'UoSAT - An Investigation into Cost-Effective Spacecraft Engineering' Sweeting M.N., The Journal of the Institution of Electrical & Radio Engineers, Vol. 52, No. 8/9, pp. 363- 378, Aug/Sept, 1982

'The University of Surrey UoSAT-2 Spacecraft Mission' Sweeting M.N., The Journal of the Institution of Electrical & Radio Engineers, Vol.57, No. 5 (Supplement), pp. S99-S115, Sept/Oct, 1987

'Microsatellites for Global Electronic Mail Networks' Ward J.W., Electronics & Communications Engineering Journal, Vol.3, No.6, Dec. 1991, pp. 267-272

'Space Science and Microsatellites - A Case Study: Observations of the Near-Earth Radiation Environment Using Cosmic-Ray Effects and Dosimetry (CREDO) Payload On- Board UoSAT-3' Underwood C.I., Ward J.W., Dyer C.S., Sims A.J., Farren J., Stephen J., 5th Annual AIAA/USU Conference on Small Satellites, Logan, Utah, August 1991

'UoSAT Microsatellite Missions' Sweeting M.N., IEE Electronics & Communication Engineering Journal, Vol. 4, No. 3, June, 1992, pp. 141-150.

'Earth Observation Using Low Cost SSTL Microsatellites', Fouquet M, Sweeting MN, IAF-96-B.3.P215, 47th IAF Congress, Oct 1996, Beijing, PR China

'Global Email System and Internet Access in Remote Regions via Microsatellites', Sun W, Sweeting MN, IAF-96-M.507, 47th IAF Congress, Oct 1996, Beijing, PR China

'UoSAT-12 Minisatellite for High Performance Earth Observation at Low Cost', Fouquet M, Sweeting MN, IAF-96-B.3.P215, 47th IAF Congress, Oct 1996, Beijing, PR China

'Technology Transfer & Training at Surrey' Sweeting MN, IAF-96-, 47th IAF Congress, Oct 1996, Beijing, PR China

'TMSAT: Thailand's First Microsatellite' Sweeting MN, IAF-96-, 47th IAF Congress, Oct 1996, Beijing, PR China

MICROSATELLITE EXPERIENCES FROM THE UNITED KINGDOM

David S. Hall

British National Space Centre, 151 Buckingham Palace Road, London SW1W 9SS, United Kingdom

ABSTRACT

16 microsatellites have been constructed by UK organisations since 1980, and 9 more are expected during the next five years. Most have been commercial programmes. The UK experience enables specific questions on features of microsatellite programmes to be reviewed, for instance, the payload capability, the reliability and the level of standardisation of the spacecraft platforms.

INTRODUCTION

For the past 17 years the UK has led the construction of microsatellites at the rate of about 1 a year. During the next five years, this rate will increase to about 2 per year. Most of the satellites are commercially funded. Table 1 lists microsatellites launched since 1980 in which the construction has been led by a UK organisation. Table 2 lists microsatellites for which launch is expected in the next five years. Table 3 list the applications of the microsatellites. What this programme has given us is a database to judge key features of microsatellite programmes. Three questions that can be reviewed in the light of the programme are :

Do microsatellites mean the range of users will be restricted ?

Do low cost programmes mean high risk ?

Are standard spacecraft buses essential for a successful programme ?

PAYLOAD CAPABILITY

The Space Technology Research Vehicle (STRV) is a microsatellite mainly for technology proving, with the additional benefit of providing staff skilled in satellite development and construction and operation. The satellites are constructed by the UK Defence Evaluation Research Agency (DERA) which works primarily as a sub-contractor to the UK Ministry of Defence. The STRV programme has to be self-supporting, with the programme funded by the users. This is achieved by charging for payload capacity. Because of this need to encourage customers to use the payload space, there is an incentive to provide as much payload space per spacecraft as possible and to attract as many users as possible. This runs counter to the argument that low-cost is most easily achieved by simple payloads with a restricted number of users. The consequence is that for the DERA STRV c there are 14 separate experiments from 10 lead groups on a single 90 kg microsatellite. In addition to UK groups, users include major organisations such as ESA, NASA, USAF and BMDO.

Table 1. Microsatellites Led By UK Organisations

Year of launch	Satellite	Lead Organisation
1980	Firewheel (UK)	Rutherford Appleton Laboratory
1981	UOSAT-1	University of Surrey
1984	AMPTE-UKS	Rutherford Appleton Laboratory
	UOSAT-2	University of Surrey
1990	UOSAT-3	Surrey Satellite Technology Limited
	UOSAT-4	Surrey Satellite Technology Limited
1991	UOSAT-5	Surrey Satellite Technology Limited
1992	KITSAT-1	Surrey Satellite Technology Limited
	S80/T	Surrey Satellite Technology Limited
1993	HealthSat-2	Surrey Satellite Technology Limited
	PoSAT-1	Surrey Satellite Technology Limited
	KITSAT-2	Surrey Satellite Technology Limited
1994	STRV 1a	Defence Evaluation Research Agency
	STRV 1b	Defence Evaluation Research Agency
1995	CERISE	Surrey Satellite Technology Limited
	FASat-A	Surrey Satellite Technology Limited

Table 2. Future UK Microsatellites

Satellite	Lead Organisation
Humble	Space Innovations Ltd
FASat-Bravo	Surrey Satellite Technology Ltd
TMSat	Surrey Satellite Technology Ltd
TiungSat-1	Surrey Satellite Technology Ltd
UOSAT-12	Surrey Satellite Technology Ltd
Clementine	Surrey Satellite Technology Ltd
PicoSat	Surrey Satellite Technology Ltd
STRV-1c	Defence Evaluation Research Agency
STRV-1d	Defence Evaluation Research Agency

Table 3. Applications of UK Microsatellites

Technology Demonstration/Verification
Education and Training
Space Science
Earth Observation
Communications
Space Environment Monitoring
Spacecraft Autonomy
Technology Transfer

Surrey Satellite Technology Ltd (SSTL) is the UK organisation which has had by far the most extensive programme, with 12 microsatellites launched since 1981. The first microsatellite had a mass of 68 kg, including payload, but constraints of a piggy-back ARIANE launch meant there was pressure to reduce the mass of subsequent satellites to nearer 50 kg. However with relaxation of the constraints the total mass can now be increased to 68 kg with most of the additional mass being assigned to the payload. Payload masses above 40% of the total spacecraft mass are possible.

Reliability

The UK programme gives evidence for reliability involved in low-cost microsatellite programmes. Of the 15 spacecraft launched successfully, all achieved the anticipated number of operating hours. For the SSTL series most are still operational more than 4 years after launch, including UOSAT-2 which has been operational for 12 years. This seems to refute the concern that low-cost means high-risk satellites.

Key factors in achieving a successful microsatellite project are :

highly-innovative technical staff in close physical proximity with technically-competent project managers
small motivated teams with responsibility for work rigour and quality delegated to sub-system engineers from the spacecraft conception to launch
testing is far more effective than analysis
plan to re-load software in orbit
inclusion of redundant systems is valuable, especially if redundant pathways use alternative technologies rather than duplication
documentation need not be eliminated but should be at the minimum level consistent with good project management practices, for instance that associated with ISO9000. Increased documentation may increase assurance during build but does not increase reliability
short timescale projects which prevents escalation of objectives

Commercial-of-the-Shelf (COTS) components have been found to be acceptable if accompanied by sound engineering practices, this means extensive test and burn-in, including radiation testing, and derating during design. In addition to cost reduction, COTS components give:

reduced procurement complexity,
reduced delivery time,
increased packaging density,
increased range of available devices,
increased technical performance.

Adaptability

The third UK organisation involved in microsatellites is Space Innovations Ltd (SIL). SIL |日 main role up to now has been in provision of sub-systems and, to serve a broad range of users, have to provide sub-systems that can be easily adapted to suit diverse customer requirements. The same is true for microsatellite buses, the customer-driven approach means that the buses have to be adaptable to cope with a range of diverse requirements, including diverse orbits and launchers. A microsatellite programme is most effective if, rather than a "standard" bus, there is a core set of modules/sub-systems, any of which can be developed separately in response to the requirements of new users.

Conclusions

The UK programme demonstrates what can be achieved in response to commercial pressures. Considering the questions posed in the Introduction:

The UK programme demonstrates that it is possible to cope with a wide range of diverse users
There is no evidence from the UK microsatellite experience that low-cost programmes mean high risk.
Spacecraft buses should be kept as adaptable as possible. Rather than a "standard" bus, the key is to have a core set of modules/sub-systems, any of which can be developed separately in response to the requirements of new users

Acknowledgements

Grateful thanks for providing material and helpful discussions are due to
Professor Martin Sweeting and his colleagues of Surrey Satellite Technology Ltd (SSTL), Centre for Satellite Engineering Research, University of Surrey, GU2 5XH, UK
Commander Richard Blott and Nigel Wells of the Defence Evaluation Research Agency (DERA), Space Department, Farnborough, GU14 6TD, UK
Kim Ward of Space Innovations Ltd. (SIL), The Paddock, Hambridge Road, Newbury RG14 5TQ, UK.

ØRSTED - A DANISH MICROSATELLITE WITH A HIGH SCIENTIFIC PROFILE

E. Friis-Christensen

Danish Space Research Institute, JuLiane Maries Vej 30, DK-2100, Copenhagen, Denmark

ABSTRACT

The first Danish satellite, Ørsted, is scheduled for launch as a secondary payload on a Delta II launch vehicle during the Summer of 1998. The Ørsted satellite is a geomagnetic research satellite that will provide very high precision measurements of the magnetic field in a polar elliptical orbit at altitudes from 520 to 850 km.

The magnetometer measurements will be used for a precise global survey of the magnetic fields and its variation, which is of importance for a wide spectrum of geo and space science studies. In spite of the internationally expressed interest in such data and in spite of many attempts by the major space agencies to conduct such a mission, the Danish enterprise is only the second of its kind.

In this presentation focus will be on the untraditional concept of this mission. This includes its limited but well-focussed instrumentation, the organization of the project comprising several research institutions and industrial companies, its international recognition, and its worldwide user interaction.

THE ØRSTED IDEA

The Danish geomagnetic research satellite, Ørsted, has a strong background in Danish science history, instrument developments and research activities. The Danish physicist Hans Christian Ørsted is known as the discoverer of the relationship between electric currents and magnetism, electromagnetism, in 1820. H.C. Ørsted did not distinguish between basic and applied research. His philosophy regarded all aspects of life as elements that were connected and that were all reflections of one divine entirety. It was quite natural that he became the founder of the first Danish magnetic observatory on February 1st, 1842. The magnetic observatory served basic scientific purposes but it also served direct societal needs in particular related to the navigation of ships.

During the next century Danish observers, instrument developers, and scientists played an active role in international geomagnetic research. They participated in the First International Polar Year 1882-1883 by conducting expeditions to Greenland to observe the magnetic field and its variations from hour to hour. Director of the Danish Meteorological Institute, D. la Cour, was instrumental in advancing international collaboration in geomagnetic research in his capacity as the President of the Commission for the Second International Polar Year 1932-33. Geomagnetic instruments designed by him and built in Denmark have been used throughout the World and have been sold to more than sixty countries.

During the recent years, Solar-terrestrial science has attracted much attention, among scientists as well as in the public. The distinction between man-made and natural causes of "Global Change" has become an important issue. Also the concept of a steady Sun, expressed for example in the term "the solar constant", has been gradually abandoned and transformed into a broadly accepted concept of a constantly varying outer environment dominated by processes in the Sun. The magnetosphere and the upper atmosphere of the polar regions are the locations where the effects of the variable Sun are most clearly seen. At present, our understanding of the global processes that determine the coupled interactions between the electromagnetic and corpuscular emissions from the Sun and the neutral and ionized species in the Earth's environment is too poor to predict the response of the system to changes in the solar output. As a consequence, several international research programs have recently been initiated to improve our knowledge of the solar-terrestrial system.

Denmark, in many ways, is in an excellent position for being involved in this type of research. This is due to the research traditions and due to the fact that Denmark can base its research on excellent scientific and logistic facilities in the polar regions of Greenland close to the magnetic pole.

The geomagnetic field is the main controlling factor of the processes in the ionized environment of the Earth. A good knowledge of the field and its variations is therefore crucial. Such knowledge may only be acquired by spacecraft that can provide the necessary global coverage. The highly successful MAGSAT mission (Langel et al., 1982) provided an 8-months "snapshot" of the geomagnetic field. Due to its short life-time, it was not possible with this single mission to acquire sufficient information about the dynamics of the field. Therefore a second mission of global mapping with comparable accuracy will not only improve the knowledge of the magnetic field but will also provide the possibility to detect the secular variation with an unprecedented spatial resolution. This will lead to major advancements in our knowledge regarding the origin and dynamics of the geomagnetic field. The International Association of Ceomagnetism and Aeronomy (IAGA) strongly supports and recommends such a renewed satellite mission to be undertaken.

For these reasons a magnetometry mission has also been part of the European Space Agency (ESA) research objectives for Earth Observation as demonstrated by the intended Aristoteles gravity and magnetometry mission, which the purpose of conducting a thorough investigation of the main field as well as the crustal and external fields. The Aristoteles mission did not happen but several national initiatives have been launched to limit the time span without adequately updated geomagnetic field values and, in consequence, a deterioration of the geomagnetic field models used in various research and application areas.The Danish satellite project Ørsted is the first of these national initiatives to be realized.

The selection of a geomagnetic research satellite as the first Danish satellite evolved from an idea, which created a lot of enthusiastic work and support among a group of Danish scientists, engineering specialists, and industry. This group obtained a research grant to conduct a System Design Phase-A study. Included in this study were two reviews by international panels of experts to evaluate the scientific merits and the overall technical feasibility of the Ørsted satellite mission. Both panel reports were very favorable. Based on the support in these reports, the Ministry of Industry, the Ministry of Research and the Ministry of Transport in June 1993 decided to establish a consortium to build and operate the Ørsted satellite and to make scientific use of the data.

The fact that the funding is provided by three different ministries representing basic research as well as applications is a splendid realization of the philosophy of H. C.

Ørsted that Natural Science should not be conducted in an ivory tower but should involve society at large as well.

MISSION SUMMARY

In 1601 William Gilbert published the first modern scientific monograph "De Magnete". His conclusion was that the Earth behaves like a very large magnet. Since then, many speculations about the causes of the geomagnetic field have been presented but current ideas are based on the idea that the major part of the field is produced by electric currents through some form of self-sustaining dynamo operating in the fluid outer core of the Earth. The magnetic field produced by this dynamo varies with time. This slowly and steadily varying field is known as the secular variation.

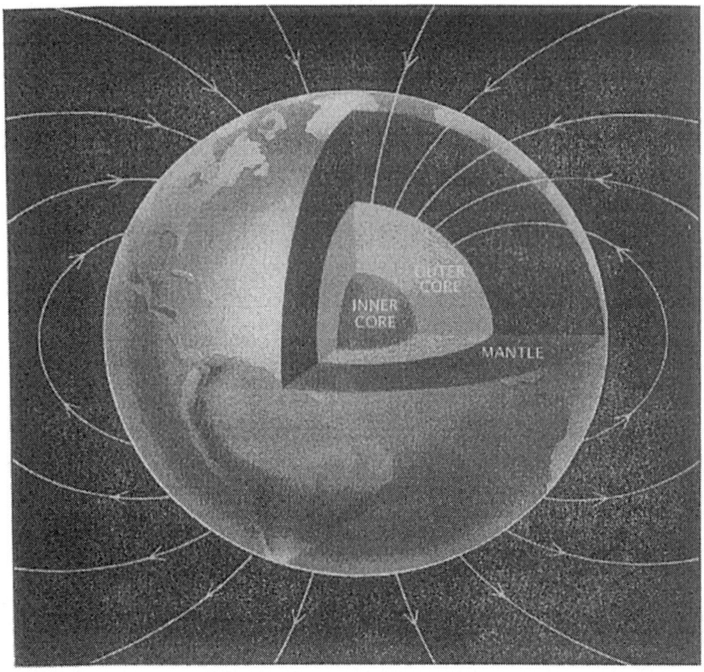

Figure 1. The Earth's internal structure. At this scale the outer layer of rock (the crust) cannot be depicted. The mantle is solid on short time scales but convects over geological time. The outer core consists of an electrically conducting fluid and is where the main geomagnetic field is generated. The inner core is solid. (From Bloxham & Gubbins, 1989).

What we measure at or near the surface of the Earth is a composite of this main field and fields caused by the varying magnetization in the Earth's lithosphere including the upper mantle and the outermost part of the Earth, the crust. Another part of the measured magnetic fields has its origin in electric currents flowing in the conducting part of the atmosphere and in the magnetosphere, and in currents induced in the Earth by these time-varying external fields.

The real value of magnetic measurements is associated with the possibility of interpreting them in terms of their sources. The benefit and the challenge are related to the sophisticated separation of these various sources which each have their specific characteristics in terms of spatial and time variations. Few other measurements, if any, of a single physical parameter can be used for such a variety of studies related to the Earth, its formation, its dynamic and its environment, stretching all the way to the ultimate source of life on Earth, namely the Sun.

The geomagnetic field is, however, not only an issue related to scientific research regarding its origin and related scientific questions regarding the evolution of our planet Earth. It is, in fact, of primary importance for the external environment of the Earth through its role in controlling the radiation belts, and its control of energy and momentum transport from the solar wind to the Earth including near-Earth effects like induced surges in long power lines. Numerous reported, but still poorly understood, correlations (e.g. Friis-Christensen and Lassen, 1991) between various solar activity parameters and climate variations have recently been related to the flux of high-energy cosmic ray particles from Space and from the Sun, possibly by affecting the formation of clouds (Svensmark and Friis-Christensen, 1997). The motion of these charged particles is controlled by the magnetic fields, both the geomagnetic field and the heliospheric magnetic field, which, in turn, is varying according to the level of solar activity.

PRIMARY SCIENCE OBJECTIVES

The Earth's Main Magnetic Field

The main field and, in particular, the secular variation are among the very few means that are available to us for probing the properties of the outer core and how these change with time.

The core is a highly conductive medium (conductivity = 10^5-10^6 Sm^{-1}) where large electric currents are driven in the core by a self-sustaining dynamo process: fluid flowing across magnetic lines of force of the field generates electromagnetic forces which drive electric currents which maintain the magnetic field.

On short time scales, less than about a century, the core may be considered as a perfect conductor. The main effect is that the geomagnetic field appears to be frozen into the material in the core. The temporal variations of the main geomagnetic field observed at the surface of the Earth, namely the secular variation, is a reflection of the fluid flow in the outermost layer of the core. Therefore, the secular variation provides a unique observational constraint on dynamo theory and the geodynamo mechanism.

The Lithospheric Anomaly Field

Apart from the magnetic field created by the currents in the outer core, a permanent contribution to the magnetic field is provided by the magnetization carried by rocks in the upper layers of the Earth, the Lithosphere. This magnetization is partly induced by the ambient field, and therefore proportional to this field and to the susceptibility of the rock. It can also be a remanent magnetization acquired during the formation of the rock. Remanent magnetization is characterized by blocking temperatures which may range from a few hundred degrees up to the Curie temperature. This magnetization may also be reset after the formation of the rock by thermal or chemical alteration. In that sense, remanent magnetization is a good recorder of any metamorphic, chemical or mechanical event in the past history of a rock.

The remanent magnetization of crustal rocks provides information about the growth and evolution mechanisms of the Earth's crust. Using this parameter, geophysicists were able to demonstrate the concept of continental drift in the 1950's and that of sea- floor spreading in the 1960's. It was, therefore, instrumental in establishing the theory of Plate Tectonics. It is still widely used today to study intra-plate deformation and mountain-building processes.

The Magnetic Field of External Origin

The question of how solar wind plasma, momentum and energy are transferred to the magnetosphere is one of the fundamental problems of magnetospheric physics. A necessary requirement for solving this problem is the understanding of the magnetopause and boundary layer processes. It is today realized that a major obstacle in reaching this understanding is the lack of a sufficient number of simultaneous observations of these regions and their ionospheric projections.

The geomagnetic field lines act as the link between the outer part of the magnetosphere and the lowest part, namely the ionosphere. This link is sustained by electric currents that can easily flow along the geomagnetic field lines because of the highly anisotropic conductivity in the magnetosphere. The only direct measurements of these currents are in the form of magnetic field measurements on board low-altitude polar-orbiting satellites. Measurements from such satellites have provided indispensable information that has made it possible to describe in gross terms, but not yet to understand, the coupling between the magnetosphere and the ionosphere.

The ionosphere is conducting and is therefore able to carry electric currents. These currents have their origin in electric fields generated by a number of different mechanisms. In the polar regions these fields are primarily of magnetospheric origin as described above. At lower latitudes there exists a permanent system of electric currents that is caused by the motion of the neutral atmosphere, the Sq system. The measurements on Ørsted will be of sufficient accuracy to provide measurements of the ionospheric part that can be incorporated in the sophisticated modeling with the purpose to separate all the different sources of the measured magnetic field better than has been possible before.

SATELLITE SUMMARY

Configuration

The Ørsted satellite is shaped as a box with a long boom carrying the magnetic instruments and the star imager, with solar panels on its five sides. It has a height of 72 cm, a width of 45 cm and a depth of 34 cm.

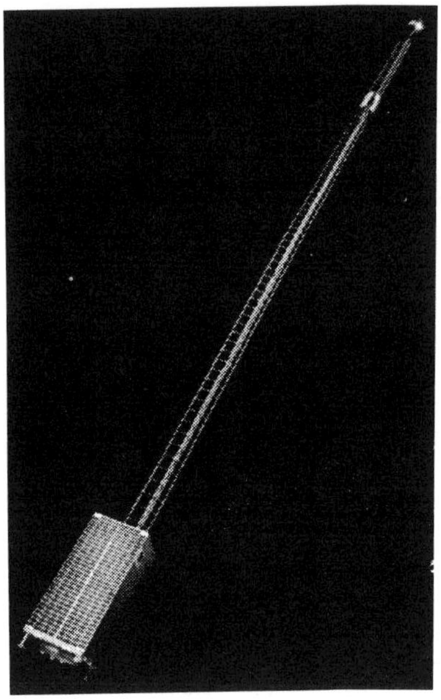

Figure 2. The Ørsted Satellite

Structure

The main structural "H" beam configuration forms two vertical compartments, one on each side of the center web of the "H" beam. Two rows of modular electronic boxes for the satellite and payload functions are mounted in one of these compartments. The other contains the boom assembly, which during launch is packed in a canister. All main structures are fabricated in aluminum.

The deployable boom consists of three coilable longerons, separated by radial spacers and tensioned by cross-wires. Redundant release mechanisms are provided, and a restraining lanyard controls the deployment. The total length of the boom is 8 meters, divided into two sections of 6 and 2 meters respectively.

Power Subsystem

The power subsystem includes a solar array of five solar panels consisting of GaAs solar cells. The array delivers 54 watts average at End-Of-Mission. Two power control units are included, which provide outputs at +5 V, ~8 V, +15 V, and an unregulated power bus. The batteries include two NiCd units, each with a 6 Ah capacity.

Communication

The satellite uses S-band communication (down link 2039.6 MHz, up link 2215 MHz) with two turnstile communication antennas mounted on the earthpointing side of the satellite. Science data accumulation rates are approximately 1 kbit/sec in normal mode, and 6 kbit/sec in burst mode. Telemetry down link is equipped with Reed- Solomon concatenated and convolutional encoding. Telecommand up link includes Bose-Chaudhuri - Hocquenghem encoding. ESA packet standards PSS-04-106 (Telemetry) and PSS-04-107 (Telecommand) have been adopted for the Ørsted mission .

Command and Data Handling

The CDH subsystem is comprised of two Central Processing Units each equipped with an Intel 80186, 16 MHz processor. It contains 16 Mbytes of RAM data storage which is sufficient for 13 hours of continuous operation. Data error detection and correction circuitry are included in the CDH.

Attitude Control

Attitude control is provided to perform detumbling after separation from the launch vehicle and to ensure stabilization during the operational phases of the mission. Attitude determination is based on inputs from the star imager, sun sensors, and the CSC magnetometer. Attitude control is performed in a way that creates minimum disturbances to magnetic field measurements.

Stabilization of the satellite attitude is accomplished by passive and active techniques. The passive technique employs gravity-gradient stabilization using the deployed boom with a tipmass of 3. 1 kg from the two magnetometers and the star imager. The active technique uses three-axis magnetorquer coils interacting with the Earth's magnetic field. The ACS maintains a yaw angle variation of +10 degrees to optimize the power output of the solar panels.

GPS receivers

The satellite position is determined by two GPS receivers. The primary GPS receiver is a six channel TANS GPS from Trimble Navigation. It uses C/A mode on L1 frequencies, and gives an accuracy of position better than ~ 50 m. The secondary GPS receiver is a TurboRogue receiver from JPL. It uses P-mode on LI and L2 frequencies for GPS occultation measurements, but uses C/A mode on LI frequency for position back-up. Universal Time Coordinated accuracy is within 1 microsec, and velocity determination is within 0.5 m/sec.

THE SCIENCE INSTRUMENTS

A total of five scientific instruments are carried onboard the Ørsted Satellite . Three of the sensors are mounted on an 8 meter long boom to minimize the disturbances from the electrical systems of the satellite.

CSC Fluxgate Magnetometer

This instrument measures the magnetic vector field (strength and direction). This instrument is stable within 0.5 nT over time spans of several days.

SIM Star-Imager

With this instrument it is possible to determine the orientation of the fluxgate magnetometer sensor. It is accurate to less than 5 are seconds.

Overhauser Magnetometer

This instrument provides an independent value of the strength of the magnetic field (not direction sensitive). It has an absolute accuracy better than 0.5 nT. The main purpose of this instrument is the calibration to an absolute scale of the measurements of the fluxgate magnetometer. It is provided by the French Space Agency, CNES.

Charged Particle-Detectors (CPD)

The purpose is to measure the flux of fast electrons (0.03-1 MeV), protons (0.2-30 MeV), and a-particles (1-100 MeV) around the satellite.

Turbo-Rogue GPS Receiver

In addition to providing an accurate position of the satellite this instrument may be used scientifically during specified intervals to investigate the atmospheric pressure, temperature, and humidity beneath the satellite. It is provided by the US National Space Agency, NASA.

Orbit and Operation

An elliptic orbit of heights between 520 and 850 km and with an orbit period of approx. 100 minutes will be used. The main receiver station is located at DMI in Copenhagen with backup stations at IKT in Ballerup and at AAU in Aalborg. The technical functionality of the satellite is monitored at the Control Center in BirkerØd.

Preliminary data reduction and calibration are performed at the Science Data Center at DMI.

THE SCIENCE DATA CENTER

The Science Data Center shall support the scientific analysis of data from the Ørsted satellite and serve as a focal point for data and information exchange between the Ørsted project and investigator teams. In particular the following tasks will be performed:

- calibration of the vector magnetic field data from the CSC instrument
- general data reduction and preparation of data sets for the scientific investigations
- distribution of data to investigator teams
- archiving of the raw telemetry data, telecommand data, and the various processed and calibrated data sets delivered to investigators
- support of visitors who wish to work with the data as well as data analysis workshops

The data organization and services reflect two different science needs:

Internal and Lithospheric field investigations

-low time resolution
-high-precision calibration
-no real-time needs

External field investigations

-high time resolution
-"best" calibration
-"real-time" needs

Date products are classified into levels according to accuracy and time resolution:

Level 0 : Summary plots and files. Free access over WWW and free use.
Level 1 : Data useful for external field investigations. Data are calibrated at least to a standard of "conventional" magnetospheric-type satellites. The best available calibration coefficients are used. Data are only for Ørsted Investigators. Password is required for access over WWW.
Level 2 : Data useful for internal, lithospheric, and external field investigations. High- precision calibration coefficients are used. These are produced by a Calibration Working Group in collaboration with the Science Data Center. Data are only for Ørsted Investigators. Password is required for access over WWW.

GROUND-BASED MAGNETIC DATA COLLECTION

The ground based magnetic data provides an essential frame of reference interpreting the magnetic observations from the Ørsted Satellite. The collection I set of magnetic observatory data, with global coverage and of high quality, therefore been given high priority in the project. Similarly the collection dissemination of a wide set of timely geomagnetic indices.

Magnetic Observatory Data

It has been decided to collect data from all available absolute measuring observatories. Because of limited resources it was decided not to collect variometer data, but possibly to produce a list of existing variometer stations. The time period 1995-1998, both years inclusive, was selected for the collection. It was fur decided to collect data of various time resolution, in the hope that observation which, for technical or other reasons, could not provide data of higher time resolution at least would be able to provide the lower time resolution data. The responsibility the collection of data was divided into different organizations. The World Data Center (WDC) CI for Geomagnetism in Edinburgh in cooperation with WDC A in Boulder offered to do a special effort to collect timely yearly averages during the Ørsted period. Institute de Physique de Globe in Paris had already initiated a collection monthly averages in Paris, also for use in global field modeling to support Ørsted. The Ørsted Science Data Center together with WDC C1 for Geomagnetism Copenhagen will collect and distribute to the Ørsted scientific investigator worldwide set of observatory 1-minute values and hourly values. This collection planned in close cooperation with WDC A in Boulder, WDC C2 in Kyoto and WDC C1 in Edinburgh.

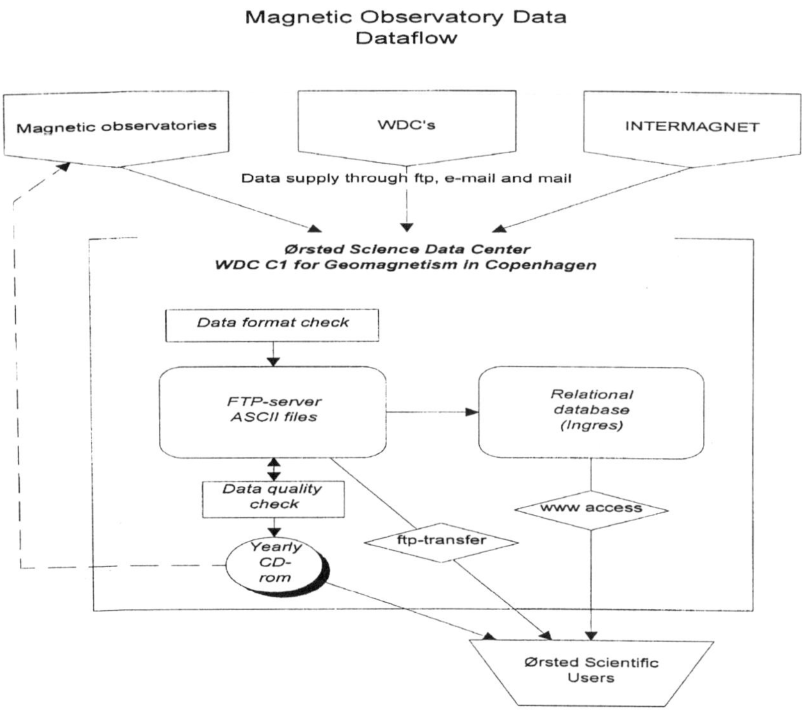

Figure 3. Data flow diagram for observatory hourly values and minute values.

INTERNATIONAL COOPERATION

From the very beginning, the project has benefited from a large international scientific attention; it could not have been realized without a clearly identified and expressed international interest for the unique data set expected to result from this enterprise.

A very important contribution is the offer from NASA to provide the launch of the Ørsted satellite. This offer is not only of significant financial value. It also demonstrated the level of the international scientific interest and confidence in the Danish team. This has proven to be of considerable value in the process of providing funding for the project within Denmark. Similarly, the offer from CNES in France to supply a scalar magnetometer demonstrates the large international interest in acquiring the most accurate data ever of the Earth's magnetic field. Finally, ESA has provided technical advice and considerable financial support of the ground segment by incorporating it in its Prodex program.

Denmark is a small country and even this small satellite is a huge investment in comparison to most scientific projects in Denmark. The number of scientific users in Denmark is rather limited and, in fact, too small for an appropriate and effective utilization of the satellite data. The project therefore announced open and extended international scientific participation in the project. In co-operation with NASA and CNES the project selected about 50 co-investigator groups from 14 countries to be included in the Ørsted International Science Team, ØIST. To assist the Project Scientist in the coordination of the international efforts an Ørsted Science Advisory Committee has been established representing the scientific key topics that can be addressed with the high-precision data from the Ørsted satellite. Another international partner is the geomagnetic observatory community, which provides the collection of simultaneous ground based magnetic data.

Two ØIST meetings have been held. The first ØIST meeting was held in Copenhagen, June 6-8, 1995. Originally, this meeting was supposed to take place less than one year before launch. A second meeting was planned to take place about three months prior to the launch. However, due to subsequent delays of the primary payload, the US Air Force satellite, ARGOS, it was decided to have the second ~IST meeting in May 1997. An Ørsted Compendium based on the first meeting and updated information available until the second meeting has been prepared by Friis-Christensen and Skøtt (1997).

ORGAN IZAT ION

The Ørsted project has been based on a partnership between an industrial/technical consortium and a science consortium. This is reflected in the financing of the project, which is provided from various sources and which have been dedicated to the basic activities, namely the production of the satellite and its subsystems and the manufacturing of the science instruments and the science data handling and interpretation. From the very beginning there has been allocated dedicated resources to the scientific part independently of the resources available for the industrial part. This was a firm requirement set in order to secure adequate use within Denmark of this unique data set.

This financial structure is also reflected in the organization of the project as sketched below. There are two main responsible units, the Project Office, headed by the Project Manager and the Science Office, headed by the Project Scientist. The instrument team is represented in both areas because they supply the hardware for the satellite and because they represent the knowledge about the instruments necessary for the scientific interpretation of the data.

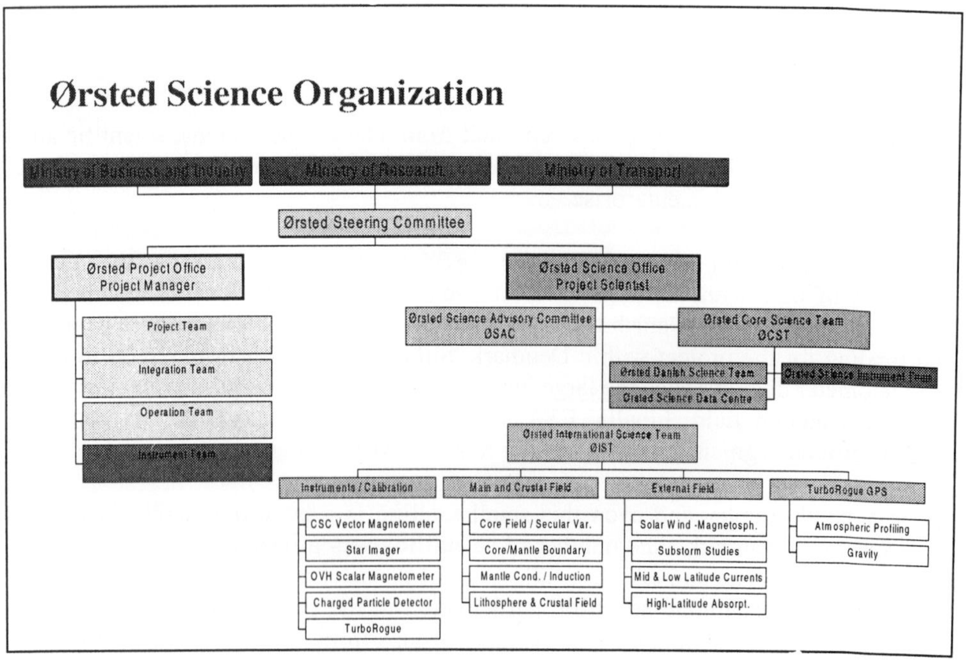

Figure 4. Organization of the Ørsted Project reflecting the three sponsoring ministries, the steering committee, and the two project lines, project management and science coordination.

REFERENCES

Bloxham, J. & Gubbins, D., The evolution of the Earth's magnetic field, Scientific American, 261(6), 30-37 (1989).

Friis-Christensen, E., and K. Lassen, Length of the solar cycle: An indicator of solar activity closely associated with climate, Science, 254, 698-700 (1991).

Friis-Christensen, E., C. Skøtt, Ørsted Compendium, Contributions from the International Science Team, Danish Meteorological Institute, Scientific Reports 97-1(1997).

Langel, R. A., Ousley, G., Berbert, J., Murphy, J. & Settle, M., The Magsat mission, Geophys. Res. Lett., 9, 243-245 (1982).

Svensmark, H. and E. Friis-Christensen, Variation of cosmic ray flux and global cloud coverage - a missing link in solar-climate relationships, J. Atmos. Sol-Terr. Phys., 59, 1225 (1997).

NEW TECHNOLOGIES FOR MINIATURIZED SPACECRAFT AND MULTI-POINT CLUSTER MISSIONS

E.L. Reynolds and M. K. Flaherty

Space Department, Applied Physics Laboratory, The Johns Hopkins University, Johns Hopkins Road, Laurel, MD 20723, USA

ABSTRACT

The Johns Hopkins University Applied Physics Laboratory is currently preparing for its next generation of low Earth orbiting and deep space missions. Auroral Multi-Scale (AMS) is one of those missions. The mission uses four miniature (36 kg each) spacecraft to perform high resolution multi-point electric and magnetic field measurements of auroral phenomena at various altitudes and latitudes. AMS uses a single Pegasus launch vehicle to inject the cluster into its operational orbit. Once deployed, the cluster maintains a tight formation--using GPS for position determination, an imager star tracker for attitude determination, and by taking advantage of atmospheric drag for position control. To implement missions like AMS, new technologies are required that up until recently did not exist. APL has been recently developing a new spacecraft architecture that is extremely small and low power yet adaptable to a variety of different mission types. This architecture uses several new technologies: 1) a modular, miniature, low power, avionics architecture; 2) a flight qualified bare die packaging technique called "chip-on-board"; 3) a technique to mold the spacecraft structures, instead of machining them, that lends itself to mass production, and 4) aerodynamic-drag station keeping. This paper focuses on the implementation of the low cost, but highly capable, Auroral Multiscale multi-point mission, and the technology that enables this highly capable science mission.

INTRODUCTION

The implementation of cluster and multipoint data collection missions are being enabled by mass production and miniaturization techniques. Mass production techniques allow multiple copies of a spacecraft design to be produced at economical costs. Miniaturization techniques allow multiple spacecraft to be launched on a single inexpensive launch vehicle. While the production of a handful of scientific spacecraft differs greatly in scale from the hundreds and thousands of spacecraft envisioned for the communication constellations, there are mass production techniques that can benefit even small constellation missions. One such mission is Auroral Multiscale, or AMS.

Over its lifetime, AMS will study Earth's auroral region using four identical spacecraft to simultaneously collect magnetic field, plasma wave, electron/ion data over different spatial scales. The nature of AMS science requires very precise position (100 m), time (10 μs), and attitude knowledge(0.01°). As each space- craft passes through the auroral latitudes and altitudes, instrument data are recorded and stored along with precision timetags and position information provided by onboard GPS receivers, and precise attitude information is provided by successive star field images from an onboard imager. The stored data are later downloaded from each spacecraft, processed, and merged with science data from the other three space- craft. These data collection sorties continue over a one year period at separations that range from tens of kilometers to thousands of kilometers.

Figure 1 shows the four spacecraft within the Pegasus fairing and Figure 2 shows a single AMS spacecraft in its orbital configuration. The AMS mission requires a highly elliptical orbit with an apogee close to 4600 kilometers and a perigee near 360 kilometers and an inclination of 83". A stock Pegasus XL can launch 173 kilograms to this orbit. When divided between the four spacecraft and the deployment sys- tem, there is only 36 kilograms of mass allocated to each spacecraft.

AMS is currently targeted for a NASA Small Explorer or Mid-Explorer program -- SMEX is cost capped at 38 million dollars, plus a baseline Pegasus XL launch vehicle. To implement a high science multi-point mission like AMS within the cost cap, the traditional approach to spacecraft design must be abandoned in favor of new methodologies that produce highly capable spacecraft at low cost. Major approaches to low cost science missions are directed toward using the smaller, less expensive launch vehicles and toward modular designs and architectures in which the modules are efficiently produced to keep the fabrication costs low and also reusable keep the non-recurring engineering costs low for later programs.

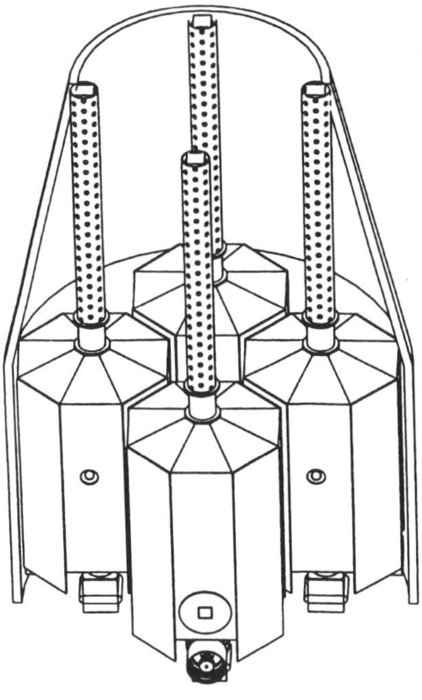

Figure 1. Launch Configuration, Four AMS spacecraft shown stowed within the Pegasus launch vehicle fairing.

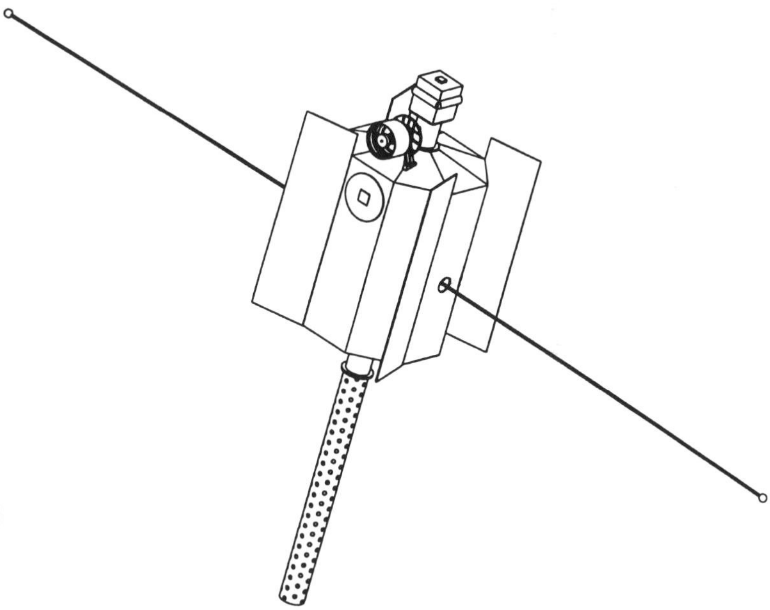

Figure 2. Orbital Configuration. A single AMS spacecraft shown with plasma wave booms and drag flaps deployed.

The spacecraft are designed to power up the instruments and collect science data simultaneously four times per day beginning at previously up-loaded start times. Thirty minutes are allocated to each collection sortie with a subinterval often minutes devoted to burst data (130.2 kbps) and the remaining twenty being devoted to lower time resolution data (3.78 kbps). Before and after each 30 minute data collection sortie, the spacecraft will collect and store GPS measurement data that is processed later on the ground to provide precision time and position (100 meter requirement). During a sortie, Sun angle, star field, and magnetometer data is collected and sorted for attitude and stability knowledge during ground processing.

Each AMS satellite will down-load the collected data to the JHU/APL ground station once per day. Start times for several days observation sorties together with the instrument modes and burst mode event trigger values can be uploaded once per day as well. The position, time, and attitude data will be processed and merged with the science data packets and then sent to the Science Data Center.

FNABLING TECHNOLOGIES

A major technology thrust of JHU/APL in recent years has been the development of systems and processes that have robust functionality (enabling cutting edge missions) at lower mass and power than conventional implementation allows. These technologies are at the heart of the avionics proposed for AMS.

Mini-IEM Technologies

The mini-IEM was developed to port circuit designs from NEAR, ACE, and TIMED into a single miniature unit using ASIC and chip-on-board technology. The mini-IEM offers a miniaturized, modular, open architecture that is perfect for micro-satellites. Mission requirements can be met by stacking together existing module designs. Figure 3 shows an integrated four module mini-IEM. Redundancy can be added by adding multiple modules. The spacecraft's RF and GPS functions are also integrated into the mini-IEM.

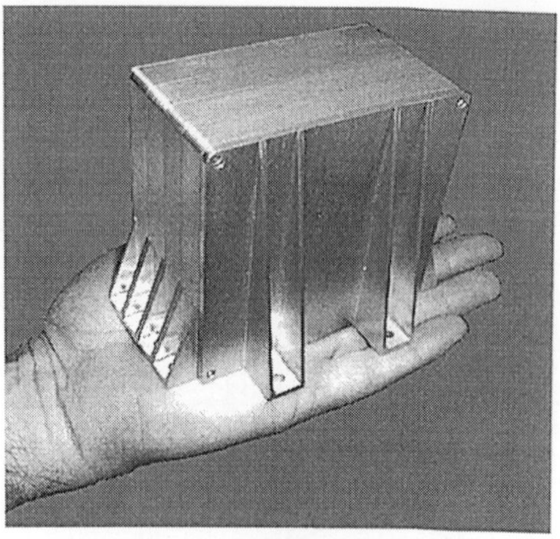

Figure 3. Integrated mini-IEM

<u>Spacecraft Electronics.</u> Most of the electronics reside in the spacecraft's miniature Integrated Electronics Module, or mini-IEM. Within the mini-IEM are six modules, analogous to cards in computer architecture. Each module is four inches square (Figure 3). Most modules contain one circuit card that mounts inside a stackable titanium frame. The frame provides structural support, thermal mass, and shielding from the high radiation environment. Each card has an integral Fuzz button/hard-hat feed through connector that serves as the mini-IEM bus and allows the modules to be stacked. The six modules communicate with each other through this connection using a IEEE-1394 (FIREWIRE) serial data bus.

Figure 4. Mini-IEM Modules

Processor Module. A single RTX-2010 performs all spacecraft processing. It divides it's time between command execution, science data time stamping and formatting, torquer control, and GPS processing.

Data Storage Module. Forty-two megabytes of storage are required for a 24 hour data collection period. This module contains 24 dynamic RAM chips for an effective capacity of 64 megabytes of triple voting data memory (64x3 megabytes of storage). Onboard voting is used in place of EDAC circuitry.

Interface Module. All instrument interfaces are low-power CMOS driver and receiver buffers. Their close proximity to the central electronics - < 1.1 meters - and their low rate digital signals make CMOS ideal.

Receiver Module. A 149 MHz command uplink was selected because of its low power consumption (0.7 Watts vs. 6 Watts for S-Band). Usable satellite communications allocations for this frequency still exist. The receiver design is functionally identical to the 1/8 watt receivers flown on the JHU/APL de- eloped Transit Navigation Satellites.

Transmitter Module. The transmitter module is a one watt implementation of the S-band RF card developed for the NASA TIMED mission. The card performs all encoding, modulation, and power amplification and interfaces directly to the down-link antenna.

GPS Module. The GPS module is based on an existing TIMED design. There are six tracking channels for each antenna enabling that antenna to track six GPS spacecraft. Once per second, each channel provides GPS satellite range and phase data along with a time tag over the 1394 bus for data storage (1200 bps total). The RTX-2010 module controls the phase lock loop and delay lock loop for each of the tracking channels over the 1394 bus.

IEEE-1394 Serial Bus. JHU/APL has developed radiation hardened IEEE-1394 serial bus chip using IRAD funding. This bus interface chip (BIG) is at the heart of the mini-IEM architecture. It autonomously transfers data between modules within the mini-IBM without requiring any special control from a processor; all protocol generation and bus arbitration is handled by the BIC chip network. The IEEE-1394 standard was just approved last year and is quickly becoming the favorite of the commercial world for networking workstations and peripherals (IEEE Spectrum Magazine, April 1997). The serial bus transfers data at rates up to 50 Mbps at lower power-watts per bit-than other architectures such as 1553 and VME .

Remote I/O Chip. This chip was also developed at JHU/APL using internal funding. It is a mixed analog-digital ASIC that allows multiple measurements of voltages, currents, discretes, temperatures, value comparisons, and timing to be made remotely instead of routing signals back to a common telemetry unit. Re- mote I/O chips are used in each of the modules as well as the Power System Box. The chips have a standard I2C serial bus interface that connects to the BIC chip using two wires. Multiple chips can be connected in parallel in a 'party-line' configuration. They are each independently addressable. When queried by the processor (over the 1394 bus) they can return a single channel value or all monitored channel values. When not actively reporting measurements, the chip is static and uses extremely little power. The chip is radiation hard to well over the 150 kilorad anticipated for AMS. This chip saves harness mass and simplifies parts selection for housekeeping type functions.

Chip-on-Board Technology

Current conventional packaging uses surface mount or through hole component technology. This approach increases the board size and the overall system weight owing to the larger sizes of the board and the components mounted on it. Chip-on-Board (COB) technology, where active dice are directly mounted onto a substrate without the need for an intermediate package, provides the basic vehicle to miniaturize electronic hardware.

COB technology utilizes the interconnections of bare dice on a substrate without the need for the component's package. Eliminating the component package reduces the required substrate area and assembly weight as much as 90% in some cases. With conventional high density printed wiring board and standard wire bonding technology, COB technology can yield a factor of ten or more in weight and volume saving.

Using bare die rather than packaged device can accommodate more components on a given board area, thus reducing the inductance created by the next level of interconnection between components. It also reduces the thermal resistance and the number of interfaces between the active die and the substrate (i.e., the pack- age pins). This can potentially improve the speed of the circuit and the reliability of the design. In terms of thermal management, un-packaged devices offer a shorter thermal resistance path than their packaged counterparts thereby improving thermal sinking and heat removal from the parts. APL has been developing and improving the chip-on-board process for spaceflight applications since 1994. This work has included techniques for known-good-die, die mounting, interconnect bonding, and post process coatings. APL's qualification standard is currently under review at NASA for approval and adoption as the industry standard.

Resin Transfer Molding

With resin transfer molding, once a mold is made, multiple copies are easily fabricated. In addition to it's mass production advantage, resin transfer molding also has the advantage of minimizing parts counts and their associated drawings. Currently, the resin transfer molding approach reduces the structure parts count from forty, using an aluminum frame construction, to three. With multiple cluster constellation missions becoming more prevalent, in some missions where constellations of tens of spacecraft are being considered, an economical but flexible design and fabrication technique is critical. We believe that for the quantities of spacecraft being considered, resin transfer molding is the best approach. The mold design process uses existing CAD tools. resin transfer molding provides flexibility to develop molds quickly and to iterate the structure design several times using several different techniques with the required tolerances. Vanguard has developed a process that is stable for high radiation environments. Available resins include Bryte and ICI/Fiberite cyanate ester. Fiber candidates range from IM-7 PAN to P-100HTS.

AMS requires only three molds to create the structure design: a main body mold, an end-cap mold, and a drag flap mold. The main body mold creates the eight sided structure with a central equipment deck. The end-cap mold creates the pyramidal shapes that taper down to the collars that support the spacecraft during launch. Solar cells will attach directly to the outer surfaces of portions of the main body, the drag flaps, and one of the two end-caps. When assembled, a central support deck molded into the main body divides the spacecraft interior into two cavities. Inserts used for mounting the avionics to the deck will be integrated into the mold tooling prior to curing. If necessary, inserts can be field installed to accommodate designchanges.

Drag Flap Station Keeping

Using drag flags for cluster station keeping has never been done before on any NASA program. With formation flying and constellation missions becoming more and more prevalent the use of drag flaps offers a more economical and long life alternative to using propulsion. McDonnell Douglas's Advanced Simulation Development System (BG-14 Integrator) code was used for our mission design development and analysis. This analysis includes: variable gravity; atmospheric drag; Sun activity; Sun, Moon and planet gravity; etc.

For AMS, four drag-flaps allow the surface area of the spacecraft, projected into a plane containing the spin axis, to increase up to 53% over the stowed configuration. The drag flaps hinge on one end to the spacecraft body. A stepper motor/ ball screw mechanism allows the flaps to deploy up to 60° away from the body. Each flap has its own separate motor/screw mechanism that can step the flap position between 0-60° in over 100 steps. The four flaps on a given spacecraft will always be deployed at the same angle which maintains the center of pressure within allowable limits of the center of mass.

The direction of deployment is radially away from the frame, lateral to the rocket's velocity vector. During deployment, the thrust axis of the rocket points along the velocity vector to within 1°. Figure 5 shows the affect of different along track components and a spacecraft's ability to stop the separation and maintain the cluster with the differential surface area control. An along track component of up to 12 cm/second (3.4° pointing error) can be overcome and corrected during the 60 days from launch. The pointing tolerance of the cluster deployment is therefore quite loose.

During the attitude stabilization period that follows launch, the along track separations are monitored and the drag flaps deployed differentially as required to achieve the desired along track separations. Even though the spacecraft are not initially in the cartwheel mode the drag flaps will still provide differential drag for along track correction. The cross track separations will accumulate with time, a range of spatial scales are studied as the mission progresses. The along track geometry is maintained by taking advantage of atmospheric drag that occurs during orbit perigee. The drag of an individual spacecraft can be varied by changing the surface area exposed to the orbit ram direction by means of the adjustable drag flaps. The along track separations can be monitored on the ground and changes in the drag flap deployment angle commanded every few days as needed.

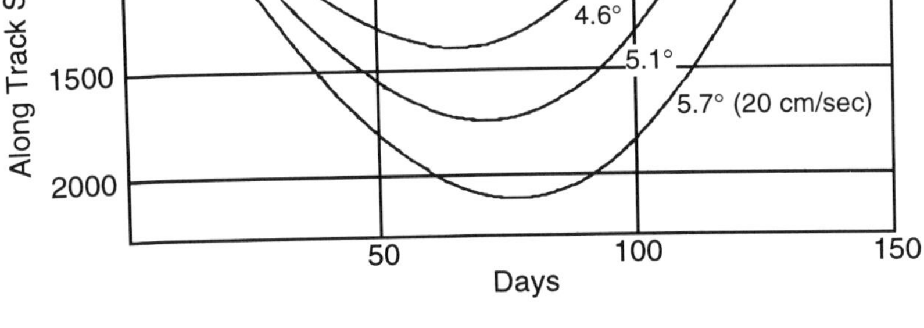

Figure 5. Deployment Error Correction Using Drag-Flaps

AMS IMPLEMENTATION

All four AMS spacecraft are identical single string designs. Overall, the system has a mass margin is 22% and power margin of 31%. Figure 6 shows the system block diagram of a single AMS spacecraft.

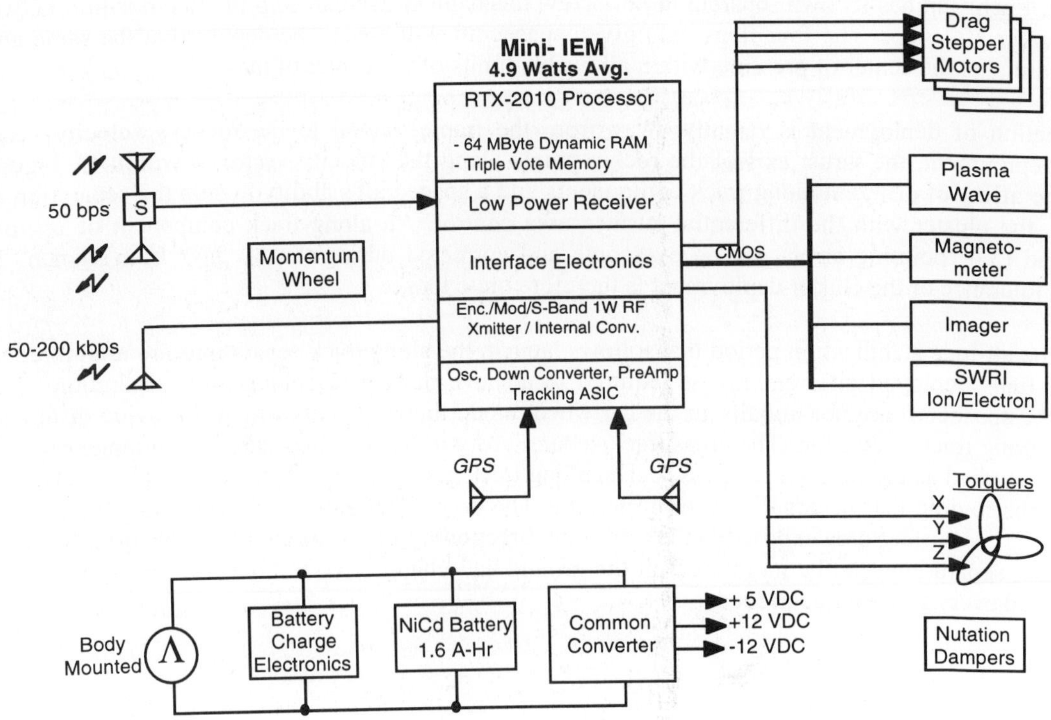

Figure 6. AMS Spacecraft System Block Diagram

Attitude Stability. In orbit, each spacecraft will rotate slowly about its central axis. An off-the-shelf spinning Sun sensor coupled with magnetometer data will be used for coarse attitude determination and attitude stability knowledge. The Sun sensor generates a Sun crossing pulse and a nine bit Sun angle to provide 0.5° coarse knowledge. A momentum wheel will store angular moment to insure the required stability during a data collection sortie. Nutation will be dissipated using two nutation dampers. Periodically during a sortie, several star field images will be captured during a spacecraft rotation to provide precise (0.01°) attitude determination and stability knowledge during ground processing.

After deployment, magnetic torquers are used to move the spin axis to the stable "cart-wheel" mode - spin axis normal to the orbit plane. The plasma wave antenna booms are then deployed and the torquers are used to set the spin rate.

Power Subsystem. With small spacecraft, with fixed, body-mounted arrays, power is very limited. In the 'cart-wheel' mode, the Sun favors different sides of the spacecraft during the year due to orbit precession. Most of the time, the power system will support an orbit average load of 10.8 watts. For power conservation, the instrument, transmitter, and torque coil loads are duty cycled and the onboard computer re- sources are shared. The orbit average load that must be supported is 7.5 watts. About half-way through the mission, the spacecraft will be flipped 180° so that the Sun once again favors the top end-cap, and not the radiator. The power bus is a 7.5 volt unregulated bus that uses a direct energy transfer topology. A 7.0 ampere-hour NiCd battery powers the spacecraft during the 42.8 minute worst case eclipses. The spacecraft contains only two power converters; a common converter located within the power system box and the on-module dedicated transmitter converter.

RF Telecommunications. The spacecraft uses simple monopole antennas for both uplink and downlink. The S-Band monopole is 1.3 inches long and mounts to the tip of the magnetometer boom. The field of view of this antenna is ±40° about the equator of the spacecraft. The VHF monopole is 20 inches long and stands on end on the electron/ion sensor. This whip antenna is captured by the lower support arm of the deployment frame during launch. With this antenna arrangement, and the spacecraft 'cartwheel' orientation, reliable ground station contact is assured. At an altitude of4000 kilometer (close to AMS apogee of 4600 km), a spacecraft can downlink its entire 24-hour collected data set in fourteen minutes.

Thermal Control. AMS has three power dissipation modes: cruise, data dump, and sortie. Cruise is by far the predominant mode consuming 22 hours of each day. The other two higher power modes last about 30-minutes each and are bounded by long periods of cruise mode. AMS thermal design keeps the spacecraft cool during cruise (-5°C) and warms (up to 30°) during the other two modes. The interior electronics is blanketed and the large thermal mass of the mini-IEM housing keeps temperature rises gradual. Spacecraft spin helps to keep temperatures constant. The bottom end-cap is used as a thermal radiator.

Ground Systems. An existing JHU/APL ground station transmit/receive antenna will be dedicated to AMS. The ground station receiver will be directly connected to a personal computer server that will automatically process the attitude, position, and time data and merge the processed data with the science data packets. This data will remain on the server as part of a time history data base. The server will also process housekeeping trending and the orbit configuration management software. These processes will be automated with defined products that will be used by the Mission Operations Manager and the Principle Investigator as decision making tools. Using these tools, command sequences are assembled and transferred to the ground station transmitter.

Communication Approach. Each spacecraft will be interrogated individually. Ground to satellite communications consisting of real-time commands, delayed commands, and table parameter uploads will all be sent to the cluster using the JHU/APL station's existing 149 MHz system. When the orbit cluster's orbit line-of-apsides causes very short passes over the JHU/APL station, the mission will use a data receive station in the southern hemisphere. The southern hemisphere station would receive and transfer the collected data to the JHU/APL mission operations center over the Internet.

On average, there are seven ground contacts with a contact time of 30.1 minutes. A long contact pass can be as long as 47 minutes. Long term cluster station keeping will be tracked and adjusted periodically. Adjustments to the drag flaps will occur as simple step values that will change no more than two or three times per week. The post processing of the time, position, and attitude stability will all be automated on a server that collects and stores the level one data products.

ACKNOWLEDGMENTS

The APL IEM was designed by a large number of staff members from the JHU/APL Space and Technical Services Departments. A committee chartered to insert advanced technology into our space programs, developed the original IEM concept. It has since been refined and enhanced through the efforts of many other staff members, and the work reported here is due to those contributions and the sup portive environment provided by our management team. We would like to thank Rich Conde, Bob Jenkins, Susan

Lee, Jim Perschy, Ed Reynolds, Kim Strohbehn, and Tom Zaremba for their contributions to working out de- tails of the APL IEM architecture. We also wish to acknowledge the efforts of Mike Kreitz, Jason Jenkins, Bob Bokulic, Barry Mauk, and Brian Anderson of JHU/APL for their application of the mini-IEM architecture toward the next generation of space missions. This work was funded by the JHU/APL Independent Research and Development Program.

REFERENCES:

Fraeman, M., "A Fault Tolerant Integrated Electronics Module for Small Satellites" 11th AIAA/USU Conference on Small Satellites, Logan, UT., September 1997.

Conde, R.F., B.Q. Le, J.F. Bogdanski, A.L. Lew, J.A. Perschy, and M.A.G. Darrin, "Command and Data Handling in Your Palm", 11th AIAA/USU Conference on Small Satellites, Logan, UT., September 1997.

LESSIONS LEARNED ?

Wing H Ip

Max-Planck-Institut fuer Aeronomie, Max-Planck-Strasse 2, Postfach 20 , Katlenburg-Lindan, D-37191, Germany , Tel: 49-555-697-9416 Fax: 49-555-697-9240 , E-mail$_1$: ip@linmpi.mpg.de , E-mail$_2$: ip@linax1.dnet.gwdg.de

ABSTRACT

The rapid economic growth and technical drives in the Asia/Pacific region over the last two decades have led to the establishment of several new space organizations with high potentials of scientific and technical contributions. Lessons learned from the recent trend of promoting space research via low-cost small satellite missions and the financial crisis in 1997 indicate that there is a need to initiate a cooperative space program in this region to pool the resources so that major environmental and scientific issues in the 21st Century may be addressed at first hand by the scientific community in Asia/Pacific. In addition to safeguarding the integrity of the natural environment and resouces, such a collaborative effort will lead to the enhancement of economic growth and regional stability. Scientific microsatellite projects under the auspice of a multi-national university-based consortium could play a key role in this enterprise.

INTRODUCTION

When the COSPAR Colloquium on "Microsatellites as Research Tools" was first planned, there was little inkling that 1997 was to become a fateful year in the Asia/Pacific region. It was not just that a major forest fire occurred in Indonesia inflicting environmental damages of both regional and planetary scales, the concurrent storms in the Asian financial markets were even more devastating. Because of the close commercial links between Asia and the rest of the world, this sequence of events have only helped to underline the basic fact that regional economic and environmental stabilities in East Asia are crucial to the global economy. The monetary loss incurred so far in this episode has been estimated to be on the order of one trillion US dollars. In comparison, the developmental cost for a scientific microsatellite which might be as low as a few million dollars is quite negligible. The question we wish to ask is: how the lessons we have learned from this scientific colloquium and those from the recent financial disasters can be synthesized such that they can be eventually used to safeguard the social security and economic developments in the Asia/Pacific region and, by implication, the world-wide community. In addition, we wish to establish the case hereby that a regional university-based program of joint microsatellite missions could be used as a cost-effective way to reinforce international space peaceful cooperation.

LESSON FROM TAIWAN

Even without the onslaugth of the economic crisis in 1997, space agencies in the Asia/Pacific regions have at times encountered difficulties in articulating the national investments in space research. This is because few of these organizations in this area are in the position of maintaining an aggressive space research program at a similar level like those of NASA, ISAS and ESA. It is therefore difficult to produce exciting scientific results by projects at the entry-level as possible in the beginning of space exploration. At the same time, high expectations of the potential scientific and technological achievements of the space programs could become rampart with tremendous social, commercial and political pressures such that the existance of a fledgling space program could be tied to the success or failure of the very first mission. A case in point was the early phase of the national space program in Taiwan which had gone through many cycles of uncertainties and endless debates. It is thus quite clear that instead of an erratic approach a systematic roadmap and a clear mandate must be followed if the vigour of a space program is to be sustained.

From past experiences, there are several recipes. First, the start-up program should not be over-ambitious. That is, the first few space missions should be limited in scope such that the costs and complexities of the mission operations could be minimized. For example, the inclusion of scientific instruments, while of high standard, must be commensurate with the local expertises. Otherwise, there will be little benefits to be reaped from the scientific returns. Second, the near-term investments in the infrastructure and human resources must be consistent with the long-term goals which in turn ought to be of major importance to the society. Third, a strong outreach and educational program and the training of new blood at universities must be introduced to ensure the continuity of the space research activity. Finally, for the reasons of cost-sharing, and pooling of scientific and technical facilities and know-hows, international space cooperations should be attempted as soon as possible. Scientific microsatellites can serve as a stepping stone in this regard.

LESSON FROM EUROPE: REGIONAL AND GLOBAL COOPERATION

Because of the European Space Agency (ESA), space researchers from different countries in Europe are enjoying many opportunities to make discoveries in astronomy and astrophysics and to make fundamental contributions to environmental studies and space technologies. Nearly everything in ESA has no national boundaries. This must have sounded like an impossible dream for young people who just came out of the shadow of the Second World War fifty years ago. One can thefore argue that ESA and other similar multi-national organizations are the very first peace dividents. On the other hand, it is appropriate to say that the long years of peace and prosperity in Western Europe have been the results of the cooperative efforts of the national leaders. Instead of wasting the most precious resources and human spirit in never-ending conflicts and warfares, European nations have jointly established the standard for high social conscience and welfare system which are supported by science and technology at the cutting edge. As a matter of fact, space technology has been identified as a crucial element in maintaining the competiveness of European industries; and this is one of the underlying reasons for the European drive towards the establishment of a lunar base in the early part of the 21st Century. In the post-Cold War era, long-term financial and commercial interests weight heavily in such large-scale space projects.

From this point of view, it is most essential that a Cold War mentality be avoided in space programs of the newly industrialized nations. For various reasons, the rush into acquiring space capabilities runs the danger of becoming a component of an expensive arms race against each other. This is like reversing the course of the modern European history with a view to consume a tremendous amount of humanity and resources for mutual destruction. Instead, we should follow the example of Europe in redirecting the "Asian values" and energies towards joint programs of peaceful cooperation. In the last analysis, mutual understanding and reduction of tension can only develop on the basis of close interactions and peaceful collaborations. The establishment of a multi-national space organization a la ESA in East Asia will go a long way in securing the regional stability at a very tiny fraction of the expenditure to be spent on advanced weapons systems.

LESSON FROM CHINA: ENVIRONMENTAL PROTECTION AND RESOURCES MANAGEMENT

While high-techs may be acquired at an ever-increasing pace, the same probably holds true for the fast disappearance of natural resources such as clear-water lakes and rivers, unpolluted coast lines and forests. These environmental problems accentuated by rapid urbanization and industrialization of the Asian nations and elsewhere are further compounded by the probable onset of global warming. We wonder, for example, can the Indonesian great fire in 1997 be repeated in future under the influence of El Nino? can such environmental dangers be forecasted and hence be prevented? and finally, how to limit the damages to natural and man-made disasters driven by energy-needs?

Many of the large-scale civil constructions would be dwarfed by the mammoth project now ongoing in the Three Gorges Dam at Yangtse River. As is well-known this project for hydraulic power generation and flood-control will be the driving force for China's economic growth. In manay ways, the future of the economies of East Asia and the world as a whole will be affected by the operation of the Three Gorges Dam in several aspects. The economic factor is obvious. The potential environmental impacts - which have made this hydraulic project a major controversy since its first proposal - must now be carefully monitored not just by China herself but also collectively by the international community. This point can be underlined by the fact that the final decision to go ahead with the Three Gorges Dam was partly triggered by the great flood in China in 1991. Almost 80\% of the densely populated area in central China was affected with heavy loss in human lives and property damages which must have amounted to tens of billions of US dollars. A similar flood in the next decade will not just cause major setbacks in the Chinese economy but also those of East Asia and beyond. With the Domino Effect in the collapse of the Asian financial markets in 1997 fresh in mind, it is quite clear that the progress and maintainence of the Three Gorges Dam will be a major concern to many governmental agencies at different disciplines and levels.

The above examples ranging from Borneo to the Three Gorges, two of the most beautiful landscapes on Earth have shown that break-neck industrialization and mounting population pressure have made environmental protection policy difficult to implement in many developing countries. If this trend continues, the geography of Asia will be drastically changed by rapid deforestation and industrial pollutions in less than thirty years (e.g., ca 2025) which has been prophesized to be the time when the economic outputs would reach 50\% of the world total. It is, however, difficult to imagine how such industrial growth can be sustained if the whole Pacific Asia region is constantly under the threats of the thick smog like in Indonesia and Malyasia in 1997 and killer-floods like in China in 1991?

Because of the integration of individual national markets into the global economy, the minimization of social disturbances and economic crises is no longer the responsibility of a single country. Conservation of natural resouces and environmental protection must be efficiently enforced at least at regional level. Take for example, Singaapore and Taiwan which have substantial commercial interests in China will probably benefit significantly from the Three Gorges Dam project because of its great potential in promoting industrial growth. The associated environemental protection mechanisms should hence in principle be shared by people across the Taiwan Strait and the Strait of Malacca. The same can be said on many other inter-related economic activities with important environmental consequences in Asia. From this point of view, there is an urgent need to initiate a joint cooperative program to map and to monitor the rapidly changing landscapes and geomorphologies of the Asia/Pacific region in the next millennium. This is exactly where space research can be of service.

LESSON FROM HONG KONG: SCIENTIFIC AD TECHNICAL DEVELOPMENTS NEEDS

Many useful results and interesting ideas were presented at the Tainan COSPAR meeting on scientific microsatellites. These included the applications of satellite communications to medical problems and the developments of nano-satellites of just a few kg for university education. They demonstrated that a wide range of technical possibilities exist to address the needs we have identified before. The question is only how to synthesize all these inputs into a coherent framework and make it a workable plan. The aim is to provide the rationales for the founding (and funding) of an international space organization in Asia/Pacific to coordinate space activities of regional interests.

We note that one million USD is likely to be a hefty sum in the budget of the National Science Council or Ministry of Education even in a country like Taiwan. But in the financial chaos experienced in 1997, many Asian establishments and business concerns – which had steadfastly refused to invest in university education and basic research - must have lost billions of dollars in a matter of days. Hong Kong is a good example. But if one does not catch up with scientific developments and technical innovations, places like Hong Kong will quickly run out of not just space, but also time, in space exploration.

We can also look at the issue from a different perspective. That is, even with the pause caused by the financial meltdown in recent months, substantial economic growth in Asia is still to be expected. Suppose in 2025, the GNP of East Asia as a whole would be comparable to that of USA and Europe. This would mean that countries in the Pacific rim should also venture to participate in space activities at a level comparable to Europe, Japan and USA. How would the space agencies in Malaysia, Thailand, Indonesia, Korea and Taiwan take part in space projects like space stations, lunar bases and human colonization of Mars? According to this projection, the Southeast Asian nations should embark on an aggressive drive towards next-generation technologies even amid the terrible economic setbacks.

Judging from the infrastructures and facilities now in place, a six-year plan can be formulated with a reasonable chance of success. The first two years will see the formation of a university-based remote-sensing research centre. This will serve as a focal point for data procesing and interpretations. The format is simply that researchers from different member-universities will come as either resident or visiting scientists to work on selected projects (i.e., land surveys, coastal pollution charts, etc) using facilities provided by the host centre. A prototype is the international cooperative program of the Centre of Space Remote-Sensing Studies at the National Central Univerity in Chung-Li, Taiwan. A robust program can probably be maintained with an annual budget of about one million USD.

The next stage will call for the establishment of a consortium of university-based organizations at circa 2002 to pursue microsatellites and small satellites in space science and engineering applications. A prototype is the University Space Research Association (USRA) of USA which is now running a successful program of student satellites for NASA. As can be found in this Proceedings, there are many examples of low-cost missions which can be attempted by such a regional consortium. An annual budget of 5 - 10 million USD should be adequate to support one to two student satellite projects per year. If there are five members, the cost-sharing will be about 1 - 2 million USD for each partner with participations in different technical components in the planning and operation of the space missions.

Depending on the progress of these two phases, the participating members may decide to upgrade the univeristy-based consortium to an inter-agency organization for cooperative space research and applications. The prototype is ESA which has a few centralized facilities as well as receiving stations and launch site (s) at separate locations. At this point, some large projects of higher complexities could be carried out to address the needs of users in Asia/Pacific and other regions. With the prospect of a strong rebound of the Tiger economy, the annual budget in 2005 might be projected to be 20 – 50 million USD or 4 - 10 million USD per partner. As in the case of ESA, the obvious advantage to participants in this space organization is that opportunities to advanced science and technologies could be open to scientists and engineers in Asia with the possibility of keeping pace with their colleagues in America, Europe and Japan.

There is a concern that the possible creation of a new organization as postulated here would only mean more bureacracy and hence less efficiency and higher cost in producing the results. To some extent, this is certainly a valid argument. However, the current problem facing Asia/Pacific is actually the lack of effective inter-governmental organziations to coordinate matters of mutual national interests. This deficiency has been fully demonstrated by the run of the Asian stock markets in the economic woes of 1997.

CONCLUSION

The discussions presented here are probably somewhat unexpected in a meeting on scientific microsatellites. This is basically a personal reflection from lessions learned that the thriving of space research activities in Asia/Pacific requires a clear definition of short-term and long-term goals and interests to the tax players. Basic research in space science and astronomy can be supported only as long as the space program as a whole brings important benefits to the society.

Because of the cultural differences, social values and technological gaps between Asia/Pacific and the West, many factors taken for granted in the space programs of the US and Europe have to be often examined and debated. This problem is compounded by the redundancy and technical/financial limits if individual newly industrialized nations are (rightfully) to establish space programs of their own. There is also a hidden danger of being mistaken for a new form of regional arms race. It is therefore suggested that a multi-national university-based space program be initiated with the following priorities:

(1) monitoring of environmental and global climate changes under the effect of rapid urbanization in East Asia,
(2) prevention and damage control of natural disasters (i.e, earthquakes, floods, and forest fires),
(3) outreach and education by processing and dispensing scientific information from previous space missions,
(4) cooperation in earth science, space science and astronomy, and applications to communications and medicine using microsatellites and small satellites.

The first two areas can be partially addressed by the establishment of a research centre for data processing and analysis using observations obtained by spacecraft from NASA, ESA and other space agencies. The third point would also benefit significantly from the experiences and vast amount of data from the outreach program of NASA. The fourth point will require the launches of satellites from the inter-agency organization which itself can be modelled after USRA. This four-points program emphasing peaceful cooperation can be supported at a low cost and can serve as a stablizing factor in the regional security and economic growth of Asia/Pacific and the global village.

ACKNOWLEDGMENTS

I thank the LOC of this COSPAR Colloquium and the Ministry of Education of Taiwan for financial supports, and Profs. F. B. Hsiao and K. Yong of NCKU, Prof. A. J. Chen of NCU, Prof. S. T. Wu of UAH and Dr. Miriam Baltuck of NASA for useful discussions.

ADVANCED MICROSAT TECHNOLOGY

Yusuke Matsumura, and T. Iwata

National Space Development Agency of Japan , 2-1-1,Sengen ,Tsukuba-city ,Ibaraki , JAPAN

ABSTRACT

Due to the progress of technologies in recent years, especially in the commercial field of miniaturized and lightened devices, the development of the small and high performance satellites will be soon made possible. In this paper, for the purpose of contributions to various missions supposed in the next generation, the system concept of the advanced microsatellites that utilized these advanced technologies are reported.

INTRODUCTION

Until now, in NASDA, large satellites have been mainly developed as shown in Figure 1., However the development cost have been enormous and long development times have been needed in order to develop such complicated large satellites. And as a result of the development cost being so expensive, a lot of importance have been placed on avoiding risks. In addition, the costs tend to increase in order to ensure sufficient reliability. As a consequence of the demand of such risk avoidance and the cost reduction, the continued use of existing technologies have resulted, and the use of new technologies have been avoided. Moreover due to the long development period required, the adoptive technologies have already been out of date at the time of the launch. During this period, base technologies such as electronics, computer, and material on the ground have advanced rapidly, and lately the space technologies are considerably older than the ground technologies.

In such situations, the satellite technologies of our country are left behind not only in cost but also in function or performance as the '80's technologies. Thus it must be tried that: ground technologies are positively adopted, and new space technology developments are accelerated more and more on the basis of the existing satellite technologies.

This paper reports on the concept of the next generation satellites and possibility of new missions.

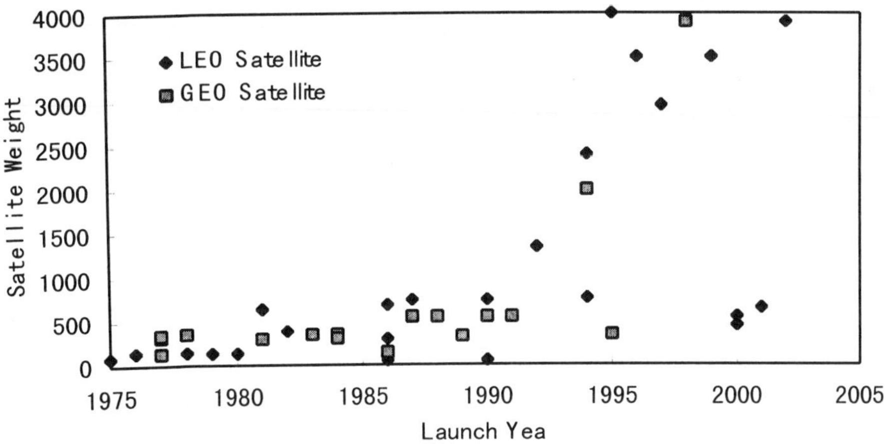

Figure 1. Satellites Developed by NASDA.

TARGET

Weight Target

We have three kind of satellite weight targets. 1) low earth orbiting satellite: around 50kg, 2) deep space explorer: 200kg, 3) service spacecraft on orbit: under several kg. Particularly in the case of the deep space explorer, we have only a launch vehicle which has the ability to carry less than 500kg to Jupiter, even if the most suitable solid kick motor can be utilized.

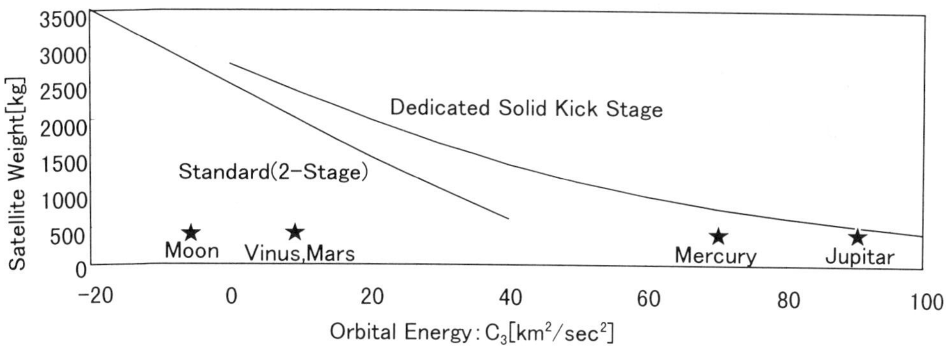

Figure 2. Capability of H-II rocket.

Cost Target

Recently as space commercialization has advanced, the spacecraft need to be evaluated from the view point of having similar economical value as ground infrastructures. As a result, drastic cost reduction has been required. Here, the goal is assumed that the development cost of the flight model are reduced to about 1/10 for the future satellite with the function which is equivalent to the present.

The possibility of this goal itself is under study now, but may be sure considering from: 1)the contribution of miniaturization and weight reduction to the cost reduction in the phase of manufacturing, integration, handling, and testing, 2)the various launch selections such as the dual launch, low cost dedicated small rocket, and 3)the effect of the mass production by constructing the system by the constellation. In addition, operating costs may also be reduced by the advancement of autonomous satellites.

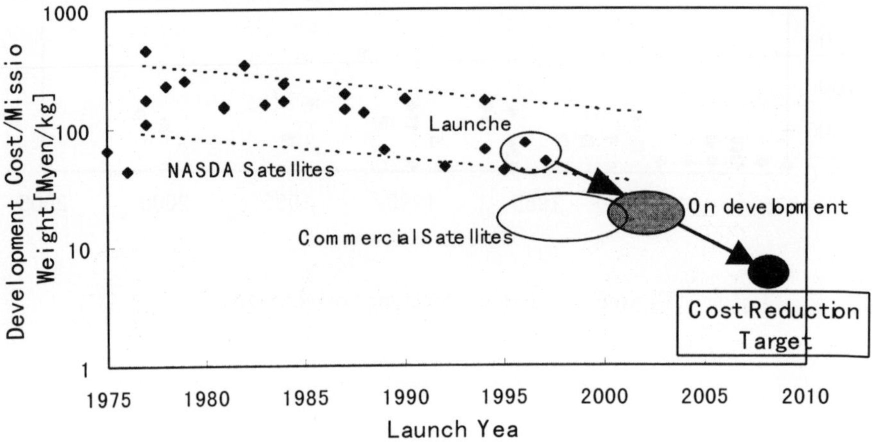

Figure 3. Cost reduction target.

SYSTEM CONCEPT

Some of the most successful miniaturized and high-performance products today are from the electronics industry. This miniaturization is realized by integrating a lot of functional parts to a small size package or chip.

Applying such method of miniaturization to satellites' electric or mechanical equipments enables miniaturization of a whole satellite possible. At first, the parts or equipments are miniaturized into chips, and next, chips are mounted on a structure board. The functions of each satellite systems including Thermal Control System, Harness, etc. are taken on an integrated board. Finally, it becomes 'system on board' as shown in Figure 4.

Approaches to the system integration on board seem to be effective concept for the satellite system as it requires small in size, light in weight, and efficient stowing, and consists of many systems. When this concept were applied to satellites, it is possible to construct various satellite systems and shapes flexibly by combining these function chips or boards depending on the system requirements. For example, on the scalable function such as antennas, power supply, propulsion, etc., a system requirement can be satisfied by standardizing every basic function board unit such as communication system board, electric power supply system board, propulsion system board, etc. and adding the electric power supply board in proportion to the required handling electric power.

The whole shapes of the satellite are considered to be extended on two-dimensions, to be constructed as the conventional type which is with polyhedron combined boxes, and so on. Further they are also considered that the systems are reconstructed on orbit by rearranging the shape depending on the conditions. While it is possible to keep the altitude more efficiently by reducing the effective sectional area with the basic structure of the board. Further more, it can be expected that these measures effectively overcome the problem of the heat concentration by the miniaturization, since the surface area per volume or weight increases, so does the security of the radiator surface become easier. And the heat conduction pass to the thickness direction shortens, so it is possible to attempt the simplification of the thermal control system.

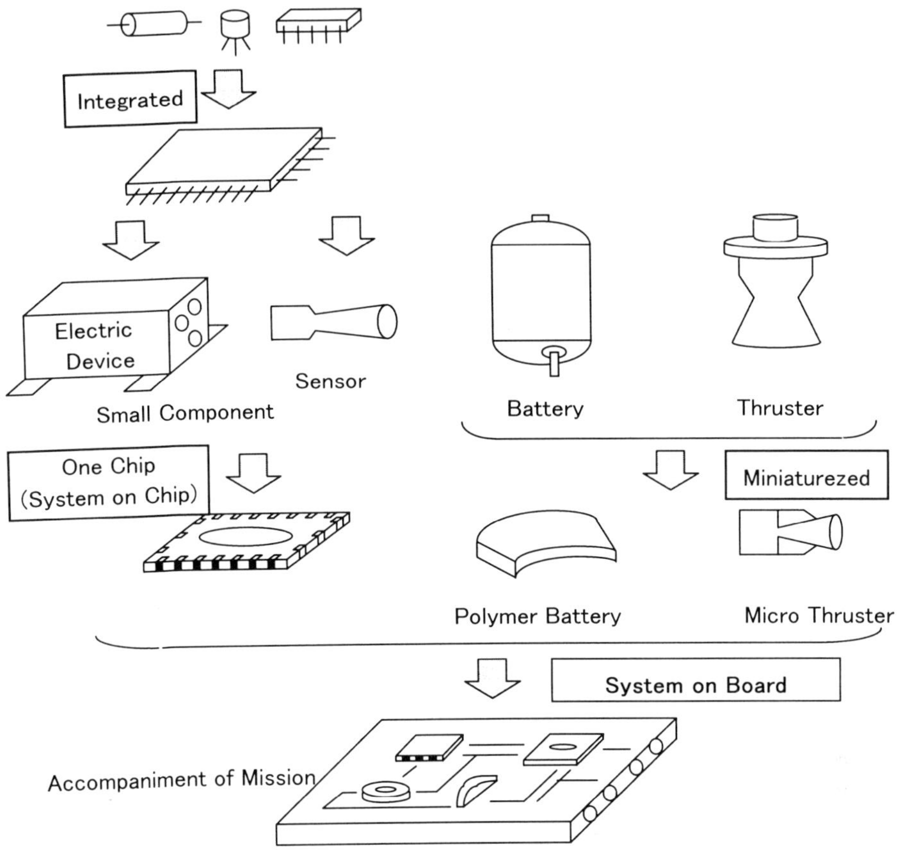

Figure 4. Image of system on board

EXAMPLE BUS

Here, we estimated the feasibility of the 10kg bus system in the 50kg LEO satellites. It is assumed that the elements adopted below will be able to developed by the beginning of the 21st Century.

System Characteristics

Main characteristics of the system and the subsystems are shown in Table 1.

Table 1. Main Characteristics

SYSTEM	Size Weight Power Consumption Mission Life	0.4m×0.4m×0.4m 50kg 100w 1Year	Altitude500kmCircular Orbit Inclination45degree
BUS	BUS Weight BUS Power Consumption	10kg 40W	
C&DH	Data Rate Data Storage	1kbps(TLM/CMD) 15Mbps(Mission) 10Gbit	 Mission Data Rate100kbps
AOCS	Control Accuracy Determination Accuracy	0.1degree 2×10^{-4}degree	
EPS	Battery Supply Power	150wh/kg 250w	DOD:40%
TCS	Radiator Surface	0.32 m^2	
RCS	Adjust Orbit Unloading	Thrust0.1N ISP200sec	

It was assumed that the function of a single mission satellite of about 500kg (which will be expected the demand increase) was realized at satellite weight 50kg and bus weight10kg. So this microsatellite has the equivalent performance on communication capability and attitude control accuracy, etc. to present small or mid-size satellites.

Communication and Data Handling(C&DH)

In the C&DH, the miniaturization and power saving are remarkable. Though the weight distributed to this system is small, the most drastic miniaturization and function enhancement can be expected. However, in order to use the commercial parts in the space, more intelligent fault tolerance systems and the reliability adequate for the mission life are required in addition to strengthening of the radiation resistance.

The miniaturization of the electronic equipment advanced like this changes hardware design into more efficient one in volume to performance, that is, from conventional hardware type of uni-function-to-one-hardware into the new and unified hardware type of multi-function-to-one-hardware. It is well-known example to attempt the integration of data processing equipments distributed into some subsystems, C&DH, AOCS, and so on.

Low supply voltage of logic IC, etc. contributes to electrical power saving and to easing the concentration of the heat source

Table 2. The Elements of The Advanced C&DH(Example)

	ELEMENT	SPECIFICATION
LSI Packaging	MCM	
Power Saving IC	logic IC	Operating Voltage:1.2V
Large Scale IC	DRAM	Capability:16Gbit
	G/A	Capability:1.5M
Very Small Parts	MMIC	
	MMCM	
Integrated Control Unit	Integrated control system	

Atitude and Orbit Control System(AOCS)

The advanced AOCS is divided into the three following functions on board:

- The observation/measurement of the reference
- The position/attitude determination
- The position/attitude control

It is supposed that the computer section which carries out the position/attitude determination is included in the integrated control unit above. On the sensors for observation, simplifying the equipment composition of the system contributes to the weight reduction, in addition to the miniaturization of each equipment. Here, the system by star tracker (SST) / gyro is adopted, but in LEO/MEO satellites the system by DGPS also becomes a candidate.

The performance of RCS and small reaction wheel which control the position/attitude has close relationship with their weight. So they are one of the hardware with much difficulty in improving volume and weight to performance.

Table 3. The Elements of The Advanced AOCS(Example)

	ELEMENT	SPECIFICATION	REMARKS
Simple Configuration Attitude Determination System	The system consist of SST and GYRO	Weight:1kg Accuracy:0.01deg	
Small Actuator	RW	Weight:1.5kg	Angular Momentum Storage ± 0.05Nmsat1000rpm
Micro Thruster	Cold gas thruster	Weight:8g	Thrust: 5mN
Integrated Control Unit			Included in DH

Electrical Power System (EPS)

From orbital parameters in Table 1., the eclipse period is about 33 minutes max and then the sunlit period is about 62 minutes per 1revolution, and beta angle is 40 degree max. When the depth of discharge is 40%, 6Ah battery capability is required, and when the power consumption of the satellite is 100W, the required power generation capability is 240W. It is necessary to achieve the performance value as indicated in Table 4., at least in order to satisfy these requirements and to realize the bus of about 10kg.

Table 4. The Elements of The Advanced EPS(Example)

	ELEMENT	SPECIFICATION	REMARKS
High Efficient Battery	Li-Ion	Energy Density:150Wh/kg	Depth of Discharge: 40%
High Efficient Cell	GaInP2/GaAs/Ge	Exchange Efficiency:25% Cell Unit Weight: 1kg/m2	

Thermal Control System and Structure(TCS & STR)

Though the reduction of the total power consumption by power saving at each equipments, and the equalizing of the satellite temperature by the miniaturization of the satellite can be expected, the improvement of the heat ejection ability is also desired

The miniaturization of the satellite contributes to the improvement of the rigidity and strength. In addition, more lightening can be expected by attempting the reduction in the mechanical environmental conditions. And the total system by the multi-functional structure board which integrates heat transfer path, wire harness, and piping, etc. will become one of the structure design in the future.

Table 5. The Elements of The Advanced TCS & STR(Example)

	ELEMENT	SPECIFICATION
Low Shock Separation Mechanism/Device	Shaped Memory Alloy(SMA)	Structure Weight:
Multi-Functional Structure	Integration of Heat Transfer, Wire Harness, and Piping, etc.	Weight:

Bus System

Using the technologies investigated above, the satellite system is examined. The most remarkable change point is to take in the control function of each subsystem by integrated control unit(ICU). Table 6. Weight and Power Distribution shows that ICU occupies 2.8kg weight, which has a higher weight ratio to bus total weight than the conventional DH subsystem, AOCS 3.1kg(except propellant), EPS 2.8kg, TCS&STR 2.1kg, Communication Subsystem 0.7kg, and finally 11.5kg bus system is possible. Payload efficiency is about 76%, and they almost satisfy the initial goal, bus weight 10kg and payload efficiency 80%. Through this result, we had the prospect of the realization of the goal.

Table 6. Weight and Power Distribution

TARGET Subsystem	Component	Weight (kg)	Power (W)	CURRENT Component	Subsystem
TT&C (0.7kg)	ANT-1,2	0.4		ANT-1,2	TT&C
	HYB/DIP	0.3		HYB/DIP	
ICU (2.8kg)	TRXM	0.3	5.0	TRX	
	DHM	0.3	1.0	CU	DH
	DBM	0.3	1.0	RIU	
	DRM	0.3	1.0	DR	
	OBCM	0.3	1.0	AOCP	AOCS
	IFM	0.3	1.0	VED	
				PDE	EPS
	PCU	0.5	10.0	PCU	
				BCCU	
	DC/DC	0.5	1.0	DC/DC	
AOCS (2.1kg)	GYRO	0.1	0.2	IRU	AOCS
	STT	0.5	1.0	FSS	
				CSS	
				ESA	
	RW-1,2,3	1.5	7.5	RW-1,2,3	
				MTQ	
				VDE	
EPS (2.8kg)	BAT	1.0		BAT	EPS
	PDL	0.7		PDL	
	CELL	0.8		CELL	
	PDM1,2	0.3		PDM-1,2	
RCS(1.0)	RCS	1.0		RCS	RCS
TCS (0.5kg)	HTR	0.5	10.0	HTR	TCS
STR (1.6kg)	STR	1.3		STR	STR
	WH	0.3		WH	WH
BUS		***11.5***	***39.7***		
PROP.		2.0		Mission Life: 1 Year	
MISSION		36.5	60.3	Payload Efficiency(Dry): 76%	
TOTAL		50.0	100.0		

ASSUMPTION OF MISSION MODEL FOR CASE STUDY

The case study with using this technologies mentioned above was carried out on some assumed mission model.

Monitoring Spacecraft

When some troubles on a mother spacecraft happen, this nano-sat separates from the mother, and monitors the trouble point. This nano-sat is required remarkable miniaturizing in order to board it on mother spacecraft, so the system is made to be simple and not to have excessive function like solar array, wheel, etc. The weight is supposed to be less than 4kg, and life time 1-2days.

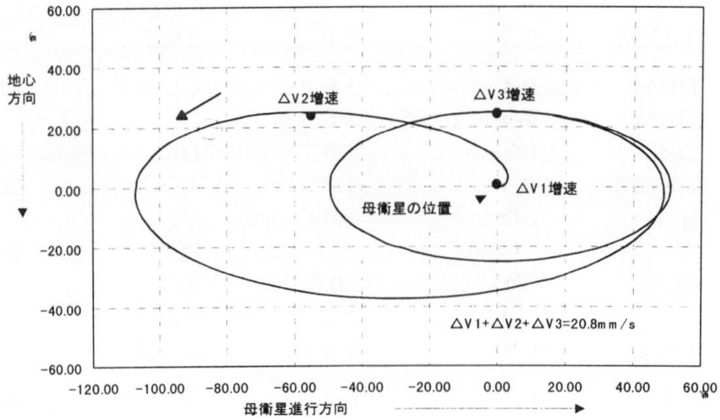

Figure 5. Trajectory of the Nano-Sat.

On the assumption of the case in which the support of the mother spacecraft can not be received, this nano-sat has remote control function from the ground or very advanced autonomy function.

Fixed Point Sun Monitoring At Lagrange Point

This mission carries out the Sun observation from the fixed point at the Lagrange point. Especially L4 and L5 are stabilized in comparison with the other points, and it is proper to observe from the fixed point. And it is possible to observe the sun from about the fourth day before the observation from the earth becomes possible, since L5 is in the rear position of the earth revolution. This is the mission which monitors the solar flare and coronal mass ejection by these utilizing features at L5.

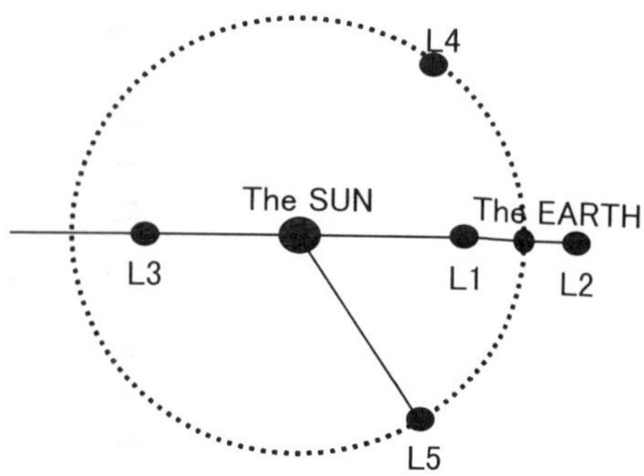

Figure 6. Location of L5 point.

As the main candidate of the mission equipments, the telescope of visible or X-ray is considered, and the weight is made to be 100kg.

The possibility of realizing this satellite which consists of the bus of 25kg, propellant of 25kg,and the satellite weight of 100kg(dry) was estimated, when the satellite was constituted using the above mentioned technology. The mission life is made to be over 5 years.

In addition even in the earth orbiting, very low orbit operations, etc. are examined in which features of miniaturization are utilized

CONCLUSIONS

The concept of 'system on board' was presented. This make it possible to improve the efficiency of the design and manufacturing of satellites, and give flexibility to the configuration in the launch and operation by constructing the system by combining enhanced function boards. And it can constitutes a minimum configuration.

However it is currently the study phase, and more detailed study on elemental technologies, system technologies, and missions, etc. will be planed in the next phase.

THE CONCEPTUAL SYSTEM DESIGN OF THE USERS SPACECRAFT

Koichi Ijichi[1], Tetsuo Yamaguchi[2], Masao Sato[2], Kotaro Kiritani[2], and Kenichiro Sato[2]

[1]*Institute for Unmanned Space Experiment Free Flyer, Shinko Building 2-12 Kanda-Ogawamachi Chiyoda-ku, Tokyo 101 JAPAN*
[2]*Kamakura Works, Mitsubishi Electric Corporation, 325 Kamimachiya Kamakura-shi 247 JAPAN*

ABSTRACT

The objective of USERS (Unmanned Space Experiment Recovery System) project is the establishment of the space experiment system under micro gravity and space radiation environment. The USERS Spacecraft will be launched in 2001 is an element of USERS, and consists of two parts, Service Module (SEM) and Reentry Module (REM). In this paper, the function of Integrated Spacecraft Controller (ISC) are focused on. ISC is the main subsystem of USERS spacecraft, and the new functions for the advanced small satellite are concentrated on ISC.

THE OVERVIEW OF THE USERS SPACECRAFT

USERS is the abbreviation of Unmanned Space Experiment Recovery System which is promoted by the Ministry of International Trade and Industries (MITI) and the New Energy and Industrial Technology Organization (NEDO). And USERS is implemented by the Institute for Unmanned Space Experiment Free Flyer (USEF). The Objective of USERS includes the following two aspects. One is the establishment of the cheap and efficient Recovery system, which will bring the result of space experiments such as the material manufactured under the micro gravity environment. The second is to establish the CFB (Cheaper, Faster, and Better) process concerning the space development system, such as spacecraft system, ground control system and so on.

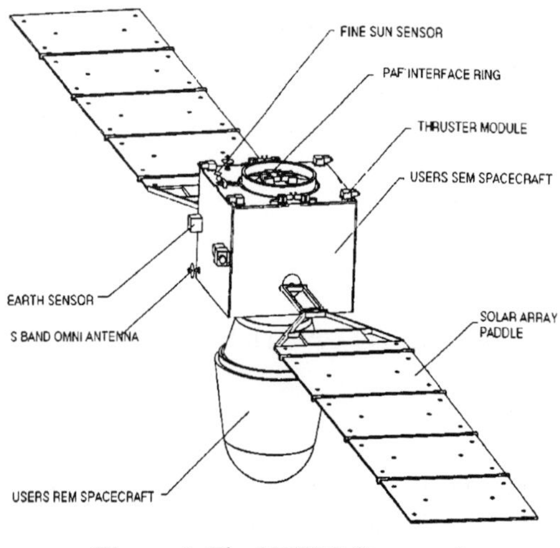

Figure 1. The USERS Spacecraft.

The USERS Spacecraft is shown in Figure 1. The USERS Spacecraft mainly consists of two parts. One is the Reentry Module (REM) , in which the Superconductivity Gradient Heating Furnace Experiment (S GHF) will be performed under good micro gravity environment. This experiment is the most important experiment in the USERS Project. After the SGHF experiment, the manufactured superconductivity material in REM will return to the earth.

Another part of the USERS spacecraft is the Service Module (SEM) which provides REM/SGHF and the other payloads in SEM with power, communication support, and the other resources. The payloads in SEM are called CFB Mission Experiments, which are the near-future space technology experiments promoting CFB process in the space development. The USERS Spacecraft System block diagram is shown in Figure 2. Major Performance of USERS Spacecraft is shown in Table 1.

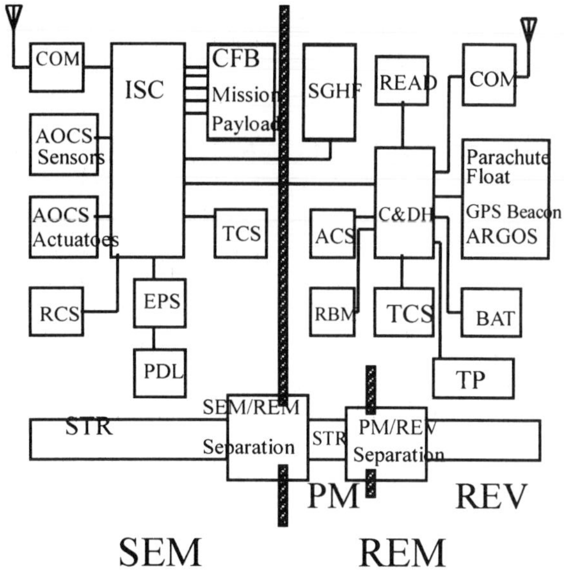

Figure 2. The USERS spacecraft system block diagram.

Table 1. Major Performance of the USERS Spacecraft

Nomenclature	Performance
Launch	A.D. 2001, launched from Tanegasima Space Center with H-IIA launch vehicle
REM Recovering	recovering at the Open Sea around the Ogasawara Islands after 6 months from launch
Mass at Launch	1,600 kg (in case SEM : 800 kg, REM : 800 kg, injection height :450 km)
Operation Orbit	Circular Orbit (Altitude : about 500 km, inclination : 30.4 deg)
Life in-orbit	6 months in case SEM/REM combination mode, 2.5 years in case SEM single mode
Micro Gravity Environment	< 1.0E-5 G (Sun Pointing Mode)

THE ADVANCED FUNCTIONS OF ISC

Over View

In this paper, the functions of Integrated Spacecraft Controller (ISC) are focused on, because the other subsystems also have the unique functions for USERS's Mission but the new functions for the advanced small satellite are concentrated on ISC. ISC consists of four kinds of components, spacecraft controller (SC), Drive Interface Unit (DIU), Data Recorder (REC) and Bus Interface Unit (BIU). SC is the main computer and performs a lot of functions by software. The ISC integrates 3 functions into 1 subsystem, and all functions are performed by 1 CPU. One is the data handling function. The core of USERS Data Handling (DH) function is the command and telemetry processing according to CCSDS recommendation. The Attitude and Orbit Control function maintains the attitude of the spacecraft and changes the orbit of it. And the last and the most important functions in the ISC are system management functions. The details of these system management functions are described in the following sections.

System Management Functions

These functions include the software function, which is formerly realized by hardware circuits and the new automated functions which make the maintenance of the spacecraft easier. And these functions are useful for CFB process of space development.

Heater Control Function. In the USERS, the heaters are designed to be controlled by the software installed in the CPU of the SC instead of the hardware electric circuits or mechanical thermostat. The most useful feature of the software controlled heaters is the capability to change their switching temperatures by commands even after they are set in the satellite. Therefore the switching temperatures can be optimized in response with the real on-orbit thermal environment. This means that the calibration of the thermal model is not necessarily very strict because the error between the thermal model and the real satellite thermal condition can be compensated by changing the switching temperature of each heater. The software heater control lessens the load of the thermal vacuum tests and the thermal analysis based on the test data and reduces the cost of the test and analysis. And of course, the cost of hardware circuits and/or mechanical thermostat for heater control can be deleted.

Payload Power Control Function. The limit of the power for each payload in the USERS Spacecraft determined by ISC is always monitored. And ISC shutdown the power for each payload if ISC finds that the power exceed the limit. This method makes the before-launch analysis of power budget very easy.

The shutdown sequence is shown in Figure 3. The most important feature of this algorithm is that each payload has chance for recovery from the over power condition even if its power exceed the limit.

Battery Charge Control Function. The battery charge condition is generally controlled by so-called V-T Curve which indicates the switching voltage from the constant current charge mode to the constant voltage charge mode corresponding to battery temperature. One battery needs several V-T curves which can be discretely selected according to the battery condition, for example in case of one cell failure.

The battery charge control function is implemented by the software algorithm. One V-T curve is defined by only two parameters because it is actually a line. Therefore it is possible to calculate the switching voltage with the two parameter and battery temperature. The ISC outputs the signal which indicates the switching voltage to the Battery Charge Control Unit (BCCU). The BCCU switches the charge modes according to the signal.

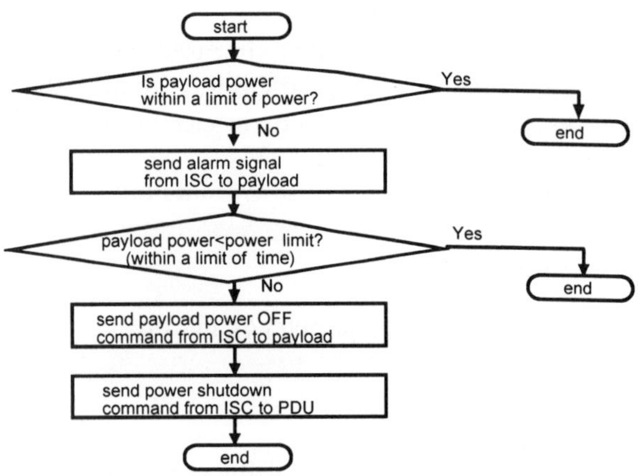

Figure 3. The shutdown sequence.

Changing the battery charge control method from hardware to software makes the hardware size of the BCCU circuits smaller and the test of the BCCU hardware circuits corresponding to V-T curve deleted. In addition, the V-T curve can be set more optimally than discretely selected V-T curve .Hence the condition of the battery is maintained better than controlled by hardware circuits. This software battery charge control function will make the test cases reduced, and the life of the battery longer, because the battery condition about both thermal environment and charge mode is kept better than the conventional hardware control. And these improvement will result in cost reduction.

THE OTHER SCHEMES FOR COST-REDUCTION

The concept of ISC contributes to reduce the total cost of designing, manufacturing and testing advanced small satellites. But there are some following ideas realized in USERS project to reduce the total cost of space development other than the ISC concept.

Automated System Test. The key technology in the system test is the automated system test process, in which the stored test procedure described as electrical text is interpreted and conducted. And the response corresponding to the command is verified automatically.

Infrastructure of System Design. Full system design analysis for the spacecraft can be conducted in a desktop computer and the result can be transferred to the other subsystem designer through the LAN.

Electronic Data Interface. The Interface Control Document (ICD) and other technical information exchanged among the constructors are transferred in form of electronic data instead of complicated drawing data.

3D-CAD Design. 3D-CAD data is utilized to make drawings of the USERS spacecraft. This method will delete the complicated and costly interface between system design engineering and manufacturing process and realize the consistent mechanical design process

CONCLUSION

In this paper, we presented the conceptual design of the USERS spacecraft, especially focusing on the ISC.

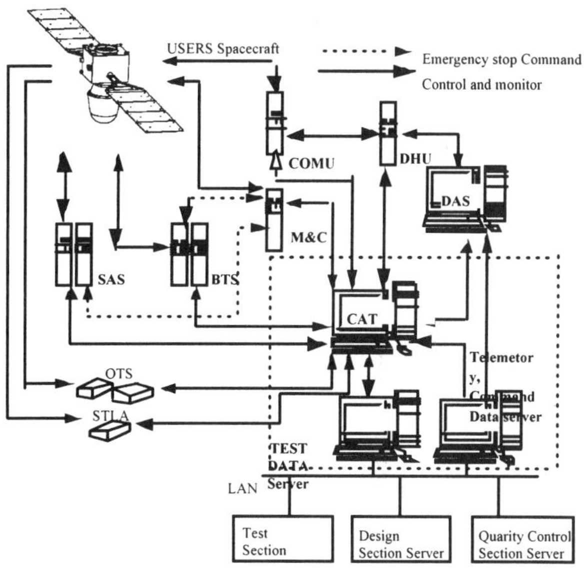

Figure 4. The configuration of automated test system.

The preliminary design of the system has already been started since April in 1997. ISC is the most important subsystem in the USERS spacecraft and many spacecraft engineer are making much effort to develop the first challenging subsystem in Japan. We expect that USERS spacecraft bus system will be the main infrastructure of small satellites for future missions.

REFERENCES

K. Ijichi et al , Unmanned Space Experiment Recovery System, ISCOP-97,C-10 (1997)

M. Sumida et al , Advanced Small Satellite System - USERS - And the Next, 11 the AIAA/USU Conference on Small Satellites (1997)

K. Ijichi et al , Unmanned Space Experiment Recovery System (USERS), 48th International Astronautical Congress, IAF-97-U.1.03 (1997)

AN ATTITUDE CONTROL SYSYTEM DESIGN BASED ON THE TURKSAT-1B GEOSTATIONARY SATELLITE

Ozan Tekinalp[1] , Ö. Uslu[2], Yurdanur Tulunay[1]

[1]*Aeronautical Engineering Department, Middle East Technical University, 06531 Ankara, Turkey*
[2]*Turk Telecom A.S., Satellite Ground Control Station, Golbasi, Ankara, Turkey*

ABSTRACT

Attitude control algorithms for a geostationary satellite which employs a strap down momentum wheel gyro for pitch control and thrusters for roll control is developed. Pitch control is realised using a ladder network type Schmitt triggers with hysteresis and a feedback lag circuit. An effectiveness of the design is demonstrated. Roll control law uses thruster on/off control scheme. The effect of thruster pulse amplitude and pulse duration on fuel consumption is investigated, and a proper pair are given.

INTRODUCTION

It is important to keep a geostationary satellite (for example a communication satellite) at a predetermined position and at a predetermined attitude for the accomplishment of its tasks (Kaplan, Chobotov). Since a satellite is subjected to various disturbing forces and moments, depending on the requirements on acceptable drift limits, there is always a need for station keeping manoeuvres and attitude control. For many satellites attitude control tasks are accomplished through on board actuators such as momentum wheel gyros, and/or thrusters.

In the past many different techniques have been proposed and employed for satellite attitude control (Kaplan, Chobotov). The purpose of this work is to propose control algorithms for satellites similar to TURKSAT-1B (Uslu). TURKSAT-1B satellite uses strap down momentum wheel gyro for pitch control and thrusters for roll control. Consequently, the satellite is assumed to have a strap down momentum wheel gyro for pitch control and thrusters for roll control.

In the next section, equations for rigid body attitude dynamics are given. Then, attitude control algorithms based on the simplified attitude equations are proposed. Results are given and discussed. Possible improvements and future work are given in the conclusions section.

ATTITUDE DYNAMICS

In this section the attitude equations of motion of a geostationary satellite similar to TURKSAT-1B will be developed. Consequently, the satellite is assumed to have a strap down momentum wheel for pitch control and thrusters for roll control.

Three sets of reference frames are used in the development of the attitude equations. An Earth centred inertial reference frame X, Y, Z, where XZ is in the equatorial plane, Earth centred, Earth fixed reference frame, X_E, Y_E, Z_E, and satellite centred satellite fixed body frame, x, y, z. The latter frames are shown in Figure 1. The equations of motion are developed in the body fixed reference frame (x, y, z) which are also along the principal directions. Here, x is in the direction of the motion of the satellite, and y is directed to south. Body angular rates are indicated by $\omega_x, \omega_y, \omega_z$. Then, the angular momentum vector of the satellite is:

$$\mathbf{h} = I_x \omega_x \mathbf{i} + (I_y \omega_y - h_w)\mathbf{j} + I_z \omega_z \mathbf{k} \tag{1}$$

where, **i, j, k,** indicates the body fixed unit vectors, I_x, I_y, I_z are the principal mass moments of inertias, and h_w is the angular momentum of the strap down momentum wheel with respect to the satellite body. From Newton's 2nd law following equations are obtained:

$$\begin{aligned} T_x &= I_x \dot{\omega}_x + (I_z - I_y)\omega_y \omega_z + h_w \omega_z \\ T_y &= I_y \dot{\omega}_y + (I_x - I_z)\omega_x \omega_z - \dot{h}_w \\ T_z &= I_z \dot{\omega}_z + (I_y - I_x)\omega_x \omega_y - h_w \omega_x \end{aligned} \tag{2}$$

In these equations $\dot{h}_w$ indicates the control torque generated by the acceleration or deceleration of the strap down momentum wheel.

Euler angles are generally used to indicate the attitude of a satellite with respect to an inertial frame (Rimrott). A rotation sequence, θ_1, ϕ_1, ψ_1, indicating pitch, roll and yaw respectively, is employed in this work. Similarly θ, ϕ, ψ, indicates the sequence of rotations between Earth fixed frame and body fixed frame. Between these Euler angles the following identities hold:

$$\phi_1 = \phi \quad ; \quad \theta_1 = \theta - \omega_0 t \quad ; \quad \psi_1 = \psi \tag{3}$$

where, ω_0 is the Earth rotational speed (7.28 e-5 rad/s).

It is well known that the transformation between Euler rates ($\dot{\phi}, \dot{\theta}_1, \dot{\varphi}$) and body angular rates ($\omega_x, \omega_y, \omega_z$) are non-linear. However, since the satellite is to be kept at a fixed orientation with respect to Earth, that is, ϕ, θ, ψ are small, these transformations may be linearized.

$$\omega_x = \dot{\phi} - \psi \omega_0 \quad ; \quad \omega_y = \dot{\theta} - \omega_0 \quad ; \quad \omega_z = \dot{\psi} + \phi \omega_0 \tag{4}$$

Externally applied torques are control torques, as well as disturbance torques. In this work only gravity gradient and solar radiation pressure torqes are considered. Detailed expressions for these disturbances as well as others are given in various references (Chobotov, Rimrott). In this work, following simplified expressions are used (Kaplan):

$$\mathbf{T}_{sp} = 2 \cdot 10^{-5}(1 - 2 \cdot \sin \omega_0 t)\mathbf{i} + 10^{-4} \cdot \cos \omega_0 t \cdot \mathbf{j} - 5 \cdot 10^{-4} \cos \omega_0 t . \mathbf{k} \tag{5}$$

$$\mathbf{T}_{gg} = -3 \cdot \omega_0^2 (I_y - I_z)\phi \cdot \mathbf{i} - 3 \cdot \omega_0^2 (I_x - I_z)\theta \cdot \mathbf{j} \tag{6}$$

Similarly, nonlinear equations of motion given in Eq. 3 may also be linearized using Eq. 5 and neglecting powers of Euler angles and rates. Together with the gravity gradient torque given in equation 7 linearized equations of motion are:

$$
\begin{aligned}
T_x &= I_x\ddot{\phi} + ((I_y - I_x - I_z)\omega_0 + h_w)\dot{\psi} + (4(I_y - I_z)\omega_0^2 + h_w\omega_0)\phi \\
T_y &= I_y\ddot{\theta} + 3\omega_0{}^2(I_x - I_z)\theta - \dot{h}_w \\
T_z &= I_z\ddot{\psi} - ((I_y - I_x - I_z)\omega_0 + h_w)\dot{\phi} + ((I_y - I_x)\omega_0^2 + h_w\omega_0)\psi
\end{aligned} \tag{7}
$$

These last set of equations are used for the development of the control laws given in the next section. However, note that under the assumptions used, the pitch dynamics is uncoupled from the roll-yaw dynamics. Consequently, the control laws for pitch and roll/yaw dynamics will be developed separately.

ATTITUDE CONTROL SYSTEM

For the development of control algorithms certain hypothetical values are selected which are representative of a geostationary satellite such as TURKSAT-1B. These values are: $I_x = 3600$ kg.m^2 ; $I_y =$ 600 kg.m^2 ; $I_z = 4000$ kg.m^2; $h_w = 60$ N.m.s.

Pitch Control Algorithm

A close analysis of the linearized pitch dynamics (Eq. 7) shows that, since the coefficient of θ is small and negative, we obtain a symmetric pair of real roots very close to the origin. Such systems are usually stabilised employing rate feedback. However, for this geostationary satellite, only angle information is available through infrared earth sensors. Consequently, there is a need to derive the rate information from attitude measurements. In the literature derived rate modulators are proposed for this purpose (Anthony et al., Kaplan, Piper and Kwathney). Here a Schmitt trigger with hysteresis and a feedback lag circuit is used. The feedback lag circuit has the following form:

$$
\frac{1}{K(\tau s + 1)} \tag{8}
$$

It is shown that this kind of derived rate modulator is basically a phase lead device (Anthony et al.). However, it is customary to use a single Schmitt trigger with a given activation level. In this work a ladder network type Schmitt trigger with different activation levels is used to provide a smoother response to the system. Response of the derived rate modulator, with a Schmitt trigger that has ten different activation levels, to a sinusoidal input is given in Figure 2. The Simulink block diagram of the system and accompanying controller is given in Figure 3.

Roll Control Algorithm

The linearized equations above (Eq. 7), predicts a marginally stable roll/yaw dynamics for the satellite under investigation. Due to the weak coupling between roll and yaw dynamics, a beating phenomena occurs between roll and yaw in an uncontrolled satellite. In TURKSAT-1B satellite roll control is achieved using onboard thrusters. Since thrusters cannot generate continuously varying thrust values it is customary to use thruster on/off control technique. To increase stability a phase lead compensation technique is required. Many investigators proposed derived rate modulators for roll/yaw control as well (Chobotov, Kaplan).

It is well known that the life of a geostationary satellite is mainly determined by the amount of fuel available for the thrusters. Consequently, proper choice of thruster amplitude and duration has an important bearing on fuel consumption. To demonstrate this situation, a double integrator roll dynamics is assumed here. Then a simple roll control algorithm is used. Thus, proper thruster pair is switched on for a predefined duration when the satellite roll attitude error exceeded prescribed limits. This situation will be discussed in the following section, and proper values that results in sufficiently low amount of fuel consumption will be given. The Simulink block diagram for the thruster on/off control law and the satellite dynamics is shown in Figure 4.

RESULTS AND DISCUSSION

Pitch Control System Results

TURKSAT-1B satellite is located at the 42° longitude. A nominal pitch angle bias of –0.8336° is required due to the requirements on the coverage zones. It is also required that the satellite is to be kept within ±0.05° of this bias value. As indicated above pitch dynamics transfer function is unstable with roots very close to origin. Sufficient amount of phase lead compensation will be required to make the system stable. Since there are two design variables, time constant τ and gain K, after some trail and error, τ is set to 100s and gain K is varied. The amplitude of the response was lower when higher gain values were used as expected. Two such simulation examples for gain values of 0.35 and 0.08 may be observed from Figures 5 and 6 respectively. In these plots the sinusoidal variation of the output is due to the cyclic variation of the solar radiation pressure disturbance (Eq. 6) included in the simulaticns. While the simulation in Fig. 5 shows that the pitch angle dynamics is within the limits, at a lower gain value (0.08) (Fig. 6) the amplitude of the response exceeds –0.8336±0.05° limit. A gain value of 0.17 keeps the system within –0.833±0.04° limit (Fig. 7). At this gain value corresponding speed fluctuation of the momentum wheel is observed to be between 4425 rpm and 4675 rpm and decided to be acceptable. The same gain value (0.17) is also tested for transient response. Fig. 8 shows that an initial 0.05° pitch angle error settles down in about 300 s with a maximum overshoot of 0.01°.

Roll Control System Results

Roll control of TURKSAT-1B is realised using thruster on/off control. The requirement is such that the roll angle is to be kept within 0.0±0.05° limit. To prevent overshoots thrusters are fired whenever roll angle exceeds ±0.04°. Simulation results for pulse amplitude 0.5 N.m. and pulse width of 1 s is given in Figure 9. The simulation shows that there is a two much thruster activity and fuel consumption. Total fuel consumption is proportional to the sum of all the impulses generated in a given period of time. The accumulated control impulse for a 5000 s simulation is plotted in Fig. 10. A similar simulation for roll angle with a pulse amplitude of 0.005 N.m and pulse duration of 10 s is given in Fig. 11. Corresponding accumulated control torque is also plotted in Figure 12. Between these two extremes other torque impulse and duration pairs are also simulated and tested. The results are given in Table 1. From the table it can be observed that for the same torque impulse value (ie. 0.08 N.m.s) the pair with lower torque pulse amplitude gives smaller accumulated control torque. However, it should be noted that, a thruster selection is not only based on attitude control considerations. Thrusters are also used for orbit correction, such as East or North manoeuvres. Consequently the thruster pulse amplitude should also be suitable for such manoeuvres.

CONCLUSION

An attitude control system based on the TURKSAT-1B geostationary satellite is designed. The pitch control of the satellite is realised using a strap down momentum wheel. Pitch control algorithm uses derived rate modulator with a Schmitt trigger. In this design instead of a single trigger, a ladder network type Schmitt triggers are used in the feed forward path. It is shown that a rather smooth and effective roll control is achieved under the cyclic solar radiation pressure torque. Roll control system uses on board thrusters. Thruster on/off control scheme is proposed for roll control. It is shown that the pulse amplitude and pulse duration has direct bearing on the fuel consumption of the satellite. Proper pair of pulse amplitude and pulse duration is selected and the effectiveness of the algorithm under the cyclic solar radiation pressure is demonstrated through simulations.

In this work, simplified attitude dynamics is used. Nonlinear dynamics together with flexible modes should also be considered, and proper control algorithms shall be developed. Especially for roll/yaw dynamics, controller gains should be optimised to reduce fuel consumption. These issues shall be addressed in a future article.

ACKNOWLEDGMENTS

This work is a part of an integrated research on satellite orbit and attitude dynamics supported by the State Planning Office of Turkey. The authors would also like to mention the support of the Satellite Ground Control Station of Turk Telecom A. S..

REFERENCES

Anthony, T. C. , Wei , B. , and S. Carroll, Pulse Modulated Control Synthesis for a Flexible Spacecraft, Journal of Guidance, Control and Dynamics, **13**, 6, pp. 1014-1022 (1990).

Chobotov, V. A. , Spacecraft Attitude Dynamics and Control, Krieger Publishing Co., Malabar, Florida, (1991).

Kaplan, M. H. , Modern Spacecraft Dynamics and Control, John Wiley and Sons, (1976).

Piper, G. E. , and H. G. Kwatny, Complicated Dynamics in Spacecraft Attitude Control Systems, Journal of Guidance Control and Dynamics, 15, 4, pp. 1018-1028, (1992).

Rimrott, F. P. J. , Introductory Attitude Dynamics, Springer-Verlag, New York, (1989)

Uslu, Ö, Orbit Dynamics, Attitude Dynamics and Control: Investigation into Possible Applications to TURKSAT, MS. Thesis, Middle East Technical University, Aeronautical Engineering Department, (1997).

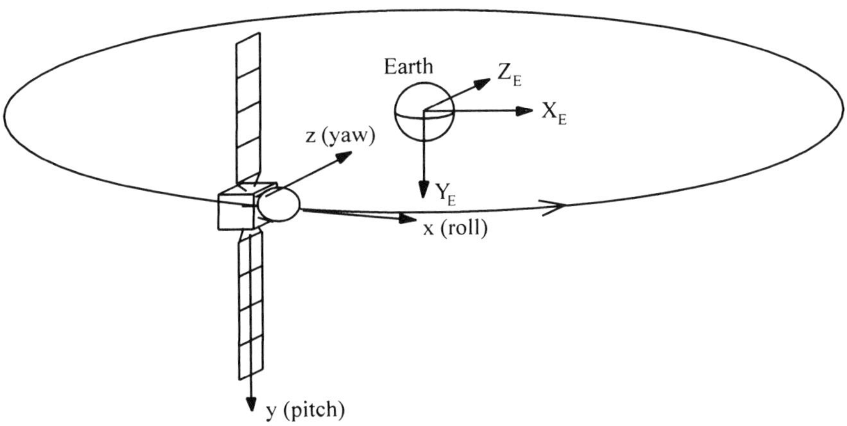

Figure 1. Earth fixed Earth centred reference frame and body fixed reference frame.

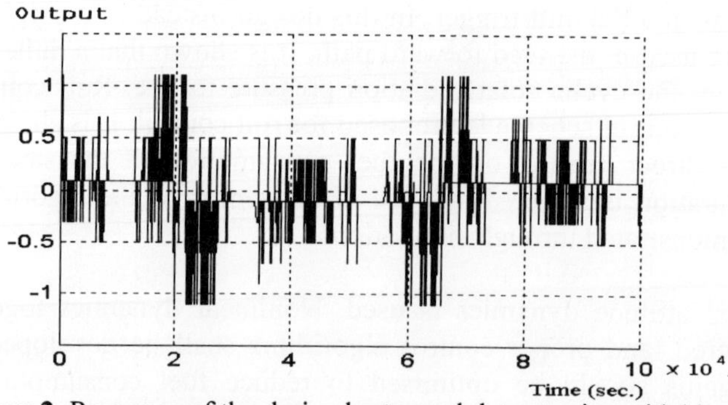

Figure 2. Response of the derived rate modulator to sinusoidal input.

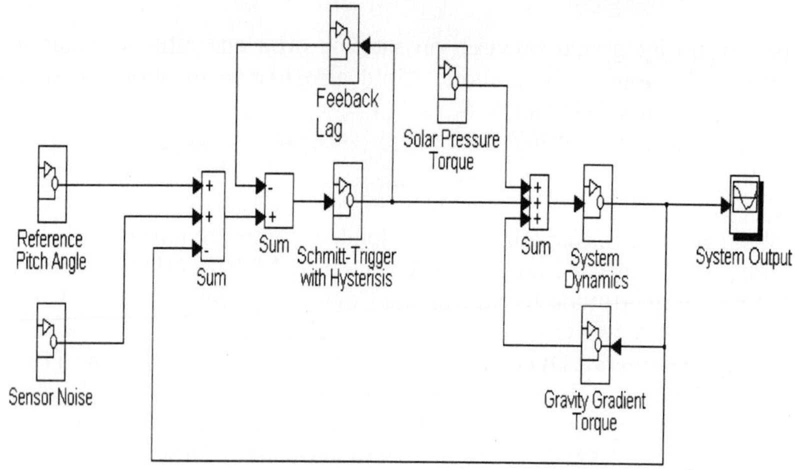

Figure 3. Pitch control system block diagram.

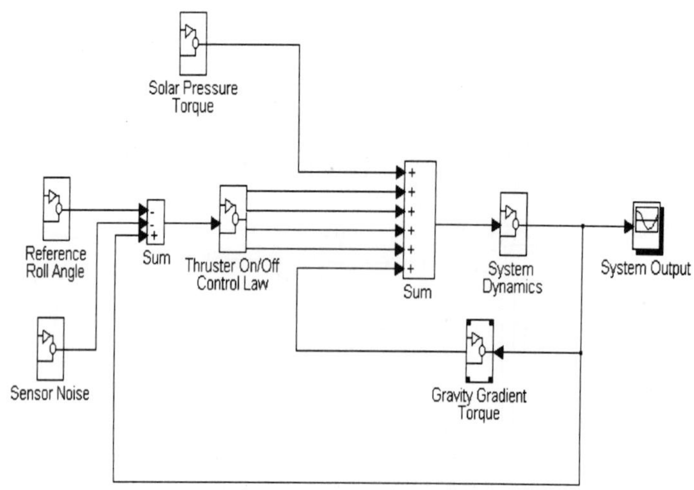

Figure 4. Roll control system block diagram.

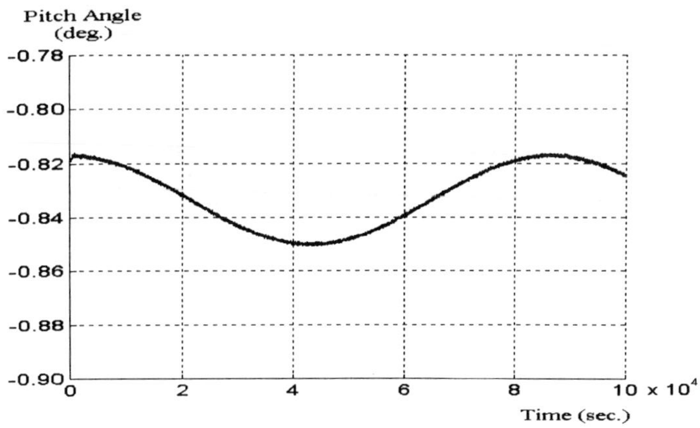

Figure 5. Time response of the pitch control system ($K = 0.35$).

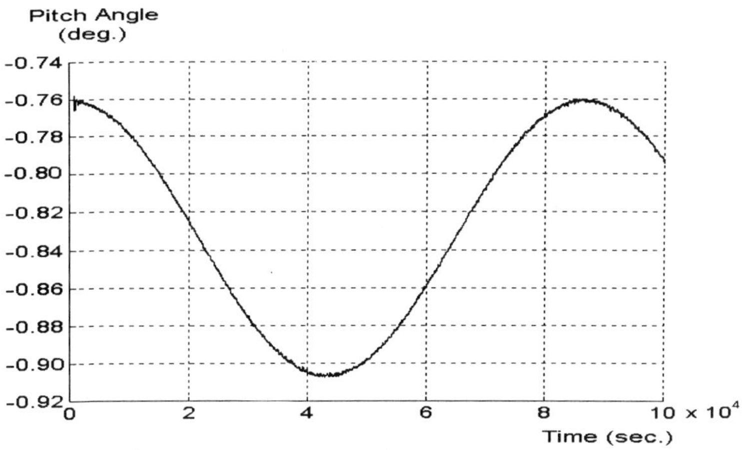

Figure 6. Time response of the pitch control system ($K = 0.08$).

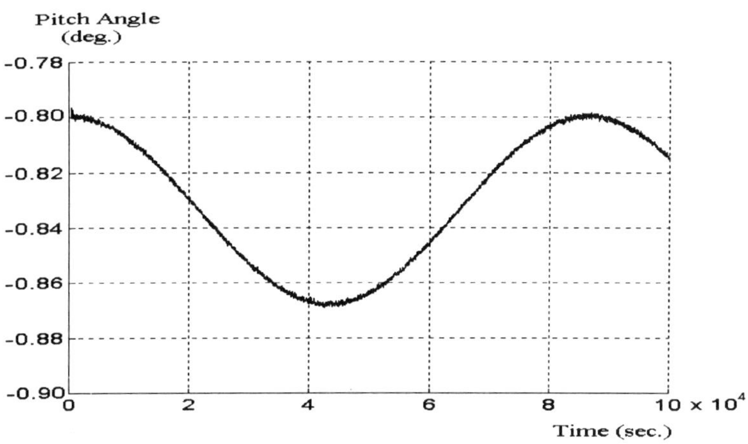

Figure 7. Time response of the pitch control system ($K = 0.17$).

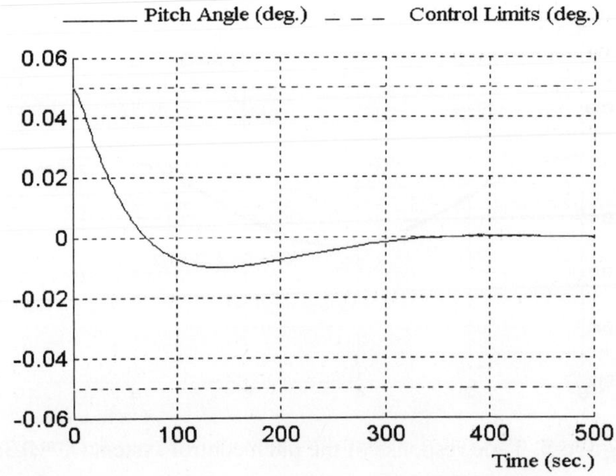

Figure 8. Time response of the pitch control system to initial pitch angle error of 0.05° ($K = 0.17$).

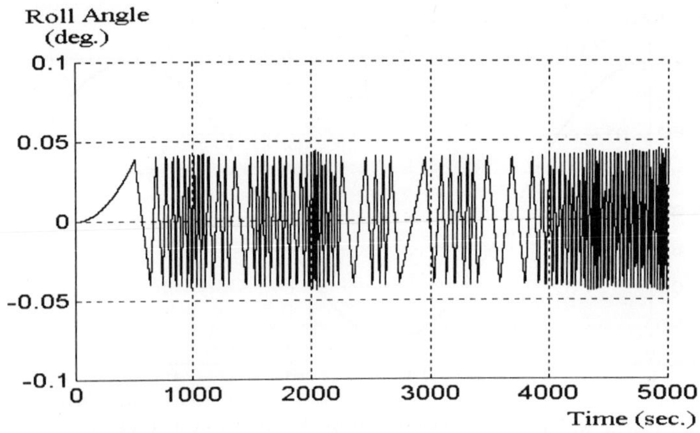

Figure 9. Time response of the roll control system (pulse width=1 s, pulse amplitude=0.5 N.m).

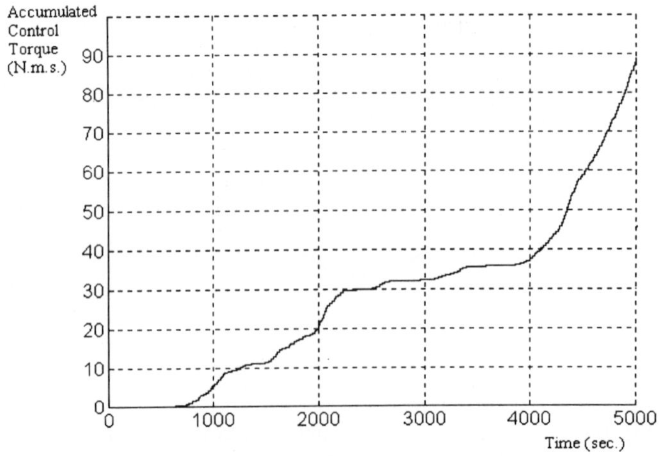

Figure 10. Accumulated control torque (pulse width=1 s, pulse amplitude=0.5 N.m).

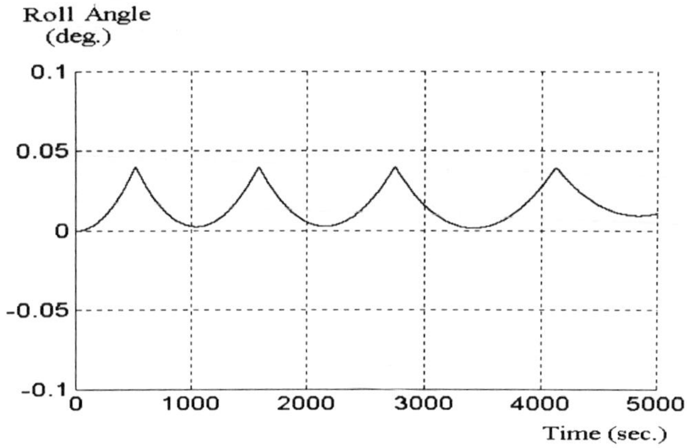

Figure 11. Response of the roll control system (pulse width=10 s, pulse amplitude = 0.005 N.m).

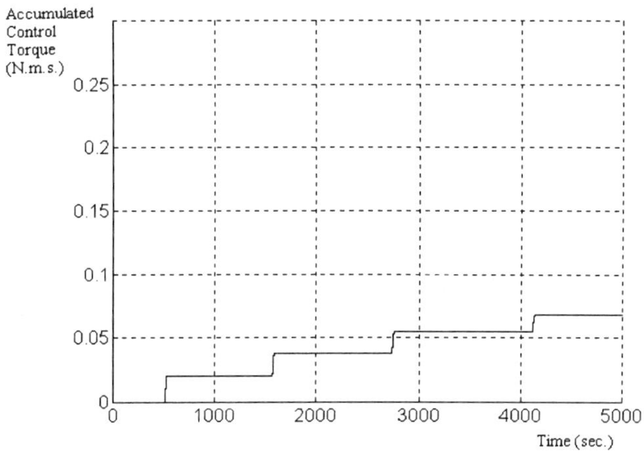

Figure 12. Accumulated control torque (pulse width=10 s, pulse amplitude=0.005 N.m)

Table 1. Accumulated Control Torque for Different Pulse Widths And Pulse Amplitudes

Control Torque Pulse Width (s)	Control Torque Pulse Amplitude (N.m.)	Torque Impulse (N.m.s)	Accumulated Control Torque (N.m.s.)
0.01	0.15	0.0015	0.2283
0.01	0.5	0.005	0.1798
0.01	0.6	0.006	0.1474
0.02	0.04	0.0008	1.3990
0.02	0.3	0.006	0.7946
1	0.003	0.003	0.0739
1	0.08	0.08*	0.6075
1	0.5	0.5	88.711
2	0.0015	0.003	0.0613
2	0.05	0.1	0.0592
2	0.07	0.14	3.9805
2	0.04	0.08*	1.2987
5	0.016	0.08*	0.0846
5	0.012	0.06	0.0736
10	0.008	0.08*	0.073
10	0.005	0.05	0.0678

MICROSATELLITES POWER SUPPLY SYSTEMS,PECULIARITIES AND METHODOLOGY OF DESIGNING

Yurii N. Agafonov[1], Vladimir V. Khrapchenkov[1], Sergey P. Shpakov[2] and Jaroslav Vojta[3]

[1] *Space Research Institute, Russian Academy of Sciences, Profsoyuznaya Str., 84/32, 117810 Moscow, Russia. E-mail: yagafonov@romance.iki.rssi.ru, vkhr@mx.iki.rssi.ru*
[2] *Moscow Power Engeneering Institute, Krasnokazarmennaya Str., 14, 111250, Moscow, Russia*
[3] *Institute of Atmospheric Physics, Academy of Sciences of Czech Republic, Bochni II, No. 1401-14131 Prague 4, Czech Republic. E-mail: jvo@ufa.cas.cz*

ABSTRACT

The present paper describes the peculiarities and methodology of small space vehicles electric power supply system (PSS) optimum designing which allows to decrease the time and cost of designing as well as to improve the operation characteristics of the system as well. The suggested solution of this task is based on the developing, manufacturing, testing and operating experience of the power supply systems for spacecraft (s/c) of the Moscow Power Engineering Institute (MPEI), Space Research Institute of Russian Academy of Sciences (IKI) and Institute of Atmospheric Physics of the Academy of Sciences of Czech Republic (IAP ASCR). The paper presents some aspects of PSS development, PSS computer models development and application, aspects of solar array (SA) parameters and buffer batteries (BB) calculation and of modes of their optimum operation.

The main materials of this paper were obtained from working up the PSSs for Interball international space project microsatellites: Magion-4 and Magion-5, namely. The first microsatellite operated successfully in the high elliptical orbit for more than two years. These microsatellites were manufactured in the IAP ASCR, all information is available on: http://www.iki.rssi.ru/interball.html (from IKI RAS) and http://www.ufa.cas.cz (from IAP ASCR).

INTRODUCTION

For the last years one of the perspective direction in the field of the space applications becomes the development and the manufacture of small (by mass and sizes) artificial satellites for different purposes (communcation, navigaion, scientfic) and vehicles of a light class rather cheap and capable to eject these small satellites into a requred orbit. In particular, there is a tendency to expand the manufacture of small satellites - microsatellites with the mass of up to 100 kg and minisatellites with the mass of 100-500 kg. As the practice shows, such s/c turn out to be rather economic while realizing low-scale research missions, different monitorng programs, etc.

Lower expenditures on development, tests and manufacture, and also on launching into space, the possibility to eject several s/c into the orbit simultaneously and also a high reliability of orbital systems of different purposes consisting of a great number of satellites refers to the advantages of small satellites in comparison with traditional large-size and expensive s/c. Besides, the cost of the manufacturing and launch of minisatellites does not exceed $7-10 mln., and of microsatellites - $2-2.5 mln. with rather high reliability and quick time of development (12-18 months), not mentioning a hitch-hike launch which cost may turn out essentially lower than the given figers [1].

DESIGN PECULIARITIES OF THE POWER SUPPLY SYSTEM OF SMALL SATELLITES

The restrictions on sizes and mass dictate the certain technology of the development of systems and parts of small satellites. As a rule, small satellites are being manufactured in the form of polyhedron, prism or cylinder with stationary solar arrays (SA) mounted on the body or, with opening and then fixing solar panels if there are restrictions on sizes during the launch.

Silicon photocells with the thickness from 20 up to 200 μm and nickel-cadmium sealed buffer batteries are used aboard the majority of modern small satellites. The uncontrolled structure with a direct transfer of energy to the load (the so called "buffer structure") based on the parallel switch of SA, BB and the load is typical to the power supply system of small satellites. The output voltage of such PSS is defined by the range of the BB voltage change with its charge and discharge, which can be sufficiently high. The control of the charge current and the protection against the BB recharge are achieved commonlz due to switching off or shunting of the whole SA or its part, more seldom - due to the use of special continuous or discrete regulators.

The simplicity of the design predetermines low specific indices of PSS. Consequently, the finding of optimum solutions becomes especially important at the stage of the PSS design, based on the detailed analysis of the totality of initial data on purposes of launch and goals of s/c, the orbit parameters, required resource, cyclogram and the load type. On this basis there should be based the choice of the structure, composition and characteristics of elements, algorithms of the PSS operation and parameters. The use of SA, BB and electronic unites, manufactured by modern technologies and having the best characteristics at present is also very important.

Military, scientific and some communication (e.g. passive retranslation) tasks can be solved with the use of small satellites. The scope of solved tasks defines the parameters of the orbit into which the s/c is launched.

Small satellites are launched into low near-Earth circular orbits, middle and high elliptic orbits with a large eccentricity. The authors do not have data on the using of small satellites in geostationary orbits.

The most wide-spread orbits of small satellites are low circular near-Earth orbits (300-600 km) and middle elliptic orbits with a large eccentricity (the perigee is 250-600 km, the apogee - 4000-6000 km or up to 200000 km). The orbit inclination depends on the solved task.

The peak power of the load for the majority of microsatellites is in the range of 50-75 Wt, the load type is variable part with a constant part (intrinsic consumption of support systems). The consumption power of support systems is, as a common, 10-15% of a peak one.

The described general features of microsatellites make it possible to realize the certain approach to the creation of its different systems including PSS [1, 2].

SCHEMES OF MICROSATELLITES` PSS

The buffer PSS refers to the group of uncontrolled PSS with the direct energy transfer to the load. Figure 1 shows the simplest structural scheme consisting of solar arrays (SA), buffer batteries (BB), switch (S), power switching unit (PSU) and load (L). The main drawback of this system is a low coefficient of the use of SA power, as for some time of the period solar arrays turn out to be switched off from the load that is irrational for high orbits with a low duration of a shadow area.

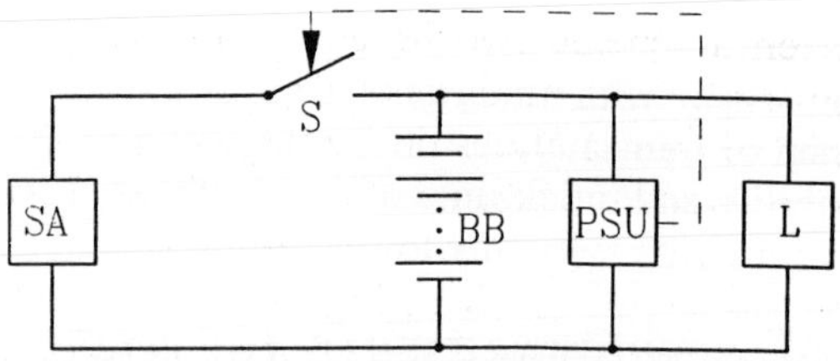

Figure 1. The simplest structural scheme.

The scheme shown in Figure 2 provides the best use of solar arrays. In this case solar arrays are manufactured sectional. Depending on the load current and the charging status of the buffer battery the sections of solar arrays are switched on or off.

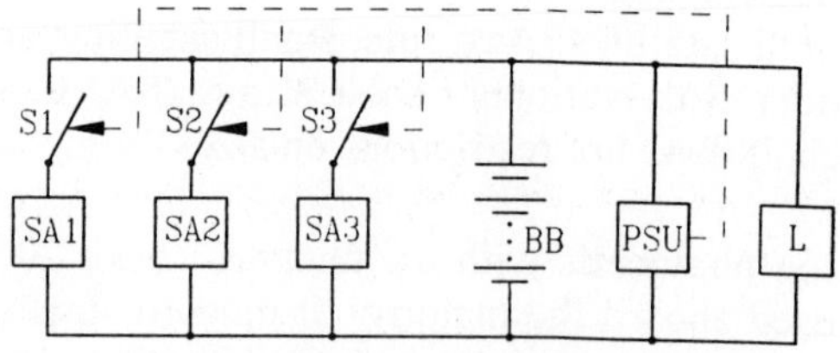

Figure 2. The simplest structural scheme with sectional solar arrays.

Figure 3 shows the structural scheme of PSS where the switching off of solar arrays with a full charge of accumulators occurs not because of the switching off but because of its short-circuit by S key. Such a mode is not dangerous for solar arrays as a short-circuit current exceeds a nominal one by 10-15%. The diode prevents the buffer battery short-circuit if the solar array short-circuit. Instead of S key the controlled shunt can be used which is switched at the moment of the end of the charge and which supports the voltage on the bus in such a way that accumulators are in on-duty mode.

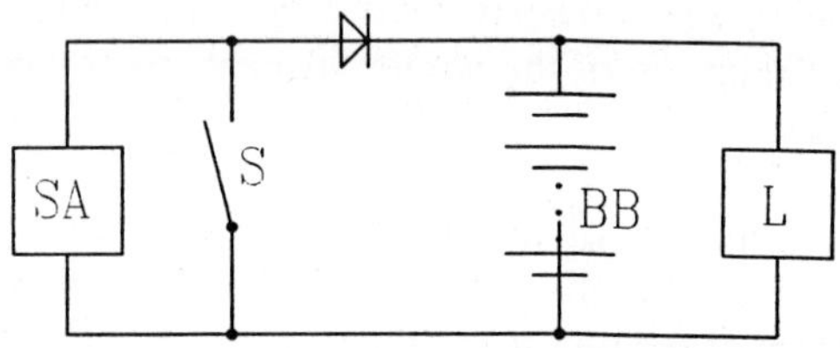

Figure 3. Solar arrays are switching off by short-circuit.

In the scheme given in Figure 4 the protection of accumulators against the recharge and consumers against the excess of voltage with the switched off buffer battery is achieved by the shunt regulator (SR). With a correct choice of the restriction voltage such a system makes it possible to support accumulators at low loads in the mode of recharge by low currents providing a high degree of its readiness. To improve the operation in a wide range of the temperature changes or to provide the forced charge of accumulators the shunt regulator can have two or more levels for the restriction of the voltage at different levels.

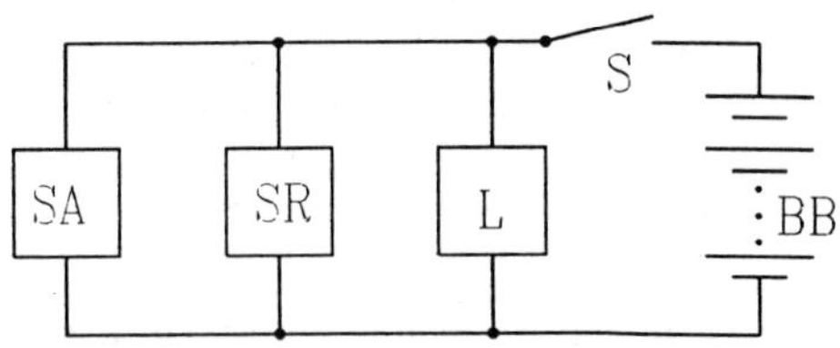

Figure 4. The scheme with the shunt regulator (SR).

The given structures are intended for the work with one buffer battery, in this case the redundant battery is switched on only in case of the main battery failure. The certain advantages can be obtained using the scheme of Figure 5 in which two buffer batteries operate simultaneously.

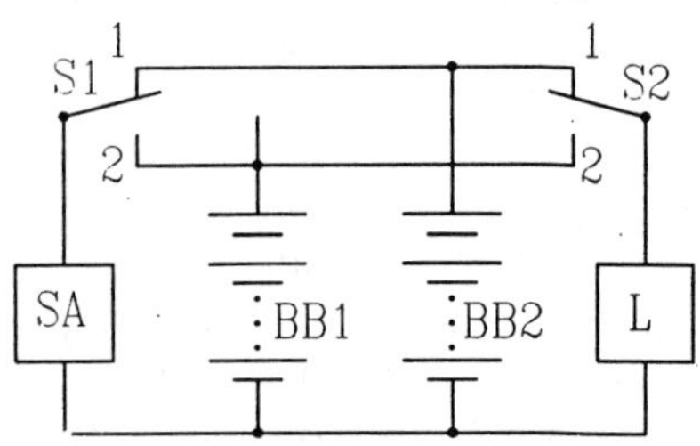

Figure 5. Two buffer batteries operate simultaneously.

In such a scheme two buses of load are generated and with each of them their storage batteries are connected. The connection of solar arrays and load with this or that bus can control the modes of the storage battery, support the voltage at the load in the certain limits, provide a high degree of the system readiness.

Such a scheme provides a high coefficient of the use of solar arrays and storage batteries. The following modes are possible in it:

- the SA operation for the load and the charge of one BB;
- the simultaneous operation of SA and one BB for the load;
- the charge of one BB from SA and the load power from the second BB. The given mode has the advantages, e.g., it gives a possibility to carry out a deep discharge of the accumulator for restoring its characteristics, but in general case it is not energetically advantageous as all SA power passes via BB.

The structures given above do not exhaust the variety of PSS creation, however, they are more simple and can be successfully used in low-powered systems with low requirements to the quality of the power voltage. Due to the studying of the PSS versions for satellites of the MAGION type there was recommended the scheme described in the next section (see Figure 6) .

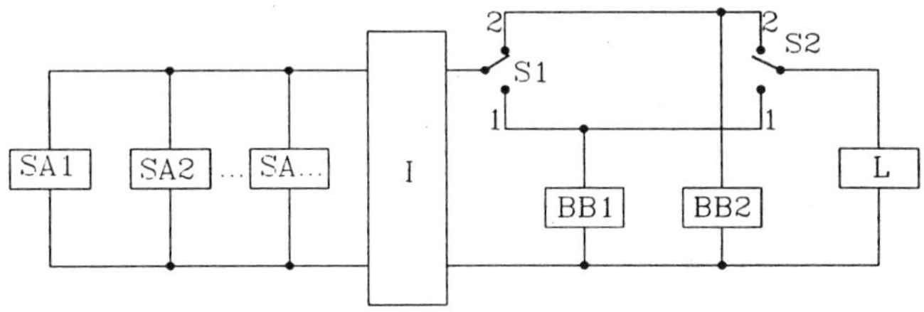

Figure 6. gives the structural scheme of such PSS.

CALCULATION OF PSS PARAMETERS OF SMALL SPACECRAFT

For about 10 years the Space Research Institute and Moscow Power Engineering Institute are working on the analysis of the different s/c PSS operation in orbit and of the PSS design automation. In particular, the experts of these institutes developed the packet of PC programs using at the design stage and automatizing the small satellites PSS development.

This packet of programs uses the standard set of initial data given below as an example. It realizes not only the mathematical approach but the many years' experience of experts of these institutes in the choice of scheme solutions and the following analysis of the PSS operation in a great number of missions, and it allows obtaining the set of parameters in the output required for the choice of PSS elements, such as, for example:

- the duration of one revolution including the duration of a shadow area;
- the required square of SA;
- the number of parallel and successively switched photo-cell SA;
- the maximum depth of BB capacity cycling;
- the dependence of SA power on time;
- the dependence of the BB capacity cycling depth on time;
- the dependence of BB current and voltage on time;
- the dependence of SA current and voltage on time.

As an example, for the initial data on the PSS parameters calculation there was taken a hypothetic microsatellite launched into the circular orbit with the height of 350 km (widely used) with panels of solar arrays fixed on the satellite body switching on periodically the whole complex of equipment to perform simultaneous measurements of parameters of the environmental plasma (the version of the microsatellite for the solution of a limited scientific problem):

- the orbit height is 350 km;
- the orbit inclination is 50°;
- the load cyclogram is shown by Table 1.

Table 1. The load cyclogram.

Interval #	Duration (min)	Power (Wt)
1	10	2.5
2	5	35
3	10	2.5
4	5	35

and so on during the orbit period.

- the SA optimum voltage is Uopt = 18 V;
- the SA operating voltage Uopr = 16 V;
- the Earth albedo is 0.35;
- the absorption coefficient of the photo-cell working side is 0.84;
- the absorption coefficient of the photo-cell back side is zero;
- the radiation coefficient of photo-cell working side is 0.87;
- the radiation coefficient of the photo-cell back side is 1.00;
- the mass heat capacity is 3600 $J/m^2 \cdot °C$;
- the coefficient of the SA degradation by current and voltage is 0.95.

Following the calculation by the given above initial data there are calculated the additional parameters of the orbit revolution (in the given case the duration of one revolution is 91.5 min with the duration of a shadow area of 33.1 min) and the following parameters of primary and buffer sources of the PSS electrical power:

Solar Array:

- the SA active square is 0.36 m^2;
- the SA full area (taking into account the coefficient of panels filling 0.9) is 0.4 m^2;
- the number of successively (parallel) switched in photo-cell chain is 42(18).

Buffer Storage Battery:

- the most suitable type of accumulators is Ni-Cd hermetic;
- the number of accumulators in the battery is 10;
- the maximum depth of BB cycling is Q = 3.2 A·h;
- the BB nominal capacity is Q=7 A·h .

For a similar scheme of PSS there are proposed three main modes of operation:

1. One buffer battery is operating (BB2 is a main battery, BB1 is a spare one). This operation mode is used when keys S1 and S2 are in position 2.
2. The alternative operation of buffer batteries: one battery is switched on only when another one is switched off as a result of a charge or discharge prohibition. Let us assume that there is available a

charge mode and S1 and S2 are in position 1, the power SA goes to BB1 and to the load. If BB1 is completely charged then keys S1 and S2 are switched on to position 2 and BB2 starts being charged till it is completely charged or till the mode of discharge comes. Then in the mode of discharge BB2 will be discharged first until its capacity or voltage comes to a minimum allowed value. After this keys S1 and S2 are switched on to position 1, thus disconnecting BB2 from the charge and connecting BB1 for the charge.

3. The simultaneous operation of buffer batteries: this is a mode of operation when one battery is being charged from SA and another one is being discharged for the load. Let us assume that key S1 is in position 1 and key S2 is in position 2. Then, BB1 is being charged from SA till it is completely charged or till BB2 is completely discharged. When the capacity or voltage BB2 reaches its minimum allowed value, the switching on of the batteries, BB1 - for a discharge, and BB2 - for a charge, takes place.

To provide BB charge and parallel operation of BB and SA it is required to introduce a control unit that will fulfil the following:

- limit the voltage of BB charge;
- control PSS operation characteristics and telemetry signals generation;
- send control signals for PSS switch devices;
- protect BB against recharge or redischarge;
- level parameters of BB elements (accumulators) during multitime cycling (one of the possible schemes of a levelling device is presented in Figure 7);
- provide restoring cycles.

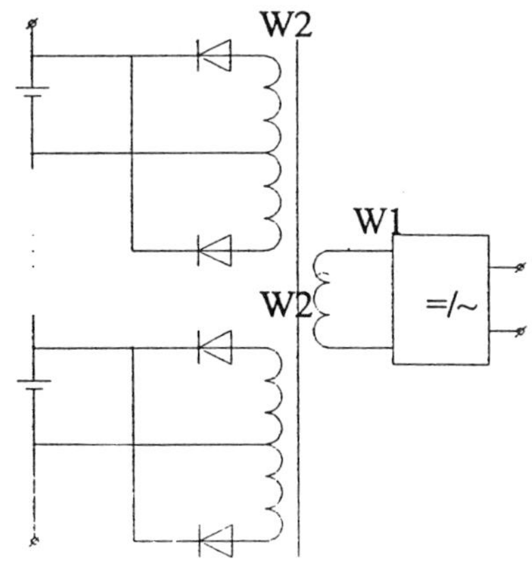

Figure 7. One of the possible schemes of a levelling device.

BUFFER DEVICE BASED ON NICKEL-CADMIUM HERMETIC STORAGE BATTERIES (NCHSB)

The wide-spread use of nickel-cadmium germetic storage batteries as buffer power storage devices is provided by lifetime, resource, specific power capacity and power, as well as by rather simple requirements to the charge-discharge control high-rate parameters.

The NCHSB lifetime and use efficiency are greatly determined by the mode of its operation. Mainly the resource is determined by the depth of the cycling as well as by a number of some important parameters, such as:

- points of charge and discharge;
- working temperature;
- choice of a charge method;
- choice of the limitation moments of charge and discharge;
- choice of the method to decrease degrading or restoring of parameters.

Below there are presented the most interesting aspects of the choice of nickel-cadmium accumulators for microsatellites MAGION-4 and MAGION-5 of the Interball international space project [3, 4].

There were performed tests of storage batteries of RSQ and RSH type (Germany), SAFT type (France) and NKGTz (Russia) for the Interball project. The purpose of these tests was to check the operation ability of the storage batteries, to investigate and compare parameters for different modes of charge and discharge and of working conditions to choose storage batteries for buffer PSS of the microsatellites. The tests were performed in IKI thermal-vacuum chamber under the conditions near to real flight ones.

Figure 8 shows the dynamics of the storage batteries properties changing based on the tests results. The Figure present charge-discharge storage batteries characteristics according to the cyclogram of Table 1.

The Figure 8 shows that the best parameters during the tests were presented by the RSQ (VARTA) type storage batteries. They were recommended to be used on the Interball subsatellite.

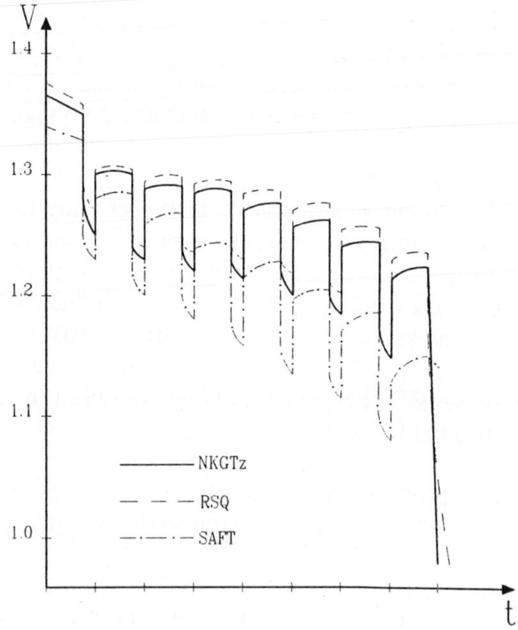

Figure 8. The dynamics of the storage batteries properties changing based on the tests results.

Below one can see the most general conclusions of the comparative tests results:

- The charge mode of a falling current, closest to the BB operation mode in a buffer PSS, provides a BB charge, corresponding to the discharge capacity of not less than 85% in the whole range of nondischarging currents and environmental conditions. After this, the further increase of the charge time is practically nonefficient;
- The storage batteries of RSQ and NKGTz types can operate under the conditions of up to minus 40°C with a short time impact of charging currents of not more than 7A and 4A, respectively, and with the discharge capacity of 85% and 75%, respectively;
- The protection against recharge on voltage is not recommended for the storage batteries under investigaton, as the increase of the voltage limitation level for a completely discharged accumulator causes the growth of a charging current, that may cause the increasing of pressure up to the seal failure;
- To provide the complete charge of BB out of 10 storage batteries the PSS voltage level should be 15 V in the range of BB working temperatures of (-10)-(+40)°C and 16V - in the range of (-40)- -(-10)°C;
- The difference in storage batteries parameters which appears due to their operation as a part of a buffer battery requires that the method of the accumulator choice for the battery should be changed. To be more precise, a preliminary cycling should be fulfilled with the purpose to choose an accumulator for a battery based on the feature of similar charging of the dynamics of the main operation parameters.

CHOICE OF ACCUMULATORS TO BE COMBINED IN A BUFFER BATTERY (MAKING UP A SET)

The point of accumulator choosing for their unification in batteries is of great importance. It is known that in the accumulator cycling process the parameters of different accumulators gathered in one battery and operating together start to differ greatly. This leads to the decrease of a total battery capacity and to repoling (repolarization) of separate accumulators. This is true for all storage batteries (including the batteries of "military standard" class), which are gathered into groups according to plant parameters, but the dynamics of these parameters changes during the operation of a battery is not taken into account.

The process of making up the set is the following. When the accumulators are in a working position and a test cycle is over the accumulators are sorted according to their capacity. The capacity spread in every group should not exceed 5-6% of a mean value. After this the cycling of accumulators as a part of battery is fulfilled and accumulators with minimum divergence of the last discharge voltage from the mean voltage are chosen. We have full technical documentation for this melthod.

CONCLUSIONS

It is not enough room to make a review of all necessary elements of PSS microsatellite designing within the frames of this paper. Briefly, below there are the main recommendations on microsatellite PSS designing:

A. To improve the efficiency of SA it is necessary:

- to choose correctly a working point of volt-ampere SA parameters taking into account the orbit parameters, illumination conditons and panel temperature changes;
- to take into account temperature modes of SA operation and to provide its optimum temperature;
- to determine exactly an SA degrading stage during its operation time and to create a flight program that makes it possible to take this factor into account completely.

B. The BB operation algorithms which allow to use their determined capacity completely and to provide long life-time of operation should envisage:

- the protection of accumulators in BB against redischarge and repoling by the control for minimum voltage of each accumulator and BB;
- the protection of accumulators against recharging by the limitation of BB maximum charging voltage and the control for the accumulator temperature;
- the levelling of accumulator capacity parameters during the operation by using special devices;
- the execution of periodic procedures with a deep element discharge of BB to restore capacity parameters.

C. It is recommended to use the buffer structural scheme with direct parallel switching on of SA, BB and loads for low power PSS. The voltage converters included between SA and BB can be used in such PSS to invert the SA voltage poling, to make the galvanic separation and to limit the voltage.

REFERENCES

D. A.Medvedchikov , Small Satellites / *Aviatsiya i kosmonavtika*, N 4, 1994 (in Russian)
Development of the Structure and Algorithms of the Subsatellite PSS Operation for the Interball Project./ Final Report, 1, 2509900,Shpakov S.P., Moscow Power Engineering Institute, M.: 1991, (in Russian). N G.R. 01900060125.
Yu. N., Agafonov, J. Vojta, P. Triska V.V. Khrapchenkov. The Interball Project Subsatellites/ *Kosmicheskie issledovaniya*, vol. 34, N 4. July-August, 1996 (in Russian).
P. Triska ,J. Vojta,Yu. Agafonov. The Interball Subsatellites S2-A and S2-T/ *Interball Mission and Payload*. RSA-IKI-CNES, 1995.
Yu. N.Agafonov, J. Vojta, S.A.Gruzkov, S.P. Shpakov. Small Satelltes PSS. Information review. /IKI RAS Preprint, 055(02)2, Moscow, IKI RAS, 1997 (in Russian).

MICROACCELEROMETRIC SATELLITE MIMOSA (MICRO-MEASUREMENT OF SATELLITE ACCELERATIONS)

L. Sehnal, Radek Perestý and P. Dostál

Astronomical Institute, Czech Academy of Sciences, Fri ova 1, Ond ejov, 251 65, Czech Republic

ABSTRACT

The aim of the small satellite MIMOSA is to measure the satellite surface forces of different origin (atmospheric, radiative, thermal, etc.) by a highly sensitive 3-axial accelerometer MACEK with cubic proof-mass. This instrument, several time tested in the space, will be held in the centre-of-gravity position as the only instrument on the board. The satellite of regular polyhedron shape and of diameter 60-cm should revolve on a highly inclined elliptical orbit to cover upper atmosphere as much as possible. At altitudes above 800 km the direct solar radiation pressure will be measured to calibrate the instrument. The acceleration will be measured as a vector, i.e., the attitude of the satellite has to be known with accuracy higher than 1 deg. The launch of MIMOSA satellite is planed to be at the beginning of 2000.
The data gathered from MIMOSA mission should improve our knowledge of the distribution and variations of the Earth upper atmosphere total density and enable us to model the atmospheric properties to predict the corresponding orbital and attitude perturbations more precisely than ever before. However, the data analysis should deal with all others sorts of perturbations, especially those of the radiative origin as solar radiation pressure, Earth albedo and infrared radiation.

ATMOSPHERIC MODELS TO IMPROVE

Does anybody have any doubts about an increasing role of close-Earth artificial satellites in today's world? Telecommunication, remote sensing of Earth, research of Earth's gravitational field and Earth's atmosphere. There we have only some well known examples of their utilization here. Knowledge and prediction of position and lifetime of these satellites are then obviously important.

The motion of close-Earth satellite is affected mainly by the gravitational field of the Earth. Its description is known quite well today, so we can calculate its influence on satellites with sufficient precision written in centimetres. But the orbit of satellite is influenced also by another natural forces, which are highly dynamical and with almost the same size as gravitational. It means forces of different origin as an atmospheric drag, direct solar radiation pressure, reflected solar radiation pressure, Earth‘s thermal radiation pressure, thermal radiation of satellite and effects coming from satellite-Earth's electromagnetic field interaction. These forces we call non-gravitational, they are highly time-dependent and uncertainty in orbit determination thanks to them is in hundreds of meters. To improve prediction of orbits of satellites with respect to these force fields we need to model their space and time variations. At low heights (250 – 850 km) above Earth's surface the atmospheric density and Earth's albedo plays the most important role. So both, the model of Earth's atmosphere total density (TD-88) and the model of Earth's albedo were worked out at the Astronomical Institute of Czech Republic.

The TD-88 model describes a distribution and variation of the total atmospheric density not considering the individual constituents included in the atmosphere. It works only with the total atmospheric density, that is practically made by a number of coefficients in rows of spherical harmonic equations. Therefore it is much more simple than usual models and on its base the analytical theory of Earth's artificial satellite motion was accomplished. The other models (for example COSPAR International Reference Atmosphere CIRA-86) are derived from a density of individual elements and take into account number of atmospheric components as H, He, O, O_2 and N_2. Compared best of these with our model, example for height 500 km shown in Figure 1., TD-88 gives similar results within sufficient accuracy for orbit calculations. However, in addition as mentioned TD-88 is much easier to use and more precise for longer orbit prediction (Ref. [1] [2]).

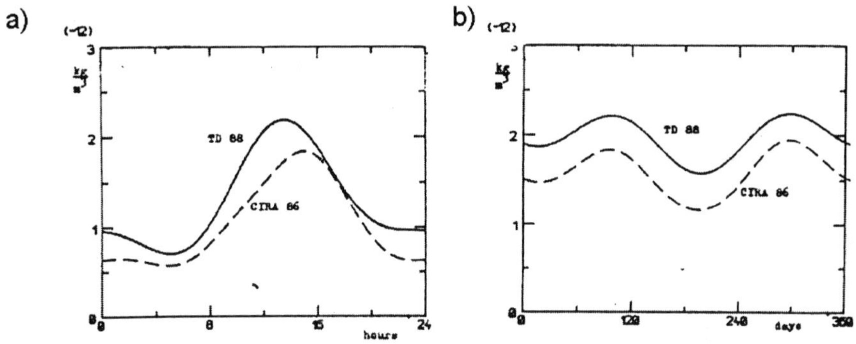

Figure 1. a) Diurnal variation of density at 500 km, F=F=150, K_p=3, day=80, φ=0.
b) Annual variation of density at 500 km, F=F=150, K_p=3, LST=14, φ=0.

Similarly the model of winter and summer Earth's albedo distribution was created and the theory of close-Earth satellite motion was built up (Ref. [3]). Graphical representation of this model for June is in Figure 2.

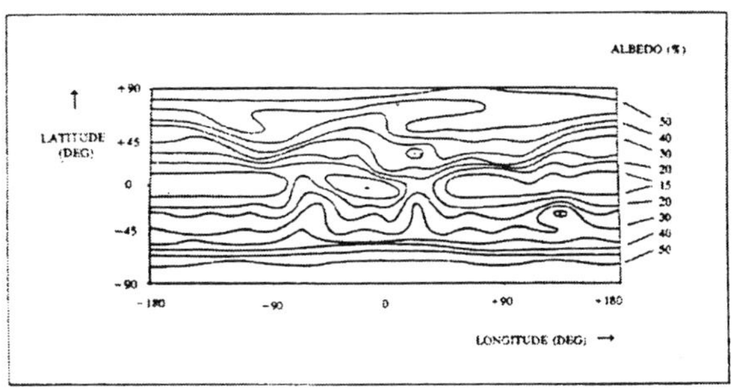

Figure 2. Earth's albedo diffuse re-radiation for June. Values are in percent of incidence solar radiation.

Since the satellite motion precision requirements are constantly increased, we have to improve current models to fulfil these demands and try to decrease satellite orbit calculation errors under 1 meter. The only way now is to obtain an amount of new data, especially by "in - situ" measurements of dynamical effects, i.e. of the accelerations caused by non-gravitational forces. Typical magnitudes of expected external accelerations vary from 10^{-4} ms^{-2} to 10^{-9} ms^{-2} (see Table 1.).
These measurements can be achieved by a highly sensitive accelerometer with resolution up to 10^{-9} ms^{-2}

and to get it as a vector, i.e. with direction definition. Such an accelerometric device called MACEK was designed, made and tested by the Astronomical Institute and to accomplish tasks mentioned above it is necessary to put it inside a special accelerometric micro-satellite.

Table 1. Expected Magnitudes Of External Disturbing Forces

Atmospheric drag	10^{-4} to 10^{-9} ms^{-2}
Solar radiation pressure	2.9×10^{-8} ms^{-2}
Earth's albedo	10^{-8} to 10^{-9} ms^{-2}
Earth's infrared radiation	4×10^{-9} ms^{-2}

THE MIMOSA SATELLITE

The last and in fact the only one highly sensitive accelerometric measurement intended for improving of atmospheric models was realized in years 1975-1979. The French CACTUS satellite had the accelerometric sensitivity 10^{-7} ms^{-2}, its attitude was made out only roughly and atmospheric rotation velocity was not measured.

The MIMOSA project, prepared by us, should overcome this old achievement, measure accelerations with sensitivity of 10^{-10} ms^{-2}. The main aims of the whole project are then to:

a) build up a new, more precise model of atmospheric total density distribution and variation
b) build up a new, more precise model of Earth's albedo distribution and variation
c) detect an atmosphere rotation velocity at heights 250 – 500 km (superrotation)
d) theoretically describe and measure influence of cloudness on satellite orbit
e) theoretically describe and measure influence of Earth's shadow entering and exit on satellite orbit
f) theoretically describe and measure influence of Earth's thermal radiation on satellite orbit
g) theoretically describe all non-gravitational influences on close-Earth satellite orbit

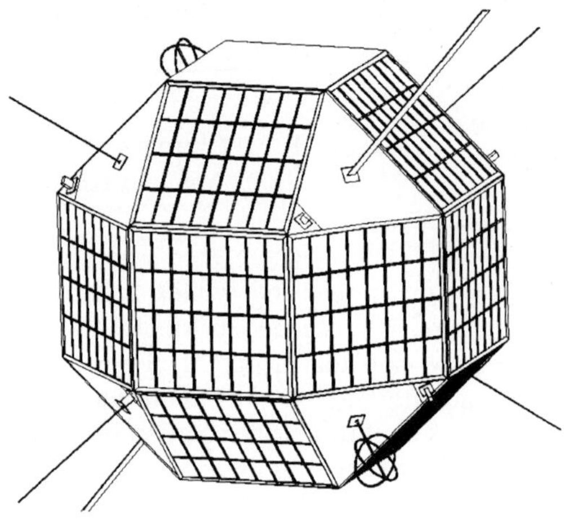

Of course, so much of tasks will need a lot of strict technical requirements on the satellite. It is necessary to avoid any undesirable interference with a sensitive accelerometer inside. So it must be the only payload. The satellite shape should be as much symmetrical and it should have a maximal size (area) and minimal weight. We have to know a satellite residual rotation and its attitude and location in any time very precisely. It is necessary to choose and launch it to a suitable orbit to measure an atmospheric drag at lower heights and a direct solar radiation pressure at higher heights. A satellite lifetime should be at least 2 years for good description of annual changes of the Earth's atmosphere. We have to build and maintain a ground station to control satellite and to collect measured data. But the most important challenge was to prepare its heart – the accelerometer.

On the next Figure 3 you can see how the whole project MIMOSA is scheduled.

MIMOSA	1996	1997	1998	1999	2000	2001
Project start						
A – Definition						
B – Design						
C – Construction						
D – Launch						
E – Operation						

FIGURE 3. MIMOSA PROJECT SCHEDULE.

THE ACCELEROMETER MACEK, THE HEART OF MIMOSA

The device for measurements of small accelerations MACEK is highly sensitive instrument that enables to measure accelerations in the space. MACEK was designed at the Astronomical Institute of Czech Republic at the department of Near-Earth Physics and built at the Space Devices Company. Its history dates back to early eighties, so it represents nearly 15 years of improving and testing of the device.

Design And Main Properties

The basis of MACEK accelerometer is a free moving cubic proof-mass. This cube of dimension 29.96 mm is placed inside a cubic cavity with size of 30 mm. This cubic cavity is created by 6 prism-blocks (see Figure 4). The cube and prism-blocks are made of Homosil glass slightly chrome-plated in such a way, that appropriate patterns on prism-blocks together with the cube build up the electrodes. By an appropriate voltage producing an electrostatical field we can keep the cube in the center of a cavity. The cube position is during measurement detected by a capacitive sensors and stabilized in all 6 degrees of freedom.

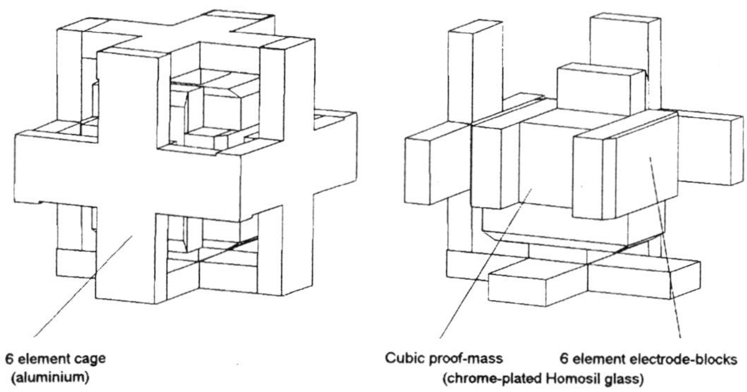

Figure 4. MACEK accelerometer body, whole and unfolded

These stabilization forces given by a control voltage are then used to analyse actual impacting accelerations along with their directions. Knowing satellite shape, area, weight, attitude and other parameters we can with help of some model (TD-88) deduce an atmospheric density.
Very important part of MACEK is its control and evaluation electronics. Only a long time of electronics developing led to MACEK's unique parameters (see Table 2.). There is a space qualified 8-bit microprocessor in its heart, that control all the process mentioned above. Maybe, the device measurement

frequency 0.5 Hz does not seem high compared to the other common accelerometers, but only until we compare their sensitivity.

Table 2. MACEK Main Measurement, Physical And Interface Properties:

Measurement range	4×10^{-4} ms^{-2}	Sensor mass	2.0 kg
Resolution	10^{-10} ms^{-2}	Sensor (cage) size	90 x 90 x 90 mm
Accuracy	10^{-9} ms^{-2}	Power	8.5 W
Scale factor	25 kV / ms^{-2}	Data interface	RS 232
Proof-cube	60 g, a=30 mm	Data rate nominal	100 bps

MACEK Space Qualification

Since a character of accelerometer MACEK does not make it possible to proof its complex function on the Earth, several test space mission were planned. The first time it has been flown on the board of a Russian satellite RESOURS-F1 in 1992 and a reliability of the construction was proved (Ref. [4]).

Improving functions of MACEK in time, we have lately been looking for a space vehicle which would provide better and more precisely defined environment for the accelerometer. So, in September 1996 the Space Shuttle Atlantis carried on board the MACEK on its STS-79 mission (Ref. [5]). The two boxes of device and electronics were placed inside the SPACEHAB module as shown in Figure 5.

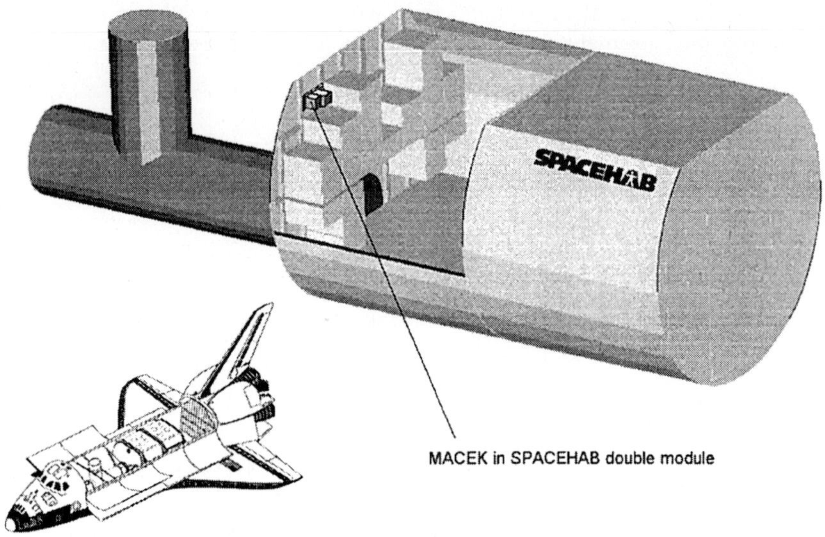

Figure 5. MACEK / STS-79 place.

Such a big spacecraft as the Space Shuttle is certainly not too favorable to gather the scientific data of dynamical perturbations so small as we would wish. But to know that output data are real accelerations does not matter if they are external or internal. However the result of Shuttle / MIR docking was – a great noise. But besides there were still some periods during which we got remarkably good records and which allowed a detailed analysis. (Ref. [6] [7]). Naturally, the best data obtained were connected with crew "sleeping periods".

During the mission MACEK has been measuring 8 days and 8 hours. The results revealed very good function without any drop-out and were of excellent quality. It means we could compare some measured data with Shuttle manoeuvres or crew activities according to NASA event timetable. You can see an example of such an event, that is jet burn of OMS (Orbital Maneuvering System), on Figure 5. After reduction of a temperature dependency we investigated the fit of the reduced data into the theoretical bandwith assuming that the accelerations are mainly caused by an atmospheric drag. For the known Keplerian elements of the Shuttle state vector all quantities can be calculated to determine density with thermospheric model TD-88 and finally the drag acceleration in terms of measurement time. The result can be seen in Figure 6 drawn for one revolution.

The Space Shuttle experience has given us an warranty of a good performance of our accelerometer and has prepared us for even more challenging task. To set up, to launch, to maintain and to process data from an independent accelerometric micro-satellite MIMOSA.

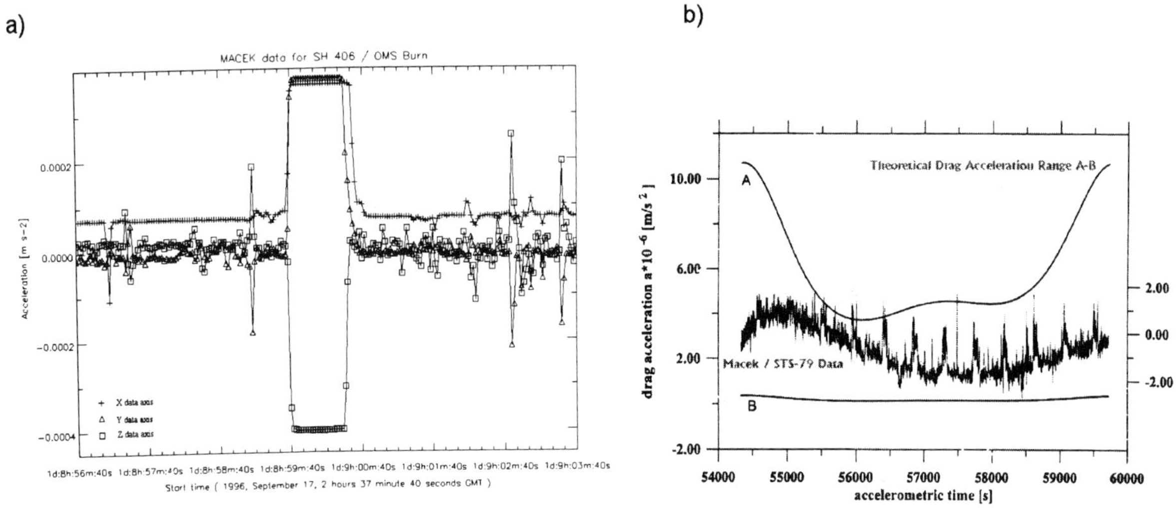

Figure 6. a) MACEK / STS-79 three-axes data example;
b) MACEK / STS-79 data comparison with the theoretical drag values .

CLOSER LOOK AT MIMOSA

After some general facts and requirements written in the chapter above, it is now the time to look closer into and over MIMOSA satellite. Since the dream of the first Czech accelerometric satellite is not really new, the MIMOSA project joins to and take experience from the old developing work. Although the project is supported by a grant awarded in 1995, a lot of work has been already done. However a progress in every technical branch is nowadays so fast, that to build a modern satellite we have to reflect many of new methods and technologies.

What is no longer object of study, is a character of a satellite, e.g. main parameters, shape and size of its casing, main components and function. The technological functional model of MIMOSA satellite was already built and is now used for testing and the design of various components lay-out.

The size and composition of the satellite with respect to the main parts is shown in Figure 7.

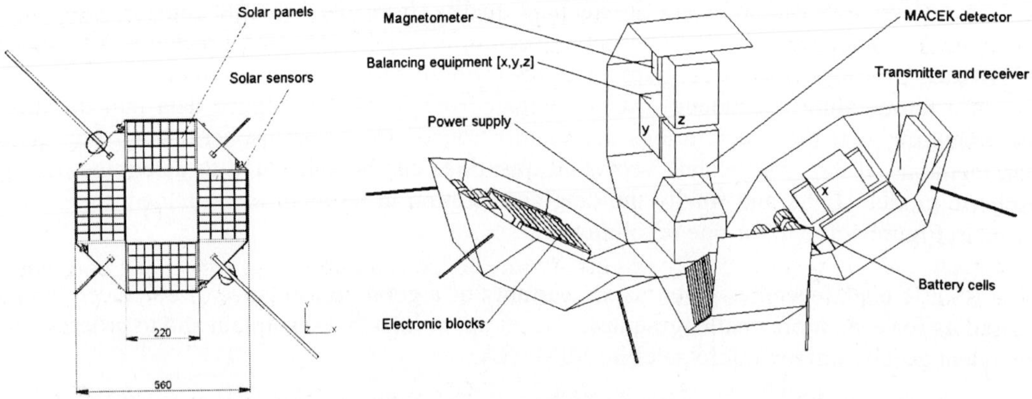

Figure 7. Sized and unfolded view of MIMOSA satellite with main components.

The satellite shape is not an optimal sphere, but a convenient 26-side symmetrical polyhedron with the greatest size of 560 mm (without antennae). Its surface is covered by solar panels as much as possible and they are as well the only power source of the satellite. For telecommunication it uses a number of antennae with different purposes and shapes (described later). The last part on the surface is a few of solar sensors for attitude detection. Inside we find several systems and control boards necessary to achieve the scientific goals as drawn in Figure 8 and describe below.

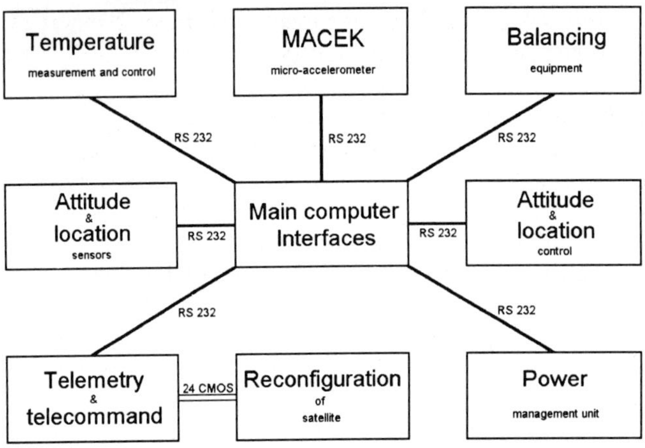

Figure 8. The main satellite parts with their star-connection done by a common RS 232 interface.

Satellite Power Management Unit controls the whole process of energy distribution of the satellite. First, it manages the utilization of solar panels by checking their illumination and individual efficiency. Then it controls the flow distribution of power to the batteries according to their actual capacity. System deals with each energy consuming part of satellite and distributes stored power to it. Of course, there is a big difference between this distribution at a normal working mode and at a possible emergency mode. In such cases, the energy flows mainly to the instruments able to inform a ground station what is happening and all efforts are concentrated to get away from this state.

An average energy consumption of the satellite is planned to be less than 11 Watts. With the use of a modern GaAs solar cells we can calculate with an average total energy input of 13 Watts. It seems satisfactory, but naturally we are trying to find out several ways to increase this difference by decreasing the energy consumption of all satellite parts.

Data Collection System is represented by a part of the main computer, interfaces, memories and appropriate software. It governs over the data collection from the accelerometer and over the operational and error data of every connected satellite part. It controls the data flow into the memories, too.

Telemetry And Telecommand is a very important part of the satellite. It guarantees the communication between a ground station and the satellite. It includes telemetry transmitters, telecommand receivers, decoders and antennae.

For a telecommand, i.e. uplink, we plan to use a 450 MHz band with the output of 150 Watts. On the satellite, it uses two rod-antennae with length of 188 mm.

For a telemetry, i.e. downlink, we intend to use two bands. The first 2.2 GHz band with the output of 2 x 1 Watt will be used as a normal mode link. It uses two cross-circular antennae with diameter of 35 mm. The purpose of the second 137 MHz with 50 mW output is to emit life-important data in case of emergency. It is transmitted by two rod-antennae 547 mm long.

Satellite Attitude And Location Control System includes mainly solar sensors, magnetometer and some device for location measurements. To achieve the scientific goals quoted above, it is a must to know precisely the satellite's attitude toward the Earth. It is arranged by a number of solar sensors of cubic shape with five solar cells on every top side. They can locate (with the precision of 1 deg) the direction of the greatest ligth source – the Sun. Then the satellite attitude is known in any time.

After the launch and separation from launcher, the satellite can still keep some residual rotation and fast satellite rotation could undesirable influence the accelerometric measurements. Therefore we would need some system to decrease this rotation. Using a magnetometer aboard, we can avoid any of the expensive and complicated jet systems and the satellite can be controled with respect to the direction of Earth's magnetic field tendency. Desirable residual rotation of the satellite is of about 0.1 rad/sec. It should not be less because of an appropriate warming up of the satellite and its parts.

The same condition as for a satellite attitude determination is valid for a satellite position determination. However, it is still uncertain task. We intend to use the Doppler measurements from the ground station(s) or GPS aboard the satellite.

Satellite Temperature Sensors measure a temperature at several places inside the micro-satellite body. It is absolutely clear, that a temperature plays a great role with the use of high-tech electronics. So, we use these sensors only for a pasive measuring and off-line processing of measured accelerometric and other data.

Balancing Equipment is one of the heaviest systems onboard. But when you imagine its function, you would understand all three weights it uses.

One of the most important thing, for a good function of the MACEK accelerometer, is its location inside the satellite body. Thanks to parasitic centrifugal forces originating from a residual rotation of the satellite, the function of the off-axis accelerometer (remember sensitivity of 10^{-9} ms^{-2}) is endangered. The must then is to locate the accelerometer into the centre of gravity of the satellite, at least with a precision of 1 μm. We use off-line control, thus we process measured accelerometric data and then we derive the position of the accelerometer and the correction. Then we uplink commands to the balancing equipment control board and it moves some of its three weights to change a position of the satellite's centre of gravity to be at the centre of the MACEK's cavity.

MIMOSA AT THE ORBIT

To define the satellite orbit is in projects like this fundamentaly important. The orbit will influence the atmospheric layers and other radiational effects we can measure and analyse. Of course we would like to explore as much as possible, but the only limitation is in availability of an appropriate launcher.

To fulfil the scientific goals mentioned above, the best orbit should be highly elliptical to reach the heights of 1500 km above Earth's ground and to descend to 300 km. Nevertheless, we have selected several other orbits, advantageous of different reasons and fitted to the eventual launcher. On Figure 9, you can find all possible orbits with their definitions in the first column. On the next one (Figure 10.) you can see the one-day projection of the second orbit.

In agreement with the table above, we plan to launch the MIMOSA satellite at the beginning of the year 2000. If everything goes well, the launcher would be either Ariane 4/5 (operated by ESA and CNES) or some Ukrainian launcher (operated by the Ukraine Space Agency).

Of course, nothing ends by the launch. It is necessary to maintain the satellite, downlink the data and uplink commands. Quite carefully we have been looking for a good ground station to operate our satellite. We have found one that operates the Czech ionospheric satellites MAGION 1-5, the location of which is $\lambda = +14° 34' 10''$, $\varphi = +50° 31' 42''$ - Panská Ves in Czech Republic.

We plan to use its 10-m parabolic antenna for downlink on bands 2.2 GHz and 137 MHz, one YAGI antenna for 137 MHz emergency downlink. For a uplink there is a spiral antenna that operates on 450 MHz with a power 150 Watts.

Orbit	Orbit Features	Perigee - Apogee Height [km]	Eccentricity	Inclination [°]	Period [h:m:s]	Visibilities per day	Visibility Time per Day [approx.h:m]	Launcher
1	-	400 – 1400	0.0688	70	01:42:49	8	1:32	t.b.d.
2	-	300 – 1500	0.0825	60	01:42:49	7	1:27	t.b.d.
3	Sun – synchronous and circular	822	0	98.7	01:41:10	7	1:14	Ariane 4/5
4	GTO	300 – 35794	0.727	7	10:32:51	3	3:23	Ariane 4/5
5	Best illumination & calibration conditions	300 - 3550	0.196	63.4	02:05:19	8	3:45	t.b.d.

Figure 9. MIMOSA orbit summary.

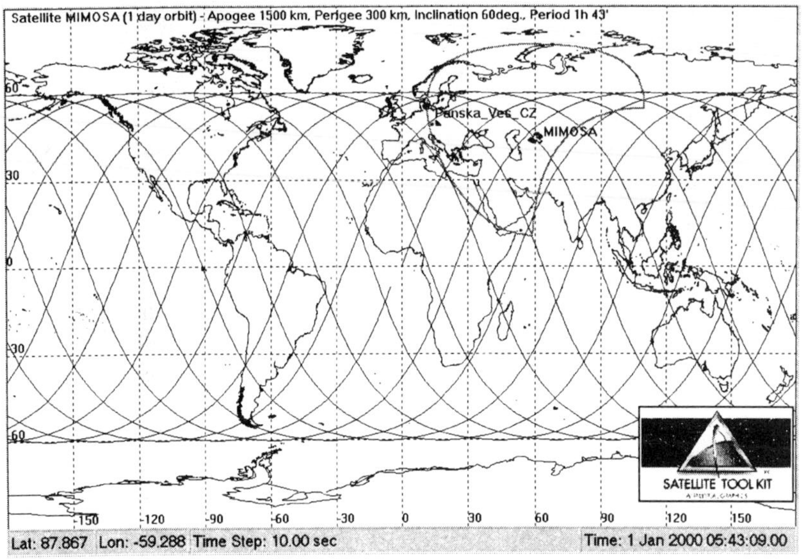

Figure 10. The proposed orbit No.2 one-day projection.

ACKNOWLEDGMENTS

This work is supported by grant No. 205 / 96 / K 119 of the Grant Agency of the Czech Republic.

REFERENCES

Sehnal L., Vokrouhlický D., *Model of non-gravitational perturbations for Cesar experiment with Macek accelerometer*, Advances in Space Research, Vol.16, No.12, pp (12)13-(12)13, 1995

Sehnal L., *Dynamics and Measurement of the attitude of a near-Earth satellite*, Mécanique Spatiale, MS 95/012, pp 785-792, 1995

Sehnal L., *The Earth albedo model in spherical harmonics*, Bull.Astron. Inst.Czechosl., Vol.30, No.4, pp 199-204, 1979

Sehnal L., Pe estý R., *In orbit testing of the MACEK microaccelerometer,* Geodesy and Physics of the Earth, IAG Symposium No.112, Springer Vrlg, pp 129-132, 1993

Pe estý R., Bijvoet J., *MACEK Experiment*, Proceedings of the 16th Meeting of MicroGravity Measurements Group, Gainesville FL, U.S.A., 1997

Sehnal L., Pospíšilová L., Pe estý R., Kohlhase A., *Accelerometric measurements of satellite dynamics*, Proceedings of the 12th International Symposium on "Space Flight Dynamics", ESOC, Darmstadt, Germany, SP-403, 1997

Pe estý R., Sehnal L., Lundquist C.A., Bijvoet J., *Measurements of microgravity environment by three-axial electrostatically compensated microaccelerometer*, Proceedings of the 48th International Astronautical Congress, Turin, Italy, 1997

AN OVERVIEW OF ROCSAT-1 ELECTRICAL GROUND SUPPORT EQUIPMENT DESIGN

Jene Chun-Hsien Wu

Spacecraft Project, National Space Program Office, Hsinchu 300, Taiwan, ROC

ABSTRACT

This paper presents the NSPO ROCSAT-1 *Electrical Ground Support Equipment (EGSE)* design which contains the system requirement development, element descriptions, design technology, trade studies, design verification and future plan. During the constructional phases of a satellite, support equipment and services will be required to enable test activities to be performed. EGSE is then designed to provide the capabilities for testing and monitoring performance of the satellite. Ethernet interface, Integrated Satellite Automated Test System (ISATS) software, VXI/GPIB, and RTworks/VxWorks technology are used to design the ROCSAT-1 EGSE. Several trade studies were performed in support of the detailed design. Each test set had been checked and tested separately to ensure that it met its functional and performance requirements.

INTRODUCTION

The space program is one of the most exciting and significant endeavors in human history. ROCSAT-1, the first Republic of China (ROC) satellite, is being developed by the National Space Program Office (NSPO). ROCSAT-1 has been designated as a scientific satellite to be injected into a 600 km Low Earth Orbit (LEO). The payload instruments of these space-science experiments include the Ionospheric Plasma and Electrodynamics Instrument (IPEI), the Ocean Color Imaging (OCI) and Experimental Communication Payload (ECP). The prime contractor of ROCSAT-1 is TRW, a major US defence and commercial aerospace corporation, was awarded by NSPO in the spring of 1994 to develop ROCSAT-1 spacecraft. An integrated team of TRW and NSPO engineers have been working together to meet the ROCSAT-1 challenge. The ROCSAT-1 Electrical Ground Support Equipment (EGSE) is an important subsystem which supports satellite testing and provides satellite configuration and control, stimulus, monitor and performance measurement. The ROCSAT-1 EGSE design is based on the Eagle Test Bed EGSE that was used on TOMS-EP satellite test program in TRW. This paper is a condensed version of Reference (NSPO,1995). Details of more ROCSAT-1 related EGSE designs can be found from the Reference (NSPO,1995) report.

EGSE DESIGN PROCESS

Using the system engineering process methodology (Hossny,1995) we translated the electrical design integration, subsystem engineering and test engineering's actual needs for EGSE into a set of requirements and specifications. The following are the six main steps by applying the system engineering methodology to design the EGSE:

System Requirements Development

The first step of EGSE design is to develop the system requirements. Using the system engineering process methodology, the satellite specification, specific flight hardware design, test requirements, COTS availability, and lessons learned from heritage are translated into the system requirements.

System Design

A system architecture was developed that supports the specified product requirements. The equipment specification were developed that document system's architecture and the functional requirements of each subsystem. Some trade studies were performed to support the detailed design. We used testbed and analytic techniques to verify that the conceptual and preliminary design supported the specified system requirements.

Detailed Design

Hardware and software were designed to meet the equipment specification. Hardware drawings including the electrical and mechanical portions were prepared to identify the physical size and signals path. Software flowchart or operation procedure was documented to meet the requirements. and each subsystem was built and tested to verify that it meets requirements.

Product Design

After the fabrication and assembly process, each subsystem was built and tested to ensure that it met its functional requirements. Then, each subsystem was integrated into a final product and performed the acceptance test procedure (ATP) to verify its functional and performance requirements.

System Integration

Each subsystem was checked and tested separately to ensure that it met its functional and performance requirements. After successful testing, the components and subsystems are integrated into a system and verified its functionality.

Design Verification

All performance specifications were verified as met under the specified operating conditions. The final system's configuration was evaluated to ensure that it complied with the original functional requirements specifications.

ROCSAT-1 EGSE ELEMENTS

The main elements of the EGSE are as follows:

1) System Test Controller (STC)

2) Power and Control/Monitor Test Set (PCMTS)

3) Telemetry and Command Test Set (TCTS)

4) Radio Frequency Test Set (RFTS)

5) Command and Data Handling Simulator (CDHS)

6) Radio Frequency Compatibility Tester (RFCT)

7) Propellant Control and Monitor Unit (PCMU)

8) Battery Simulator

The STC is used to control testing and record all test results. The STC performs these functions by use of the Integrated Satellite Automated Test System (ISATS) software package. The STC control and monitors the three major test sets (PCMTS, TCTS and RFTS) via an 10BASE5 Ethernet interface. The EGSE system block diagram is shown in Figure 1.

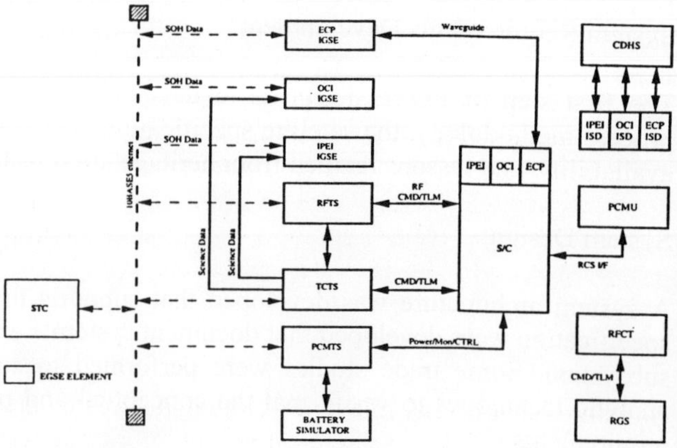

Figure 1. ROCSAT-1 egse block diagram

The STC generates spacecraft commands and passes the commands to the TCTS. The TCTS then formats the commands and routes them to the spacecraft on either the hard-line interface to spacecraft Transponder Interface Electronics (TIE) or to the RFTS. The RFTS modulates the commands onto a carrier and transmits them to the spacecraft. Selection of hard-line or RF as the source of both commands and telemetry is done under control of the ISATS in the STC. The telemetry generated in the spacecraft is output via an RF link and TIE hard-line. The RFTS demodulates the telemetry data and passes the baseband signal to the TCTS. The TCTS, under ISATS control, select either the hard-line or baseband from the RFTS. The TCTS decommutates the selected telemetry stream and passes the science data to the appropriate Instrument Ground Support Equipment (IGSE) and either the real-time or playback state of health (SOH) data to the STC. Selection of real-time or playback data is under control of STC. The PCMTS supplies DC power to the satellite, monitors satellite hard-line test points, and provides stimuli to the satellite sensors. Power also can be supplied via the external battery simulator or solar array simulator (SAS). Selection of the power sources is under control of STC. The satellite hard-line test points are monitored with measurement results passed back to the STC. Control of the stimuli to the satellite is under control of STC.

DESIGN TECHNOLOGY

Ethernet

Ethernet is the most popular Local area networks (LAN) technology in use today. Other LAN types include Token Ring, Fiber Distributed Data interface (FDDI) and LocalTalk. Each has its own advantages and disadvantages, Ethernet strikes a good balance between speed, price and ease of installation. The Ethernet standard is defined by the Institute for Electrical and Electronic Engineering (IEEE). IEEE Standard 802.3 defines rules for configuring an Ethernet as well as specify how elements in a network interact with one another. The immunity to common level of electrical noise and the integrity of the network signals makes the Thickwire, or 10BASE5 Ethernet, is selected to create “backbones” for communication of ROCSAT-1 EGSE computers.

VXI/GPIB Architecture

In 1987, a consortium of test and measurement companies introduced VXIbus, a new standard instrument architecture. VXIbus was developed to meet the needs for portable application and to provide a standard modular architecture for integrating into the traditional GPIB test system and for standalone application. VXI mean smaller, faster test systems that can be configured for a variety of applications. Based on VMEbus and GPIB, VXI is actually a combination of two existing standards that improve test system performance and simplify integration-two of the biggest hurdles in test system development. The popular VMEbus architecture provides an efficient computer backplane for VXI, including high-speed data rates (up to 100 MB/s local bus) and all necessary communication protocols to assure high throughput. The design of PCMTS, RFTS and CDHS is based on VXI/GPIB technology.

XTM/LabVIEW Software

Software plays a vital role in developing automated data acquisition and instrument control systems. With XTM or LabVIEW, the graphical programming and user interface can be designed quickly. XTM, product of WaveTest, creates program with flowcharts. It is easy to understand the overall program. The control of PCMTS and RFTS is based on the XTM software. LabVIEW, product of National Instruments, graphically assembles the program as a block diagram. It is used to design the CDHS and RFCT.

RTworks/VxWorks

RTworks has been selected as core COTS technology to develop and deploy the ground operations systems that monitor and control satellites for major NASA, DOD and commercial space programs. RTworks is a modular, integrated environment for building and supporting a wide variety of command and control software applications. It provides central interface for user and distribution of real-time data. VxWorks is the premier development and execution environment for complex real-time and embedded application on a wide variety of target processors. Three highly integrated components are included with VxWorks: a high performance scalable real-time operating system which executes on a target processor; a set of powerful cross-development tools which are used on a host development system; and a full range of communications software options such as Ethernet or serial line for the target connection to the host. It provides multi-tasking with preemptive priority, semaphores, message queues, sockets and remote procedure calls. The design of TCTS and RFCT is completed under the control of RTworks and VxWorks.

TRADE STUDIES

Downlink Data Rate Increase

In order to accommodate the proposed increase of the spacecraft downlink telemetry rate to approximately 1.4 Mbps the processing capabilities of the spacecraft EGSE must be enhanced. Specially the TCTS currently planned to process the 1.0 Mbps telemetry data rate. If the telemetry data rate increases to 1.4 Mbps, there is only 14% CPU margin available. This is not meet the 50% design requirement. Therefore, the upgrade of the TCTS host and target computer is necessary. However, even in the faster computers, the margin is still less than 50%. The addition of a high speed recorder to the current TCTS design provides the added capability to record and playback the telemetry data as a slower rate thus increasing the CPU margin to meet the design requirement.

High Speed Tape Recorders

A high speed tape recorder system is to be used with TCTS during playback from the Solid State Recorder (SSR). The high bit rate spacecraft telemetry (1.3981013 Mbps) is to be temporarily recorded on a high speed tape recorder medium. The same data is then played back at a slower bit rate to the TCTS for processing of spacecraft and instrument SOH. The recommendation is to use Metrum buffered VLDS for TCTS design. (NSPO, 1995)

Solar Array Simulator

The TOMS solar array simulator AT8000 has thermal and power requirement problem. Also, AT8000 is a custom module. The recommendation is to use ELGAR Fast Profiling Current Sources (FPCS) for solar array simulator. (NSPO,1995)

DESIGN VERIFICATION

Acceptance Test

The Acceptance Test is to verify that the test set design meets the functional requirements. In order to run the acceptance test, the Acceptance Test Procedure (ATP) is necessary. After successful testing, the test set design is fully completed. And it can be used in I&T activities.

Validation Test

Before connects the test set to the satellite, the validation test must be performed to ensure the no interface problems. The validation test is done with specific equipment, validation unit. The design of this equipment is fully relied on the feature of the test set. The PCMTS and CDHS design comes with the validation unit design.

FUTURE PLANS

The ROCSAT-1 EGSE elements, design process, and trade studied have been presented. ROCSAT-1 EGSE design has been completed with the participation of both TRW and NSPO engineers. This experience is a successful example of technology transfer due to the deep involvement of ROC engineers in the actual ROCSAT-1 design, analysis, and testing activities. Currently, the ROCSAT-1 satellite integration and test (I&T) is underway in NSPO highbay. The major efforts are to maintain/modify the EGSE during the satellite testing. With the new coming satellite project, how to take the advantage of the existing ROCSAT-1 EGSE and maximize the technology transfer will be a big challenge to NSPO.

REFERENCES

NSPO, Electrical Ground Support Equipment Design Report, DRL 14, 61738.DRL14.002A, 25 October 1995.

NSPO, ROCSAT-1 Power and Control Monitor Test Set, DRL 14, 30 August 1995.

NSPO, ROCSAT-1 Telemetry and Command Test Set, DRL 14, 25 October 1995.

NSPO, ROCSAT-1 Radio Frequency Test Set, DRL 14, 18 October 1995.

NSPO, ROCSAT-1 C&DH Simulator, DRL 14, 25 October 1995.

NSPO, ROCSAT-1 Radio Frequency Compatibility Tester, DRL 14, 3 November 1995.

Hossny El-Sherief, A Systems Engineering Approach for the Design of Digital High Resolution Ambulatory ECG Recorder, TRW, Winter 1995.

NSPO, Data Rate Increase RFP Inputs, ROCSAT-1.95.320.010, TRW, 23 February 1995.

NSPO, High Speed Tape Recorders Trade Study, ROCSAT-1.95.322.036, TRW, 12 June 1995.

NSPO, PCMTS: SAS TRADE SUMMARY, ROCSAT-1.95.321.014, TRW, 2 February 1995.

MULTILAYER INSULATION FOR SPACECRAFT APPLICATIONS

Che-Shing Kang

Spacecraft Project, National Space Program Office, Hsinchu 300, Taiwan, ROC

ABSTRACT

Multilayer Insulation (MLI) blankets provide a lightweight insulation system with a high thermal resistance in vacuum. MLI blankets are utilized to reduce heat loss from a spacecraft to the cold space, or to prevent excessive heating of the surroundings from an internal component with heat dissipation. MLI blankets consist of a number of highly reflecting radiation shields interspaced with a low thermal conductivity spacer material or separated by crinkling the radiation shields themselves. The radiation shields are generally a plastic film metalized on either one side or both sides of the film. The principle of an MLI blanket is to use multiple layers of radiation shields to reflect back, in the opposite direction of heat flow, a large percentage of the radiant heat flux reaching each radiation shield. MLI is therefore very effective if solid conduction through the spacers and gaseous conduction through the gas medium can be minimized.

MLI HEAT TRANSFER

Heat transfer modes through an MLI system include thermal radiation between the radiation shields, heat conduction and radiation through the gaseous medium in the spacing, and heat conduction through the spacer or the contact points between the crinkled MLI film surfaces. Radiation through the spacing gas medium must be considered if the medium is a participating gas, such as water vapor or carbon dioxide. Otherwise radiation of the gas medium is negligible. For launch vehicle or satellite MLI applications, air is the medium which is almost transparent to thermal radiation. However, gaseous heat conduction can be neglected only if the air pressure is less than 10^{-5} mm Hg (Torr), such as in space or upper atmosphere for most satellite applications.

Radiation Heat Transfer Between the Radiation Shields

Radiation heat transfer between two adjacent parallel radiation shields is governed by:

$$Q_{rad} = \sigma \Im A\left(T_1^{\,4} - T_2^{\,4}\right) \qquad (1)$$

where T_1 and T_2 are the absolute temperatures of the adjacent radiation shields, A is the open surface area of the radiation shields, σ is the Stefan-Boltzmann constant and $\Im$ is the radiative interchange factor (*Script-F*). In general, the radiative interchange factor between two opposing surfaces is defined in terms of the *geometric form factor* (F_{12}) for two black surfaces, the surface areas (A_1 and A_2), and the surface emissivities (ε_1 and ε_2):

$$\Im_{12} = \frac{1}{(1/F_{12}) + (1/\varepsilon_1 - 1) + (A_1/A_2)(1/\varepsilon_2 - 1)} \quad (2)$$

Solid Conduction through Spacers

Solid heat conduction through the spacer may be expressed as:

$$Q_{solid} = k_s A_s \frac{T_1 - T_2}{d} \quad (3)$$

where k_s is the thermal conductivity of the solid spacer material, A_s is the solid area through the spacer, T_1 and T_2 are the temperatures of the two opposing radiation shields, d is the thickness of the spacer (also spacing between two shields). If contact resistance between the shields and the spacer is also considered, k_s represents the combined thermal conductivity.

Combined Normal Heat Transfer

If heat conduction through a thin metalized film is negligible, the total normal heat transfer between n similar radiation shields closely placed, in a steady state, is expressed by :

$$Q_{total} = \sigma A(1-\phi)\frac{\varepsilon}{(n-1)(2-\varepsilon)}(T_1^4 - T_n^4) + k_s \frac{A\phi}{(n-1)d}(T_1 - T_n) \quad (4)$$

where $\phi = A_s/A$ is the solid fraction of the spacer. In practice, if we consider an MLI system with $T_h = T_1$ and $T_c = T_n$ to be the hot and cold boundary temperatures, Equation 4 can be expressed by :

$$Q_{total} = \sigma A \varepsilon^* (T_h^4 - T_c^4) \quad (5)$$

where ε^* (*E-Star*) is the *effective emissivity* which is a measure of the combined effect of radiative heat transfer through the radiation shields and the solid conduction through spacers. Therefore, ε^* may be calculated from measured data of Q_{total}, T_h and T_c under vacuum conditions for a particularly designed MLI system.

MLI BLANKET CONSTRUCTION

Radiation Shields

Radiation shields are low emissivity metal foils or metal coated plastic films used to attenuate the incoming thermal radiation. Aluminum, gold and silver are the most commonly used metals either for coatings or to form thin metallic foils.

Plastic films can be metalized (through a vacuum deposition process) on either one side or both sides. For single-sided metalized films, the metal can be vacuum deposited on either the top side or the back side of the film. The top-sided metalized films (usually called *First-Surface Mirrors*) and the double-sided metalized films are good solar absorbers (high solar absorption over IR emissivity α_s / ε ratio) and infrared reflectors. They are commonly used as inner layers (infrared heat shield) in MLI blankets and can also be used as the outermost MLI layer (with thicker plastic structure) if the covered area is shaded from the Sun for most portion of the orbit cycle. The back-sided metalized films (usually called *Second-Surface Mirrors*) are good solar reflectors (low α_s / ε ratio) and are commonly used as radiators to dump the spacecraft internally generated heat to the deep space.

Among the available plastic films, *Mylar, Teflon* and *Kapton* (trade names of DuPont Company)are commonly selected by space industries as the shield materials for MLI.

Mylar, from the family of polyester, is inexpensive, and has been mass produced and widely used in the past. However, it disintegrates under prolonged UV exposure. Therefore *Mylar* is usually used as inner layers of an MLI system, or test MLI for a spacecraft in a thermal vacuum test.

Kapton, from the family of polyimide, is a gold-colored plastic. Its color becomes more darker by increasing the thickness, thereby increasing its emissivity. *Kapton* is slightly heavier than *Mylar*, but more expansive (in the past). *Kapton* is more rugged and high-temperature resistant (non-flammable) than both *Mylar* and *Teflon*. Therefore it can be installed as the outermost layer of an MLI blanket to provide protection during installation and handling, and also offer desirable α_s / ε ratios when used as a *second-surface mirror* (*SSM*). The α_s / ε value does not change with thickness because both ε and α_s increase with increasing thickness. For an aluminized *Kapton*, the α_s / ε value remains at about 0.5 for thickness from 0.5 mil to 5.0 mil. Embossment of *Kapton* is fairly troublesome that a general rule is to use long cycles to produce the molded shapes from this polymer.

Teflon, from the family of fluoroplastics, is expensive than both *Mylar* and *Kapton*, but most stable under UV and charged particle exposure. *Teflon*, vacuum deposited with silver, offers the best for *SSM* (with lowest α_s / ε ratio). Its α_s / ε ratio also decreases with increasing Teflon thickness because ε decreases with increasing thickness while α_s remains unchanged. Therefore it is usually applied as the outermost surface of an MLI blanket. Its drawback is that it does not adhere easily to spacecraft surfaces.

Metal Deposits on Plastic Films

Available metals for vacuum deposition are silver, aluminum and gold to offer low solar absorptance, and Inconel and chromium to offer high solar absorptance. The infrared emissivities of silver, aluminum and gold are also lower than those of Inconel and chromium.

Aluminum is the most popular metal for vacuum deposition, followed by gold and silver. Aluminum is inexpensive and readily available as a coating on a variety of plastic surfaces. It vaporizes at a lower temperature than gold, making the aluminum deposition process easier to control. As a result, aluminum coated film are of better average quality than gold or silver-coated films. The emissivity of aluminum ($\varepsilon = 0.03$) is only slightly higher than that of clear silver ($\varepsilon = 0.02$), but, whereas silver tarnishes in air, aluminum forms a very thin layer of oxide which prevents further degradation of the surface.

Vapor deposited gold appears very attractive for reusable spacecraft radiation shields because of the lower emissivity ($\varepsilon = 0.02$) and no degradation effects after exposure to moisture or contaminated air with salt and pollutants which would corrode aluminum metal away.

Silver has the lowest emissivity value ($\varepsilon = 0.02$) as compared to that of gold. *Silverized Teflon*, usually used as *Second-Surface Mirrors*, offers the lowest α_s / ε value of less than 0.1 for a layer of 5.0 mil type A *Teflon*. To protect silver from contamination, it is normally over-coated with silicone oxide or vacuum deposited by an additional layer of Inconel.

Thermal Control Tapes and Laminates

The various types of thermal control surfaces can be converted into a variety of products, e.g., combining different metal deposits on opposite sides of a substrate may provide one type of surface on one side and another type on the other side.

Thermal Control Tapes have pressure-sensitive adhesive coated on the opposite side for easy and positive attachment. Typical *thermal control tapes* are aluminum or silver Teflons with acrylic of silicone pressure sensitive adhesives. Standard tape widths from manufactures are 1.0 to 4.0 inches and the standard roll length is 108 ft (Shedahl 1989).

Thermal Control Laminates are formed by laminating a reinforcement scrim or film to the back side for added toughness as for the covers of an MLI blanket. A typical example is a Dacron scrim sandwiched between two Mylar films (by polyester adhesive) and vacuum deposited by aluminum on the outside of the Mylar films (Shedahl 1989).

A Complete MLI System

MLI blankets typically consist of three to thirty layers of metalized plastic sheets. The inner layers are typically as thin as practical (0.25 mil) to minimize weight and are usually made of aluminized *Mylar* or *Kapton*, separated by spacers. The innermost and outermost layers of the blanket, however, are usually made of thicker 1- or 2- mil aluminized *Kapton* which provides protection during installation and handling. Thermal control laminates can also be used as the outermost layer for additional toughness.

A standard MLI layup of Lockheed-Martin (former GE Astro Div.) is shown in Figure 1 where 8 layers of 0.25 mil, double aluminized *Mylar* are separated by *Dacron* polyester nets. A 2-mil aluminized second surface *Kapton*, coated with *Indium Tin Oxide (ITO)*, is placed on top to face the space UV radiation and atomic oxygen environment. A 1-mil double aluminized *Kapton* is located at the bottom to shield the spacecraft.

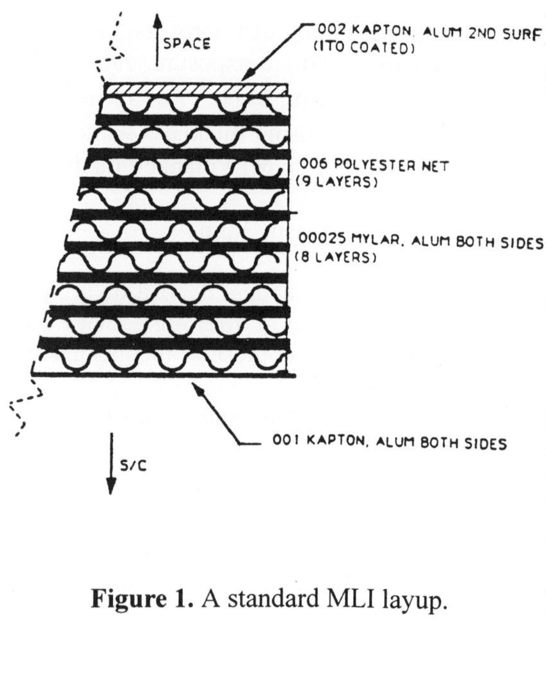

Figure 1. A standard MLI layup.

MLI DESIGN EXAMPLE

ROCSAT-1, the first Republic of China satellite, has been designated as a scientific satellite to be injected into a 600 km Low Earth Orbit (LEO) to conduct three scientific experiments. Eclipse time of the

ROCSAT-1 orbit can be as high as 30%, therefore, to protect the electronic components from thermal radiation heat loss to the deep space, MLI is applied to cover most of the exposed areas, except those places designated for radiator (Kang 1996).

The MLI blankets used on ROCSAT-1 include two types of external blankets, one backside and internal blanket, and one spiral wrap tape. Each layer of the ROCSAT-1 MLI blanket is composed of aluminized-Kapton, either plain for the inner and outer layers or crinkled for the filled layers. The external ones have either regular outer layer or carbon filled electric conducting outer layer. The backside MLI has a thinner outer layer and less crinkle filler layers. The regular external MLI blankets cover most exposed surfaces while the electric conducting ones cover the front area of the payload enclosure which is facing the ram direction. The thinner MLI blankets cover the backside of some external components and separate the high dissipation internal components from neighboring components. The spiral wrap tapes cover cables, waveguides and struts.

Since the ROCSAT-1 600 km orbit typically passes through the Earth's ionosphere, all layers of the MLI blankets shall be grounded between the outer layer's metalized surface to the spacecraft structure with a D.C. resistance less than 1000 ohms.

REFERENCES

Sheldahl Corp, "Thermal Control Material & Metalized Films," July 1980.

Kang, C. S., "An Overview of ROCSAT-1 Thermal Control Design," Proc. of AASROC Annual Meeting, Taipei, Taiwan, Dec 1996.

ATTITUDE DETERMINATION AND CONTROL SYSTEM FOR A SMALL LEO SATELLITE

Ying-Wen Jan (ADCS Senior Engineer) , Lee H. Ting (Spacecraft Project Manager)

Spacecraft Project, National Space Program Office, Hsinchu 300, Taiwan, ROC

ABSTRACT

This paper describes the Attitude Determination and Control System (ADCS) for ROCSAT-1. The nominal ROCSAT-1 ADCS mission scenario and an overview of the ADCS design are presented. This paper summarizes design considerations in selecting the hardware that can fulfill the ROCSAT-1 mission requirements. After separation from the launch vehicle (Athena-1, Lockheed Martin), the ROCSAT-1 satellite uses four one-pound thrusters to perform a sun acquisition in a 600 km, 35-degree inclination circular orbit. Then the satellite transits to Science Mode through Maneuver Mode, in which the spacecraft attitude is three-axis stabilized with the z-axis nadir-pointed. Upon reaching its Science Mode, a wheel-based control system is designed to keep the ROCSAT-1 instrument accurately nadir-pointed for four years. Two fail-safe modes add robustness to the design. This paper shows how a small LEO satellite can be implemented with proven sensors and actuators resulting in low power usage, low weight and affordability.

INTRODUCTION

ROCSAT-1 is a three-axis stabilized light-weight satellite which will be launched by a small expendable launch vehicle (Athena-1). It is designed for a two year mission lifetime with four years as a goal. The operational orbit is circular, inclined at 35 degrees, and at 600 km altitude. Figure 1 illustrates the fully deployed bus configuration.

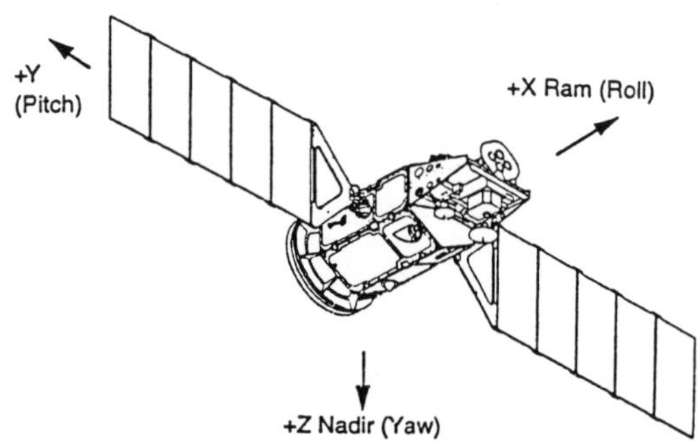

Figure 1. ROCSAT-1 Satellite.

The Attitude Determination and Control Subsystem (ADCS) performs attitude determination, control and stabilization during all operating phases of the ROCSAT-1 satellite, from launch vehicle separation until mission termination. During early orbit operations the ADCS provides firing signals to the hydrazine thrusters. During normal science operations in a 600 km, 35 deg inclination orbit the ADCS provides attitude control (<0.5 deg for roll, pitch and yaw) and knowledge (onboard accuracy wrt LVLH < 0.15 deg per axis) as required for the science instruments. Three-axis attitude control is achieved with speed-biased wheels for pitch momentum bias and a gyro based reference with earth and sun sensor updates.

)CS HARDWARE IMPLEMENTATION

gure 2 shows the overall block diagram of the ADCS equipment. In this diagram, the spacecraft ocessors are part of the Command and Data Handling (C&DH) subsystem and the thrusters and plation valves belong to the Reaction Control Subsystem (RCS). The remaining assemblies are ADCS nsors and actuators with associated electronics. Sufficient standby and functional redundancy has been ovided in the subsystem configuration to achieve a reliability of 0.97 over 2 years. Table 1 summarizes e ADCS hardware weight and power. The total weight is 58 pounds and the average power for science erations is 30 watts. A brief description of each ADCS component is given below.

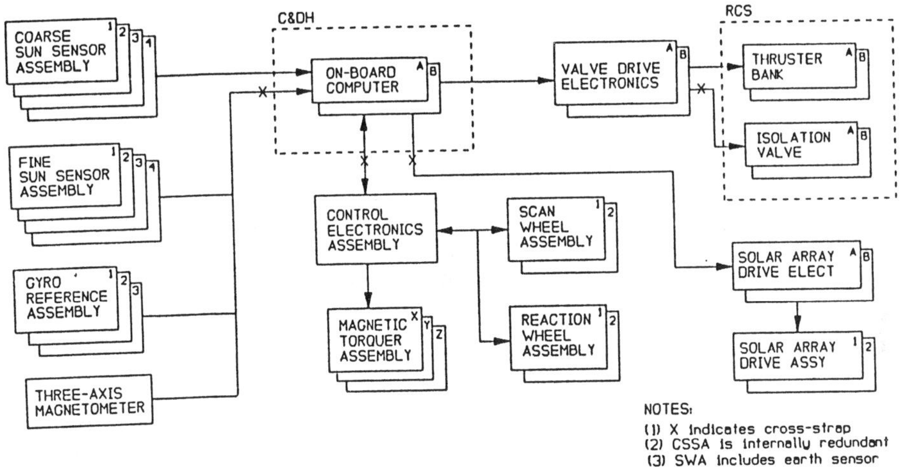

Figure 2. ADCS Equipment Block Diagram.

Table 1. ADCS Hardware Summary

Assembly	Number per Spacecraft	Unit Weight (1b)	Total Power (w) in Science Mode
Coarse Sun Sensor	4	0.1	-
Fine Sun Sensor	4	0.4	-
Gyro Reference	3	3.4	14.2
Three-Axis Magnetometer	1	1.1	0.3
Scan Wheel	2	6.9	9.0*
Reaction Wheel	2	5.6	-
Magnetic Torquer	3	2.0	1.5
Control Electronics	1	11.9	5.0
Valve Drive Electronics	2	6.5	-
Solar Array Drive Electronics	2	3.6	13.0
Solar Array Drive Assembly	2	11.3	-

*Includes motor drive power in CEA

ADCS OPERATION DURING EARLY ORBIT

Solar Array Deployment

After separation from the launch vehicle, the onboard computer (OBC) boots up in Maneuver Mode with the thrusters disabled. As the stowed spacecraft is tumbling slowly, a sequence of commands is automatically initiated to deploy the two solar array wings. The solar array panels on both sides unfold in an accordion-like fashion and cause the vehicle rates to slowdown due to the increasing inertia. Before attitude control can be activated, this solar array deployment has to be verified. Since no limit switch or latchup telemetry is available, the solar array deployment is verified indirectly via gyro information. First, two gyros have to be selected for this test. All three gyros are powered since liftoff and are still running. The gyro selection logic is executed to pick two of the three gyros for rate sensors used by the flight software. Once this selection is made, the solar array deployment check can be performed.

The basic concept behind the solar array deployment test is to check for the change of roll inertia due to the array deployment. An alternating sequence of positive and negative roll thruster firings of fixed length are commanded and resulting rate changes are measured by the gyros. If the rate changes are as expected for a fully deployed vehicle configuration, a relay is set to indicate deployment. Next, the two solar array wings are slewed to their null position with the cell side facing the -x direction. Active attitude control can now be enabled. By this time the gyros and thrusters have been checked out and a command sequence is initiated to perform a sun acquisition. This command sequence changes the operating mode from Maneuver to Sun Mode.

Sun Acquisition and Earth Search

Sun Mode uses four coarse sun sensors mounted on the outward corners of the two solar array wings to generate sun pointing errors. Thrusters, fired in an effort to null these errors, orient the spacecraft -x axis toward the sun. Rotation about the sunline is controlled to within 0.25 deg/sec by a roll rate control system. At this point, the automatic command sequence halts and the spacecraft waits for further instructions from ground. After checking out the health of spacecraft, ground operators initiate an earth acquisition. This process involves spinning up the two scan wheels to 1000 rpm and the pitch reaction wheel to 1880 rpm near 4 pm local satellite time in preparation for earth search. A command is sent to roll the spacecraft at about 2 deg/sec in order to sweep the earth sensor field-of-view across the earth. When the mode transition logic detects the proper earth sensor signals, a switch is made from Sun Mode to Maneuver Mode.

Earth-Referenced Attitude Control in Maneuver Mode

In Maneuver Mode the spacecraft attitude is three-axis stabilized with the z axis nadir-pointed and the X-axis along the sunline. Subsequently, a small yaw maneuver is commanded to align the x axis with the velocity vector and the solar arrays are put into track mode to stay sun-pointed.

Maneuver Mode uses a gyro-based reference with attitude updates from earth and fine sun sensors. Since gyro rates are propagated in ECI coordinates, an onboard orbit ephemeris is required to relate the ECI and LVLH frames. When the spacecraft is nadir-pointed, the onboard attitude determination filter begins to work. This six-state filter determines roll, pitch and yaw estimates with respect to LVLH and also updates the gyro rate biases. Earth sensor roll and pitch updates are available continuously while the fine sun sensor yaw updates occur only once per orbit around 6 am LST. Convergence of this attitude determination process takes about 1-2 orbits.

After the vehicle attitude control is achieved with two thruster-based modes: Sun and Maneuver Mode, the next mode transition is to Science Mode which uses three wheels and three magnetic torquers for actuators. First the wheel speeds have to be changed for Science Mode operation. The two scan wheel are spun up to a bas speed of 2000 rpm and the pitch reaction wheel is spun up to 4000 rpm. Solar array control in Maneuver and Science Mode is the same. It uses solar ephemeris data to generate a tracking command for the two Solar Array Drive Actuators (SADA) per wing. The Alpha actuator has sliprings and it rotates 360 degrees each orbit. The Beta actuator has a flexible cable for its limited travel and it makes small adjustments in the array beta angle once per orbit near midnight LST.

ADCS OPERATION FOR SCIENCE MISSION

ADCS operation in Science Mode utilizes a simple, low-cost design implemented for high reliability and robustness. Two scan wheels, canted 20 degrees from the pitch axis, and the reaction wheel along the pitch axis provide the pitch momentum bias and control torque as well as redundant roll and pitch attitude information. A backup reaction wheel along the yaw axis is available for wheel redundancy. The attitude reference is propagated using gyros. Yaw attitude updates are provided by four fine sun sensor assemblies. Wheel momentum accumulated due to environmental disturbance torques is unloaded using three orthogonal magnetic torque rods and a tri-axial magnetometer. The fail-safe logic in Science Mode has two tiers. The first tier of the fail-safe response, Safe Hold Mode, relies on a thrusterless B-dot control law using magnetic torque rods. The second tier of the fail safe response, Sun Mode, relies on thruster-based sun pointing with full redundant equipment.

Attitude Control

In Science Mode, attitude control is provided by a pitch momentum bias system with continuous momentum unloading. Active control is provided by the wheel torques in pitch and roll which cause slight variations of the wheel speeds about the set points used to maintain the pitch momentum bias. The yaw axis is passively controlled using the roll/yaw coupling due to the momentum bias. The nominal spacecraft attitude in Science Mode maintains the positive spacecraft yaw axis pointing at nadir, the positive spacecraft roll axis along the spacecraft velocity vector and the spacecraft pitch axis normal to the orbit plane.

Momentum Management

Continuous momentum unloading is provided in Science Mode by employing three orthogonal magnetic torque rods in a bang-bang manner. The momentum unloading control system uses the well-known cross product law. Measurements of the magnetic field and the wheel speeds are inputs to the control law. In typical mission scenarios, the torque rod duty cycles in Science Mode are maintained under 10%.

Fail-Safe Response

The first tier of the fail-safe response is designed to catch problems in the attitude determination and control system based on a detected loss of nadir pointing. The earth sensor roll and pitch signals are continuously monitored and compared against threshold values. If a threshold value for either signal is exceeded, the flight software triggers an autonomous transfer to a wheel-based Safe Hold Mode using a B-Dot control law.

The objective of the B-Dot control law is to maintain the pitch axis of the spacecraft normal to the orbit plane and to induce an inertial pitch rate equal to twice the orbit rate. the spacecraft, therefore, pitches once per orbit with respect to the earth. Maintaining the pitch axis normal to the orbit pane insures high solar array exposure to the sun while allowing a quick return to Science Mode without the need to fire thrusters. While in safe Hold Mode, the wheel speeds are controlled to maintain the nominal pitch momentum bias for the current wheel configuration.

If the nadir pointing loss in Science Mode was caused by a wheel failure, the fail-safe logic first switches to Safe Hold Mode but then autonomously enters the thruster-based Sun Mode and maintains sun pointing. Safe Hold Mode requires a constant pitch momentum bias an therefore, acceptable performance cannot be assured with a failed wheel. The switch to Sun Mode is made automatically whenever a wheel control loop cannot maintain a constant wheel speed. All detected power anomalies in Science Mode trigger a direct transfer to Sun Mode.

CONCLUSIONS

The ROCSAT-1 ADCS attitude control and determination subsystem has been designed to meet the weight and power constraints of a small satellite. It was shown that low-cost hardware components can be integrated onto a reliable system architecture to satisfy the mission requirements. A two-tiered safing approach was implemented instead of adding more hardware components for standby redundancy. The ADCS is used during the early orbit phase and for science operations in the final orbit. Sensors and actuators have been selected to accommodate both mission phases. As a result, the final ROCSAT-1 ADCS design is versatile, robust and affordable.

REFERENCES

TRW Report 61738, ADCS Subsystem Design Report, Dated October 20, 1995.

AN OVERVIEW OF ROCSAT-1 OCI SCIENCE TEAM AND SCIENCE DATA DISTRIBUTION CENTER

Hsien-Wen Li, Chung-Ru Ho, Nan-Jung Kuo, Wei-Peng Tsai and Shih-Jen Huang

Department of Oceanography, National Taiwan Ocean University, Keelung, Taiwan

ABSTRACT

The ROCSAT-1, the first Taiwan's scientific experimental satellite, will be launched by February 1999 to a 600 km circular orbit. It will carry an ocean color instrument on board to monitor ocean color. This space borne ocean color instrument is called Ocean Color Imager. It is a multi-spectral push-broom imager that is designed to map six reflected spectral radiances from ocean surfaces. The ROCSAT-1 is on a 35° inclination orbit that is different to most of operational remote sensing satellites which are placed on a sun-synchronous orbit to cover the entire earth. The selection of the orbit was based on the need of maximum contact time with the ground station in Taiwan. There are two major groups for the scientific purpose of Ocean Color Imager, the science team and the science data distribution center. The aims of science team are to do sensor calibration, to correct atmospheric errors, to develop the bio-optical algorithms, and to work on data validation. Data from the ocean color instrument will be processed and then distributed by the data distribution center. Data are free to the scientific and educational communities. They can request the data through internet or regular mail after the satellite launch. The web site of the data center and science team is at "http://www.oci.ntou.edu.tw."

INTRODUCTION

Ocean color is generally affected by the concentrations of phytoplankton, suspended matters and gelbstoffe. The change of ocean color means the change of concentrations of such materials near the sea surface and it may also relate to marine productivity and pollution. From remote sensing point of view, we mean that ocean color is the relative amounts of water-leaving radiance in the various portions of the visible spectrum. Measuring water-leaving radiance can derive concentrations of pigment or chlorophyll a. Because of the role of phytoplankton in the global carbon cycle, to understand its distribution and variability can help us to assess the role of ocean in the global change. The distribution and variability of ocean color can also help scientific communities to study the dynamics of ocean, the evolution of eddies, coastal currents and the physics of mixing. From a scientific perspective, acquisition of ocean color data from space in the 1990s is a high priority goal that has been recognized in reports of the National Research Council of the National Academy of Science (Hooker et al., 1992).

The Republic of China on Taiwan plans to launch her first experimental satellite by February 1999 to a 600 km orbit. The satellite is dubbed "ROCSAT-1" which is a 400 kg, three-axis stabilized, low-earth orbit satellite with a 35° inclination. The period is about 96 minutes. The ROCSAT-1 is designed to carry out three scientific experiments, including ocean color imaging, ionospheric plasma and electrodynamics, and Ka-band communication. The ocean color imaging is trying to follow the success of Nimbus-7 Coastal Zone Color Scanner (CZCS, 1978-1986) and the Sea-viewing Wide Field-of-View Sensor (SeaWiFS, 1997-) to better understand environmental changes. The payload instrument of ocean color imaging on ROCSAT-1 is called Ocean Color Imager (OCI) which is used for mapping chlorophyll a concentration distribution and investigating influence of atmospheric aerosols in remote sensing.

DESCRIPTION OF OCI

Most of operational remote sensing satellites are placed on a sun-synchronous orbit to cover the entire earth but the ROCSAT-1 is on a 35° inclination orbit. The selection of the orbit was based on the need of maximum contact time with the ground station in Taiwan. Since the ROCSAT-1 does not broadcast data in L band, it transmits science data through its Telemetry, Tracking, and Command (TT&C) channel on S band (Yang, 1995). It means that only the selected sites can the data be downloaded, and that the data volume is severely limited. The situation is exasperated by the fact that only one ground station in Taiwan can the data be dumped. Therefore, in order to download the data as many as it can, the contact time with the ground station has to be as much as it can.

The OCI is a multi-spectral push-broom imager that is designed to map six reflected spectral radiances from ocean surfaces. The six spectral bands (443 nm, 490 nm, 510 nm, 555 nm, 670 nm and 865 nm) are identical to the six of the eight spectral bands on the SeaWiFS. A comparison of SeaWiFS and OCI is shown in Table 1. Of the six bands, the first four short wavelength bands are expected to reveal spectral contents of the scattered matters. The red (670 nm) and the near infrared (865 nm) bands are primarily used for atmospheric corrections because the incident radiations are mostly absorbed on ocean surfaces. Since this push-broom instrument does not have any moving parts, this makes design and handling much easier. An 800-m square footprint on the ground is formed by the "push-broom" action. The OCI's fixed optics has a total field of view of 60 degrees. This translates to about 700 km swath. The OCI is fixed and does not have any mechanism to tilt for solar glint avoidance. It is always nadir-pointing. A detail description of the OCI can be found in Yang (1995).

Table 1. Comparison of characteristics of SeaWiFS and OCI.

	SeaWiFS	OCI
inclination	98.25°	35°
altitude (km)	705	600
period (min)	98.9	96.6
orbital repeat time (days)	16	52
spectral bands (nm)	B1 402-422	B1 433-453
	B2 433-453	B2 480-500
	B3 480-500	B3 500-520
	B4 500-520	B4 545-565
	B5 545-565	B5 660-680
	B6 660-680	B6 845-885
	B7 745-785 B8 845-885	B7 545-565
nadir pixel (m^2)	1130 × 1130	800 × 800
swath width (km)	2801	704
redundancy	no	555 nm
color sensing	scanner	push broom
crossing equator time	noon (local time)	9:00 ~15:00
bits	10	12
tilt	-20°, 0°, 20°	no
gain selection	0.5, 1, 2	band 1~6: 0.5, 1, 2 band 7: 0.5, 1, 6
launch date (year/month)	1997/8	1999/2

THE SCIENCE TEAM

To have valid OCI data products, the sensor calibration has to be performed first. There is no on-board solar calibrator on the OCI. Therefore, the possibility of using a biologically inert area to calibrate the radiometric stability of OCI is considered. The 35° inclination orbit provides an interesting opportunity to try some experimental imaging. As the satellite crosses all local times, it is believable that the OCI can be switched on to image the biologically inactive scenes right after the local dawn, and to repeat the same measurement at various local time as the satellite passes the same scene overhead. If the scene is truly Lambertian, then the signal should follow the cosine law with respect to the solar elevation angle. In this case, it would be possible to calibrate OCI's radiometric stability using the biologically inactive area as a reference source which can not be afforded by a sun-synchronous imager.

The second proposed method is inter-comparison. As OCI crosses a given ocean at a selected local time, image data can be used to compare against information from other instruments, such as SeaWiFS. This type of symbiotic inter-instrument calibration activities near Taiwan waters is being considered. However the cross-calibration method is not good enough because of the uncertainty of sensors. Therefore, a vicarious calibration method described in Muller and Austin (1992) has to be employed in accordance with the cross-calibration method. Some experimental regions near to Taiwan for calibration are selected. When the OCI and the SeaWiFS pass those region at the same time, the OCI will take images simultaneously. We will compare OCI images with SeaWiFS images for the inter-calibration. Meanwhile, in-situ optical properties and water samples at that region will be taken for vicarious calibration.

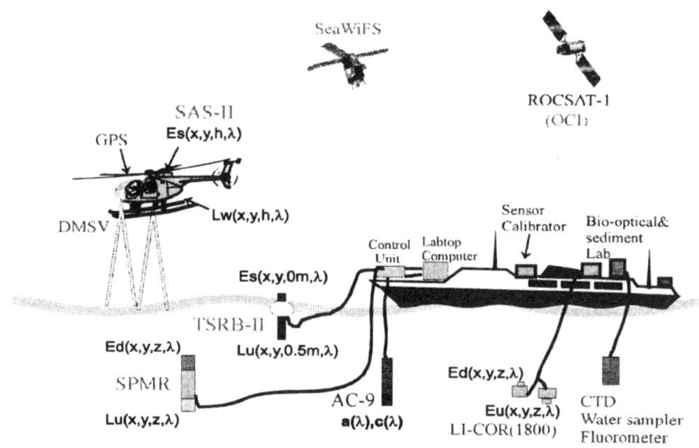

Figure 1. OCI cal/val optical system.

The OCI data products have to be validated before they distributed to users. The region for a validation program of OCI will be on the adjacent waters to Taiwan. Three research vessels, Ocean Researcher I, II, and III, can support the program for ship measurements. To correct the atmospheric effects, a modified LOWTRAN-7 maritime and coastal aerosol models will be employed for aerosol correction. Theoretic model will be applied to Rayleigh correction. After doing atmospheric correction, the retrieval of the water-leaving radiance from measurement of the total radiance will be validated by field measurements. A Satlantic ROCSAT-1 OCI cal/val optical system including a SAS II (SeaWiFS Aircraft Simulator), a TSRB II (High Resolution Tethered Spectral Radiometer Buoy) and a SPMR (SeaWiFS Profiling Multi-channel Radiometer) will be applied to work on the water-leaving radiance validation. Figure 1 illustrates the cal/val optical system. An aircraft to measure sky irradiance and water leaving radiance from atmosphere and research vessels to measure downwelling irradiance, upwelling radiance, and sky irradiance from ocean simultaneously are employed in these in-situ measurements. The CZCS pigment concentration algorithm will be applied to derive pigment concentration from those optical properties (for example, Morel and Gordon, 1980). We will also apply SeaBAM (SeaWiFS Bio-optical Algorithm Mini-workshop) algorithms to derive chlorophyll a concentration.

When we measure spectral radiation, we also take water samples to determine pigment and chlorophyll a concentrations. The HPLC (High Performance Liquid Chromatography) and fluorescence methods will be used to analyze pigment and chlorophyll a concentrations. The chlorophyll a or pigment concentration derived from the system will compare with the in-situ measurements to validate the algorithms. The coefficients of those algorithms will be corrected if they are necessary. When the OCI on its orbit, the data products derived from the sensor will be used to compare with the products from the cal/val optical system and be compared with the in-situ measurements in order to validate the OCI data and products. The science team participants are listed in Table 2.

Table 2. OCI science team participants.

Calibration/Validation	Hsien-Wen Li (National Taiwan Ocean University)
	Wei-Song Lin (National Taiwan University)
	Chung-Ru Ho (National Taiwan Ocean University)
	Nan-Jung Kuo (National Taiwan Ocean University)
Atmospheric Correction	Gin-Rong Liu (National Central University)
	A.J. Chen (National Central University)
	Wann-Jin Chen (Chung Cheng Institute of Tech.)
	Chung-Yi Tseng (Academia Sincica)
	Shih-Jen Huang (National Taiwan Ocean University)
Bio-optical Algorithm	Chun-Te Chen (National Taiwan Ocean University)
	Tisn-His Fang (National Taiwan Ocean University)
	Wei-Peng Tsai (National Taiwan Ocean University)

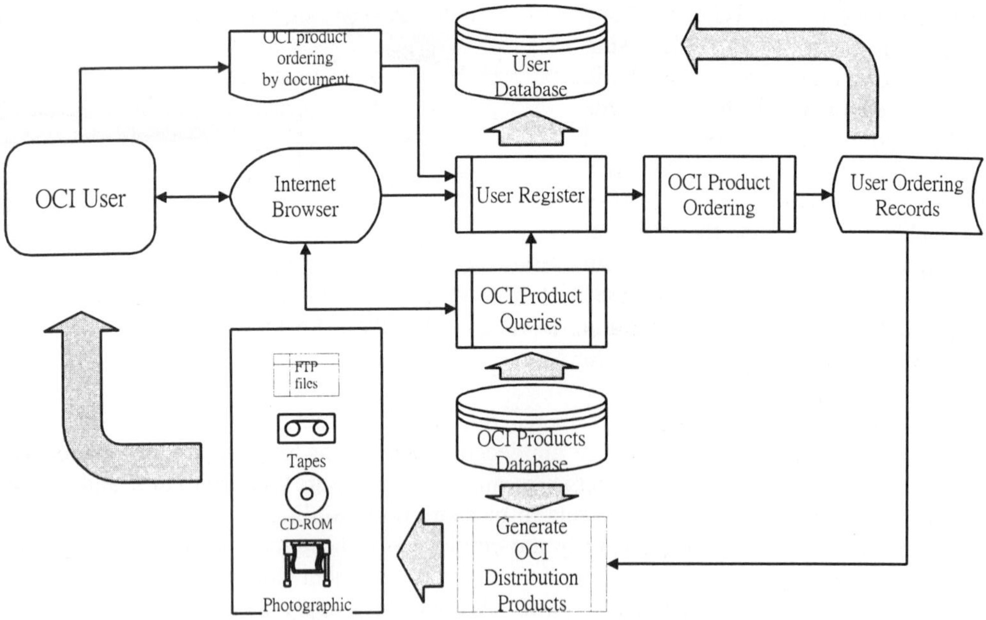

Figure 2. OCI data distribution flowchart.

THE SCIENCE DATA DISTRIBUTION CENTER

Data from the OCI instrument will be processed and then distributed by the science data distribution center (SDDC). The SDDC is conducted by the Department of Oceanography of National Taiwan Ocean University (NTOU). Scientists from universities and research institutions in Taiwan as well as that from abroad can request the OCI data from SDDC through electronic network. The data distribution flowchart is shown in Figure 2. Major functions of the SDDC include OCI science operation support and science data management. Specifically, the SDDC is responsible for organizing the OCI user groups and for developing software that will be requires for processing and analyzing the OCI data. During the mission operation phase, the SDDC is also a day-to-day operating and mission planning center. Upon receiving the OCI Level-0 data sets and the satellite orbit and attitude information from the Science Control Center (SCC) at the National Space Program Office (NSPO) of Taiwan, the SDDC will perform data decoding and further processing to generate the Level-1 data files and Quick-Look plots for monitoring the instrument performance. Moreover, the SDDC will compile the higher level (Level 2 and 3) OCI data according to various parameters in response to the requests of scientists and users. Throughout and after the mission operation phase, the SDDC is responsible for storing, archiving, querying, and distributing the OCI data and algorithms. Furthermore, the SDDC and science team will actively communicate with the scientific and educational communities to promote the maximum usage of OCI data. A web site related to the OCI SDDC and science team is located at "http://www.oci.ntou.edu.tw/."

CONCLUSIONS

The ROCSAT-1 is Taiwan's first scientific experimental satellite that will carry an ocean color sensor to monitor ocean colors. The OCI science team and data distribution center have the responsibilities to prepare quality OCI data and make them available. In order to have contribution to the international scientific communities, OCI data are free to the scientific or educational users of the world. Anyone who has interest on the OCI data for scientific or educational purposes can request the data from OCI SDDC through electronic network. The OCI data products will include normalized water leaving radiance (at 443 nm, 490 nm, 510 nm and 555 nm, respectively), CZCS type pigment concentration, and chlorophyll *a* concentration. The detail information about OCI and data policy are also available on the OCI world-wide-web homepage.

ACKNOWLEDGMENTS

We thank Mr. T.-M. Lin and Miss W.-J. Su for preparing the plots. This work is supported by National Space Program Office through grant NSC83-NSPO-A-RDD-019-002, NSC84-NSPO-(A)-OCI-001-02, NSC85-NSPO-(A)-OCI-019-02, NSC86-NSPO-(A)-OCI-019-01 and NSC86-NSPO-(A)-OCI-019-02.

REFERENCES

Hooker, S. B., W. E. Esaias, G. C. Feldman, W. Gregg, and C. R. McClain, An overview of SeaWiFS and ocean color. *NASA Technical Memorandum 104566*, Vol. 1, ed. S. B. Hooker and E. R. Firestone, 24 pp (1992).

Morel, A. Y. and H. R. Gordon, Report of the working group on ocean color, *Boundary Layer Meterol.*, **18**, pp. 343-355 (1980).

Mueller, J. L. and R. W. Austin, Ocean optics protocols for SeaWiFS validation, *NASA Technical Memorandum 104566*, Vol. 5, ed. S. B. Hooker and E. R. Firestone, 43 pp (1992).

Yang, B. T., The first ocean remote sensing payload of the ROC: an introduction, *COSPAR Colloquium, Space Remote Sensing of Subtropical Oceans*, Sept. 12-16, 1995, Taipei, Taiwan, pp. 13A2-1-13A2-9 (1995).

POWER, PROPULSION, AND COMMUNICATIONS FOR MICROSPACECRAFT MISSIONS

Wim De Groot, T. M. Maloney, and M. J. Vanderaar

Nyma Inc., 2001 Aerospace Parkway, Brook Park, OH 44142, USA

ABSTRACT

The development small sized, low weight spacecraft has the potential to reduce scientific mission costs through lower fabrication and launch costs. An order of magnitude reduction in spacecraft size can be obtained by miniaturizing components. An additional reduction in spacecraft weight, size, and cost can be obtained by utilizing the synergy that exists between different spacecraft systems. State-of-the-art of three major systems are discussed; spacecraft power , propulsion, and communications. Possible synergies between these systems or between these systems and the payload are identified. Benefits of several of these synergies are discussed.

INTRODUCTION

In an era of continually shrinking budgets it is prudent to investigate means of reducing spacecraft mission costs. These costs can roughly be divided among three categories: a. spacecraft cost (mission payload and spacecraft bus); b. launch costs; c. mission operations and data analysis cost. Reducing the spacecraft size and mass impacts both spacecraft- and launch costs. Proposed spacecraft definitions are indicated in Fig. 1. which shows spacecraft mass and power. Most existing spacecraft follow the 1 W/kg energy density line, which will be used as general guideline. Because of improvements in photovoltaics, this value may increase for additional power. Microspacecraft are approximately defined as spacecraft smaller than 100 W, 100 kg. Few existing spacecraft in this category are equipped with propulsion systems. Fig.1 should not be seen as a spacecraft comparison because Cassini includes a large propulsion system wet mass, whereas the Hubble and Microsat do not posses propulsion systems.

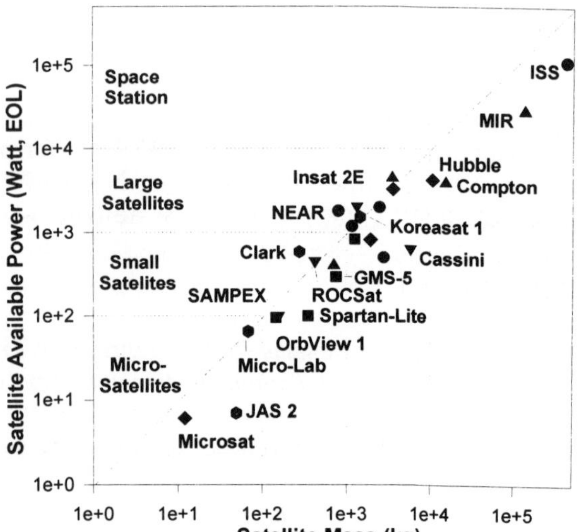

Figure 1. Spacecraft Classification

An order of magnitude mass and volume reduction requires radical changes in component technology. Janson (1994) provides an approach towards obtaining microcomponents with micro-electromechanical systems (MEMS) technology and describes the inherent MEMS advantage of easy subsystems integration as a result of the silicon base material properties. His bold vision of an armada of sub-kg spacecraft, composed out of a three-dimensional stack of silicon wafers, and lofted in orbit with a single launch vehicle, is a challenging goal for future spacecraft builders.

However, one of the most important spacecraft components that does not lend itself to be downscaled easily is the propulsion system. One scenario used to circumvent this problem is the use of a "mothership" that brings the microspacecraft in the desired orbits. This eployment bus" will have the propulsion system needed to deploy the microspacecraft and have the high data rate communications capability lacking the microspacecraft. A weak point in this approach is the single point failure potential of the mothership to be deployed. Although it is very likely that this vision will, at sometime in the future, become a reality, technical hurdles and limitations in R&D funding will delay the realization.

For many smaller countries and for scientific institutions with spaceflight aspirations but limited budget, a different, low cost approach towards miniaturization needs to be followed to obtain significant mass and cost savings. Such a strategy, which includes incremental reduction in size and weight of spacecraft components, calls for more efficiently designed satellites, which are smaller in size and budget. Several subsystems, such as sensors, electronics, and communications lend themselves readily for miniaturizing. Due to the usually large required consumables, however, propulsion systems are harder to miniaturize. One approach to reduce propulsion system volume and mass is to leverage multi-functionality of spacecraft systems. An example of such multi-functionality was proposed by McElroy (1994). In this proposal, propulsion, power, satellite structure and tankage were integrated to provide maximum performance for a minimum in system weight and volume. In this scenario, the consumables (water) doubled as energy storage medium.

This paper describes the current state of the art in the three disciplines of propulsion-, power-, and communications. A number of possible integration options for these three subsystems that could reduce satellite mass and volume, while retaining mission capabilities, are reviewed. Both satellite bus subsystems and payload instrumentation can be integrated with any of the satellite subsystems. However, that option depends on the type of instrumentation used.

POWER-ENERGY CONVERSION AND STORAGE TECHNOLOGY

A number of different electrical power systems can be assessed for application to spacecraft payload power. In addition to solar cells, non-rechargeable (primary) and rechargeable (secondary) batteries, fuel cells, semi-cells, flywheels, and capacitors are candidate components for integrated power systems. Microwave and laser beamed power systems can also be considered subject to severe operational constraints as can nuclear power sources. Future spacecraft will require smaller and cheaper components than those that are presently in use so the space industry as a whole will benefit greatly from the ongoing worldwide efforts in electric vehicle component development and from the development of new military power systems. A comprehensive survey of energy source technologies is presented which includes state-of-the-art performance parameters as well as performance parameters projected to be achieved in the year 2000.

It is essential that all the components of candidate power systems are included in the assessments and comparisons among those systems. The ancillary components are especially important for space power systems where harsh environments are encountered. The nickel-hydrogen battery system designed for the International Space Station, for example, has an overall specific energy of 6.5 Wh/kg even though the batteries are nominally rated as 50 Wh/kg. In addition to the total power system mass and volume, the availability, reliability, environmental compatibility, and cost must be addressed during spacecraft design. An excellent survey of the current and projected electrical power system technologies was reported by Power Computing Solutions, Inc. (1997)

Solar Cells

Solar cells are a major component of virtually every spacecraft power system. There are three main types or classes of solar cells under production today, thin film, single junction and multiple junction. Solar cell integration with a specific spacecraft requires that the individual cells be mounted on a suitable support structure with the ability to cool the solar cell modules and/or to shield the solar cells from a space plasma environment. In general, the multi-junction cells are most efficient, reported as 23% compared to 15% for single junction and 10% for thin film. But the multi-junction cells are also the most expensive, and most difficult to handle and install. While each type of solar cell has already been flight demonstrated on solar electric unmanned aerial vehicle (UAV) platforms, the multi-junction cells are generally deemed uneconomical. Even though the lowest cost thin film cells have relatively low efficiencies, they may be the most attractive spacecraft system option if a lightweight mounting structure is used (PCS 1997).

Batteries

A myriad of primary and secondary batteries are available today ranging from the low cost primary commercial alkaline batteries that are used in ordinary flashlights to the higher cost lithium and nickel based batteries. Excellent performance has already been demonstrated in primary battery systems and no further advancements in performance are anticipated in the near future since no significant development programs are underway to improve the primary battery performance. In contrast, there are significant ongoing secondary battery development programs that will advance the state-of-the-art for lead-acid, Ni-Cd, NiH_2, Ni-MH, and several different lithium based designs.

The rechargeable lithium batteries exhibit the most promise for superior energy density and specific energy but the nickel based chemistries are still considered the baseline battery systems for space missions today. This may change before the turn of the century if the goals of the major battery development programs are achieved. One of the most significant development programs is being conducted by the United Sates Advanced Battery Consortium (USABC). While the long term goals of the USABC shown in Table 3 were targeted for 1998, it is reasonable to consider those battery level performance targets as the upper limit for secondary batteries for the year 2000. The system level performance will obviously be less than the cell level performance. A comprehensive discussion of secondary battery technology characteristics is presented by Pellerin (1996).

Table 1. United States Advanced Battery Consortium Battery Goals

	Mid-Term (1995)	Long-Term (1998)
Specific Energy (Wh/kg)	> 80	200
Energy Density (Wh/l)	> 135	> 300
Peak Specific Power (W/kg)	150	400
Power Density (W/l)	250	600
Cycle Life (cycles)	600	1000
Calendar Life (yrs.)	5	10
Efficiency (%)	75	75
Recharge Time (hrs.)	< 6	3-6

Fast Recharge Time (hrs.)	0.25	0.25
1 hr Continuous Discharge	75	75
Power and Capacity Degradation (%)	20	20
Operating Environment (°C)	-30 - 60	-40 - 85
Cost (US$/kW-hr)	< 150	< 100

Fuel Cells

Non-regenerative fuel cells were first used as a primary on-board electrical power source for the NASA Gemini Program in the 1960`s and today, the Space Shuttle Orbiter extracts all needed electricity from fuel cells. In addition to the NASA program that was initiated in early 1997 to investigate implementing new fuel cells on the Space Shuttle (Warshay 1997), there is approximately $200M per year being invested worldwide to develop fuel cell systems for military and commercial stationary power and transportation applications. As a result of these programs, the state of fuel cell technology will advance significantly in the next three years. For the short charge/discharge power system duty cycles prevalent in LEO spacecraft and for the large charge/discharge time ratios prevalent in GEO (as well as LEO) spacecraft, secondary battery systems have been better suited than fuel cells. If the spacecraft power demands are increased or if the discharge time period of the non-photovoltaic power source increases to more then several hours, fuel cell systems would be a better power system options than batteries. For a 5 kW regenerative fuel cell system with twelve hour charge and discharge times, over 230 Wh/kg can be achieved today and over 500 Wh/kg is projected to be achieved in the year 2001 (PCS 1997).

Flywheels

Several flywheel development programs are being carried out under support from the Partnership for a New Generation of Vehicles (PNGV) where the objective is to integrate flywheels into hybrid vehicles. They can also be considered as a dual purpose energy storage system for future satellites. There are several beneficial features of flywheel systems. Round trip efficiencies of over 90% have been demonstrated in system tests at Oak Ridge National Laboratory. Flywheels are projected to have long service lives (especially if magnetic bearings are used) and the bearings and power electronics should provide adequate life for the flywheel system. The principal life limiting factor is fatigue of the flywheel rotor although most of the performance predictions in the flywheel literature are based on a life of at least 100,000 deep cycles. The primary risks associated with energy storage in flywheel systems arise from a rotor failure where explosions and/or disintegration could lead to a loss of the spacecraft. Flywheels can be designed for high specific power or for high specific energy and the maximum specific energy achievable is about 400 W-hr/kg energy densities of over 100 Wh/liter are achievable.

PROPULSION

Many satellites and spacecraft require a variety of low thrust propulsion functions, including orbit insertion, attitude control, station keeping, repositioning, and primary propulsion for planetary spacecraft. As a general rule, chemical propulsion options are for high-thrust, low power applications. Electric propulsion, and its high performance levels, can be used when there is sufficient power available and where low thrust levels are acceptable. With the drive towards smaller spacecraft, miniaturized electric and chemical propulsion systems are gaining attention, especially those with dual use potential. Leveraging spacecraft subsystems for more than one function may make it easier to meet demanding mission scenarios with smaller spacecraft.

Propulsion needs for microspacecraft depend on the launch system used and the mission scenario. The most affordable launch option is to piggy-back" on an existing launch. In this case, the microspacecraft mission is constrained by the primary payload`s mission and the availability of an on-board propulsion system for orbital transfer. More expensive options are small launch vehicles or using a getaway special canister on board of the shuttle. These missions are less constrained but might still require primary propulsion for orbit transfer and/or orbit insertion. Fine pointing requirements determine the minimum impulse bit specifications of the ACS system. At the same time, scientific instruments might limit the firing frequency during deadbanding, and slew rate required for planetary observations might determine thrust level. A propulsion system that can provide optimum impulse bits satisfying fine pointing, firing frequency limitations, and slew rate needs to be found. In many cases, such a system is not available and a compromise needs to be made.

A number of low thrust propulsion options are available to carry out specific small missions (de Groot 1997). Volume and mass constraints dictate system selection. Cold gas propulsion is commonly used for tasks where the propulsion requirements are limited and where low cost and system simplicity are decisive factors. Cold gas systems are volumetrically inefficient and the relative high leakage of gaseous storage may make them unsuited for anything but small ΔV, short duration missions. Monopropellant hydrazine (N_2H_4) has been the preferred option for orbit insertion of smaller satellites. They have a higher specific impulse (Isp) than cold gas systems but are more costly and complex. Storable bipropellants, utilizing nitrogen tetroxide (NTO) as oxidizer and either monomethylhydrazine (MMH) or N_2H_4 as fuel, have been used extensively for orbit insertion of medium to large satellites and primary propulsion for planetary spacecraft. Bipropellant systems operate most effectively at thrust levels which are too high for microspacecraft and can only be considered for some primary thrust functions. Monopropellant N_2H_4 and bipropellant MMH/NTO have been used for satellite on-orbit functions, but the trend is now toward electric thruster systems to perform these duties. Resistojets have been used to augment monopropellant performance , and arcjets have recently been utilized for large satellite station keeping. These systems require too high power level to be of use for micropropulsion. Low power ion and Hall thrusters are being developed for orbit transfer and primary planetary propulsion missions (Myers 1994). However, ion and Hall thrusters operate most efficiently at power levels that can not be supplied by microspacecraft. Pulsed plasma thrusters are poised to be flight tested for precision on-orbit functions for smaller satellites.

New developments may facilitate microspacecraft missions and in some cases enable them. In the chemical arena, tri-constituent gas mixtures, with nitrogen forming the bulk of the gas and diluted with a non-combustible mixture of hydrogen and oxygen, can provide more than twice the performance of cold gas when passed over a catalyst bed. Subliming solid thrusters designed specifically for microspacecraft, have the advantage of high propellant density, simplicity, and restartability, but the low performance and potential toxicity are a drawback. reen" monopropellant systems, utilizing more benign propellants, could improve performance and will result in lower cost as the result of improved ground operations. Downscaling electric thrusters for low power requirements could make them less efficient, but still quite competitive. Pulsed plasma thrusters are relatively simple and can operate at very low power levels. The pulse frequency, however, and thus the average thrust depends on the available power. Resistojets can be build for spacecraft with higher available power, but the high weight and power requirements makes them less attractive. Field Emission Electric Propulsion, currently under development, has all the characteristics attractive for micropropulsion ACS functions. A high (~8000 s) Isp, low mass and volume, and low thrust level combine for an attractive package. However, some major technical hurdles, such as neutralization of the ion plume, spacecraft contamination by the metallic plume, and feed system problems still need to be resolved.

COMMUNICATIONS

The communications capability (data transmission rate) of a microsatellite is closely related to the amount

The communications capability (data transmission rate) of a microsatellite is closely related to the amount of power available to the communications electronics as well as the distance over which communications will take place. This is obviously very mission specific. Moreover, the physical size limitation of onboard antennas is also an important issue. In most cases the limit is on the return link from the microsatellite to the ground. In a very rough sense, near-earth microsatellite systems can support rates in excess of 100 kilobits per second (kbps) while deep space missions may have limits on the order of 10 kbps.

Increasing these rates and/or decreasing the onboard power consumption to levels appropriate for microsatellite systems can be approached in a number of ways. First, microsatellites can leverage the investment from the burgeoning commercial low power, high performance electronics industry. Much of this technology is commercially available or being developed for mobile applications such as cellular phone, personal communications systems, and wireless networks. In these applications, it is very important to conserve power to increase battery lifetime and thus increase airtime. This is being achieved by advances in devices as well innovative architectures and algorithms.

Advances in devices include low voltage operation, decreasing feature sizes, and increasing die sizes. However, in general, low power, small feature size electronics can be much more susceptible to effects of radiation including latchup and single event upset. In fact, some of these systems are sensitive to the radiation environment at sea level! Currently, researchers at Stanford University and the University of New Mexico are developing a low power standard cell library that is radiation tolerant for both terrestrial and space based applications (Gambles 1997).

Second, microsatellites can trade processing power for RF power. Architectural advances include a new breed of multi-processor, multi-instruction, and special function devices that enable the efficient using advanced processor architectures, and using advanced signal processing techniques. There are a number of algorithmic techniques to compress, error control, and modulate digital data to enable more data to be transmitted with less RF power (DC to RF power is roughly only 40% efficient). The limitation is the amount of DC power that required in digital processing to achieve the gains. Also, the same processors used for other on-board purposes can be shared for the communications applications, further reducing the required onboard mass.

For example, NASA has developed lossless signal and image compression techniques that can be used for scientific data (Yeh 1995). In some circumstances lossy compression may also be used if the data integrity can be compromised. In this case, higher compression ratios can be obtained using some of the standard commercial algorithms. As a dual to compression, error correction of the data is of critical importance. The standard method of error correction in space is to use a concatenation of two error correcting codes. This Consultative Committee for Space Data Systems (CCSDS) standard uses an nner" code based on convolutional coding in conjuction with an optional outer Reed-Solomon block code (CCSDS 1994). This combination provides a powerful error correction capability that reduces the amount of onboard power required to obtain an error-free link. The encoding is normally done on board the spacecraft, so radiation hardened encoder devices are available. There has also been some recent interest in hardening the decoder devices for receive space applications. Moreover, new iterative decoding techniques (Berrou 1993) are being developed that provide greater error correction capabilities to within 0.5 dB of the theoretical limit.

Third, higher transmission frequencies offer the promise of smaller antennas for the same amount gain. For example Ka Band frequencies and above are getting interest from both the antenna size aspect as well as the frequency congestion aspect. The antenna diameter decreases linearly with transmission frequency for the same beamwidth. The drawback of using the higher frequencies is that they are more susceptable to attentuation and distortion by intervening material such as water vapor / droplets in the atmosphere.

Moreover, power amplifier efficiency is generally lower at higher frequencies. Increasing Ka band power amplifier efficiency is and active area of research (Schellenberg 1995).

Finally, in a number of microsatellite missions, real time access to the data may not be critical. In these missions, a power scheduling techniques can be used to require the duty cycle of the satellite to ground link to be rather low. Such store and forward systems will then offer relatively high data rates for short periods of time. In these periods of time, most of the onboard power can be directed towards the communications package, supporting the downlink of the required data. The cost of this technique is that more onboard data storage is required and that instruments may have to be powered down during data transfer.

UNITIZED POWER AND PROPULSION SYSTEM

A powerful illustration of the benefits of the integration of propulsion and power is the *reversible* Unitized Regenerative Fuel Cell (URFC) design by Hamilton Standard, and shown in Figure 1. A number of papers have been published describing the operation of the regenerative system, with cell operation as an electrolysis and fuel cell (McElroy 1994, Butler 1995). The working fluid of the system is water, a safe, clean, storable fluid. The central unit is a H_2/O_2 solid proton exchange (SPE®) URFC design, based on the proton exchange membrane as sole electrolyte. This sulfonic acid membrane electrolyte is fashioned into electrochemical cells by bonding catalyst electrodes to both faces of the membrane.

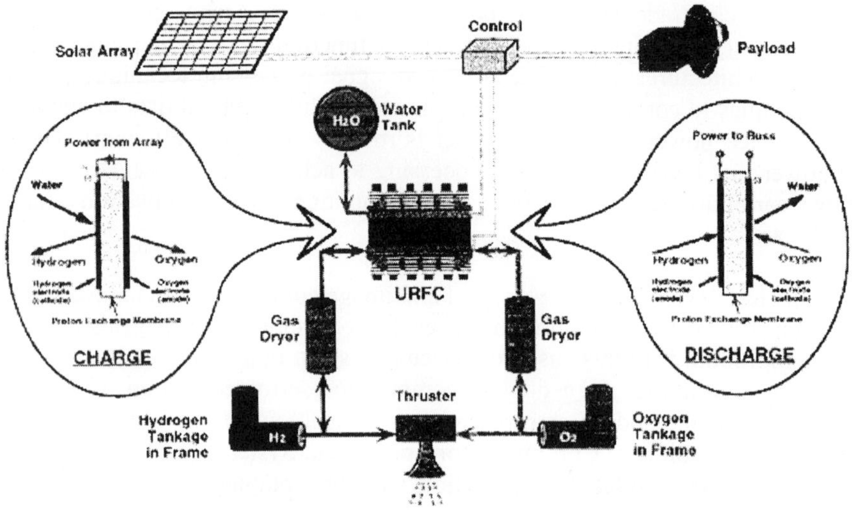

Figure 2. Integrated Modular Propulsion and Regenerative Electro-Energy Storage System.

A single electrolysis cell consists of a water feed chamber, a water permeable membrane, a hydrogen chamber, a SPE membrane, an oxygen chamber, an electrochemical hydrogen pump, and electrical insulators on both end plates. Hydrogen and oxygen are produced on either side of the SPE membrane with the application of DC power. The water feed chamber is separated from the hydrogen gas chamber by water permeable membranes which allow osmotic water transport into the hydrogen chamber. Because water is being consumed to produce propellants, a water gradient is established across the water feed barrier and more water from the storage tank enters the cell. An electrochemical hydrogen pump, drawing a few milliwatt of power, assures that no hydrogen builds up in the water feed chamber.

Surplus power from the photovoltaic array, not needed for payload and housekeeping functions, is supplied to the URFC. Water electrolysis inside the URFC generates hydrogen and oxygen. These gases are stored inside spacecraft structural components lined with lightweight bladder liners providing a permeation barrier for gas storage. State of the art hydrogen bladder technology developed at Lawrence Livermore National Laboratories safely allows such storage.

Several spacecraft functions can be serviced by the stored gases, as shown in Figure 2. The cold gases can be used for attitude control (ACS) and fine pointing, by providing low thrust, small impulse bit pulses through small cold gas thrusters. The gases can also be used as propellants in gaseous bi-propellant thrusters for primary propulsion or larger delta-V maneuvers. This dual mode potential allows replacement of conventionally used combinations of hydrazine/cold gas systems at a significant weight and volume savings. For periods during which the solar arrays can not provide power, the URFC, operating as a fuel cell, produces DC power and water, returning the water to its thin-walled, light weight storage container. For many missions, the fuel cell by itself compares favorably to state of the art batteries in terms of power vs. weight ratio. Integration with the propulsion system provides additional reduction in spacecraft weight and volume.

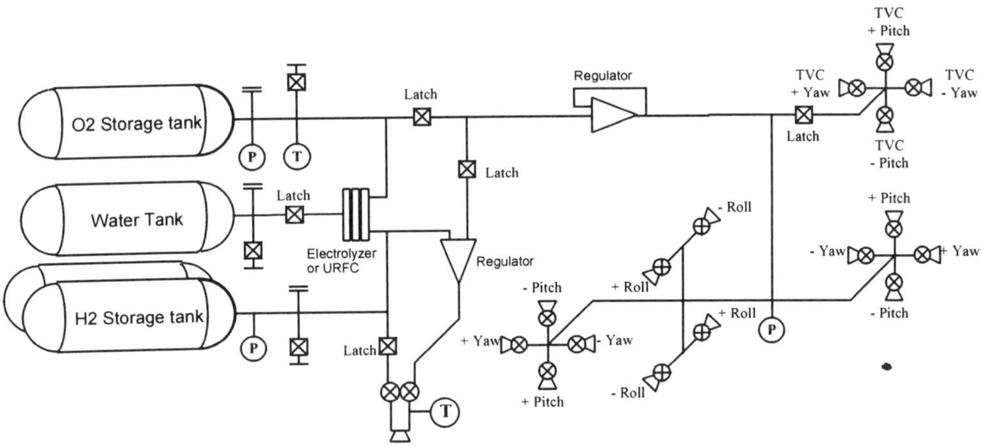

Figure 3. Schematic of Dual-Mode Electrolysis Propulsion System (Lawrence Livermore Labs.)

INTEGRATED POWER AND ATTITUDE CONTROL

One opportunity which should be explored during spacecraft design phases is to utilize a flywheel system for both energy storage and spacecraft control. By appropriate positioning and control of flywheels, a net torque can be placed on the spacecraft by a combination of spin up/spin-down of the flywheels used for energy storage which could, as a best case, eliminate the need for any control surfaces on the spacecraft. This Integrated Power and Attitude Control System (IPACS) stores kinetic energy in the motion of the flywheel, and utilizes the angular momentum of the rotating flywheel for spacecraft attitude control, such has historically been done with reaction wheels (control moment gyros).

Developments in flywheel design that have made the IPACS possible are: 1. the use of magnetic bearings which reduce friction losses, increase lifetime, improve balancing, and reduce noise effects (vibrations) that can introduce pointing jitter and affect sensor instruments, and 2. the use of composites which made it possible to concentrate the mass on the edge of the flywheel for a maximum of energy storage with a minimum of mass.

The main advantages of the magnetic bearing based IPACS system are: 1. the high energy storage roundtrip efficiency (85-90%) as compared to batteries (75-78%) or fuel cells (50%), which saves energy; 2. the small volume required; 3. the long life which increases mission life and reduces cost; 4. increase in specific energy, which reduces mass; 5. the large control torques which reduces the need for propulsion systems; and 6. the reduction in vibrations which improved instrument performance. Maximum benefits are expected for LEO.

Flywheels utilizing magnetic bearings have been investigated for energy storage systems in space (Santo 1995). At the same time, a Japanese on-orbit experiment, named Magnetic Bearing Flywheel Experiment System (MABES) showed (Nakajima 1990) that magnetic bearings exhibited sufficient stability in a microgravity environment for spacecraft attitude control. Most component issues that needed to be addressed to make the IPACS a reality have been resolved (Gondhalekar 1991). Issues pertaining to component integration are currently being addressed.

INTEGRATED PAYLOAD, COMMUNICATION, AND PROPULSION

For a number a scientific experiments, cryogenic dewars carry liquid helium to cool the instrumentation to a temperature where noise levels are low enough to obtain desired accuracy. Missions planned that include cryogenic helium are the Gravity Probe B (GP-B), STEP, and GOCE. Despite good insulation, boil-off of the liquid helium will occur. If the gaseous helium leaks, an undesirable, constant, microscopic thrust will occur.

However, it is possible to utilize the gaseous helium to the spacecrafts' advantage. The same cryogenic liquid can improve Ka band reception and reduce reciever size by improving antenna pointing. The improvement is not sufficient to warrant a dedicated cryogenic system, but leveraging cryocooling of an available system will reduce power consumption and reduce errors.

Depending on the orbit and solar activity, some drag will occur on the spacecraft. This drag causes a microgravity environment in the spacecraft. To make the spacecraft drag free, the boil-off helium can be used to compensate for the drag. Because the boil-off is extremely low with good insulation, the available gaseous helium might not be enough to compensate for the drag encountered. In that case the insulation needs to be reduced until a level of boil-off occurs sufficient to make-up for the predicted drag. However, this will also reduce the experiment lifetime.

To make up for the average drag of approximately 1 mN encountered by the GP-B, helium thrusters are designed to operate at a thrust range from 73 μN to 30 mN. The comparable Isp are 150 s and 135 s. These thrust levels are in the desirable range for micropropulsion applications.

MISCELLANEOUS

Other opportunities are leveraging subsystem components for spacecraft structure. Examples are the use of flywheel containment structure as part of the spacecraft structure or using batteries as spacecraft structural components.

SUMMARY

In order to reduce mission costs, spacecraft have to be designed smaller and more efficient. A reduction in spacecraft size can only be accomplished with size reductions in systems and subsystems. Several subsystems, such as sensors, electronics, and communications lend themselves for miniaturizing. However, propulsion systems, due to the usually large required consumables, are harder to miniaturize.

One approach to reduce weight and volume is to combine hardware for several spacecraft systems, such that the same hardware performs several subsystem functions.

State-of-the-art in the power, propulsion, and communications area as it relates to small and miniature components have been described. Technology directions to obtain lighter, less expense systems have been highlighted. Individual approaches to reduce volume and weight of each of those systems have been discussed.

Several integrated systems and subsystems have been discussed. An integration of primary and ACS propulsion with power and spacecraft structure has been described. This system which is based on water electrolysis and fuel cell technology for bipropellant thrust and energy storage utilizes state-of-the-art membrane technology for propellant storage inside spacecraft structural components.

Flywheels have been discussed as a means of combining high specific energy storage with accurate ACS functions, primarily for LEO spacecraft. The utilization of cryogenic boil-off for drag make-up thrusters has been discussed.

REFERENCES

Berrou, C., A. Glavieux, and P. Thitimajshima, Near Shannon Limit Error-Correcting Coding and Decoding: Turbo Codes, *ICC*, pp 1064-1070 (1993).

Butler, L.B., Integrated Modular Propulsion and Regenerative Electro-Energy Storage System (IMPRESS), Hamilton Standard Proposal to New Millennium Program, (1995).

CCSDS 101.0-B-3: *Telemetry Channel Coding. Blue Book.* Issue 3. This recommendation has been adopted as ISO 11754:1994. Available at http://www.ccsds.org (1992).

de Groot, W.A., Propulsion Options for Primary Thrust and Attitude Control of Microsatellites, COSPAR Colloquium on Scientific Microsatellites, Tainan, Taiwan (1997).

Gambles, J. W., A Path Towards Low-Cost Rad-Tolerant CMOS, 6th NASA Symposium on VLSI Design (1997).

Gondhalekar, V., J.R. Downer, D.B. Eisenhaure, R.L. Hockney, and B.G. Johnson, B.G., Low Noise Spacecraft Attitude Control Systems, *Proc. 26th Intersociety Energy Conversion Engineering Conference*, Boston, MA, Vol.4, pp 244-249 (1994).

Janson, S., Chemical and Electric Micropropulsion Concepts for Nanosatellites, Paper AIAA-94-2998, 30th Joint Propulsion Conference, Indianapolis, IN (1994).

McElroy, J., Unitized regenerative fuel cell storage system for aircraft and orbital applications, UTC Hamilton Standard Div., Rept. BD94-02 (1994).

Myers, R.M., Oleson, S.R., Curran, F.M., and Schneider, S.J., Chemical and Electrical Propulsion Options for Small Satellites, Proc. 8th AIAA Utah State University Conference on Small Satellites (1994).

Nakajima, A., On-orbit Experiments with Small Satellites , *Proc. of 17th Int Symp. On Space Technology and Science*, Tokyo, Japan, Vol. 2, pp. 1465-1470 (1990).

Pellerin, Alain, Are You Talking Batteries, Electric and Hybrid Vehicle Technology (1996).

Power Computing Solutions, Inc., Energy Source Technology Survey for High Altitude UAV, Final Report, Contract NAS2-96011, NASA Ames Research Center (1997).

Santo, G. E., S.P. Gill, J.F. Kotas, and R. Paschal, Feasibility of flywheel energy storage systems for applications in future space missions, NASA CR-195422 (1995).

Schellenberg, J.M. A High-Voltage, Ka-Band Power MMIC With 41% Efficiency, *Proc. of the 17th Annual GaAs IC Symposium*, 95CH35851, Oct. 29 - Nov. 1, 1995, pp. 284-287.

Warshay ,M. *et al.*, The NASA Fuel Cell Upgrade Program for the Space Shuttle Orbiter, IECEC (1997).

Yeh, P. and W. H. Miller, A Real Time Lossless Data Compression Technology for Remote Sensing and Other Applications, ISCAS-95 (1995).

PROPULSION OPTIONS FOR PRIMARY THRUST AND ATTITUDE CONTROL OF MICROSPACECRAFT

Wim De Groot

Nyma Inc., 2001 Aerospace Parkway, Brook Park, OH 44142, USA

ABSTRACT

Order of magnitude decreases in the size of scientific satellites and spacecraft could provide concurrent decreases in mission costs because of lower launch and fabrication costs. Although many subsystems are amenable to dramatic size reductions, miniaturization of the propulsion subsystems is not straightforward. There are a range of requirements for both primary and attitude control propulsion, dictated by mission requirements, satellite size, and power restrictions. Many of the established propulsion technologies can not currently be applied to microspacecraft. Because of this, micro-electromechanical systems (MEMS) fabrication technology is being explored as a path for miniaturization.

INTRODUCTION

During the last few decades, increasingly larger scientific satellites have been built and launched to provide powerful research tools. The expense of these large spacecraft is such that only one or two satellites can be built per decade. The single large spacecraft approach brings with it the risk that failure of the satellite to be deployed leaves a large void in scientific research.

In order to spread the risk of deployment and to reduce the costs associated with satellite launch and operation, it is imperative to reduce satellite size by at least an order of magnitude. This sharp reduction in size has far reaching consequences for all spacecraft subsystems, including the propulsion. Virtually all current spacecraft utilize propulsion systems in one way or another. The propulsion subsystem can account for anywhere from 20 to 90 percent of the total spacecraft mass. Within limits, propulsion system dry mass can be scaled down. However, for conventional systems, both electric and chemical, performance typically decreases with decreasing size. In most cases, therefore, propellants will occupy a larger proportional mass and volume for smaller satellites than for larger satellites with similar ΔV requirements. Electric propulsion systems are state-of-the art for primary propulsion for larger spacecraft with high ΔV requirements. Microsizing spacecraft and spacecraft subsystems, however, requires a re-evaluation of the propulsion system selection with spacecraft characteristics such as required mission ΔV, available propulsion system mass, available power, and plume contamination issues as prime criteria.

The simplest method to reduce weight, size, and cost of propulsion systems is to accept single string design (with few exceptions) (Santo 1996). Because a large fraction of launch costs consists of safety procedures surrounding the storage, handling, and loading of toxic and/or carcinogenic propellants used by established propulsion systems, additional cost savings can be obtained by utilizing environmentally safe propellants, with better performance and/or storage characteristics than existing monopropellants.

To enable an order of magnitude reduction in spacecraft size while retaining mission capabilities, nothing less than revolutionary developments in propulsion technology are needed. Such developments have not been forthcoming, most likely because the emphasis on microspacecraft is just starting. The need for innovative technologies is graphically represented in Figure 1. Some existing spacecraft are shown, from the 12 kg, 0.006 kW Microsat-I (no propulsion) to the 10.8 metric ton, 4.3 kW Hubble. At the low mass/power end of the graph, an area exists where, depending on mission requirements, spacecraft propulsive needs may be met with scaled down versions of conventional propulsion systems. The boundaries of this echnology barrier" are not well defined.

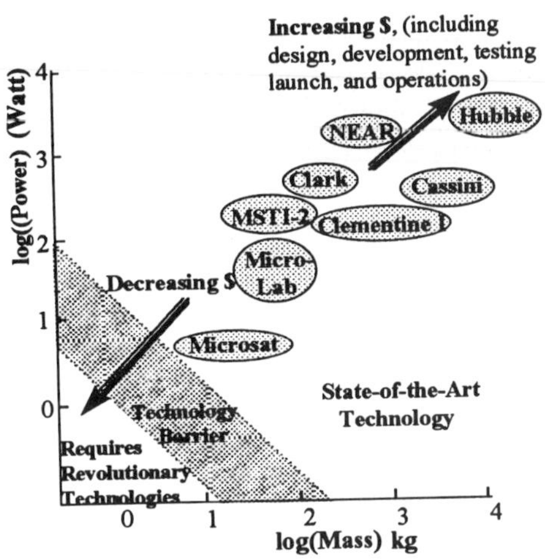

Figure 1. Technology Requirements for Microspacecraft.

One technology that has the potential to revolutionize micropropulsion is a manufacturing technique that utilizes etching and photolithography, such as used in computer chip manufacturing. Both chemical and electric propulsion concepts could benefit from this micro-electromechanical systems (MEMS) technology. Some propulsion requirements that can not be met with propulsion systems made with conventional manufacturing, could be replaced with MEMS manufactured hardware, such as MEMS resistojets or MEMS monopropellant thrusters. A major advantage of applying MEMS technology at this stage of development appears to lie in the cost savings that can be obtained with batch fabrication. An additional advantage is the potential of chip integration of thrusters with computing, communication, and sensing equipment, allowing low cost, batch fabricated microspacecraft. Several programs have been initiated to investigate MEMS technology for propulsion and power applications. For very low mass/power requirements, MEMS machined conventional concepts would not be sufficient and innovative new concepts need to be developed.

A number of papers have recently been written on the subject of propulsion options for microsatellites. De Groot and Olesen (1996), in a review paper covering chemical propulsion options, categorize systems by thrust level without regard to satellite size. Thrust level deemed relevant for micropropulsion ranged from 10^{-6} to 10 N. Several mission scenarios were described in that paper for spacecraft from 20 to 350 kg mass. An excellent review of microspacecraft propulsion options was given by Mueller (1997). Mueller defined microspacecraft as spacecraft of mass roughly smaller than 20 kg, power smaller than 0.02 kW, and dimensions approximately 0.4 m on the side. For scaling he assumed a power density of 0.001 kW/kg and subsequently subdivided these spacecraft by evaluating whether scaled down versions of conventional technology could be utilized or whether alternate (e.g. MEMS) technology need to be applied.

This paper reviews propulsion systems currently available or in development stages nearing completion. Critical performance data for minimum thrust level of full size propulsion systems are given as well as estimates of performance for a strongly scaled down version. Typical mission criteria are reviewed and the most likely propulsion systems for the microspacecraft mission are analyzed.

MICROSPACECRAFT PROPULSION REQUIREMENTS

Many of the impulse requirements laid out in this section were generated during a micro-propulsion workshop and are shown in Table 1 (JPL 1997). Typical science missions envisioned to be executed with micro-spacecraft are single spacecraft planetary or asteroid missions, or multiple spacecraft observer clusters. For many micro-spacecraft missions, the primary velocity increases (ΔV increments) required during spacecraft maneuvers are equal to those of larger spacecraft. The primary ΔVs refer here to large changes in orbit parameters (e.g. orbit transfer, repositioning, deorbit, and planetary capture/escape). Secondary ΔV refers to smaller changes in orbit parameters, such as orbit maintenance (e.g. stationkeeping and drag makeup), adjusting orbit parameters (phasing of a single or multiple spacecraft), or attitude control [ACS] (e.g. three axis control, momentum/reaction wheel dumping, or spacecraft spin control.). Microspacecraft primary ΔV missions call for lower thrust levels corresponding to the spacecraft mass and power availability. Microspacecraft secondary ΔV missions will additionally call for small impulse bits to achieve similar secondary ΔV increments as larger spacecraft. In some instances, the widely varying primary and secondary ΔV requirements of each mission type can only be met by a combination of different primary and secondary propulsion systems, whether of impulsive or long thrust duration(e.g. electric propulsion).

Table 1. Typical Microspacecraft Propulsion System Requirements (JPL Workshop):

Mission	Primary ΔV/Duration	(m/s)/yr	ACS ΔV/Duration (m/s)/yr	Thrust (mN)
	Impulsive	Non-impuls.		
Vesta Rendezvous	3400/4.3		2.5 * S/Cmass	0.5 * S/Cmass
		7000/2.7	2.5 * S/Cmass	0.5 * S/Cmass
Europe Orbiter	2500/4.8		5.0 * S/Cmass	0.5 * S/Cmass
	945	5500/5.8	5.0 * S/Cmass	0.5 * S/Cmass
DS-3 Interferometer	---	---	100-300/0.5-1.0	
Earth Observing Clust.	500/5.0	550/5.0	5.0 * S/Cmass	0.5 * S/Cmass
Heliocentric Cluster	1400/3 (from GTO)		6.0/1-2	1250
		3000/3 (from LEO)	6.0/1-2	1250

Clusters of small spacecraft are intended to replace single large observation spacecraft. Virtual apertures created by multiple, formation flying spacecraft have the potential to obtain signals that would be impossible to obtain with a single large spacecraft, and probably at a much lower cost. These clusters could be launched from either a mothership or from a launch vehicle. In both cases, most of the primary propulsion requirements could be performed by the launch vehicle. Primary propulsion might be needed for a system that iggy-backs' on an existing launch vehicle and requires orbit raising, or for an interplanetary or asteroid rendezvous mission. Typical mission requirements for impulsive thrusters are 1.2 to 4.0 km/s. The same missions undertaken with electric thrusters require 2.2 to 7.0 km/s. Depending on the scope of the planetary observation, ACS functions can be done by either electric or impulsive propulsion systems and could possibly be performed by one of the primary propulsion systems. For earth or planetary clusters, propulsion could be used to maintain formation using non-Keplerian orbits (Janson 1994). Continuous thrust is used to assume the non-Keplerian orbit. ΔVs for providing this non-Keplerian orbit depend on spacing and operational altitude. Propulsion for stationkeeping is also required to offset disturbances. For deep space interferometer missions it is assumed that the launch vehicle delivers the spacecraft in earth escape solar orbit. The only primary propulsion required is for "aperture filling" maneuvers, where the distance between spacecraft varies from 0.1 to 1.0 km. For a 3-axis stabilized spacecraft, the ΔV requirement would be 100 to 300 m/s for a 0.5 to 1 year mission.

For large spacecraft missions requiring large primary ΔV, the reduced wet mass provided by an electric propulsion system makes it attractive compared to impulsive chemical system. Figure 2. shows the ratio of propellant mass versus spacecraft dry mass (including propulsion system dry mass) as a function of specific impulse for two values of ΔV, 1 km/s and 5 km/s. A cold gas system (75 s Isp) would need 890 times the spacecraft mass in propellant in order to achieve a ΔV of 5 km/s. A high performance bipropellant engine (315 s Isp) would require a ratio of propellant mass to spacecraft dry mass of 5.8, and an arcjet (600 s Isp) would only require a ratio of 1.3. For PPTs (1500 s Isp) and Ion thrusters (3200 s Isp), these ratios would be approximately 0.4 and 0.19, respectively. Issues that need to be addressed to make a number of electric propulsion systems amenable to microspacecraft are power requirements, efficiency, weight of the propulsion system dry mass, and contamination issues. For several classes of missions, such as low earth orbit (LEO) missions, and major constellation maneuvers, thrust level is also an important consideration.

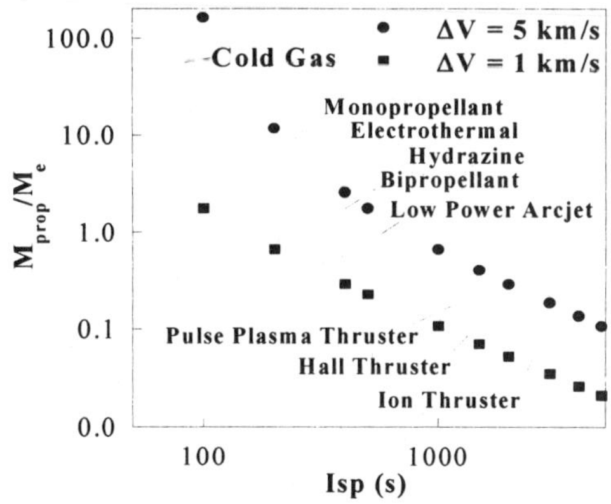

Figure 2. Propellant Mass to Achieve a ΔV of 1 and 5 km/s as a Function of Isp

In some, but not all cases the electric propulsion system will require a longer trip time. For instances where burn time must be low (e.g. thrust high) such as ballistic planetary capture, an impulsive system must be used. Secondary ΔV missions can also require large ΔVs. For these cases electric propulsion can save mass as long as slew rate and pointing accuracy requirements are within the system's capability. The availability of power on the microspacecraft can impact propulsion system selection. For primary propulsion, the payload is usually inactive, freeing up a large amount of power for electic propulsion. However, during secondary propulsion the payload is normally active, reducing the power availability unless the payload's reserve system can be used (e.g. batteries for stationkeeping of communication satellites). For very low power spacecraft with primary and/or secondary propulsion requirements power capability must be added or an impulsive system used.

STATE-OF-THE-ART PROPULSION TECHNOLOGIES

Chemical. Of the chemical propulsion systems frequently used in spacecraft, cold gas propulsion systems have the lowest complexity and cost. They can provide highly repeatable, extremely small impulse bits for accurate orbit maintenance and attitude control. However, these systems also have the lowest performance in terms of specific impulse (Isp) and total impulse for a given volume. Cold gas thruster performance likely scales better than the performance of most conventional chemical and electric thrusters. However, the low Isp still leads to storage problems. Leakage can become a liability for long mission durations, both in terms of attitude control (leaking gas could exert a force), and in terms of lifetime. For minor primary propulsion functions and ACS tasks with a relatively short mission duration and a low overall impulse, cold gas systems may work well. For these applications, the simplicity and low dry mass are a benefit, despite the low Isp. The minimum obtainable impulse bits are on the order of 10 μN–s.

Solid propellant thrusters are frequently used for orbit insertion. These thrusters are simple, reliable, and have a high propellant density, giving high density specific impulse. Because of the high density, easy storage (no leakage, no valves and/or regulators), and relatively high performance, solid propellants could become prime candidates to perform primary propulsion functions on microspacecraft. The main disadvantage is the lack of restartability which limits the use to a single, high impulse burn for each thruster used. Together with issues such as packaging, restartability need to be addressed before alternate propulsion functions are considered.

The most common propulsion system found on current spacecraft is the hydrazine monopropellant system. Hydrazine monoprops are among the best currently available candidates for micropropulsion applications for spacecraft larger than 10 kg. The storability, catalytic ignitability, restartability, and good performance of monopropellant systems are sufficient reasons to select this concept for some primary propulsion applications and many orbit maintenance functions. The system simplicity has proven to provide reliable performance for many spacecraft. The simplicity also allows miniaturization of all components without a large reduction in performance. Because of the liquid storage, leakage is less of a problem than for cold gas systems. Some loss in performance will occur when scaling down as the result of larger proportional wall area which increases heat losses and viscous losses. However, the performance is expected to only be slightly lower than larger versions. The smallest currently available hydrazine thruster can operate at a thrust level of 0.19 N with a 206 Isp. Although additional research is needed, catalyst beds can be created with sub-millimeter characteristic dimensions. Lower limits on flow velocity through the catalyst bed and heat transfer out of the catalyst will determine the smallest possible impulse bit. Impulse bit delivery spans a wide range, from 10 μN-s upwards. Issues that have to be addressed are similar to its larger version, namely material compatibility and propellant safety and handling issues. Some power is required to pre-heat the catalyst bed.

Hydrazine/nitrogen tetroxide (NTO) or monomethyl hydrazine/NTO bi-propellant thrusters have the highest performance of chemical spacecraft propulsion systems. Because of their high performance, they are frequently used for high impulse primary thrust applications, such as orbit insertion, orbit raising, divert propulsion or interplanetary propulsion. The large dry mass required can only be justified for high total impulse applications. Because of the large proportional dry mass, bipropellant systems do not lend themselves well to being downscaled. An additional problem of small scale bipropellants is the difficulty in obtaining thorough propellant mixing within the small combustion chamber. This causes ignition delays and reduced performance. The smallest available bipropellant thrusters operate at a thrust level of 2 N with an Isp of 265 s compared to an Isp of 315 s for a 100 N thruster.

Electric. Three classes of electric propulsion devices are currently in use or near being used in flight. These types are referred to as electrothermal, electrostatic, or electromagnetic devices, depending on the principle by which the working fluid is accelerated to provide thrust. Electrothermal thrusters create a high temperature fluid which provides a driving force by acceleration through a conventional nozzle. The thermal energy of the fluid is partly converted to kinetic energy. Electrostatic thrusters provide thrust by accelerating a charged plasma by means of a static electric field. Electromagnetic thrusters apply an electromagnetic field to accelerate an electrically charged plasma. This electromagnetic field can be self-induced or externally generated.

The least complex electric propulsion system available is the resistojet. In a hydrazine resistojet, the heat of the products of hydrazine, decomposed in a catalyst bed, is resistively augmented in a heating coil. The increase in performance over monopropellants can range to 80 s, but this requires 0.3-0.4 kW of electric power. In scaling this concept to smaller size, previously stated concerns about the propellant flow rate and heat transfer are valid. In addition, power becomes scarcer when miniaturizing spacecraft, such that power consumption needs to be considered. Hydrazine resitojets require too much power for microthrust applications.

Arcjets, also an electrothermal propulsion concept, operate by directly heating the propellant (usually hydrazine passed through a catalyst bed to create hydrogen and nitrogen) with a stationary arc in/near the nozzle contraction. The primary application of arcjets has been stationkeeping of large satellites. Up to 600 s Isp has been demonstrated with 2.2 kW systems. Low power arcjets perform well at 0.5 to 0.75 kW. Below 0.5 kW, the performance decreases (Sankovic and Jacobson 1995) and current technology does not allow operation at microspacecraft power levels.

Ion thrusters are electrostatic propulsion devices. In an ion thruster, a plasma is created from a propellant (usually xenon) by means of an electrical discharge in the discharge chamber. The plasma is accelerated in an electrostatic field created by a set of ion grids (ion optics) placed at the exit of the thruster. After the ions leave the thruster, a neutralizer cathode is used to return electrons to the plasma. This prevents a negative charge from building up on the spacecraft. The smallest ion thrusters available are the 10 cm diameter British DERA T5 and the DASA RITA thruster. A 13 cm diameter Hughes XIPS thruster with an Isp of 2585 s, overall efficiency of 51.3%, and power of 300 Watt was launched on the ASTRA 1G satellite on December 3rd, 1997. An experimental ion thruster was launched on-board the Japanese ETS-VI, but the spacecraft failed to reach its intended orbit and the test program was severely shortened. Ion thruster loss mechanisms, such as recombination, are dominated by wall effects. Reducing the physical size will therefore reduce the efficiency and Isp. Ion thrusters could be considered for microspacecraft propulsion if issues regarding these wall losses, plume neutralization, and high voltage arcing can be resolved.

Another electrostatic concept is the Hall effect thruster. Xenon thrusters of this type have been used on Russian spacecraft (Meteor, a.o.) for orbit-raising, drag make-up, and North-South Station Keeping (NSSK). Russian Hall thrusters, such as the Stationary Plasma Thruster (SPT) and the Thruster with Anode Layer (TAL) have been adapted and qualified for American standards in a collaborative effort to develop a NSSK device for use on geostationary satellites. The SPT-100 typically provides 80 mN thrust at an Isp of 1600 s and 48% efficiency with an input power of 1.35 kW. A SPT-50 has been laboratory tested down to 0.09 W with an Isp of 700 s and efficiency of 21% (Manzella 1996). At the design point of 0.3 kW, the SPT-50 operated at an Isp of 1160 s and an efficiency of 32%.

The electromagnetic pulsed plasma thruster (PPT) has been in operation on satellites for many years. The concept is relatively simple. A solid (usually teflon) propellant bar is spring loaded inside an insulating container. A capacitor discharge strikes an arc at the propellant surface, vaporizing several molecular layers of the solid and creating a plasma. The same discharge generates an electromagnetic field. This field accelerates the plasma externally, providing a small thrust pulse. Capacitor discharge frequency is on the order of 1 Hz, with Isp of 1500 s. Typical thrust levels are between 0.05 and 2 mN, which makes this device well suited for accurate spacecraft positioning such as the multi-spacecraft interferometric experiment planned to be launched in 2001. PPTs are possibly the best candidate for many micro-propulsion tasks. The minimum impulse bits obtainable are on the order of 10 μN-s, with larger impulse bits possible up to 1 mN-s. A total impulse of up to 20 kN-s is targeted for larger systems. Performance does not significantly deteriorate with power level. Total impulse and impulse bit size can be varied by changing the fuel bar geometry.

Characteristics of currently available small thrusters, together with the manufacturer and model, are shown in Table 2. For the Hall thruster, the values given are obtained for a SPT-50 thruster tested at NASA Lewis Research Center. The PPT values are obtained from the specifications of the Deep Space 3 mission.

Table 2. Characteristics of Potential Microthrust Propulsion Systems

	Cold Gas	Solid	Mono-prop	Bi-prop	EHT	Arcjet	PPT	Hall	Ion
Thrust (mN)	4	169000	190	2000	180	150	0.05	5.3	11
Isp (s)	65	270	206	265	304	465	1150	1049	3171
I-bit (mN-s)	0.1	?	2	30			0.06	100	
Power (W)	9		9	18	350	1400	20	90	275
Pow/T (W/mN)	2.25		0.05	0.009	1.95	9.3		16	24.6
Eff. (%)					90	30-35	8	24	62
μ S/C Applic.	Prim./ACS	Prim.	Prim/ ACS	No	No	No	Prim/ ACS	No	No
Source	Moog 58x125	Thiokol STAR 5A	Primex MR-103C	Aerojet	Primex MR-501	Primex MR-507	Deep Space 3	SPT-50	DERA T5

TECHNOLOGY DIRECTIONS

An important issue that cuts across almost all propulsion systems is the feed system design. The extremely small impulse bits required for a number of secondary functions requires fast valve and regulator response times for propellant feed systems and an overall more compact feed system design. MEMS technology could provide a means to reduce feed system component size. To this end, programs have been initiated to address a number of issues pertaining to valve closing, material compatibility, thermal loads, and bonding of MEMS components to conventional components.

Chemical

A variation of the cold gas systems which has the potential to double the performance is the tri-constituent gas thruster. A non-combustible, stochiometric mixture of hydrogen, oxygen diluted with a third, inert gas is passed over a catalyst bed, where the H_2/O_2 reaction heats up the third gas, providing approximately double the Isp of cold gas thrusters. The third gas can be nitrogen or helium. For a system which is more volume then mass constraint, xenon could be the third gas because of the higher density Isp. Because of the higher temperatures with Xe as third gas, issues pertaining to catalyst bed and chamber survivability need to be addressed.

Renewed interest in subliming solid thrusters, first investigated in the sixties, has focused mainly on micro-propulsion applications. The high density specific impulse, solid storage, controllability, and capability to produce small impulse bits are all advantages that can be exploited for micro-propulsion. Subliming solid systems can be designed passively by utilization of heat from the spacecraft for sublimation, or actively with heating elements. The last type gives greater controllability, but requires power. For a higher Isp, and to eliminate problems with recondensation of the sublimation products during expansion, heat can be added with a resistive heating element.

An alternate solid proposed for micro-propulsion is the gas generator compound. A solid compound, usually a compound with high nitrogen content, is stored in solid form. An ignition source, such as laser, pyrotechnic or hot-wire, initiates decomposition. Exothermic decomposition creates a high temperature gas which is accelerated to provide propulsion. No chemical reaction takes place. Proposed uses of a solid gas generator compound is as a solid propulsion device or as a solid storage cold gas system, with the solid compound packaged to replenish the gas in storage.

One potential bipropellant candidate is hydrogen/oxygen propulsion where the propellants are stored until needed in the form of water. Electric energy is used to electrolyze the water into H_2/O_2, which can be consumed immediately or stored in propellant tanks. A single propellant could be used to provide ACS functions with cold gas accuracy and impulse bits, while the bipropellant combination could be used for high Isp, high thrust level primary propulsion. An additional advantage, described in detail in an accompanying paper (de Groot 1997), is that the same hardware can be used as a fuel cell for energy storage to replace heavy batteries.

Electric

Janson (1994) describes a MEMS fabricated ammonia resistojet. Assuming a 50% thrust efficiency and an Isp of 250 s, the thrust level of this resistojet is 0.41 mN. A proposed application is the orbit raising of sub-kg class spacecraft.

Ion thrusters are the focus of several downsizing programs. Patterson (1997) is developing an 8 cm diameter ion thruster. Estimated thruster characteristics range from an efficiency of about 37% at 1810 seconds specific impulse and 85 W input power, to an efficiency of about 54% at 2960 seconds and 300 W input power. Janson (1994) describes a micromachined ion thruster concept with a projected Isp of 1900 s.

An additional electrostatic propulsion concept is a field emission electric propulsion (FEEP) thruster. Currently being developed by European sources, this thruster concept utilizes liquid indium or cesium as propellant. The expected Isp is over 8000 s with thrust level around 10 μN. Major technology issues which still needs to be resolved are neutralization of the ions in the plume, feed system difficulties, and spacecraft contamination due to the metallic plume.

Operating Range.

A generic graph showing the operating range of the different propulsion systems applicable for microthrust propulsion is provided in Figure 3. The graphs indicated the thrust and total impulse that the different concepts cover, and does not account for the delivery of the total impulse in terms of impulse bits. It is clear that at the high total impulse and high thrust level requirements a void exists that makes a number of missions impossible with current or near future technologies.

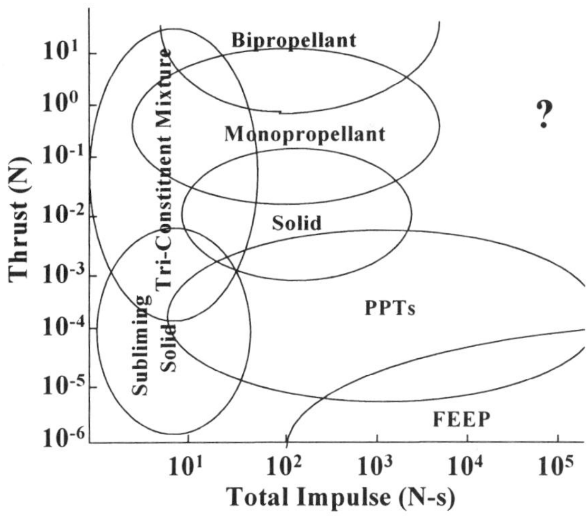

Figure 3. Operating Range for Potential Microthrust Concepts.

MEMS Technology

A cross cutting technology that has the potential to aid in the miniaturization of established propulsion systems is MEMS technology and laser lithography. This technology can provide low cost miniature components for feed system valves, orifices, catalyst bed, chamber, nozzle, and heating element.

Fabrication details can be significantly smaller than with conventional techniques. When used as a batch manufacturing process, costs of component and systems can be very low.

MEMS technology is an extension of integrated circuit (IC) technology. The base material is silicon. Silicon is an excellent mechanical material, with a strength comparable to steel (Si modulus ~ 190 GPa). However, these excellent mechanical properties start to decrease above ~ 400° C. An alternative base material is silicon carbide SiC, which retains its mechanical properties to well over 800° C. However, MEMS techniques applied to SiC are not as well understood and much harder to implement with the desired details and accuracy than for pure silicon. Common MEMS fabrication techniques are wet and dry bulk silicon etching, used to micro-machine detailed devices such as proposed for micro-thruster development, and surface micromachining, usually employed in the design of sensors. Additional techniques used include bonding, both lower temperature anodic bonding and higher temperature diffusion bonding; micromolding, which utilizes photolithography and electroform deposition to create detailed, three dimensional shapes. Laser lithography, which uses high power, accurate lasers to machine shapes from a silicon substrate is an alternate, non-etching technique.

Many satellite sub-systems can be designed and built with MEMS technology. Miniature sensors, data acquisition, photovoltaic arrays, electronics, fuel cells, and communications antenna can all be designed and build on a single silicon wafer, or on a stack of wafers, with MEMS technology. The development of MEMS-based thruster technology would improve microspacecraft integration issues (Janson 1994). Because of the fact that the current developments in microthruster technology are based on conventional concepts, these MEMS-based microthrusters need to be joined with non-MEMS based propellant storage vessels. The types of joints feasible depend on the storage vessel material and pressure. These issues need to be addressed on a case by case basis.

The transition to MEMS components, specifically MEMS valves, will probably magnify the leakage problems of gaseous storag because of the inability to provide sufficient force on valve closing and the probability of contaminating the valve seat. MEMS research in the valve area should be directed to mitigate these problems. For the short term, the easiest approach is to perform experiments on concepts that are both easy to execute as well as have some benefit to planned missions. To that end, MEMS based subliming thrusters (Mueller 1997), in which a heating element is used to provide some control over the thrust, MEMS based cold gas thrusters, MEMS based bipropellants, and MEMS based ion thrusters (Janson 1994) are being investigated.

SUMMARY

A number of established propulsion concepts have been analyzed for applicability towards microthrust propulsion. A range of requirements covering the higher thrust levels and higher required total impulse can not be achieved with the current propulsion technology, not even after MEMS technology has been introduced for components. New propulsion technology is needed to enable these missions with microspacecraft.

Candidates for primary propulsion with limited total impulse required are cold gas, solid propellants and monopropellants. PPTs can be used for primary propulsion for selected missions.

Cold gas propulsion is acceptable for ACS functions but can only be applied for very low ΔV missions with limited lifetime. Slew rate requirements can be met with cold gas thrusters. Tri-constituent gas, monopropellants, and PPTs are also candidates for selected ACS micropropulsion tasks. FEEP thrusters can be considered when certain technology issues have been resolved

New developments in fabrication technologies, such as MEMS or photolithography could improve performance of conventional concepts on microscale. Such developments are being tested for cold gas, subliming thrusters, resistojets, bipropellants and ion thrusters.

REFERENCES

de Groot, W. A., and S.R. Oleson, Chemical Microthruster Options, AIAA Paper 96-2868, 32nd JPC Conf.., Lake Buena Vista, FL (1996).

de Groot, W. A., M. Vanderaar, and T. Maloney, Power, Propulsion, and Communications for Microspacecraft Missions, COSPAR Colloquium on Scientific Microsatellites, Tainan, Taiwan, (1997).

JPL Micropropulsion Workshop, April 7-9 (1997).

Janson, S., Chemical and Electric Micropropulsion Concepts for Nanosatellites, Paper AIAA 94-2998, 30th JPC Conference, Indianapolis, IN (1994).

Manzella, D., S.R. Oleson, J. Sankovic, T. Haag, A. Semenkin, and V. Kim, Evaluation of Low Power Hall Thruster Propulsion, AIAA 96-2736 32nd JPC Conf, Lake Buena Vista, FL (1996).

Mueller, J., Thruster Options for Microspacecraft: A Review and Evaluation of Existing Hardware and Emerging Technologies, AIAA Paper 97-3058, 33rd JPC Conf., Seattle, WA (1997).

Patterson, M. J., and S.R. Oleson, Low-Power Ion Propulsion for Small Spacecraft, AIAA 97-3060 (1997)

Sankovic, J. M, and D. T Jacobson, Performance of a Miniaturized Arcjet, AIAA Paper 95-2822, 31st JPC Conf., San Diego, CA (1995).

Santo, A., NEAR Spacecraft Case Study, ATI Small Satellite Design and Technology Short Course, (1996).

COMPARISON OF MAGNETIC AND AERODYNAMIC STABILIZATION FOR A MICROSATELLITE

Zuu-Chang Hong[1]*, Wen-Hu Hu[1]**, Yen-Hsun Chen[1]**, and Da-Ban Lee[1]**; Jeng-Shing Chern[2++]

[1]*CFD and Satellite Engineering Laboratory*[+] *,National Central University ,Chungli, 32054 Taiwan, China*
[2]*Chung Shang Institute of Science and Technology*

ABSTRACT

When the orbital altitude of a satellite is relatively low, the methods of magnetic and aerodynamic stabilization can be considered. Both methods are very effective when the satellite is small. In particular, magnetic and aerodynamic stabilization systems have been installed in some satellites. Examples are UO-sat and Cosmos-320 with magnetic and aerodynamic methods respectively.

Magnetic torques acting on a spacecraft result from the interaction of the spacecraft's residual magnetic field and the geomagnetic field. Because a magnetic torque is always orthogonal to geomagnetic field vector, it is not favorable in all regions of the orbit. The aerodynamic restoring torque can be provided by a very light tail stabilizer which is several meters downstream from the main body of the small satellite. The gravity gradient stabilization method also has been used in many satellites. One of the examples is the Eole satellite, In the recent years, many small satellites have adopted this method owing to its long life and continuous Earth pointing characteristics.

The purpose of this paper is to compare the methods of magnetic and aerodynamic stabilization both based on the aid of a gravity gradient boom for attitude control. We shall consider the attitude motion of a 50kg microsatellite and the set of nonlinear equations of motion will be solved numerically with specified initial condition. The satellite operates in a circular orbit with 500 km altitude. The history of the attitude stabilization will be investigated and discussed in detail.

INTRODUCTION

Satellite becomes more and more important in the world because the application of satellite in telemetry, communication etc, are very close to life of the human. According to the progress in on-board computer and material, the satellite size is being decreased step by step. In Taiwan, we are developing a new microsatellite, called Taiwan University United Satellite No.1 (TUU SAT-1), which is designed and

---

* Professor, Department of Mechanical Engineering, Senior Member AIAA.
** Graduate Student, Department of Mechanical Engineering.
++ Senior Scientist, Associate Fellow AIAA.
\+ Develop The microsatellite TUU SAT-I, TUU SAT-1 is an university microsatellite program initiated at Satellite Engineering Laboratory, National Central University, Taiwan, in 1995

constructed since 1995 in the Taiwan University United Satellite Engineering Laboratory. It is approximately 50 kg and 0.36*0.36*0.72 m^3 in size and will operate in a sun-synchronous orbit.

In this paper we discuss the magnetic and aerodynamic stabilization both based on the aid of gravity gradient boom for TUU SAT –1, where the previous is based on fuzzy controller and the latter is based on proportional controller. In derivation, we use Euler equation to obtain the satellite dynamics with coordinate transformation matrix.

Keywords: TUU-SAT1, attitude control , fuzzy control.

SATELLITE DYNAMICS

Magnetic stabilization with gravity gradient boom

In the section, we derive the satellite dynamic equations from the Euler equations. Variables of the equation are Euler angles.. In magnetic stabilization we use 2-1-3 Euler rotation to get satellite's dynamic equation. As 2-1-3 Euler rotation, the first rotation is pitch around the Y_0 axis and this rotation angle is ϕ. The second rotation is roll around the L axis; this rotation angle is θ. The third rotation is Yaw around the Z axis; this rotation angle is ψ, and the XYZ body axis is formed, see Fig.1. From rotation formula, we can get the rotation matrix as follows:

The first rotation around the Y_0 axis: $$[R(\theta)] = \begin{bmatrix} c\theta & 0 & -s\theta \\ 0 & 1 & 0 \\ s\theta & 0 & c\theta \end{bmatrix} \tag{1a}$$

The second rotation around the L axis: $$[R(\phi)] = \begin{bmatrix} 1 & 0 & 0 \\ 0 & c\phi & s\phi \\ 0 & -s\phi & c\theta \end{bmatrix} \tag{1b}$$

The third rotation around the Z axis: $$[R(\psi)] = \begin{bmatrix} c\psi & s\psi & 0 \\ -s\psi & c\psi & 0 \\ 0 & 0 & 1 \end{bmatrix} \tag{1c}$$

With the use of the transformation matrix Eq.(1), Euler angle rate in the XYZ body axis vector components can be written as: $$\begin{bmatrix} \omega_x \\ \omega_y \\ \omega_z \end{bmatrix} = [R(\theta)] \begin{bmatrix} \dot{\phi} \\ 0 \\ 0 \end{bmatrix} + \begin{bmatrix} 0 \\ 0 \\ \dot{\psi} \end{bmatrix} + \left[R(\psi) R(\phi) \begin{bmatrix} 0 \\ \dot{\theta} - \omega_0 \\ 0 \end{bmatrix} \right] \tag{2a}$$

And the Euler angle rate in the LMZ axis vector components is $$\begin{bmatrix} \omega_l \\ \omega_m \\ \omega_z \end{bmatrix} = \begin{bmatrix} \dot{\phi} \\ c\phi\left(\dot{\theta} - \omega_0\right) \\ \dot{\psi} - s\phi\left(\dot{\theta} - \omega_0\right) \end{bmatrix} \tag{2b}$$

Assume the moment of inertia around the X and Y body axis is equivalent (IX=IY=IT), and a boom deployed along the Z axis. The angular momentum is $$\bar{h} = I \bullet \bar{\omega} = \begin{bmatrix} I_T \dot{\phi} \\ I_T c\phi\left(\dot{\theta} - \omega_0\right) \\ I_Z\left[\dot{\psi} - s\phi\left(\dot{\theta} - \omega_0\right)\right] \end{bmatrix} \tag{3}$$

Take differentiation with respect to time: $\dot{\bar{h}} = I \cdot \dot{\bar{\omega}} = \begin{bmatrix} I_T \ddot{\phi} \\ I_T\left[-s\phi\left(\dot{\theta}-\omega_0\right)\dot{\phi}+c\phi\ddot{\theta}\right] \\ I_Z\left[\ddot{\psi}-c\phi\left(\dot{\theta}-\omega_0\right)\dot{\phi}-s\phi\ddot{\theta}\right] \end{bmatrix}$ (4)

And $\bar{\omega}\times\bar{h} = \begin{bmatrix} (I_Z - I_T)c\phi\left(\dot{\theta}-\omega_0\right)\left[\dot{\psi}-s\phi\left(\dot{\theta}-\omega_0\right)\right] \\ (I_T - I_Z)\left[\dot{\psi}-s\phi\left(\dot{\theta}-\omega_0\right)\right]\dot{\phi} \\ 0 \end{bmatrix}$ (5)

Take Eq(4) and Eq(5) in Euler dynamics equation, we can get

$$\bar{T} = \dot{\bar{h}} + \bar{\omega}\times\bar{h} = \begin{bmatrix} I_T\ddot{\phi} - \left(\omega_0 - \dot{\theta}\right)c\phi\left[\dot{\psi}-s\phi\left(\dot{\theta}-\omega_0\right)\right]I_Z - \left(\omega_0-\dot{\theta}\right)s\phi I_T \\ \ddot{\theta}\, c\phi I_T + \dot{\phi}\left[2\left(\dot{\theta}-\omega_0\right)s\phi I_T - \left(\dot{\psi}-s\phi\left(\dot{\theta}-\omega_0\,\dot{\phi}\right)\right)I_Z\right] \\ \ddot{\psi}\, I_Z - \ddot{\theta}\, s\phi I_Z - \dot{\phi}\left(\dot{\theta}-\omega_0\right)c\phi I_Z \end{bmatrix} \tag{6}$$

Magnetorquing

Magnetic coils around the spacecraft can be fed with constant current to generate a magnetic dipole moment ***M***. And the dipole moment will interact with the geomagnetic field vector ***B*** to generate a torque ***N*** by taking the cross product: $\boldsymbol{N} = \boldsymbol{M}\times\boldsymbol{B}$

The geomagnetic field vector can be defined in the coordinates of the orbital plane, its components are Bxo, Byo and Bzo, where the Xo axis is along the velocity vector, the Yo axis is normal to the orbit plane and the Zo axis is toward the Earth's center(nadir direction). The vector components can be obtained from a simple dipole model. The dipole model is: $B_0 = \nabla\left[\frac{\hat{R}^T M_e}{R^3}\right] = \left[I - 3\hat{R}\hat{R}\right]\frac{M_e}{R^3}$ (7)

Where
∇ = vector gradient operator
R = radius vector magnitude
$\hat{R}$ = unit vector along radius
I = identity matrix
M_e =geomagnetic strength of dipole

In the orbital coordinates this model is expressed as $B_0 = \begin{bmatrix} B_{X0} \\ B_{Y0} \\ B_{Z0} \end{bmatrix} = \begin{bmatrix} \sin i \cdot \cos\alpha \\ \cos i \\ 2\sin i \cdot \sin\alpha \end{bmatrix}\frac{M_e}{R^3}$ (8)

Where
i = orbit inclination
α = orbit angle as measurement from the ascending node

Gravity Gradient torque

For gravitation field in an inertial reference frame, the gravitational acceleration can be expressed as follows: $\bar{a}_g = -\frac{GM\bar{R}}{R^3}$ (9)

Integration of the gravitational force in the mass center is: $\overline{T_g} = \int \bar{r} \times \overline{a_g} dm$ (10)

$$\overline{T_g} = \frac{3\mu}{R_0^{\ 3}} \begin{bmatrix} (I_Z - I_T) s\phi c\theta c\phi c\theta \\ (I_Z - I_T) c\phi c\theta s\theta \\ 0 \end{bmatrix}$$

Aerodynamic stabilization with gravity gradient boom

The geometry of the Keplerian orbit of an axisymmetric satellite with the center of mass *o* moving about the center of force *O* is shown in Fig. 2. The *OXYZ* is the inertial coordinates where *OY* is normal to the orbital plane and *OZ* passes the perigee. The *OX0YOZO* is the rotating coordinates with *Oyo* normal to the orbital plane and *OZO* along the local vertical. Therefore, *OY* and *oYo* are parallel. Then, the *oxyz* represents the principal body coordinate system.

If we rotate *OXYZ* an angle η about the *OY*-axis and then translate the origin to point *o*, we have the system $ox_o y_o z_o$. Then, by rotating a roll angle θ_1 about *oxo* -axis, a pitch angle θ_2 about *oy*1-axis, and a yaw angle θ_3 about *oz*-axis, the spatial orientation of the *oxyz* system can be completely specified. The transformation matrix from the rotating coordinate system $ox_o y_o z_o$ to the principal body coordinate system *oxyz* is

$$L = \begin{bmatrix} \cos\theta_2 \cos\theta_3 & \sin\theta_1 \sin\theta_2 \cos\theta_3 + \cos\theta_1 \sin\theta_3 & \sin\theta_1 \sin\theta_3 - \cos\theta_1 \sin\theta_2 \cos\theta_3 \\ -\cos\theta_2 \sin\theta_3 & \cos\theta_1 \cos\theta_3 - \sin\theta_1 \sin\theta_2 \sin\theta_3 & \sin\theta_1 \cos\theta_3 + \cos\theta_1 \sin\theta_2 \sin\theta_3 \\ \sin\theta_2 & -\sin\theta_1 \cos\theta_2 & \cos\theta_1 \cos\theta_2 \end{bmatrix} \quad (11)$$

The angular velocity vector of satellite with respect to the inertial frame can be expressed as

$$\vec{\omega} = \theta_1 \vec{e}_{x_o} + \theta_2 \vec{e}_{y1} + \theta_3 \vec{e}_z \dot{-} \eta \vec{e}_{y0} = \begin{bmatrix} \dot{\theta}_1 \cos\theta_2 \cos\theta_3 + \dot{\theta}_2 \sin\theta_3 - \dot{\eta}(\sin\theta_1 \sin\theta_2 \cos\theta_3 + \cos\theta_1 \sin\theta_3) \\ -\dot{\theta}_1 \cos\theta_2 \sin\theta_3 + \dot{\theta}_2 \cos\theta_3 - \dot{\eta}(\cos\theta_1 \cos\theta_3 - \sin\theta_1 \sin\theta_2 \sin\theta_3 \\ \dot{\theta}_1 \sin\theta_2 + \dot{\theta}_3 + \dot{\eta} \sin\theta_1 \cos\theta_2 \end{bmatrix} \quad (12)$$

The equation of rotational motion of a rigid body is $\vec{M} = \dot{\vec{H}} = \left(\dot{\vec{H}}\right)_r + \vec{\omega} \times \vec{H}$ (13)

where $\left(\dot{\vec{H}}\right)_r = \dot{H}_x \vec{i} + \dot{H}_y \vec{j} + \dot{H}_z \vec{k}$ (14)

$\left(\dot{\vec{H}}\right)_r$ is the relative rate of change as viewed by an observer on the moving coordinate system *oxyz*. Now, if the *oxyz* coordinate system is chosen such that the coordinate axes are principal axes at the center of mass of the rigid body, all products of inertia vanish and Eq. (3) become

$$\begin{aligned} M_x &= I_{xx}\dot{\omega}_x + (I_{zz} - I_{yy})\omega_y\omega_z \\ M_y &= I_{yy}\dot{\omega}_y + (I_{xx} - I_{zz})\omega_z\omega_x \\ M_z &= I_{zz}\dot{\omega}_z + (I_{yy} - I_{xx})\omega_x\omega_y \end{aligned} \quad (15)$$

These are known as Euler's equations of motion. The differential equations for the three components of can be written as

$$\dot{\omega}_x = \frac{M_x}{I_{xx}} - (\frac{I_{zz} - I_{yy}}{I_{xx}})\omega_y\omega_z$$
$$\dot{\omega}_y = \frac{M_y}{I_{yy}} - (\frac{I_{xx} - I_{zz}}{I_{yy}})\omega_z\omega_x \tag{16}$$
$$\dot{\omega}_z = \frac{M_z}{I_{zz}} - (\frac{I_{yy} - I_{xx}}{I_{zz}})\omega_x\omega_y$$

From Eq. (12) and Eq. (16)we can derive the three equations about θ as

$$\dot{\theta}_1 = \frac{1}{\cos\theta_2}(\omega_x \cos\theta_3 - \omega_y \sin\theta_3 + \dot{\eta}\sin\theta_1 \sin\theta_2$$
$$\dot{\theta}_2 = \omega_x \sin\theta_3 - \omega_y \cos\theta_3 + \dot{\eta}\cos\theta_1 \tag{17}$$
$$\dot{\theta}_3 = \frac{1}{\cos\theta_2}(\omega_x \cos\theta_3 - \omega_y \sin\theta_3) + \omega_z - \dot{\eta}\frac{\sin\theta_1}{\cos\theta_2}\sin\theta_1 \sin\theta_2$$

Thus Eqs. (16) and (17) are the three-dimensional equations of motion for attitude motion of the satellite.

External Torques

The external torque acting on the satellite to be consider are the gravity gradient torque and the aerodynamic torque.

$$\vec{M} = \dot{\vec{H}} = \vec{G}_G + \vec{G}_A \tag{18}$$

Where $\vec{G}_G$ is the gravity gradient torque and $\vec{G}_A$ is the aerodynamic torque. All other torques are regard as disturbance and will be neglected.

Aerodynamic Torque

The aerodynamic stabilizer for yaw angle stabilization is shown in Fig.3. The aerodynamic torque acting on satellite due to the presence of the stabilizer can be written as [4]

$$G_{a_z} = -\frac{1}{2}\rho V^2{}_R Al[C_p \cos\delta_3 + \frac{\sin\theta_3 - \sin(\theta_3 + \delta_3)}{\cos(\theta_3 + \delta_3)}C_\tau] \tag{19}$$

where $C_p = (2-\sigma')[\frac{2}{\sqrt{\pi}S}\sin(\theta_3 + \delta_3)\exp(-S^2\sin^2(\theta_3 + \delta_3)) + (2\sin^2(\theta_3 + \delta_3) + \frac{1}{S^2} * erf(S\sin(\theta_3 + \delta_3))]$

$$C_\tau = \frac{2\sigma}{\sqrt{\pi}S}\cos(\theta_3 + \delta_3)[\exp(-S^2\sin^2(\theta_3 + \delta_3)) + \sqrt{\pi}S\sin(\theta_3 + \delta_3)erf(S\sin(\theta_3 + \delta_3))$$

Here we analyze the attitude motion of the microsatellite with the existence of external torques

$$M_x = G_{G_X} \quad M_y = G_{G_y} \quad M_z = G_{G_z} + G_{a_z}$$

where

A = area of yaw stabilizer

C_p = normal force coefficient for yaw stabilizer
$C\tau$ = shear force coefficient for yaw stabilizer
$\vec{e}_i$ = unit vector (i = x_o, y_1, z, y_o)
$\vec{G}_A$ = aerodynamic torque
$\vec{G}_G$ = gravitational torque
I = inertia matrix for satellite
Iz = principal moment of inertia about z axis
i = orbit inclination
L = transformation matrix
l = moment arm of yaw stabilizer
K3 = gain in yaw channel feedback loop
R = orbital radius
S = molecular speed ratio
T = orbital period
VR = air velocity with respect to satellite
VR = VR/VR
$OXYZ$ = inertia coordinates
$Oxyz$ = principal body coordinates
$ox_o y_o z_o$ = rotating coordinates (orbital coordinates), yo normal to the orbital plane and z0 along the local vertical
α = 90 -($\theta_3 + \delta_3$)
η = orbit true anomaly
$\theta_1, \theta_2, \theta_3$ = roll, pitch, and yaw angles
= defection angle of yaw stabilizer
ρ = atmospheric density
σ, σ' = tangential and normal accomodation coefficients
$\vec{\omega}$ = angular velocity of satellite with respect to inertial coordinates

DESIGN OF THE CONTROLLER:

Fuzzy Controller for Magnetic Stabilization

In the above section, we obtain the satellite dynamic equation is:

$$N_L = \ddot{\phi} I_T - (\omega_0 - \dot{\theta}) C\phi [\omega_Z I_Z - (\omega_0 - \dot{\theta}) S\phi I_T] + 3\omega_0^2 S\phi C\phi C\theta C\theta (I_T - I_Z)$$

$$N_M = \ddot{\theta} C\phi I_T + \dot{\phi}[2(\omega_0 - \dot{\theta}) S\phi I_T - \omega_Z I_Z \omega_0 + 3\omega_0^2 S\theta C\theta C\phi (I_T - I_Z) \quad N_Z = \ddot{\psi} I_T - \ddot{\theta} S\phi I_Z + \dot{\phi}(\omega_0 - \dot{\theta}) C\phi I_Z$$

with $\omega_Z \doteq \psi + (\omega_0 - \theta) S\dot{\phi}$

where
ω_0 : orbit rate.
I : moment of inertia around LMZ axis.
C : cosine function.
S : sine function.

If $\phi = 0$, $\theta = 0$ and $\dot{}$ = constant, it means the attitude has no error, the satellite spins in a constant speed, and Z axis is coincident with Zo axis to form a nadir pointing. Although the magnitude and direction of M can be controlled by the currents of the 3 magnetic coils, the ***B*** is not always favorable in the orbit to control the attitude of a specific axis. It is also possible that a desirable control torque around an attitude axis will generate undesired disturbance torque on the other axes. The main objective of the control effort will be to maximize the desired influence and to minimize the undesired disturbance.

Therefore, an block diagram of fuzzy controller based on fuzzy control rules is designed as shown in Fig.4. The input variables are the measured state variables of the satellite and the estimated control torque. The torque can be estimated by using magnetometer readings and computation from Eq.8 There are 14 variables used as follows.

$x_1 = \phi_{ref} - \phi$, for the coincident of Z axis and Zo axis $\phi_{ref} = 0$.

$x_2 = \phi$

$x_3 = \theta_{ref} - \dot{\theta}$, $\theta_{ref} = 0$.

$x_4 = \dot{\theta}$.

$x_5 = \psi_{ref} - \psi$, $_{ref}$ is the desired spin rate.

$x_6 = \hat{N}_{LX}$, $x_7 = \hat{N}_{MX}$, $x_8 = \hat{N}_{ZX}$ estimated control torque when $i_x=1$ 、$i_y=0$ 、$i_z=0$

$x_9 = \hat{N}_{LY}$, $x_{10} = \hat{N}_{MY}$, $x_{11} = \hat{N}_{ZY}$ estimated control torque when $i_x=0$ 、$i_y=1$ 、$i_z=0$.

$x_{12} = \hat{N}_{LZ}$, $x_{13} = \hat{N}_{MZ}$ estimated control torque when $i_x=0$ 、$i_y=0$ 、$i_z=1$.

The membership functions of these variables are shown in Fig.5, Where $X_P=0.1$ degree/sec, $X_Z=-0.1$ degree/sec, $T_P=10^{-5}$, $T_N=-10^{-5}$, and $T_Z=10^{-}$. The control rules of *ix, iy, iz* are shown in Table.1.

Proportional controller for Magnetic Stabilization

A simple feedback loop as shown in Fig. 6 is designed. The deflection angle of the stabilizer is used as the control and is designed to be $\delta_3(t) = K_3\dot{\theta}_3(t)$

From Eqs. (16)~ (19) with the given gain value K3=17.3, the numerical computation results have been shown in Fig.8. The initial attitude motion are stabilized and converged to nearly zero, as required.

SIMULATION RESULTS

Magnetic Stabilization with Gravity gradient boom

Fig.7 shows the simulation results. When the satellite is controlled in zero spin rate. The satellite can reach the control effect in about one half period of orbit with non-spin. And it's error keeps in 0.1 degree. It's a very good choice that non-spin need for gravity gradient stabilization.

Aerodynamic Stabilization with gravity gradient boom

From fig.8 we can see that roll and pitch axis will being settled in about one orbit(about 6000 sec) with accuracy of two degree, and yaw axis will being settled in about half orbit (about 3000 sec)with accuracy of 0.1 degree.

CONCLUSION

The analysis of magnetic stabilization and aerodynamic stabilization based on gravity gradient boom to control our satellite is completed. From the result we can see clearly that both magnetic stabilization and aerodynamic stabilization can reach the control requirement very fast, however, for more activity than aerodynamic stabilization, the performance of magnetic stabilization is better than the previous one except its more complexity.

Magnetic stabilization is active control, it's advantage is shorter settling time and high responsibility, but it is more complex.

The response time of aerodynamic stabilization (with only one stabilizer) is semi-active control, the response time is longer than magnetic stabilization, but it also have good performance, and relative simple.

For requirement of quickly response, and for the reason that the effect of aerodynamic torque is smaller than magnetic torque above 200 km, we recommended the magnetic stabilization.

REFERENCE

Wall, J.K., "The Feasibility of Aerodynamic Attitude Stabilization of a Satellite Vehicle," presented a ARS Controllable Satellite Conference, M.I.T., April 1959.

Schrello, D. M., "Passive Aerodynamic Attitude Stabilizion of Near Earth Satellites, Vol. I, Libratio due to Combined Aerodynamic and Gravitational Torques," WADD Tech. Report 61-133, Vol I. Jul 1961.

Sarychev, V. A., "Stability Conditions of the System of Gravitational Stabilization of Sallites Fitte with Gyro Damping," Astronautica Acta, Vol.14, 1969, pp.299-310 (in Russian).

Ravindran, R, and Hughes, P C., "Optimal Aerodynamic Attitude Stabilization of Near-Eart Satellites," Journal of Spacecraft, Vol.9, No.7, July 1972, pp.499-506.

Flanagan. R. C., and Rangarajan. R, "Liapunov Stability Analysis and Attitude Response of Passively Stabilized Space System," Astronautica Acta, Vol.8, 1973, pp.21-34

Modi, V.J., and Pande, K. C." Aerodynamic-Solar Hybrid Attitude Control of Near-Eaaai Satellites, The Journal of the Astronautical Sciences, Vol. XXII, No.1, July-September, 1974, pp.36-54

Shrivastava, S.K., and Modi. V.J., " Satellite Attitude Dynamics and Control in the Presence of Environment Torques-A Brief Survey, "Journal of Guidance, Control, and Dynamics, Vol.6, No.6 November-December 1983, pp.461-471.

Sarychev, V A,, Camelier, L A,, and Paglione, P, "Aerodynamic Stabilization System for Small Satellites." Proceedings of the 20th Congress of the International Council of the Aeronautica Science distriblited by, the AIAA, 1996, pp.325-329

Sweeling, M, "Smail Satellite at Surrey," Surrey Satellite Technology Ltd. (SSTL), University of Surrey , U. K., 1992.

J. R. Wertzs and W. J. Larson, "Space Mission Analysis and Design", London, Kluwer Academic Publishing.

C. D. Brown, "Spacecraft Mission Design", AIAA Publishing, American Institute of Aeronautics and Astronautics, Inc. Washington, DC 20024-2518,1992.

Vladimir A. Chobotov, "Spacecraft Attitude Dynamics and Control", Krieger Publishing Company, 1991

M. D .Shuster and S. D. Oh, "Three-Axis Attitude Determination from vector Observation", Journal of Guidance and Control, Vol.4, Jan.-Feb. 1981, pp.70-76.

F. L. Markley, "parameterization of the attitude", Spacecraft Attitude Determination and Control, edited by J. R. Wertz, D. Reidel Publishing Co., The Nethrland, 1987, pp.410-420.

Ai Cbun Fang. An Approach to Attitude Determination For a Spin-Stabilized Spacecraft(IMP 1).

Techinical Report, AIAA Paper 82-0070R,1982.
Mohinder S. Grewal, Angus P. Andrews, “Kalman filtering: theory and practice”, Englewood Cliffs, N.J: Prentice-Hall, c1993.
Peter C. Hughes. “Spacecraft Attitude Dynamics” ,John Wiley & Sons, Inc., 1976.
R.T. Merrill and M.W. Mcelhinny, " The earth's magnetic field", *ACADEMIC PRESS, INC.* , 1983.
M. L. Psiaki, F. Martel and P. K. Pal, “Three-axis Attitude Determination via Kalman Filtering of Magnetometer data,” J. Guidance, Vol. 13, No. 3, pp.506-514, May-June 1990.
J. Jung, N. Kuzuya, j.Alvarez, “The Design of the OPAL Attitude Control System”, 10th Annual AIAA/USU Conference on Small Satellites, Utah State University, Sept. 16-19, 1996, pp1-7.
F. Martel, P. K. Pal and M. Psiaki, “Active magnetic control system for gravity gradient stabilized spacecraft” Proceedings of 2nd Annual AIAA/USU Conference on small Satellites, Utah University, Sept. 1988.
M. S. Hodgart, “Gravity Gradient and Magnetorquing attitude control for low-cost low-earth orbit satellite-The UoSAT experience,” Ph. D. submision at the University of Surrey, Jun, 1989.
W. H. Steyn “Comparison of Law-Earth-Orbit Satellite Attitude Controllers Submitted to Controllability Constraints”Jounal of Guidance Control and Dynamics. Vol.17, No. 4, Jully-August 1994.
J. J. Rodan, “Closed-loop Magnetic Control of a Spin-stabilized Satellite”, International Federation of Automatic Control vol.20, pp.729-735. 1984.
W. Pedrycz, Fuzzy Control and Fuzzy Systems, Research Study Press LTD, 1989.
Mohammad Jamshidi, Nader Vadiee, Timothy J. Ross, “Fuzzy logic and control : software and hardware applications”, --Englewood Cliffs, N.J.:PTR Prentice Hall, c1993.

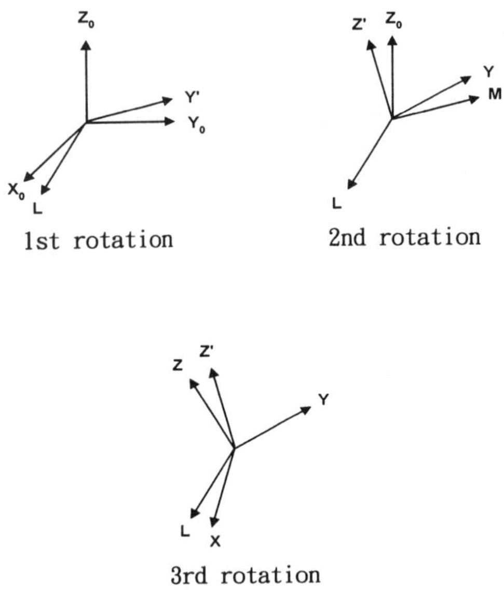

Figure 1. Euler rotation.

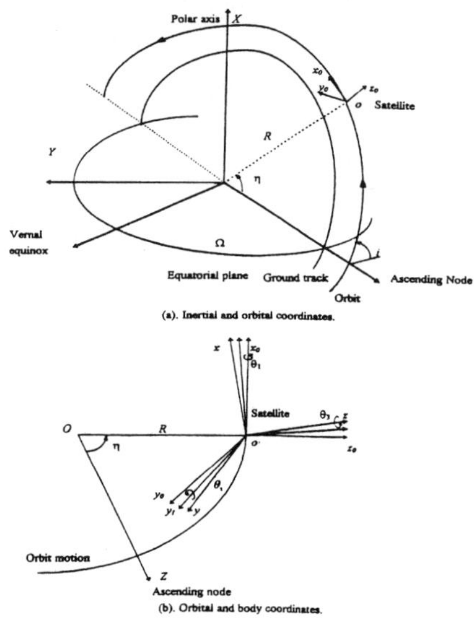

Figure 2. Geometry of orbital flight.

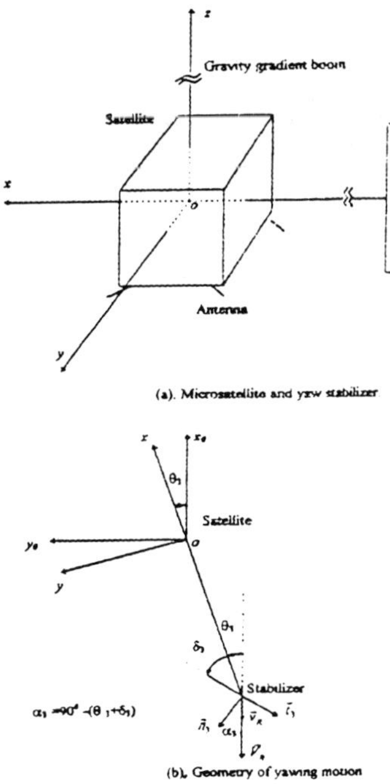

Figure 3. Geometry of yawing motion and stabilizer deflection.

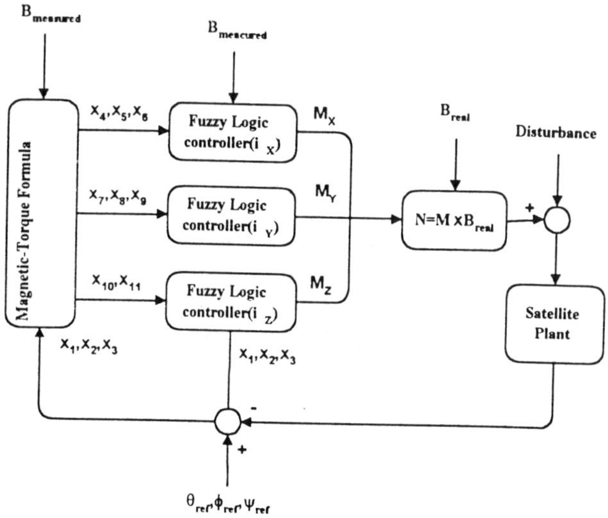

Figure 4. The simulation block diagram based on fuzzy control rule.

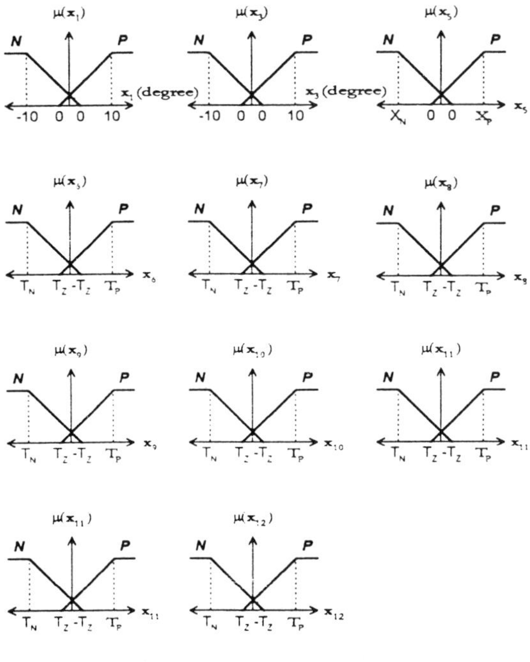

Figure 5. The fuzzy membership function sets.

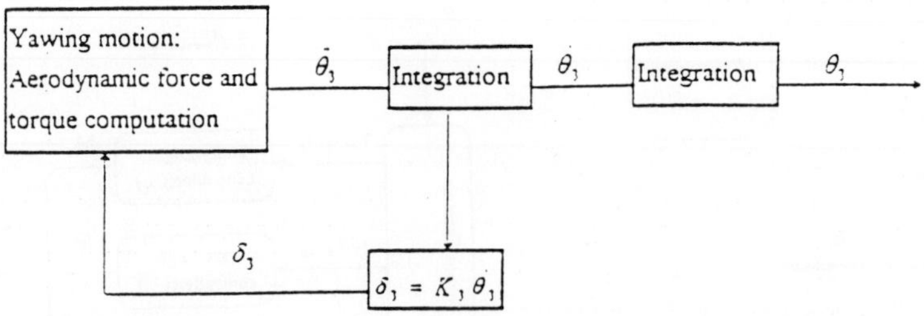

Figure 6. A simple feedback loop.

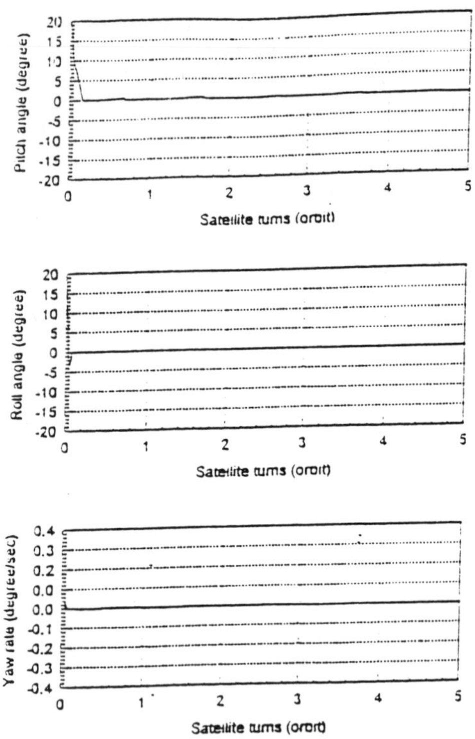

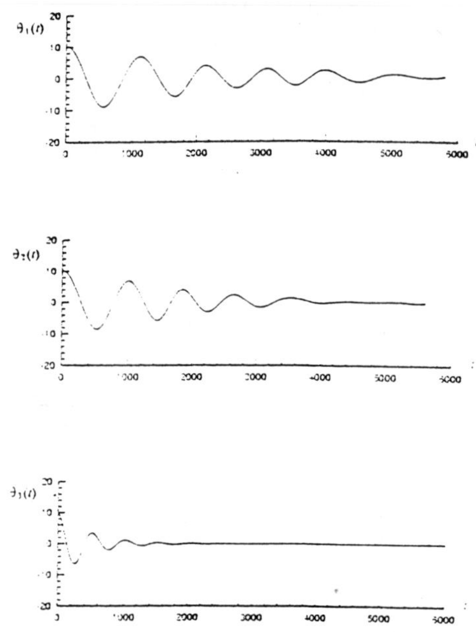

Figure 7. The simulation result when the satellite is controlled in zero spin rate

Figure 8. The time history of attitude motion with gravity gradient and aerodynamic torques

Table 1. The fuzzy control rules of ix, iy, iz

The control rules for ix on x axis

Rule	x_1	x_3	x_5	x_6	x_7	x_8	U
R_1	P	—	—	P	—	—	-1
R_2	P	—	—	N	—	—	+1
R_3	N	—	—	P	—	—	-1
R_4	N	—	—	N	—	—	-1
R_5	—	P	—	—	P	—	-1
R_6	—	P	—	—	N	—	+1
R_7	—	N	—	—	P	—	+1
R_8	—	N	—	—	N	—	-1
R_9	—	—	P	—	—	P	-1
R_{10}	—	—	P	—	—	N	+1
R_{11}	—	—	N	—	—	P	+1
R_{12}	—	—	N	—	—	N	-1

Rule	x_1	x_3	x_2	x_4	α_1
R_{13}	P	—	P	—	1
R_{14}	P	—	N	—	0.1
R_{15}	N	—	P	—	0.1
R_{16}	N	—	N	—	1

Rule	x_1	x_3	x_2	x_4	α_2
R_{17}	—	P	—	P	1
R_{18}	—	P	—	N	0.1
R_{19}	—	N	—	P	0.1
R_{20}	—	N	—	N	1

The control rules for iz on z axis

Rule	x_1	x_3	x_9	x_{10}	U
R_1	P	—	P	—	-1
R_2	P	—	N	—	-1
R_3	N	—	P	—	-1
R_4	N	—	N	—	-1
R_5	—	P	—	P	-1
R_6	—	P	—	N	+1
R_7	—	N	—	P	+1
R_8	—	N	—	N	-1

Rule	x_1	x_3	x_2	x_4	α
R_9	P	—	P	—	1
R_{10}	P	—	N	—	0.1
R_{11}	N	—	P	—	0.1
R_{12}	N	—	N	—	1

Rule	x_1	x_3	x_2	x_4	α
R_{13}	—	P	—	P	1
R_{14}	—	P	—	N	0.1
R_{15}	—	N	—	P	0.1
R_{16}	—	N	—	N	1

The control rules for iy on y axis

Rule	x_1	x_2	x_3	x_{11}	x_{12}	x_{13}	U
R_1	P	—	—	P	—	—	-1
R_2	P	—	—	N	—	—	+1
R_3	N	—	—	P	—	—	+1
R_4	N	—	—	N	—	—	-1
R_5	—	P	—	—	P	—	-1
R_6	—	P	—	—	N	—	+1
R_7	—	N	—	—	P	—	+1
R_8	—	N	—	—	N	—	-1
R_9	—	—	P	—	—	P	-1
R_{10}	—	—	P	—	—	N	+1
R_{11}	—	—	N	—	—	P	+1
R_{12}	—	—	N	—	—	N	-1

Rule	x_1	x_3	x_2	x_4	α_1
R_{13}	P	—	P	—	1
R_{14}	P	—	N	—	0.1
R_{15}	N	—	P	—	0.1
R_{16}	N	—	N	—	1

Rule	x_1	x_3	x_2	x_4	α_2
R_{17}	—	P	—	P	1
R_{18}	—	P	—	N	0.1
R_{19}	—	N	—	P	0.1
R_{20}	—	N	—	N	1

THE MINIATURIZATION OF DEEP SPACE TELECOMMUNICATION SYSTEMS

Mary Katherine Flaherty and R. S. Bokulic

The Johns Hopkins University Applied Physics Laboratory, 11000 Johns Hopkins Road, Laurel, Maryland 20723-6099, USA

ABSTRACT

As part of NASA`s Discovery program, which focuses on exploring the solar system with inexpensive scientific spacecraft, the Near Earth Asteroid Rendezvous (NEAR) was launched on February 17, 1996. It will rendezvous with the asteroid Eros in early 1999. With NEAR, The Johns Hopkins University Applied Physics Laboratory (APL) continued its tradition of building small spacecraft and demonstrated that new technologies can be included in a cost-capped program. APL is currently working on the next generation of deep space and low Earth orbiting spacecraft. Next generation missions will require the use of smaller, more capable spacecraft, and some missions will use spacecraft clusters to perform multipoint scientific data collection. To reduce the size of the spacecraft, APL has defined a new architecture that incorporates spacecraft subsystems into an integrated electronics module to reduce mass, volume, and cost. This "spacecraft in a box" not only includes subsystems that lend themselves to circuit board architecture, but also attacks the difficult problem of packaging the RF telecommunications system onto a set of cards. These cards use off-the-shelf monolithic microwave integrated circuits (MMICs) as well as field programmable gate arrays (FPGAs) to provide the functionality that previously required a separate box. An innovative technique for noncoherent Doppler tracking permits the use of a transceiver-based system instead of a coherent transponder. This paper discusses the deep space telecommunications technology that will enable future small spacecraft missions and presents recent APL technology developments and their applications to current flight programs.

INTRODUCTION

New advances in RF telecommunication system design are paving the way toward the highly integrated, l ow-cost missions of the future. These advances include new system-level ideas that permit the effective a pplication of miniaturization technologies. This paper provides an overview of The Johns Hopkins Unive rsity Applied Physics Laboratory`s (APL) recent accomplishments and their application to current flight programs.

In the recent past, the design methodology for scientific spacecraft has been dominated by the "box-and-harness" approach. This distributed approach provides clear boundaries between subsystems and has many advantages, including versatility, ease of testing, and ease of cross-strapping. However, it also comes with burdens, such as increased size, mass, and cost relative to more integrated approaches.

Historically, RF telecommunication systems have naturally fallen into the box-and-harness approach because of their special packaging requirements and because of the clear boundary has been drawn between the command and data handling (C&DH) system and the RF system.

This paper discusses the development of a new, highly integrated approach to RF telecommunications system design. Instead of incorporating stand-alone components, the RF transmitter and receiver elements are integrated with the rest of the spacecraft electronics into an integrated electronics module (IEM). This integration results in a natural blurring of the boundaries between the C&DH system and the RF system. For example, the IEM plug-in cards that contain the RF electronics also contain a telemetry framer compatible with Consultative Committee for Space Data Systems (CCSDS) standards, and a critical command decoder. In addition, a new, noncoherent Doppler tracking technique provides the high accuracy required for a deep space missions (0.1 mm/s). Flight IEM transceiver hardware is currently being developed for the Thermosphere-Ionosphere-Mesosphere Energetics and Dynamics (TIMED) mission. The Comet Nucleus TOUR (CONTOUR) spacecraft, NASA`s sixth Discovery mission, will also use the IEM approach.

THE EVOLUTION OF DEEP SPACE TELECOMMUNICATIONS SYSTEMS

The design of RF telecommunications systems for deep space historically has been centered about a coherent transponder system. At present, these systems typically operate in the X and Ka frequency bands, providing a command link, telemetry link(s), and radiometric tracking. The typical transponder system includes a microwave transponder, as well as baseband circuitry for demodulating commands and conditioning telemetry. For example, the transponder systems used for the Cassini, NEAR, and Mars Pathfinder missions included an X-band transponder, a command detector unit (CDU), and a telemetry conditioning unit (TCU). These three missions used identical transponders and CDUs, but each mission incorporated its own custom-built TCU. On NEAR, the total mass for the transponder, CDU, and TCU was 5.2 kg. The latest technology development is the small deep space transponder (SDST), which incorporates the microwave transponder, CDU, and TCU into a single chassis and weighs 3.0 kg. This unit, built by Motorola, will fly aboard the New Millennium program`s DS1 mission in mid-1998.

In response to the need for lighter weight, more compact spacecraft, APL has developed an IEM concept. In the APL IEM architecture, most core spacecraft subsystems are implemented on circuit board cards that plug into a common cardcage chassis. Communication between subsystems occurs across a high-speed, low-power bus. Significant mass savings result from the common packaging and the elimination of harness. Figure 1, which lists the masses of the NEAR transponder system, the SDST, and the IEM transceiver cards, illustrates the mass reduction. Another benefit of the common chassis is that co-located boards share resources such as oscillators and power converters. Subsystem boards can be readily reused on different missions, resulting in significant savings of nonrecurring development costs. Successful adoption of the IEM concept requires the RF communications to be incorporated within the IEM. Designing a deep space telecommunications system for the IEM is not a trivial effort; it has led to new ideas on the architecture of RF telecommunication systems.

TECHNOLOGY DEVELOPMENT AND APPLICATION

The CONTOUR Deep Space Probe

Engineers at APL will soon have the opportunity to apply these new system-level concepts to a deep

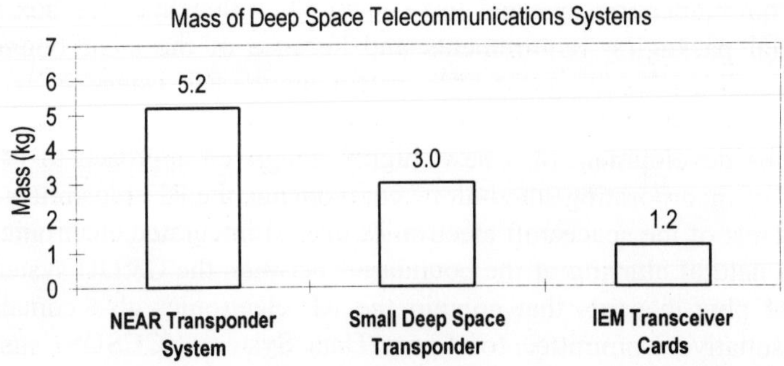

Figure 1. A comparison of mass in deep space telecommunications systems. Mass values are for a single system, excluding the external power amplifier and diplexer.

space mission. The CONTOUR mission, due to be launched in July 2002, will perform close flybys of three major near-Earth comets (Encke, Schwassmann-Wachmann-3, and d'Arrest). The spacecraft's suite of science instruments will take images, make spectral maps, and analyze dust flowing from the bodies to improve our knowledge of comet nuclei and their diversity. The core spacecraft electronics for CONTOUR will use the IEM concept.

The CONTOUR spacecraft will have a highly redundant telecommunications system. It will comply with CCSDS standards for communicating and will provide high-rate science data return, precision tracking data, and low-rate "Maintenance" communications. Figure 2 shows a block diagram of the CONTOUR IEM with emphasis placed on the telecommunications system. The heart of the CONTOUR telecommunications system will reside in two cards within the IEM. The uplink and downlink cards replace the deep space transponder, CDU, and TCU of the typical box and harness deep space probe.

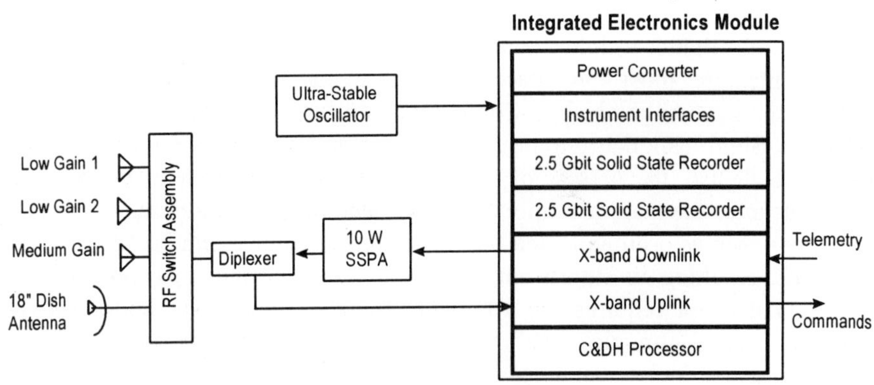

Figure 2. The CONTOUR IEM, with emphasis on the telecommunications system. Redundant IEMs will fly on this mission.

The CONTOUR X-band downlink card will accept data packets from the modified peripheral component interconnect (PCI) bus of the IEM, format and encode the data into downlink frames, and modulate the data onto an X-band carrier wave. This signal will then go to an external 10 W solid state power amplifier (SSPA) and the chosen antenna. The uplink card will be capable of receiving standard NASA commands at rates from 7.8125 bps to 2 kbps (in 7.8125 X 2^n steps). It will also feature a critical command decoder that can execute relay commands. Noncoherent Doppler tracking circuitry located on the uplink card will enable the primary means of tracking the spacecraft. An ultra-stable oscillator (USO), also built at APL, will furnish an accurate reference signal for the IEM.

The TIMED Low Earth Orbiter

The CONTOUR transceiver effort is an extension of APL`s work on the TIMED program. The NASA-sponsored TIMED satellite will be launched in mid 2000 to study the mesosphere, lower thermosphere, and ionosphere. For the TIMED program, engineers at APL have designed, built, and tested a two-card S-band transceiver that works within an IEM (Bokulic *et al.* 1997).

The TIMED IEM incorporates the S-band transceiver system, command and data handling system, solid state recorders, power converters, and Global Positioning System (GPS) navigation into each IEM chassis. Pre-engineering models of the TIMED S-band uplink and downlink cards have been built and tested (Figure 3). On both the uplink and downlink cards, the RF and analog circuits reside on the front side of a 15.2 X 22.9 cm card with digital circuitry mounted on the reverse side.

The TIMED downlink card provides a high-rate science downlink at 4 Mbps and a low-rate emergency downlink at 9 kbps. The downlink card also formats telemetry data into CCSDS frames and can encode the data with rate ½, $k = 7$ convolutional and Reed⇋olomon coding. To preserve bandwidth, differential quadrature phase shift keying (DQPSK) modulation is used for the high-rate data. The downlink card transmits 3 W of RF output power. The TIMED uplink card provides command reception capability at 2.0 kbps. It also contains a real-time command decoder that can execute relay commands. The uplink card stays powered continuously and can operate the spacecraft in the event of an emergency when all other IEM electronics are turned off. Finally, the circuitry necessary to perform noncoherent Doppler tracking resides on this card, although GPS will provide TIMED`s primary method of navigation.

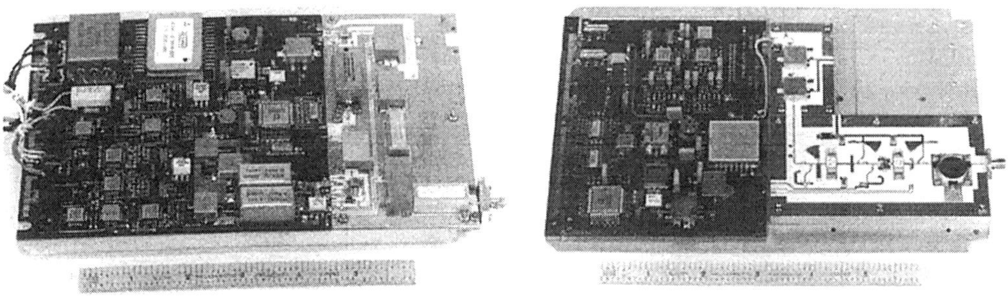

Figure 3. Photographs of the pre-engineering models of the uplink (left) and downlink (right) cards of the TIMED S-band telecommunications system.

The APL IEM Advanced Technology Effort

Several ongoing technology efforts have enabled APL to commit to the IEM concept for the CONTOUR and TIMED missions. These parallel research activities have resulted in the development of a composite-based IEM chassis, two custom ASIC chips, an S-band receiver card, and an X-band receiver card. The prototype IEM chassis is shown in Figure 4. This cardcage structure is manufactured using composites with conductive layers for electromagnetic interference control. APL has also developed two radiation-hardened ASIC chips customized for the IEM. The first is a serial bus interface chip based on the IEEE-1394 standard, which provides the interface for each card to the internal backplane bus (Fraeman 1997). The second ASIC performs remote input/output (RIO) operations.The RIO chip will remotely measure voltages, currents, and tell-tales. It also contains counters, comparitors, and analog and digital control outputs (Paschalidis 1997). Finally, in the telecommunications area, S-band and X-band receiver cards have been designed, built, and tested. These cards have significantly reduced the risk of infusing the technology into flight programs.

Figure 4. A composite-based IEM chassis.

X-Band Receiver Card Development

An X-band receiver card has been built and is being tested at APL. This prototype card provides a design basis for the uplink card for the CONTOUR mission. The experimental receiver card receives a modulated X-band signal and outputs the demodulated command data stream.

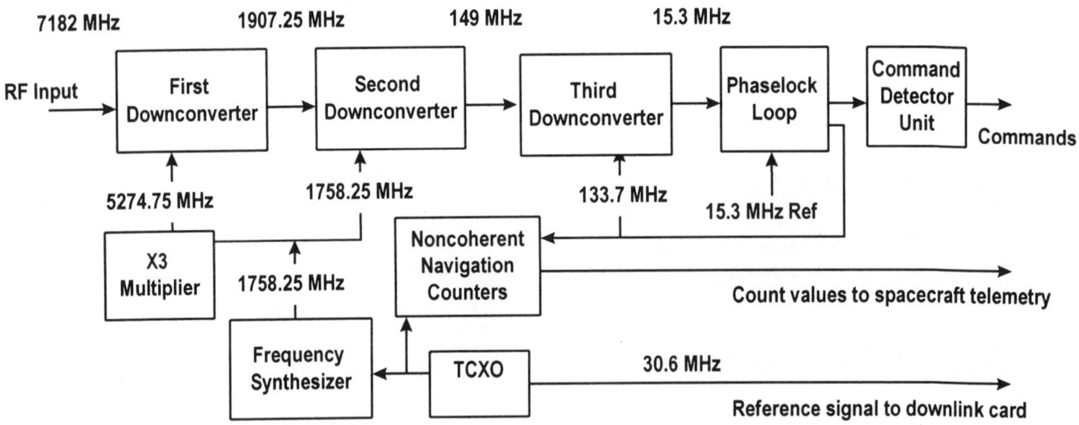

Figure 5. Block diagram of the experimental X-band receiver card.

Figure 5 shows a functional block diagram of the card and labels the input and output connections. The input signal to the receiver consists of a 7182 MHz carrier signal with a 16 kHz sine-wave subcarrier phase modulated onto it. When the X-band signal enters the receiver, it is downconverted in three steps to 15.3 MHz. The downconverter is implemented using off-the-shelf monolithic microwave integrated circuit (MMIC) amplifiers, active mixers, and AGC attenuators. Next the signal is passed to the phase-locked loop, which strips the subcarrier off of the carrier wave and provides it to the CDU. The CDU amplifies the 16-kHz signal, digitizes it, and demodulates the command data stream. The CDU is implemented in two field programmable gate array (FPGA) chips. The local oscillator frequencies for the downconverter are synthesized from a local 30.6 MHz oscillator. To facilitate noncoherent Doppler tracking, the receiver card shares its reference frequency with a downlink card. The noncoherent Doppler tracking circuitry is contained within one of the FPGA chips of the CDU. The layout of the X-band receiver mimics that of the TIMED S-band uplink card. The RF and analog circuits are mounted on the front side of a 15.2 X 22.9 cm card (Figure 6), with the CDU mounted on the reverse side. Table 1 lists the design goals of the receiver card.

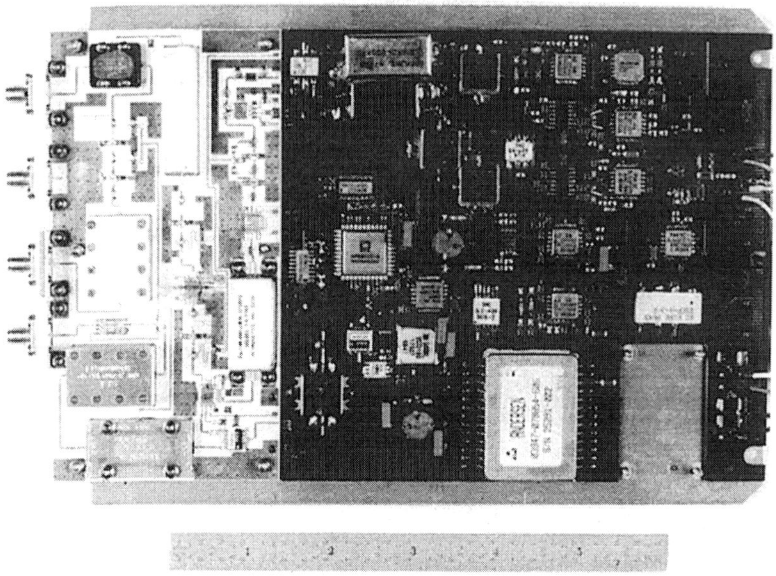

Figure 6. The experimental X-band receiver card. The card size is 15.2×22.9 cm.

Table 1. X-Band Receiver Card Design Goals.

Parameter	Design Goals
Center frequency	7182 MHz
Acquisition threshold	–155 dBm
Noise figure	< 3 dB
Dynamic range	–70 dBm to threshold
Image rejection	60 dB
Carrier tracking loop band width (2 B_L)	20 Hz at threshold
Tracking range	± 150 kHz
Command bit rate	7.8125 bps × 2^n (to 2 kbps)
Size	15.2 ×22.9 cm
Mass	≤ 600 grams
DC power	7 W

Two Way Non Coherent Doppler Tracking

One of the enabling technologies that led to the IEM-based telecommunications system is the development of a method to perform Doppler tracking using a noncoherent transceiver system on the spacecraft. This new approach, developed at APL, is compatible with existing ground station assets and provides Doppler tracking with the accuracy needed for deep space missions (0.1 mm/s). The TIMED mission will incorporate this technique on an experimental basis. The CONTOUR spacecraft will rely on this navigation technique for its primary tracking capability. Compatibility tests with the Deep Space Network are planned for early 1998.

Figure 7 illustrates the noncoherent Doppler tracking concept. The uplink signal is received and compared with an onboard oscillator in the command receiver. The results of this comparison (phase counts) are placed in the spacecraft telemetry and transmitted to the ground. At the same time, the ground station is tracking the spacecraft as if it had a coherent transponder on board. The Doppler data obtained at the ground station will have significant errors due to drifting of the spacecraft oscillator; however, these errors can be corrected using the counter values from the spacecraft telemetry.

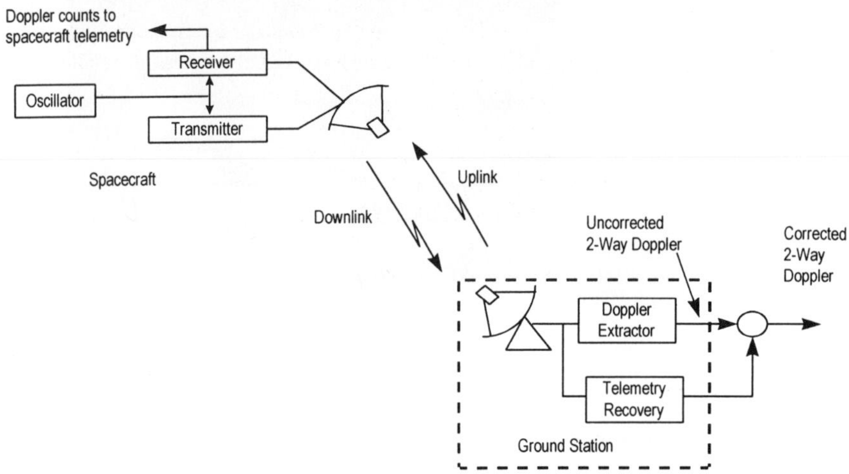

Figure 7. The two-way noncoherent Doppler tracking concept.

The consequences of using this technique are minimal, both on the spacecraft and at the ground stations. Spacecraft oscillator stability is not critical, and accurate time-tagging on the spacecraft is not needed. For deep space missions, the technique is expected to require on the order of 24 bits of telemetry per minute during cruise. The counter circuitry consumes only 0.3 W and is largely incorporated into existing FPGA chips. As mentioned before, ground station assets use their conventional methods of Doppler tracking without system upgrades. The correction is made in a postprocessing software step as part of the navigation solution.

This concept was tested at APL under "zero velocity" laboratory conditions. With the experimental hardware located in a laboratory environment, the true velocity of the uplink card was zero. Figure 8 shows that gross velocity errors up to 30 m/s occurred when the on-board oscillator was tracked without any correction. However, when the tracking data was corrected using counter values from the uplink card, the velocity errors were reduced to the 0.1 mm/s level. The correction worked even in the face of large oscillator drift, such as that produced by a thermal transient.

Another feature of the receiver is its ability to make one-way uplink Doppler measurements. When used in conjunction with a USO, this facilitates the use of hibernation modes in deep space probes. In this scenario the receiver is constantly powered during hibernation mode. A ground station with a low cost antenna can transmit an X-band tone to the spacecraft and lock up the receiver. The uplink Doppler counts are stored in the spacecraft memory. When the spacecraft emerges from hibernation mode, this Doppler history can be telemetered to Earth and used as an aid in obtaining a navigation solution.

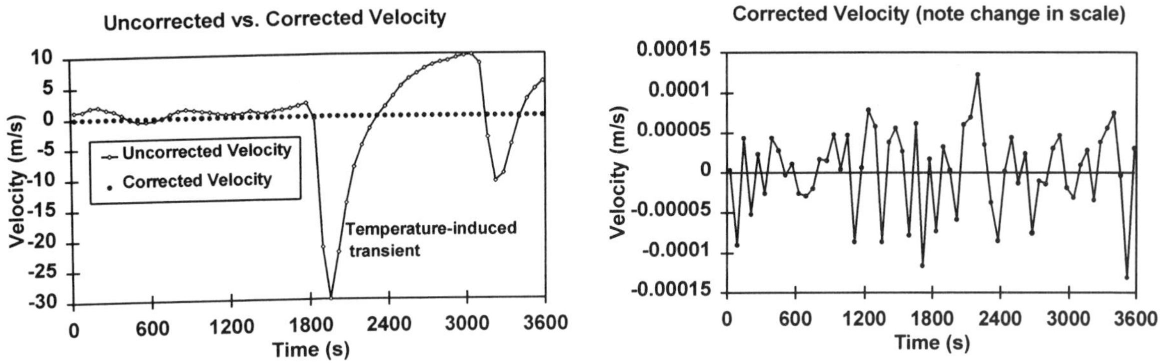

Figure 8. Results of a two-way noncoherent Doppler tracking test performed at APL. Using this technique, gross errors in onboard oscillator stability (left) are readily canceled, producing tracking data accurate to the 0.1 mm/s level (right).

Mini-IEM

The next generation in integrated electronics modules will include a further size reduction. Work has begun on a command and data handling system that leverages APL`s chip-on-board technology and fits on a 10.2 × 10.2 × 1.3 cm stackable slice (Conde *et al.* 1997). The modules currently in development are an RTX2010 processor and interface module, a 2.0 Gbit solid-state recorder module, and a Mongoose V processor and interface module. A photograph of two modules is shown in Figure 9. The next challenge is reducing the size of the telecommunications system so that it is compatible with this new architecture. Some methods that may be applied to achieve this size reduction are the use of higher-level MMIC chips, the implementation of the carrier tracking and automatic gain control loops with digital signal processing, and the development of an ASIC to replace the FPGA chips of the CDU and real time command decoder. Chip-on-board technology will be implemented in the digital portions of the telecomm-unications systems.

Figure 9. The Mini-IEM. The module size is 10.2×10.2×1.3 cm.

CONCLUSION

Significant effort has been expended at APL to support future small spacecraft missions by reducing the size and mass of deep space telecommunications systems. APL has achieved this goal by moving from a traditional box-and-harness approach to an integrated electronics module approach. New ideas for packaging and Doppler tracking have enabled us to obtain the high performance required for deep space missions using a transceiver-based telecommunications system. These innovations have circumvented the need for expensive new technology development and enabled rapid infusion into flight programs such as TIMED and CONTOUR.

ACKNOWLEDGMENTS

The authors would like to acknowledge Bob Jensen, co-inventor of the noncoherent Doppler tracking technique. We also thank Jeff Will and Lloyd Ellis of APL RF Engineering Group for their work in fabricating the experimental IEM X-band receiver card.

REFERENCES

Bokulic, R. S., C. C. DeBoy, D. B. Minarik, J. Daniels, S. Cheng, and J. H. Sinsky, A Highly Integrated S-Band Transceiver System with Two-Way Doppler Tracking Capability, Proc., 11th AIAA/USU Conf. on Small Satellites, Logan, UT (1997).

Conde, R. F., B. Q. Le, J. F. Bogdanski, A. L. Lew, J. A. Perschy, and M. A. G. Darrin, Command and Data Handling in Your Palm, Proc., 11th AIAA/USU Conf. on Small Satellites, Logan, UT (1997).

Fraeman, M. E., A Fault Tolerant Integrated Electronics Module for Small Satellites, Proc., 11th AIAA/USU Conf. on Small Satellites, Logan, UT (1997).

Paschalidis, N, A Remote I/O (RIO) Smart Sensor Chip Supporting Serial Communications Bus for Small Satellites, Proc., 11th AIAA/USU Conf. on Small Satellites, Logan, UT (1997).

THE AEROSPACE PROGRAM AT UNIVERSITI SAINS MALAYSIA

Md. Azlin bin Hj. Md. Said, B.Eng.(Hons), M.Sc.(Eng.), Ph.D

Aerospace Programme Coordinator · School of Civil Engineering · Universiti Sains Malaysia, Perak Branch Campus · Seri Iskandar, 31750 Tronoh, MALAYSIA

ABSTRACT

This paper describes the aerospace activities in Malaysia especially in the areas of space technology. The recent satellite based activities in Malaysia had made the need for human resource development and aerospace research based activities more prominent. The efforts done by Universiti Sains Malaysia to meet this need is described. Apart from the academic programs planned, research and development activities are also described in which it was designed to meet the national satellite and space programme and also support supporting aerosapce industries with the assistance from various parties.

INTRODUCTION

Malaysia is in the process of becoming an industrialised nation by the year 2020 as envisaged in Vision 2020. The Government of Malaysia has recognised that aerospace sub-sector in the areas of aeronautics and space technology is one of the critical areas to be tackled if it is going to achieved the target set for Vision 2020.

The two successful launches of MEASAT-1 and MEASAT-2 by Binariang Sdn. Bhd. recently for the purposes of telecommunications and broadcasting had spurred the nation. The two satellites had pushed Malaysia from a developing country status into one of the world's space nation. The launch of the first national microsatellite project sometime this year will no doubt focus the nation's interest in space technology. The satellite projects have provided a shortcoming that is to provide the need for the development of indigenious technology and human resource development in space engineering and related technology.

To overcome this shortcoming, several aerospace programmes were introduced in local tertiary institutions in Malaysia, but the programmes were generally aeronautics based. Several government agencies such as MACRES and BAKSA were also involved in setting up the necessary groundwork in satellite technology, but there exist a gap in the development of producing graduates competent in the area of space technology.

Recently, Universiti Sains Malaysia had started the process to introduce the program of aerospace engineering at its engineering campus, where the focus was directed towards aeronautics, astronautics and satellite technology, and the introduction of its School of Aerospace Engineering. It was hoped that the programme would produce graduates who are competent enough to provide the nations human resource requirements in aerospace technology. It was proposed that the School of Aerospace Engineering starts in 1998 and the first intake of students would be in first semester of 1998.

THE AEROSPACE ENGINEERING PROGRAMME

Universiti Sains Malaysia started its aerospace programme in 1994 by offering seven elective courses in the areas of satellite technology and astronautics. The programme was greatly assisted by Binariang Sdn. Bhd. as it had assisted in providing training to several academics at Hughes and ISRO in satellite technology. However, in 1996 efforts were made to start the process of offering a degree program in aerospace engineering.

The aerospace programme was developed after consultations and advice from prominent academics of various aerospace institutions of higher learning throughout the world. The programme was finally drawn up, which it is a three-year degree programme in where it would encompass the necessary elements for a recognised graduate programme in aerospace engineering, without compromising its quality and balanced in content. The programme is designed as a 126 unit programme where one unit is equivalent of 13hours of contact with the lecturers, and follows the norm set by Universiti Sains Malaysia for its other engineering programmes. It was recognised there would be shortcomings in the programme, but efforts were made to reduce them to a minimum.

The distribution of the various course types is presented in Table 1 and Table 2, show at least 88% engineering type courses will be covered by the students throughout the three year programme. The remainder 12% will be covered by language courses, religious courses, skill based courses , co-curriculum activities and option courses.

Table 1. Course Structure Distribution.

Course Type	Units	%
Common	19	15.08
Basic	16	12.70
Core	64	50.79
Electives	12	9.52
University Requirements	15	11.91

Table 2. Distribution of Basic, Common and Core Courses.

Course Type	Units	%
Structure/Materials	12	12.12
Propulsion/Heat Transfer	10	10.10
Mission	6	6.06
Mathematics	8	8.08
Laboratories	6	6.06
Industrial Training	10	10.10
Final Year Project	8	8.08
Engineers in Society	3	3.03
Design	4	4.04
Control Systems	19	19.19
Computer/Drawing/CAD	4	4.04
Aerodynamics	9	9.09
	99	100.00

To further enhance the knowledge of students in a perticular area of aerospace engineering, there are three elective packages being offered, i.e. astronautics engineering, satellite technology engineering and aeronautics engineering. There is a choice of subjects in each package for the students to take to fulfil the elective package requirements. The courses include courses in management, advanced design, systems and advanced materials.

The students are required to undergo two levels of hand-on training in-campus and then at least ten weeks of industrial based training. The final year project will be conducted in the final year project in which aerospace projects will be conducted.

Assistance were obtained from various countries, in the form of academic assistance and the setting-up of aerospace based laboratories. Universiti Sains Malaysia can therefore develop the program with confidence and also benefit in terms of transfer of technology.

Apart from undergraduate courses, Universiti Sains Malaysia has also plans to introduce postgraduate courses i.e. at M.Sc. and Ph.D levels. This is to encourage research to be conducted to utilise the expertise and equipments and contribute to the national research programme.

THE FUTURE RESEARCH AND DEVELOPMENT PLANS

Universiti Sains Malaysia has ambitious plans for the proposed School of Aerospace Engineering in the area of astronautics and satellite technology. It was agreed that the focus of the School would be to develop the expertise and know how to design, built, test, launch and ground control of satellites. This project would therefore focus the development of the laboratories, capability of the staff and group the various expertise available at the University into clearly defined focus groups. The University can therefore contribute in providing the expertise and professionals in the nations space programs.

A program of satellite projects was developed in which Universiti Sains Malaysia personnel will work together with fellow researchers from various countries. At present there are a variety of satellite project proposals. One of the proposed satellite project is in collaboration with the Russians where the overall mission would be earth observation or generally known as remote sensing. The satellite would be a microsatellite, where the mass would be around 40-50kg. The spectral resolution for multispectral scanner in the visible range is 0-5-0.6 μm , 0.6-0.75 μm , 0.75-0.8 μm , whilst in the panchromatic mode is 0.5-0.8 μm. The spatial resolution is around 50-100m with a typical scene of 50km $\times$ 50km. The data obtained can be used in the areas of hydrology, water resources, land-use monitoring, etc. The satellite would be designed and built in Malaysia jointly by Malaysian and Russian scientists and engineers.

Apart from the proposal stated above, several proposals were being made to jointly develop satellites from 100kg – 250kg for various missions and scientific payloads. Discussions are currently being made for such projects in which several factors would be tackled. One of them is the ability of transfer of technology to Universiti Sains Malaysia personnel in all aspects of satellite technology, i.e. from the design stage, analysis, production, integration, control and tests. With this, the development of indigenous technology can therefore be developed as the projects progresses from one satellite to another.

CONCLUSION

Universiti Sains Malaysia has taken a great step in the area of aerospace engineering. The proposed academic programme outlined above is ambitious, practical but without compromising the quality required for such a programme will greatly contribute to the nations aerospace and satellite program as it progresses towards an industrialised nation by the year 2020.

ACKNOWLEDGEMENTS

The Author wishes to expressed his sincere thanks to Universiti Sains Malaysia especially to Prof. Hassan Said (Dep. Vice Chancellor (Academic)) and Prof. Jamjan Rajikan (Dep. Vice Chancellor (R&D)) in providing the financial support, encouragement in preparing this paper. To Prof. Fei Bin Hsiao of National Cheng Kung University, Taiwan in providing support and assistance in making this paper presentation possible.

PLANETARY ATMOSPHERIC MICROPROBES

BERND HÄUSLER

Universität der Bundeswehr München, Institut für Raumfahrttechnik, 85577 Neubiberg, Germany

ABSTRACT

Microprobes can be used as research tools allowing dedicated measurements in regions near planets or moons usually not accessible for orbiting spacecraft. Here, we mainly discuss the investigation of the near environment of Jupiter`s Galilean moon Io. Of special interest are the composition and density distribution of Io´s atmosphere, also Io´s mass properties and the configuration of the current system in the Jupiter-Io magnetospheric environment. It is proposed to release the microprobe ca. one month before its arrival at Io from an interplanetary spacecraft (for example Pluto-Kuiper Express) near the begin of the gravitational swing-by manoeuver. The microprobe will be fully active and will transmit telemetry data to Pluto-Kuiper Express during the last twenty minutes before it performs a hard landing on Io´s surface or a grazing fly-by.

The proposed scientific model payload consists of a neutral gas spectrometer, a 3 axis magnetometer sensor and a radio science instrument, the mass allocated to the Io probe will be kept extremely low, not exceeding 17 kg. The probe will be powered by both primary batteries and solar cells, it will be spin stabilized and actively controlled by a cold gas system, its thermal control will be accomplished by passive means and radioactive heater units (RHU´s). We discuss in this paper the separation strategy from the main spacecraft, the design of an ultralight structure, the thermal control concept, the attitude control concept and the onboard electronics which has to operate in the extreme harsh radiation environment of Jupiter magnetosphere.

The proposed microprobe concept offers enough flexibility to equip the probe also with different payloads. We discuss the possibility to perform measurements made by mass spectrometer and surface penetration radar instruments in the near environment of Jupiter`s moon Europa.

MISSIONS TO THE OUTER PLANETS

Early ideas about a mission to Pluto, the outermost planet of the solar system, have been developed around 1970 within the framework of the original Grand Tour mission. The idea then came back to the scientific community after the successful encounter of the Voyager 2 spacecraft with Neptune in August 1989. This event left Pluto with its satellite Charon as the last planet and the meanwhile discovered disk (Kuiper Disk) of "Ice dwarfs" or minor planets as a region not yet explored at all by a spacecraft encounter. NASA/JPL has now proposed a mission "Pluto-Kuiper Express"(PKE) to be launched in 2003.

Because many of the mission scenarios led to close encounters with Jupiter as the most favorable trajectory option, it was suggested to use such a close encounter for the release of a microprobe to study the Jovian satellite Io, one of the most exotic and enigmatic objects in the solar system. This attracted also the interest of a large part of the German planetary and space physics community. In addition, it was considered favorable to interrupt the long flight to Pluto by a truly exciting scientific highlight (Neubauer 1996).

Starting in early 1995, the Pluto Express concept was further developed from a scientific and technical point of view during the work of a Science Definition Team (SDT) established by NASA. The report of the SDT was completed in September 1995 and included a section on the Russian "Drop Zond" and a German "Io-Probe" as options for international cooperation. These discussions led to the idea of a technical study of the Io-Probe. The results of the technical feasibility study (Häusler 1996) form a major part of this report.

These results can also be adopted to microprobes dedicated to studies of other bodies in the outer solar system (Pluto, Europa). Especially Europa, fourth largest of the Galilean moons, has attracted recently tremendous interest since it became known that it has an atmosphere (Hall *et al.* 1995, 1996, Ip 1996) and possibly a liquid water ocean buried underneath a very smooth ice surface several hundred meters thick (Kerr 1997). It is also speculated that this environment could even harbour biological activities.

IO MICROPROBE

Science

Due to the orbital eccentricity of Io, forced by the orbital resonance with the neighbouring satellites Europa and Ganymede, ongoing strong tidal friction provides a strong internal heat source of several 10^{13} Watts (Peale *et al.* 1979). This heat source drives volcanic eruptions releasing large amounts of gas which also constitute the source for Io highly dynamic atmosphere (Strobel *et al.* 1995). It looses about 1 ton/sec or 10^{28} SO_2 molecules per second due to a violent electrodynamic interaction with the Jovian inner magnetosphere, where the Io torus plasma is stripping the atmosphere and neutral cloud of particles that are subsequently ionised, themselves becoming part of the plasma torus (McGrath 1997; Brown and Bouchez 1997). This also leads to a system of Alfven wings launched by Io towards the Jovian northern and southern hemispheres (Neubauer 1980, Acuna *et al.* 1981). The existence of a magnetized interior is still under dispute (Frank et al. 1996). In order to investigate these effects the following scientific instruments were given the highest priority for a model payload:

- Neutral/ion gas mass spectrometer (RTOF): mass range: 1-100, mass resolution: 1, pressure: 10^{-6} - 10^{-17} bar
- Three-axis fluxgate magnetometer: range: 5000 nT, resolution: 10 pT
- Radio Science Investigation (USO): $\Delta f/f = 10^{-10}$

The Mission. For a mission to Io, the probe will be released from the Pluto-Kuiper Express (PKE) sciencecraft at a time based on considerations of mission safety, Δv-requirements, targeting accuracy, thermal and power considerations etc.. For the Io encounter trajectory, the nominal Pluto-Express trajectory from the SDT-report has been taken (JPL 1996). In this case, the distance of closest approach between Io and PE is 40.000 km. In order to traverse a maximum developed atmosphere on the trajectory approaching Io, an impact near the subsolar point on Io should be attempted as well as a distance of a few tens of kilometers from an active volcanic plume.

Data acquisition of the probe should be performed at heights below several 1000 km. In addition some time interval for experiment preparation, calibration etc. has to be taken into account. The data rate should be ≤ 2 kbps in order to achieve at least one full mass spectrometer measurement per 100 km or per scale height (at low altitudes) as well as at least one vector magnetic field measurement per 10 km.

Mass and cost constraints are the determining factors when designing the subsystems for the probe. As a first target for the feasibility study the Io-Probe system should not exceed a total of 15g. Therefore, a major subsystem such as the attitude control system was selected to be spin-stabilized. However, because of the high velocity of the probe (ca. 13 km/s relative to Io with respect to ca. 0.5 km/s thermal velocity) it must be equipped additionally with a capability to keep the spin axis of the spacecraft parallel to the direction of the velocity vector when close to Io. Further constraints imposed on the mission are given by the large distance from the sun resulting in a practically completely battery-based microprobe power subsystem and the extreme hard radiation environment in Jupiter´s magnetosphere (Häusler *et al.* 1996)

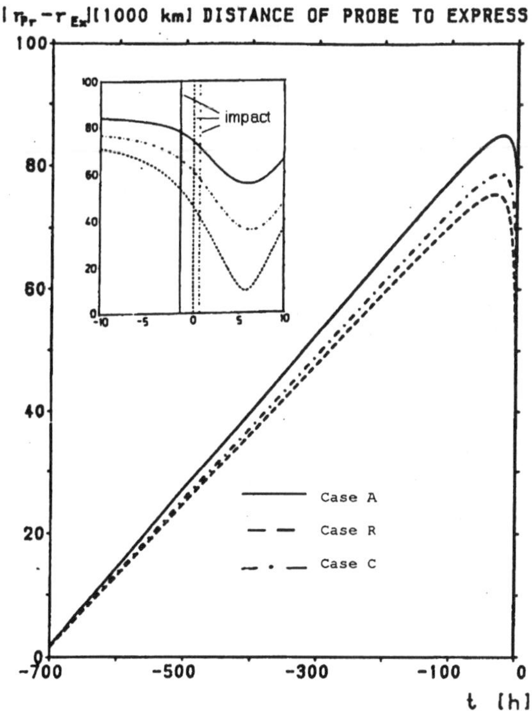

Figure1. Distance between Probe and Pluto-Kuiper-Express.

Separation Strategies. The Io-Probe will not be equipped with a propulsion system. In order to impact on Io's surface or to achieve a close "grazing" fly-by we assume that the necessary trajectory (Δv)-maneuvers will be solely done by Pluto-Kuiper Express (PKE). It shall release the probe on its trajectory towards Io and perform afterwards a correction maneuver to become aligned with the desired Jupiter swing-by trajectory towards Pluto.

The separation was selected to occur at Jupiter`s sphere of influence (t = -713h). Figure 1 shows the temporary development of the distance between Io-Probe and Pluto-Express for the cases A,R,C. First, it increases nearly linearly with time up to a maximal distance of about 80.000 km and decreases afterwards very fast to a minimal distance at time of impact (or to a local minimum distance 5 hours after passing Io) depending on the actual study case:

In case R (---), with an Io-Probe flying nearly parallel to Pluto-Express, the distance is about 50.000 km half an hour before impact. In case A (—), meaning "Io-Probe leading Pluto-Kuiper Express by 45°", the distance between Io Probe and Pluto-Kuiper Express within this time period is about 80.000 km and in case C (-.-.), where Io-Probe is located behind PKE, the critical distance is about 60.000 km.

Case A and case C fulfill the requirements set by the radio science experiment, i.e. these configurations will allow to determine possible dust impact effects due to the Doppler effect with high sensitivity. We selected case R be-cause it leads to the smallest distance between both spacecraft and provides also (not shown here) a rather constant antenna aspect angle as well for the Io-Probe as also for the PKE receive antenna. Both effects contribute considerably to an improvement of the telemetry link quality. The necessary Δv amounts to 32.3 m/s which is also the minimum Δ v, verified by an optimization algorithm.

We further analyzed the targeting accuracy vs. Δv-injection errors. The analysis shows that the deviation from the computed Δ v should be less than 4% (< 1.2 m/s), otherwise the Io-Probe would miss Io. Since the total Δ v provided by the separation spring is ca. 0.35 m/s, it will not be difficult to meet this requirement.

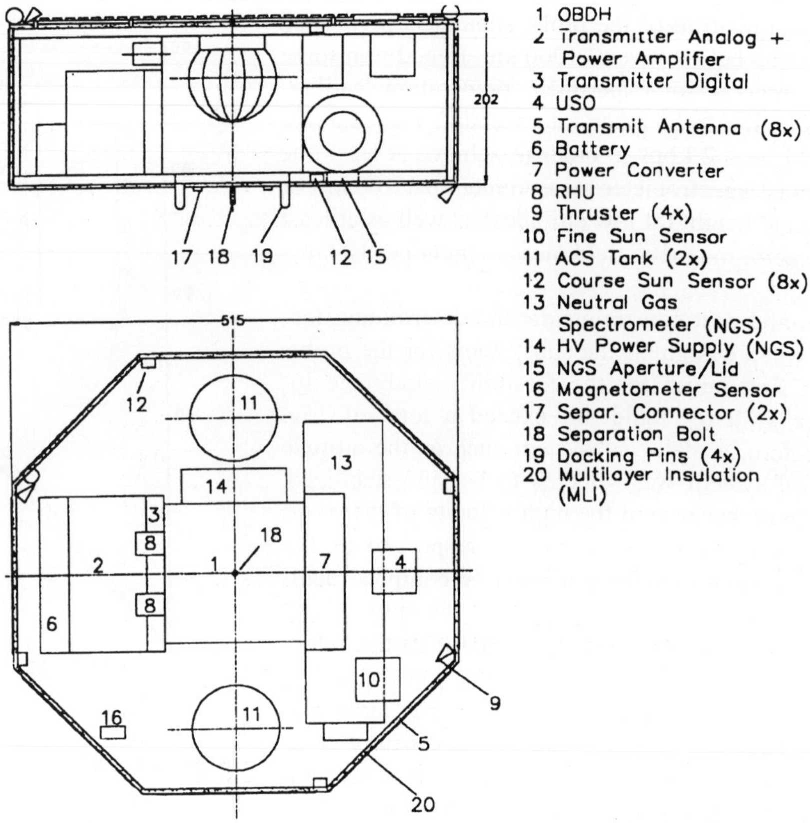

Figure 2. Structural Layout of Io-Probe

Structure

Structural Layout, Structural Analysis. The Io-Probe including subsystems and instruments is shown in Figure 2. Its diameter is 515 mm, its height 202 mm. Top and ground panel are manufactured in a sandwich construction using a PMI foam core. The two panels are rigidly connected via the housing of the central electronics box (OBDH). The housing is made out 7075 standard aluminium material (2.5 mm). Both panels carry equipment and instruments, the top panel is also loaded with solar cells, while 4 shear pins belonging to the "docking module" to the Pluto-Express spacecraft are integrated into the bottom panel. The OBDH box is the main (shear) load carrying structure. Its bottom plate of the box is an integral part of the sandwich panel. The mass of the Io Probe structure is 1.13 kg.

In order to provide for the structure as well high strength as high damping characteristics, the bottom and top panels of the probe are made out of carbon reinforced epoxy laminates with a PMI foam core. This combination was found superior in comparison to more conventional materials such as NOMEX or Aluminium honeycombs. For box mounting epoxy fillets glued into the core foam or into the sandwich beam stringers will be used. This will reduce the weight by avoiding a complex insert and potting layout. The mass budget of the Io Probe is shown in Table 1.

Table1. Mass Budget of the Io-Probe.

	Mass [kg]
Io-Probe Subsystems	
Structure	1,13
NGS Cylinder	0,25
Thermal Control	0,65
Attitude Control System (ACS)	2,03
ACS Propellant (Argon)	0,40
Power	0,94
Solar Cells	0,20
On Board Data Handling System (OBDH)	2,76
Transmitter S-band	1,75
Antenna	0,64
Neutral Gas Spectrometer (NGS)	2,87
Magnetometer Sensor (MAG)	0,05
Oscillator (USO)	0,25
USO Spot Shielding (Tl)	0,25
Harness, Connectors, etc.	0,81
Balance Mass	0,20
Margin (10%)	1,52
Total Mass	**16,70**
Dust Shield (optional)	0,40
Pluto-Express Support Equipment (shared with Drop-Zond)	
Pyrotechnics	0,10
Receiver S-band, TC-Interface	2,27
Oscillator (USO)	0,25
USO Spot Shielding (Tl)	0,25
Antenna	0,70
Harness, Connectors, etc.	0,29
Margin (10%)	0,39
Total Mass	**4,25**
Io-Probe System	
Io-Probe	16,70
Pluto-Express Support Equipment - Shared with Drop-Zond (50%)	2,12
Total System Mass	**18,82**

The following FE-analyses were performed: Static launch load simulations, thermal stress analysis, free vibration analysis. All analyses gave satisfactory results with respect to deflections and margins of safety. The thermal stress analysis showed that special care has to be given when considering first ply failure and laminate cracks at very low temperatures (delta T = -200°C, ref. glasspoint). It is expected that these problems are reduced when taking into account the non-linear behaviour of the used materials.

Mechanical Interface to PE. In order to save mass and to gain the necessary stiffness in the bottom panel, the OBDH-box will be integrated into the lower sandwich plate. Four shear pins (Al) with a length of 2.5 cm and a cross-section of 100 mm^2 are integrated into the bottom plate and are resting in the Pluto Express docking pads before separation. A titanium bolt with a crossection of 3 mm^2 will be used to attach the Io Probe to Pluto-Kuiper Express. For separation, it will be cut by a pyrotechnic guillotine. A Δv of 0.35 m/s will be provided to the probe at separation by a conical shell spring.

Thermal Aspects

Design Constraints, Thermal Analysis. The thermal design of the Io-Probe had to consider severe environmental thermal conditions (5 ears in spacecraft shadow, 15 minutes of incident solar radiation during Venus swing-by, very little solar power available when near Jupiter). In addition, the thermal design had to cope with a variety of different operational modes (NGS heating, aperture opening, high dissipated power during operational mode, no electrical power for heating purposes available). The analysis was performed using the FEM program MSC-NASTRAN for Windows.

Thermal Design. The multi-layer insulation (MLI) used will be a 15 layer superinsulation type with Dacron spacers. The inner layer will be a 50μ double aluminized Mylar (DAM) foil, the following 13 layers 6μ DAM. The outer layer (facing space or solar array) will be a double goldized Kapton foil with a high ratio of $\alpha/\varepsilon = 10$.

This concept led to a solar array which is thermally decoupled from the upper sandwich panel. The MLI is located between solar array and upper panel (Fig. 3). Further, the analysis has shown that the long cruise phase in the shade of Pluto-Kuiper Express and the 30 day free flight phase near Jupiter makes it necessary to install two radioactive heater units (RHU) with a dissipating energy of 1W each to maintain the proper thermal environment for the Io-Probe experiments and subsystems.

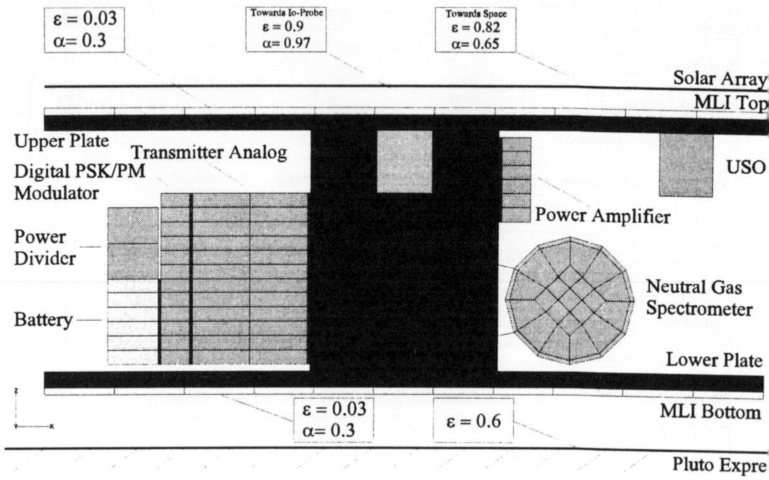

Figure 3. Thermal Model with Surface Properties.

The transient thermal analysis revealed that for operating times close to 30 minutes all temperatures will stay within their operational limits. For longer time periods the microprobe temperatures will exceed those limits resulting finally in either a thermal destruction or a disruption of operation due to the limited battery capacity, whatever is earlier. In any case, at this time a collection of scientific data is of no further use.

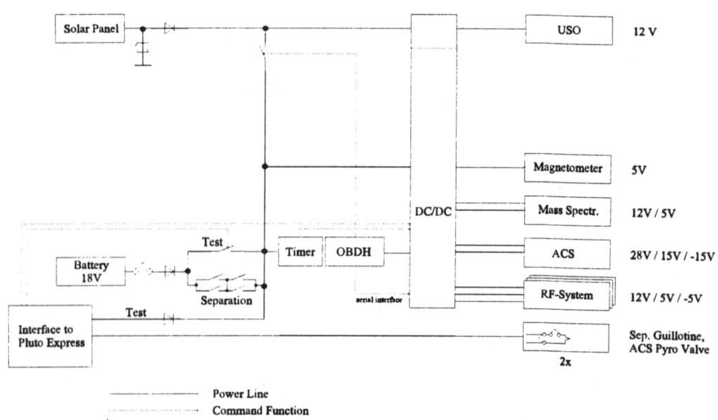

Figure 4. Power Management Subsystem.

Power Management Subsystem

The block diagram of the power management subsystem is shown in Figure 3. Because of mass savings reasons, all components (except separation switches and timer) must be non-redundant. The subsystem consists of the primary power sources, battery and solar generator, DC/DC converters including current limiters, switches, and the interface to the mothership Pluto-Kuiper Express. Because of EMC reasons all subsystems are equipped with separate DC/DC converters. The digital +5 V supply is the only supply shared among OBDH and subsystems. Switch on/off of all DC/DC-converters is controlled by OBDH via a serial interface. The power converter of the on-board computer itself is controlled by a low-power timer, which will activate OBDH at predetermined times. This timer can be programmed via the Pluto-Kuiper Express command interface.

Battery. Low solar energy input (ca. 51 W/m^2) and extreme mass restrictions are the most important design drivers for the power subsystem of the Io Probe. This applies also for primary batteries and converters. A Lithium/SO_2 battery cell (SAFT) was selected which has been already flight-qualified in the Galileo project as primary energy source. To minimize power consumption where possible, all subsystems - including OBDH - will be switched off when not needed. Furthermore, to avoid any leakage currents, the DC/DC converters have to be switched off and disconnected from the bus via relays.

Solar Generator. The USO quartz needs many days of operation to reach the required frequency stability of at least 10^{-10}. Although it needs very low power (0.7 W), it is not possible to operate it from the battery, the overall energy consumption would exceed the available battery capacity. A solar generator will be able to supply the energy. It has to deliver power at very low levels of solar intensity (51 W/m^2) at very low temperatures. It is planned to use LILT (Low Intensity Low Temperature) high efficiency, low mass Si-cells, which are presently being developed and qualified by ESA for deep space applications such as the Rosetta mission. They are optimized for an operating temperature of 150°K (the expected temperature of the Io-Probe backpanel is about 170°K). With an efficiency of 25% they can deliver a power of 2.2 W.

Onboard Data Handling Subsystem

Design Constraints and Main Functions. Design driving constraints are:

- Severe radiation environment, ≈ 80 krad total dose behind 1 mm Al.
- Very restricted mass budget, less than 3 kg for the Data Handling subsystem
- Limited energy budget because of supply by batteries (≈ 100 Wh).

The Onboard Data Handling Subsystem OBDH combines the functionality of a conventional probe data and command subsystem with

- all digital data processing functions of the Neutral Gas Mass Spectrometer (traditionally located partly in the sensor electronics box and partly in the instrument DPU),
- all electronic functions of the flux gate magnetometer,
- the logic functions of the attitude control and power management subsystem

During the pre-separation phase the OBDH can receive commands from the mother-S/C. Vice versa it transmits verification and status information to the mother-S/C. After separation, the OBDH subsystem is completely autonomous, i.e. it cannot be influenced by the mother-S/C any more.

The long-time storage of status information is sensitive to SEU-induced bit-flip errors. Therefore, all sensitive NVRAM-parameter are protected by both error coding and triple redundant storage in non-adjacent array locations of the NVRAM. The timer is triple-redundant and triggers with the occurrence of the second run-out. For reliability reasons a dual-redundant microprocessor system would be preferable. But the implied add-on mass is not tolerable. We consider the signal processors 320C40 and ADSP21020 as favourite candidates.

Radiation Environment, Shielding Thickness. The radiation environment of the Io-Probe is characterized by the intensive electron and proton fluxes of the Jovian radiation belt. Todays digital CMOS standard devices show typical tolerance doses of 10 ... 20 krad. Shielding down to 10 (15) krad for standard devices requires a shielding mass density of approx. 10 (7) g/cm^2. In contrast, rad. hard devices with 100 krad tolerance dose require 0.7 g/cm^2 only, i.e. no radiation-imposed enforcement of aluminium box walls. Assuming rad.-hardened components, we estimate the mass of the OBDH system to 2.7 kg.

Attitude Control Subsystem

The type of attitude stabilization of the Io Probe is determined by as well scientific as spacecraft operational requirements. These are: Neutral gas mass spectrometer angle of attack pointing, spin modulation of the magnetometer, high quality communications link to Pluto-Kuiper Express, the necessity for continuous solar power supply for the high stability oscillator and thermal control of the probe (Fig. 5).

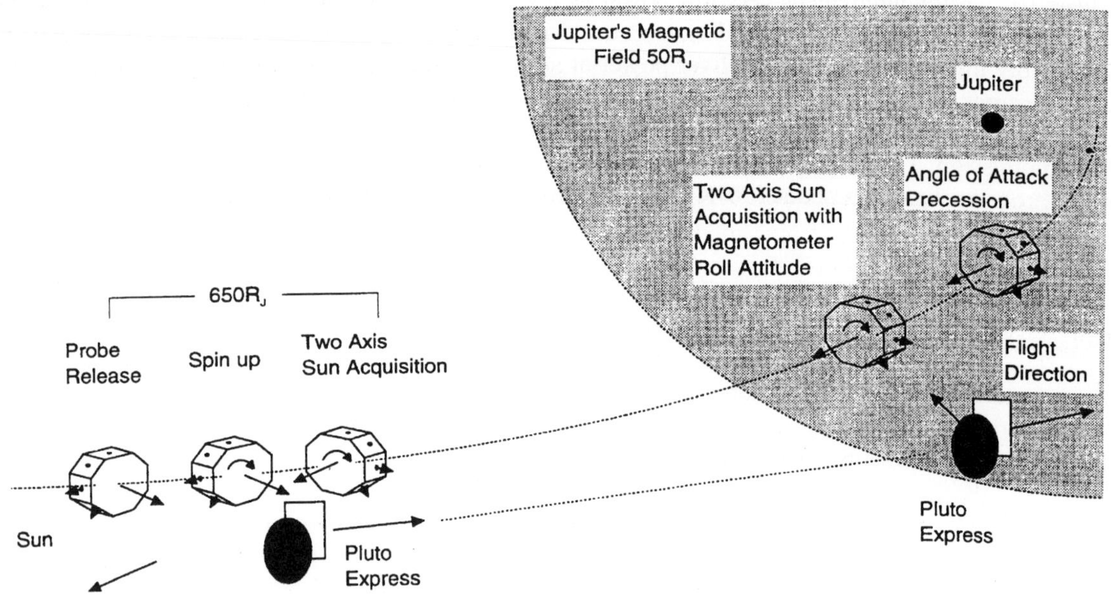

Figure 5. Operational Modes of the Attitude Control System.

Environmental Effects. Environmental effects influencing the attitude of the Io Probe are: gravity gradient, solar radiation pressure, magnetic torque, aerodynamic drag, and dust impact. The solar radiation torque can lead to a precession of the order of 2.5° out of the trajectory plane. The gravity gradient torque resulting from Jupiter to within 5 R_J is not significant for the mission. Depending on the magnitude of the residual magnetic field in the probe, the worst case resulting angular precession error caused by the magnetic torque could exceed 5° in the trajectory plane. The most significant environmental disturbing influences are aerodynamic drag and dust impact during the final 1000 km or 80 seconds to impact and particularly below 400 km. These will eventually result in sufficient deflection of the Z axis to exceed the field of view of the mass spectrometer and cause loss of telemetry contact with Pluto-Kuiper Express.

Active Spin Stabilization. Cold gas thrusters provide a simple and reliable torquing capability for the whole free flight period. The probe will contain omnidirectional sun sensing in two axes. Each sensor axis comprises a coarse and fine sensor. Each coarse sensor array comprises four small silicon cells with their optical axes at 90° spacing in the axial plane and at 45° to the Z axis. The cells are installed on the rim of the upper and lower plates to have an unobstructed field of view and avoid stray reflections from the probe body. A fine sun sensor mounted on or flush with the solar array panel provides sun attitude information in the X and Y axes within a range of ±60° to an accuracy of the order of 0.1° and is used for fine control when the sun is within its field of view.

Angle of Attack Pointing. Angle of attack pointing during the final 30 minutes to encounter requires three axis attitude information in the inertial frame. As the magnetic field of Jupiter is well known within ±2° and the sun direction and probe trajectory are roughly in its magnetic equatorial plane, the magnetometer provides a relatively accurate roll reference with no additional mass or power requirements.

The probe is spin stabilized at approximately one radian per second and requires a precession torque to control the spin axis about the sun vector or along the angle of attack of the trajectory. This torque can be generated by one or more jets with a moment arm about the longitudinal centre of gravity. The precession jets (thrust 7.5 mN) will be mounted diametrically opposed on the upper and lower plates such that the plane normal to their axes contains the probe centre of gravity. The spin up thruster pair are mounted on the extremities of the solar array plate which simplifies the mechanical construction of the probe.

Cold Gas System. Argon gas is used as the propellant because of the mass spectrometer experiment. The high pressure section is a fully welded system comprising a graphite wound tank, fill and drain valve, filter, pressure transducer and pyrotechnic valve. The specification for the proposed carbon reinforced tank with 250 bar pressure is $1 \cdot 10^{-6}$ scc/sec of He, which for the six year flight implies a loss of 190 scc which is negligible and with argon would be even less. The spin up, worst case initial acquisition of the sun and disturbance free angle of attack pointing require less than 20 Ns. At $2 \cdot 10^{7}$Pa (200 bar) pressure, this implies in the order of 0.1 litre gas.

Radio Frequency Subsystem

The S-Band frequency range was selected for the Probe-Express link due to

- the relatively high manufacturing density of S-band radio equipment which results in low to moderate mechanical dimensions and moderate mass
- the very high power efficiency of linear S-band transmitters / power amplifiers which results in low DC power consumption and power dissipation
- a good maturity of electrical parts, components and units at S-Band
- the toroidal shape of the probe transmit antenna pattern which can be implemented with a few (eight) radiators without great risk for signal reflections and multipathing caused by the probe structure.
- multiport (Butler Matrix) amplifiers which can be manufactured at S-Band without great risk. This type of amplifier allows the instanteanously changing excitation of the antenna ports without any need

for power switching. The RF signals can electronically be switched by e.g. PIN diodes in the small signal region infront of the SSPAs.

A data rate in the order of 1 kbit/s results for the Io-Probe link assuming a RF transmit power of 4 Watts beamed towards Pluto-Express via an electronically despun antenna.

Transmitter Module Description. Circuitries operated at frequencies up to approx. 1 GHz can be implemented with discrete, lumped elements on PCBs. Surface Mount Technology (SMD), together with application specific integrated circuits (ASIC) shall be applied wherever possible to miniaturize the equipment. Above 1 GHz lumped elements and microwave monolithically integrated circuits (MMIC) in coplanar thinfilm technology on soft- or ceramic substrates can be used. To save mass and DC power, the multiplexer and encoder, as well as the controller shall be implemented as part of the OBDH. A digital PSK modulator is followed by a TX-NCO which is used for linear phase modulation of the transmit 1. IF frequency. The local oscillator (L0) synthesizer module generates both reference carriers required for both frequency upconversions. All frequencies are derived from the USO output signal. The power amplifier consists of a minimum of two conventional driver amplifiers followed by an input switch network and an eight channel multiport amplifier (MPA).When the antenna is electrically despun, there is an equal distribution of the RF power between all eight power amplifiers of the MPA independently of the momentary selected antenna ports.

Ultra Stable Oscillator (USO). The USO, a subminiature crystal oscillator, is part of the Radio Science Investigation experiment. With a similar device installed in the receiver at PKE, the Doppler shift due to gravitational trajectory and plasma effects can be extracted. It has to be mentioned, however, that further analyses or tests are necessary to ensure the required frequency stability of 10^{-10} while operating in a high dose environment.

Transmit Antenna. The TX antenna consists of eight radiating elements mounted, one per plane, on the octagonal lateral areas of the Io-probe. Each rectangular sized radiating element consists of a group of four circular patches. The four patches are excited by an RF distribution network. The transmit patch antenna can be manufactured using either soft substrate materials or ceramic carriers.

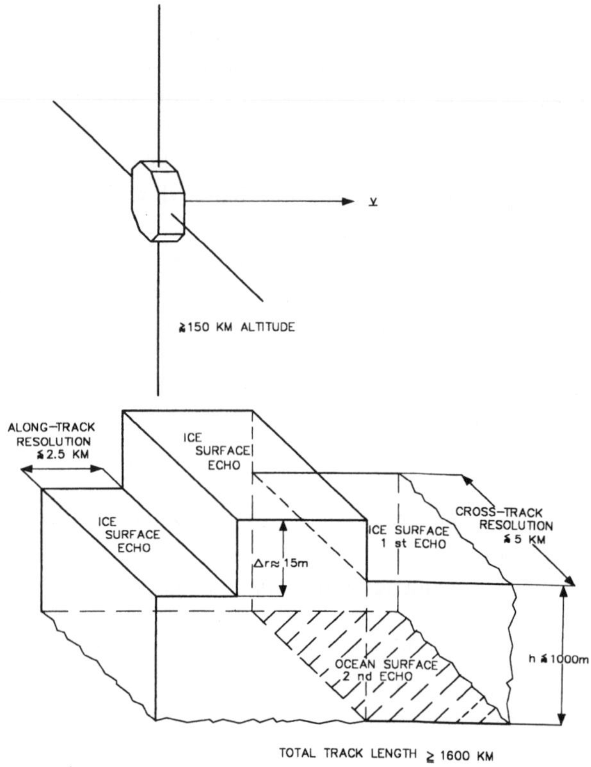

Figure 6. Spinning Europa Probe - Radar Sounder.

EUROPA MICROPROBE

The proposed Io-microprobe design can be also used for the investigation of other moons, such as for example Europa in a grazing fly-by trajectory with a minimum altitude of 150 m-200 m. Of special interest is here the measurement of the composition of the ionosphere/atmosphere (Kliore *et al.* 1997) and the experimental verification for a liquid water ocean underneath a young and smooth ice surface, possibly several hundred meters thick (Kerr 1997). Therefore, a model payload should consist of a neutral/ion mass spectrometer optimized for the detection of, H, H_2, 0H, 0_2, 0, related ions and isotopes. In the following, we will concentrate on a payload equipped with a stepped frequency radar altimeter to detect ice/water discontinuities.

The 10 MHz Ground Penetration Radar As shown in Figure 6 it is proposed to use a spinning microprobe equipped with two crossed $\lambda/2$ dipoles which transmit and receive sequentially a stepped frequency pulse pattern (PRF $\cong$ 1 kHz, 512 frequency steps), around a center frequency of 10 MHz. Attitude reference and phase angle are obtained by sun sensor and magnetometer measurements.

10 MHz are selected since at these low frequencies, the attenuation in dry ice is extremely small (loss factor $\mathrm{tg}\,\delta = 0.005$). Useful radar measurements can be expected for a time interval of ca. 4min. The surface region covered by this technique has a total length of ca. 1600 km and a cross track dimension of 5 km. This dimension and the along track resolution of 2.5 km are dependent on the size of the first Fresnel zone which constitutes the dominating back scatterarea. The range resolution in ice, depending weakly on the dielectric constant of the ice, is approx. 10 m. Under the assumption that the ice is very dry, we will be able to detect horizontal features of the ice surface by measuring the first radar echo and a horizontal layer of water by a strong reflection forming the second echo.

Current studies (PIRA, ESTEC/RST) show that such an instrument can be realized with a mass of $\leq$ 10g and a power consumption of $\leq$ 10 W (Braun 1997, Braun *et al.*, 1997), ideally suited for a battery-operated microprobe. Assuming a 12 it resolution for the I and Q channels, the data rate will be 24bit/s, resulting in ca. 5 Mbits totally. Depending on trajectory selection, it has to be decided whether the data shall be transmitted in real time to the mother spacecraft or in a memory readout mode after the encounter.

SUMMARY

It has been shown that it is possible to use microprobes as research tools in the atmosphere and close environment of moons and outer planets. The microprobes can be released from interplanetary spacecraft and can operate under extreme environmental conditions for a time period of ca. 20 min performing "snapshot measurements". Separation strategy and trajectory have to be selected carefully to allow data transmission to the mother spacecraft from distances not exceeding a few tenthousand kilometers in order to guarantee still a good quality of the telemetry link.

ACKNOWLEDGEMENTS

The author is grateful to F.M. Neubauer (Univ. Cologne) and W.I. Ip (MPAe Lindau) for the scientific guidance of the Io Probe Feasibility Study, he further thanks M. Päzold (Univ. Cologne), G. Billig, G. Bode, W. Eidel, K. Hornung, O. Zeiler, (UniBwM), K. Sauer (MPE-Berlin), H.U. Auster (MPE Berlin), S. Livi (MPAe Lindau), J. Turner (TWIG/DLR), R. Kohl (DASA-Dornier), F. Gliem (IDA-TU Braunschweig), and H.M. Braun (RST, Salem) for their contributions and helpful discussions. This study was partially supported by DARA GmbH, Bonn, Germany.

REFERENCES

Acuna, M.H., F.M. Neubauer, and N.F. Ness, Standing Afvén wave current system at Io: Voyager 1 observations, J. Geophys. Res., 86, 8513 (1981)

Braun, H.M., H. Lentz, A. Woode, A planetary into-ground radar and altimeter (PIRA), paper presented at Twelfth International Conference and Workshops on Applied Geology and Remote Sensing, Denver, Colorado, 17-19 November, 1997.

Braun, H.M., private information (1997).

Brown, M.E. and A.H. Bouchez, The response of Jupiter`s magnetosphere to an outburst of Io, Science, 278, 268 (1997)

Frank, L.A., W.R. Peterson, K.L. Ackerson, V.M. Vasylinnas, F.V. Coroniti, S.J. Bolton, Plasma observations at Io with the Galileo Spacecraft, Science, 274, 394, 1996

Hall, D.T., D.F. Strobel, P.D. Feldman, M.A. McGrath, H.A. Weaver, Detection of an oxygen atmosphere on Jupiter´s moon Europa, Nature, 373, 677 (1995)

Hall, D.T., P.D. Feldman, D.F. Strobel, M.A. McGrath, Far-ultraviolet emissions from Ganymede and Europa, Bull. Am. Astron. Soc., 28, 1071, (1996)

Häusler, B., Feasibiliy Study Io Probe, Technical Report, Universität der Bundeswehr, München (1996).

Häusler, B., W. Eidel, J. Bode, G. Billig, O. Zeiler, F.M. Neubauer, and W. Ip, The Pluto Express-Io Probe, Proposal for a small interplanetary probe for the research of Io's environment, in Proceedings Small satellites and services, International Symposium June 24-28, Annecy France, edited by CNES (1996).

Ip, W.H., Europa's oxygen exosphere and its magnetospheric interaction, Icarus, 120, 317 (1996)

JPL, Trajectory information given to the project at the Io Probe Kick Off Meeting, January 11, 1996 (1996)

Kerr, R.A., An icy world looks livelier, Science, 275, 478 (1997)

Kliore, A.J., D.P. Hinson, F.M. Flasar, A.F. Nagy, T.E. Cravens, The ionosphere of Europa from Galileo Radio Occultations, Science, 277, 355 (1997)

McGrath, M.A., Io and the plasma torus, Science, 278, 237 (1997)

Neubauer, F.M., The Pluto Express Program, in Feasibility Study Io Probe, Technical Report, Universität der Bundeswehr, München (1996)

Neubauer, F.M., Nonlinear standing Alfvén wave current system at Io: Theory, J. Geophys. Res., 85, A3, 1171 (1980).

Peale, S.J., P.Cassen, and R.T. Reynolds, Melting of Io by tidal dissipation, Science, 203, 892, (1979).

Strobel, D.F., X. Zhu, and M.E. Summers, On the vertical transport in Io´s atmosphere and ionosphere, Icarus, 111, 18 (1994).

CHARACTERISTICS OF THE KOMPSAT- I PAYLOADS AND ITS APPLICATION

Hong-Yul Paik, Hyeong-Sik Youn, Seung-Hoon Lee, Sun-Hee Woo, Hyung-Sik Shim, Kyoung-Hwan Oh, Young-Min Cho, Sang-Soon Yong, Sang-Gyu Lee, Haeng-Pal Heo, Byung-Kyo Kim

Satellite Application Dept., Korea Aerospace Research Institute, P.O. Box 113 Yuseong Taejon, 305-600, KOREA

ABSTRACT

Korea Aerospace Research Institute (KARI) is developing a Korea Multi-Purpose Satellite I (KOMPSAT-I) which accommodates Electro-Optical Camera (EOC), Ocean Scanning Multi-spectral Imager (OSMI), and Space Physics Sensor (SPS). The satellite has the weight of about 500kg and will be operated on the 10:50 AM sun-synchronized orbit with the altitude of 685 km. The satellite will be launched in 1999 and its lifetime is expected to be over 3 years.

The main mission of EOC is the cartography to provide the images from a remote earth view for the production of 1/25000-scale maps of KOREA. EOC collects 510~730 nm panchromatic imagery with the ground sample distance(GSD) of 6.6 m and the swath width of 17 km by push broom scanning. EOC also can scan ± 45 degree across the ground track using body pointing method.

The primary mission of OSMI is worldwide ocean color monitoring for the study of biological oceanography. It will generate 6 band ocean color images with 800 km swath width and 1km GSD by whiskbroom scanning. OSMI is designed to provide on-orbit spectral band selectability in the spectral range from 400 nm to 900 nm through ground command. This flexibility in band selection can be used for various applications and will provide research opportunities to support the next generation sensor design.

SPS consists of High Energy Particle Detector (HEPD) and Ionosphere Measurement Sensor (IMS). HEPD has missions to characterize the low altitude high-energy particle environment and to study the effects of radiation environment on microelectronics. IMS measures densities and temperature of electrons in the ionosphere and monitors the ionospheric irregularities at the KOMPSAT orbit.

INTRODUCTION

There are currently three kinds of satellite development in Korea. One is the Koreasat series for satellite communications and direct broadcast services (Hwangbo 1996). The second is the KITSAT series for micro-satellite scientific experiment (Shin 1996). The last is Korea Multi-Purpose Satellite (KOMPSAT) series for practical application on the Earth observation.

Korea Aerospace Research Institute (KARI) is developing KOMPSAT-I per Korea National Space

Program under Korea Government support. KOMPSAT-I is a small satellite with the mass of 500 kg to be operated at the sunsynchronous orbit which has 98 minutes orbit period, 28 days repeating ground track. 10:50 AM ascending orbit crossing time, and 98.13 degrees inclination at the altitude of 6tS5km (Lee 1996, Baek 1996). KOMPSAT-I is scheduled to be launched at the third quarter in 1999 and earth observation data acquisition from KOMPSAT-I can be available from the end of 1999 during its lifetime of at least 3 years.

KOMPSAT-I accommodates three instruments i.e. Electro-Optical Camera (EOC), Ocean Scanning Multispectral Imager (OSMI), and Space Physics Sensor (SPS) for the mission of cartography, worldwide ocean observation, and space environment monitoring respectively.

EOC has cartography mission to provide images for the production of scale maps, including digital elevation models, of the Korea peninsula from a remote earth view in the KOMPSAT orbit. EOC collects panchromatic imagery with the ground sample distance(GSD) of 6.6m and the swath width of 15km at nadir through the visible spectral band of 510 ~ 730 nm. EOC scans the ground track of 800km per orbit by push-broom and body pointing method.

OSMI mission is worldwide ocean color monitoring for the study of biological oceanography. OSMI image data can be used for ocean ecological observation, ocean resource management, and ocean-atmosphere environment analyses. OSMI is a multi-spectral imager generating 6 color ocean images with 800km swath width and less than 1km GSD by whisk-broom scanning method. OSMI is designed to provide on-orbit spectral band selectability in the spectral range from 400nm to 900nn for flexible ocean observation. The color images are collected through 6 primary spectral bands centered at 443, 490, 510, 555, 670, 865nm or 6 spectral bands selected in the spectral range via ground commands after launch

SPS consists of High Energy Particle Detector (HEPD) and Ionosphere Measurement Sensor (IMS). HEPD has missions to characterize the low altitude high energy charged particle environment and to study the effects of radiation environment on microelectronics such as single event upset, total dose effect and RAM test. IMS measures densities and temperature of electrons in the ionosphere and monitors the ionospheric irregularities in KOMPSAT orbit.

The shape and the configuration of the instruments (EOC, OSMI, SPS) are shown in Figure 1. The operational concept of KOMPSAT-I is illustrated in Figure 2 for the mission of cartography, ocean color monitoring, and space environment monitoring.

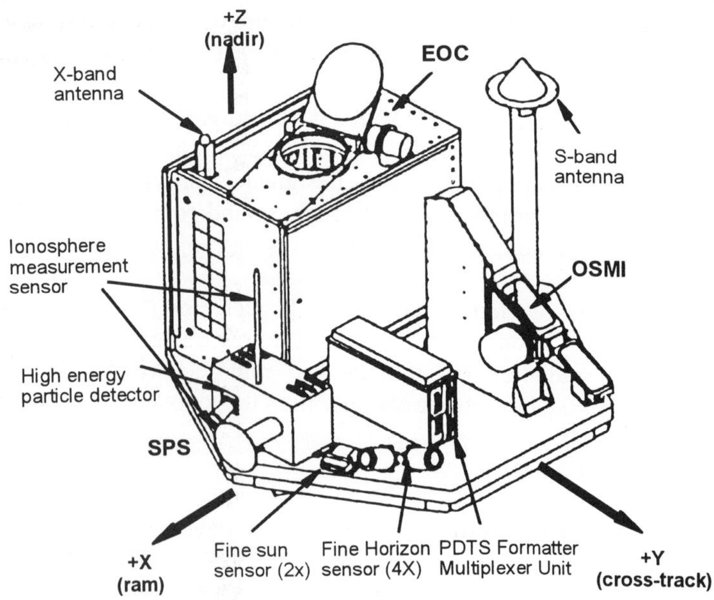

Figure 1. KOMPSAT-I payload shape and configuration.

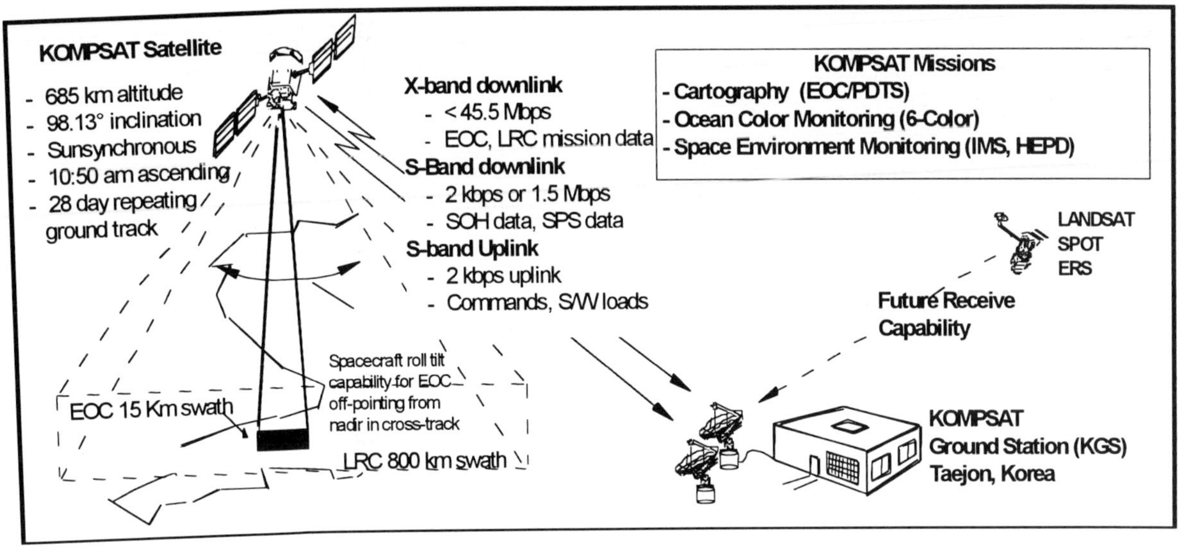

Figure 2. The operational concept of KOMPSAT-I

ELECTRO-OPTICAL CAMERA (EOC)

EOC Mission

The mission of EOC is cartography to build up Korean digital map including Digital Terrain Elevation Map (DTEM).

EOC Design

The orbit of EOC is sun synchronous and located at 685km altitude. EOC collects ground image through 510 ~ 730nm panchromatic spectral band by push-broom scanning and spacecraft body pointing: At Ground station, stereo image can be generated with images which are obtained at a few body tilt angles.

The EOC system specifications are as follows.
Duty cycle for image collection: 2 minutes/orbit(800 ion image/orbit)
Ground Sample Distance : 6.6 meters at 685 km in NADIR view
Swath width :≧15 km at 685 km in NADIR view
Mission life time: ≧ 3 years
Reliability: ≧ 0.94
MTF: ≧10% at Nyquist frequency
SNR : > 50 over entire FOV
FPA : 2592 pixels
Digitization : 8bits

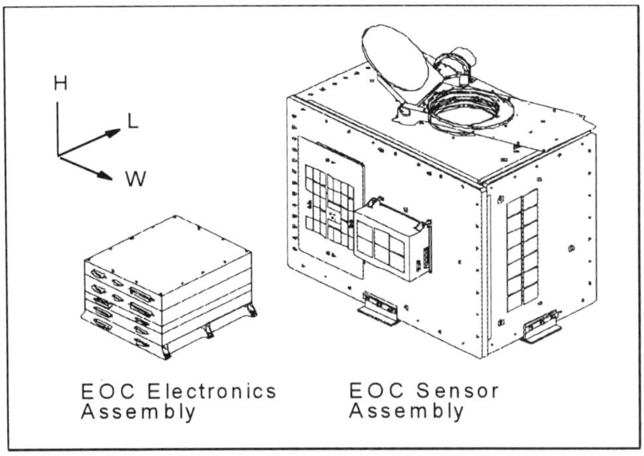

Figure 3. EOC configuration.

Total weight : 35 kg
Maximum power consumption: 52 Watt
Image data transmission rate : ≦ 25 Mbps

The on-board analog signal processing module of EOC has the function of programmable gain and offset to take the various conditions of ground reflected radiation into account.

The EOC (Figure 3) physically consists of two assemblies, the EOC Sensor Assembly (ESA) and the EOC Electronics Assembly (EEA), and each assembly is mounted on the payload platform of spacecraft. The ESA is thermally isolated from the payload platform and EEA is thermally coupled to that.

EOC Operation

For EOC mission, S-band communication link is used to transmit command from the ground station and to receive the State Of Health (SOH) data of EOC. The On-Board Computer (OBC) controls S-band data with the MIL-STD-1553B interface. EOC transmits the image data to ground station via X-band channel.

The EOC collects image for 2 minutes per 98 minutes orbit cycle, which covers about 800 km along ground track. The EOC image can be transmitted to Korean Ground Station in real time during Korean peninsular observation or be stored in the Solid State Recorder (SSR) of PDTS out of the Korean Ground Station reception area. The stored image data can be transmitted when the data reception connection is available later. It is possible to observe earth globally using the image storage scheme.

Ground station performs the radiometric correction and geometric correction on the received image data, and performs several image processing steps with ancillary data. For calibration, observed reference area such as desert or Ground Control Point (GCP) is used. Spacecraft provides ancillary data such as image time, position and attitude of spacecraft for EOC image processing.

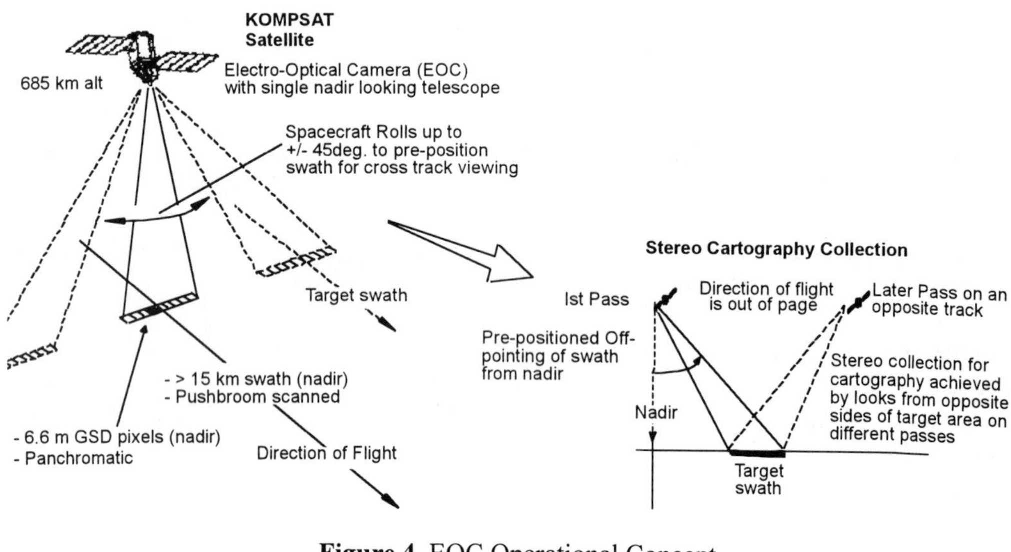

Figure 4. EOC Operational Concept

The EOC has four (4) operation modes as follows:

Off mode: Power-off
Safe-Hold mode: power-on for survival only
Standby mode: EEA power-on, ESA power-off, SOH available
Imaging mode: All power-on, Image collection

Figure 4 illustrates the EOC operation concept of image collection and stereo image collection. The EOC collects stereo images of target area from opposite sides on different passes by roll-tilting of spacecraft, then ground station can make DTEM with stereo image. It is possible to collect image data 39 times in the daytime per 28 days revisit cycle by roll-tilting of up to ± 45degree. For cartography, up to 30degree roll-tilting is used. Korean Ground Station can obtain EOC images 20 times out of 39 times in this case.

OCEAN SCANNING MUTISPECTRAL IMAGER (OSMI)

OSMI Mission

OSMI mission objective is to provide ocean color measurements for biological oceanography. The OSMI image data is collected for various researches and applications in the fields of worldwide ocean resources management and ocean environment monitoring.

OSMI Instrument Description

OSMI is designed to provide 6 spectral image data with GSD less than 1km over 800km swath width by whisk-broom scanning method at the altitude of 685km.

The KOMPSAT OSMI instrument uses 6 primary spectral bands in the 400 nm to 900 nm range to perform ocean color monitoring. The OSMI spectral band centers and bandwidths are as shown in Table 1.

Table 1. OSMI ocean color monitoring spectral channels

Ocean Color Spectral Channel	B1	B2	B3	B4	B5	B6
Band Center(nm)	443	490	510	555	670	865
Bandwidth(nm)	20	20	20	20	20	40

The OSMI ocean color spectral channels B1 through B4 provide ocean color data while channel B5 and B6 provide information for atmospheric (aerosol) corrections.

The six OSMI ocean color data channel performance requirements in signal-to-noise ratio (SNR), modulation transfer function (MTF), and ground sample distance (GSD) are as shown in Table 2. In addition to the requirements of SNR, MTF and GSD, the OSMI signal channels has a polarization requirement of less than +/-5%.

Table 2. OSMI SNR, MTF and GSD requirements

Ocean Color Spectral Channel	B1	B2	B3	B4	B5	B6
SNR Requirement	450	450	450	350	350	350
MTF Requirement(%)	20	20	20	18	18	15
GSD Requirement(km)	1.0	1.0	1.0	1.0	1.0	1.0

The OSMI instrument optics consists of the image scanner, the objective lens group and the multi-channel spectrometer. The scanner consists of a scan mirror driven by a servomotor. The scanner has a +/-30 degree scan angle with respect to nadir resulting in an 800km swath from 685km altitude. The objective lens group forms the primary image which is dispersed into its various colors by the spectrometer. The dispersed image is then focused onto and detected by the focal plane assembly (FPA).

The OSMI electronics consists of(l) the FPA electronics section providing timing and voltage bias for the CCD FPA, (2) the analog signal processing and analog-to-digital (A/D) conversion section providing analog gain and offset signal adjustment and a IO-bit AID signal conversion, (3) digital data processing and compression section providing 6 channels of data with the desired band centers and bandwidths, and uses data compression to reduce data volume, (4) a control processor section to provide control, communication and house keeping, (5) a motor drive section providing servo control for the scan motor and read out of the optical encoder for the position of the scan mirror, and (6) the power converter section to regulate DC power.

For on-orbit calibration, the OSMI design employs a two level, black and white, calibration scheme. The sensor black calibration level is provided by a dark cell while the white calibration level uses attenuated solar radiance. The solar calibration is performed over the northmost part of the satellite orbit and the dark calibration is done at the beginning and end of each continuous image scan.

Physically, the OSMI instrument consists of two components (Figure 5), the sensor unit and the electronics unit. The OSMI instrument nominal weight is about 18Kg and the peak power consumption is about 30 watts.

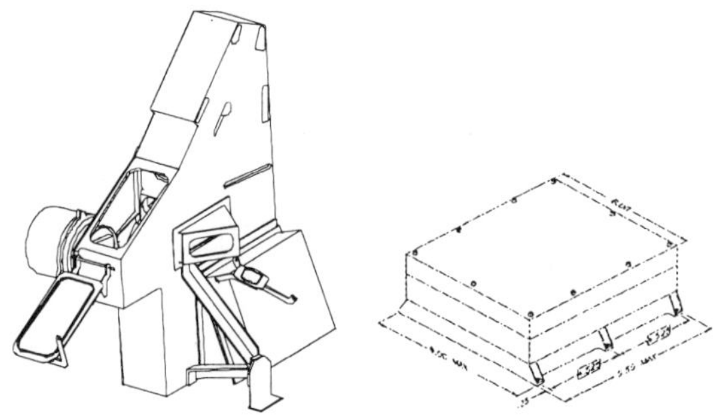

(a)OSMI Sensor Assembly (b) OSMI Electronics

Figure 5. OSMI Configuration.

OSMI Spectral Band Selection Capabilities

The OSMI is a multi-spectral instrument covering the visible spectrum from 400nn to 900nn. In the OSMI design there is also built-in flexibility to provide band center and bandwidth selectability via ground station command. Any 6 spectral band can be selected in the spectral range from 400nm to 900nm. There are always 6 bands without overlap among the bands. The band centers of the 6 data channels can be varied with the accuracy of 2.6nm and the bandwidth for each of the 6 data channels can be changed from the minimum value of 5.2nm to the maximum value of 166.4nm by the step of 2.6nm.

The OSMI's on-orbit spectral band center and bandwidth selectability provide great flexibility in ocean color monitoring. For example, the 412nm band, which is of interest in the Yellow Sea area near the Korean peninsula, can be monitored without adding instrument complexity. In addition, this flexibility in band selection provides research opportunities to support the next generation of sensor design.

OSMI Operation Description

The KOMPSAT OSMI is designed to provide worldwide ocean color data from a 685km sun synchronous orbit. The orbit crossing time is 10:50 AM and the inclination is 98.13 degrees. The OSMI instrument performs whisky-broom scan imaging operation with a cross-track ground swath of 800 km. The ground track re-visit time is 28 days.

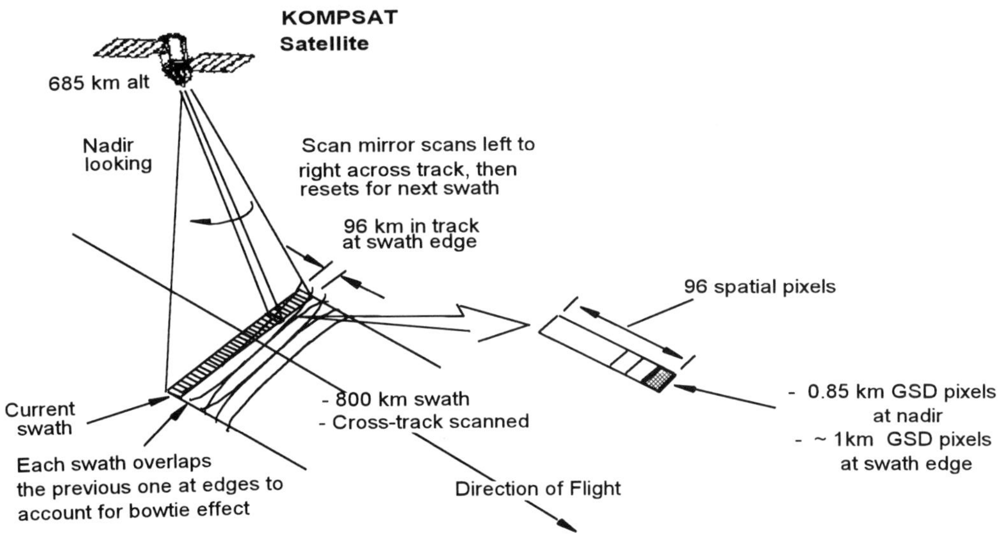

Figure 6. OSMI Operation Concept

The OSMI instrument is designed to perform imaging operation for 20% per orbit and a planned on-board solid state recorder will support worldwide imaging operation by providing data archive and downlink to the ground station at Taejon, Korea.

SPACE PHYSICS SENSOR (SPS)

SPS Mission

Space Physics Sensor (SPS) is composed of two observation sensor, High Energy Particle Detector (HEPD) and Ionosphere Measurement Sensor (IMS). HEPD characterizes the low altitude high energy particle environment. We can study the effects of radiation environment on microelectronics through HEPD. IMS measures densities and temperatures of electrons in the ionosphere and monitor the ionospheric electron irregularities in KOMPSAT orbit. We can study the effect of spacecraft charging through IMS.

SPS Design

HEPD consists of Proton and Electron Spectrometer (PES), Linear Energy Transfer Spectrometer (LET), Total Dose Monitor (TDM) and Single Event Monitor (SEM) and these are separated physically as depicted in Figure 7. PES is composed of 7 channels (three channel of Proton, three channel of Electron, a channel of alpha particle) and physically consists of Telescope and Signal Processing Unit. LET Spectrometer consists of a Solid State Detector and measures the amount of energy transferred to Silicon. TDM consists of four pairs of RADiation sensitive Field Effective Transistor (RADFET) and SEM monitors single event upsets using random access memory (RAM).

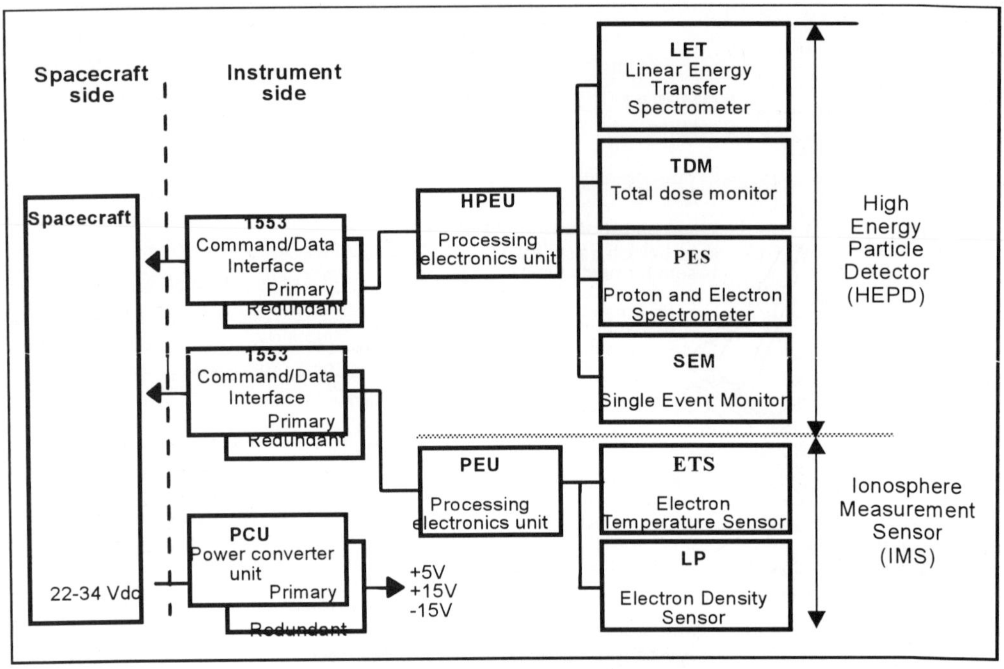

Figure 7. SPS block diagram.

IMS consists of Electron Temperature Sensor (ETS) that measures the temperature of thermal electrons in the ionosphere and Electron Density Sensor (EDS) that measures the density and temperature of electrons (Figure 7).IMS EDS is a modification of Langmuir Probe (LP). Dynamic measurement range of IMS is $10 \sim 10^6$ Electrons/cm^3 in electron density and 0~leV in electron temperature.

HEPD and IMS are installed in a SPS (HEPD+IMS) box and total weight is 7.6Kg and maximum power dissipation is 16W. Data transfer rate of HEPD is 1.5kbps and that of IMS is 5.6kbps. SPS needs time data, position and attitude of spacecraft and the intensity of earth magnetic field and

SPS can obtain time data, position and attitude of spacecraft from the spacecraft. ETS and LP face ram direction directly.

SPS Operation

SPS calibration data and SPS SOH (State Of Health) are transmitted to Ground Station through S-band after being saved to memory of OBC (On-Board Computer). The operation of SPS is controlled by ground command transferred by OBC that communicates with Ground Station through S-band. SPS has 100% duty cycle. SPS operational concept for the execution of mission is shown in Figure 8.

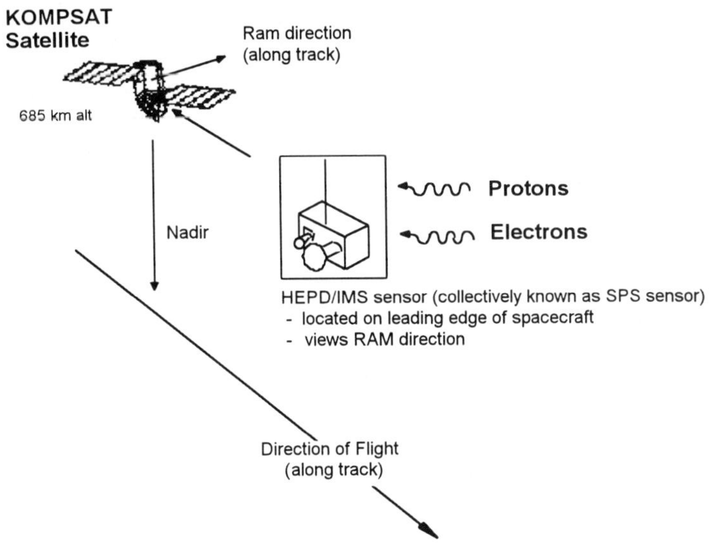

Figure 8. SPS mission operation concept

PAYLOAD DATA TRANSMISSION SUBSYSTEM (PDTS)

The purpose of the PDTS is to provide KOMPSAT-I with an X-band downlink capability to transmit image and OBC data to the KOMPSAT Ground Station (KGS). The PDTS consists of an internally redundant Formatter/Multiplexer Unit (FMU), separate primary & redundant X-band transmitter (modulator and RF amplifier functions) units, a RF assembly, and an X-band Antenna(Figure 9).

The FMU is commanded to accept image data from one or both of two imaging instrument (EOC and OSMI) and telemetry data from the On- Board Computer (OBC) which contains state-of-health, satellite reference information (time, ephemeris, and orientation). The FMU is able to receive the image data at the maximum input data rates of Table 3 (defined for the EOC and OSMI as the rate of data obtained via the interfaces).

Table 3. PDTS input data rate

Input	Maximum Rate
EOC data	25 Mbps
OSMI data	1 Mbps
OBC data	1.5 Mbps

To provide compatibility with the Korean Ground Station (KGS), the image data are formatted and encoded for transmission in CCSDS Grade 2 prior to being sent to the X-Band transmitter. The formatted data from FMU are passed to X-band Transmitter for real-time information transmission. The FMU contains Solid State Recorder (SSR) to store processed EOC/OSMI image data and OBC data for later playback downlink to KGS when KOMPSAT-I is within the data reception coverage of the KGS. The storage capacity of SSR is 2.5 Gbits at the end of life. The FMU shall format a serial CADU bit stream into two-bit wide words suitable for offset quarter phase-shift- keying (OQPSK) modulation, assigning even bits (starting with the first bit of the CADU) to one bit of the word (Q-channel bit) and odd bits to the other bit of the word (I-channel bit) in accordance with CCSDS recommendation 401-B-1.

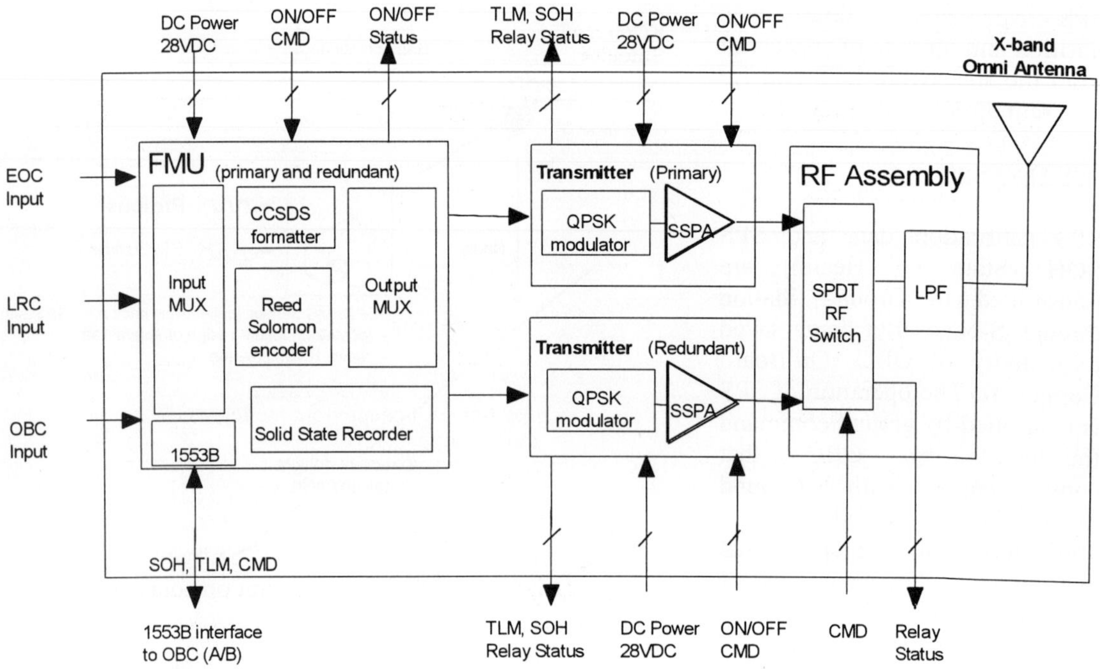

Figure 9. The functional block diagram of PDTS

The X-Band transmitter has the data transmission rate of 45 Mbps. The transmitter center frequency is 8.3 GHz and the transmitter output power is 3 watts. Transmitter output is visible to the ground station up to 1500 km from the satellite nadir point in the ground. The X-band link margin is more than 3dB with Bit Error Rate (BER) of 10^{-6}. The Mass of PDTS is 14.5 Kg and the peak power consumption is 63 watts.

CONCLUSION

Korea Aerospace Research Institute (KARI) is developing three instruments of Electro-Optical Camera (EOC), Ocean Scanning Multispectral Imager (OSMI) and Space Physics Sensor (SPS) to be accommodated onto Korea Multi-Purpose Satellite I (KOMPSAT-I) for the mission of cartography, ocean color monitoring, and space environment monitoring respectively. The current status design budget and performance of KOMPSAT-I payloads are summarized in Table 4.

The flight model (FM) of EOC and OSMI will be manufactured by the end of 1997 and SPS FM will be by April of 1998. The instrument is scheduled to be assembled with spacecraft bus and tested as a part of the satellite system at KARI in the third quarter of 1998. After the final check on the interfaces between the satellite, KGS, and the launch vehicle of Taurus, the satellite is to be launched in August of 1999 and it is expected to operate more than 3 years.

ACKNOWDGEMENTS

The development of KOMPSAT-I payload is financially supported by Korea Ministry of Science and Technology. We thank TRW and Dr. K. W. Min of Korea Advanced Institute of Science and Technology for their contribution to the development of the KOMPSAT-I payload.

Table 4. KOMPSAT-I Payload design budget and performance summary

Payload	Performance factor	Specification
EOC	Ground Sample Distance(GSD)	6.6 ± 10 % m @685km altitude, Nadir
	Life Time	3 years at least
	Spectral Band	PAN: 0.51 ~ 0.73 μm
	Swath Width	> 15 km
	Image Data Rate	≤ 25 Mbps
	Digitization	8 bits
LRC	GSD	≤ 1 km
	Swath Width	800 km
	Spectral Band	selection of 6 bands from the region of 400 ~ 900 nm
	Digitization	10 bits
PDTS	BER	$< 10^{-6}$
	SSR	2.5 Gbits(End of Life)
	Modulation	QPSK(X-band)
	Error Correction	R/S(255,223) (FEC)
	Packettizing	CCSDS II
	Data Transmission Rate	45 Mbps(Constant Rate)
IMS / HEPD	Duty Cycle	100 %

REFERENCES

Baek, M.-J. and Y.-K. Chang, KOMPSAT Spacecraft Bus System Overview, Proceedings of the third Asia-Pacific conference on multilateral cooperation in space technology and applications May 27-31 Seoul Korea, pp. 275-281 (1996).

Hwangbo, H., S.-J. Kim, S.-H. Choi, KOREASAT: The First Korean Communication and Broadcasting Satellite, Proceedings of the third Asia-Pacific conference on multilateral cooperation in space technology and applications May 27-31 Seoul Korea, pp, 175-183 (1996).

Lee, S.-R. and H.-J. Kim, The Mission Orbit Selection for the Korea Multi-Purpose Satellite (KOMPSAT), Proceedings of the third Asia-Pacific conference on multilateral cooperation in space technology and applications May 27-31 Seoul Korea, pp. 199-203 (1996).

Shin, Y.-H., D.-H. Lee, S.-M. Park, J.-J. Lee, and K.-W. Min, Space Environment Scientific Experiment (SENSE) on KITSAT-3, Proceedings of the third Asia-Pacific conference on multilateral cooperation in space technology and applications May 27-31 Seoul Korea, pp. 550-560 (1996).

TELEMEDICINE USING SMALL SATELLITE MEO/HICAT FOR INTERNATIONAL MEDICAL COMMUNICATIONS

I. Nakajima [1], Y. Sawada [1], T. Maeda [2], D. L. Martin [2], S. Nagano [2], and N. Hamano [2]

[1] *Tokai University School of Medicine, Emergency Medical Services, Boseidai, Isehara-city, Kanagawa, 259-11, Japan. Nakajima's e-mail : js2hb@is.icc.u-tokai.ac.jp*

[2] *Hitachi, ltd. Space System Division, 6 Kanda-Surugadai 4-chome, Chiyoda-ku, Tokyo, 101, Japan. Maeda's e-mail : t_maeda@cm.head. hitachi.co.jp*

ABSTRACT

Here is described the effectiveness of communication between medical facilities using satellite, in particular of international medical communications, targeted to communications between medical practitioners. Communications between these facilities are characterized by low call origination count, the range of the radio station is relatively large, and the communication time is relatively long. We belive an example of MEO is shown in consideration of this feature.

We designed HiCAT (High-performance International Communications for Advanced Telemedicine) satellite; sun-synchronous quasi-recurrent Orbit (n=7), specifically, six satellites to be launched, three orbit planes, and two satellites for each plane, and the users targeted are those living in areas from the equator as the center to the latitude of 70 degree. The average angle of elevation of HiCAT is approximately twice that of Iridium. These advantages; higher visibility and higher elevation angles can be explained to make MEO suitable for communications between these facilities. Teleconsultations with clinical data as a store-and-forward packet communications linking with Internet, clinical image diagnosis with JPEG, and broadcasting medical news distribution services using MEO are believed to be one of the optimal applications of small satellite.

OBJECT

This document describes a Medium Earth Orbit (MEG) applications, especially suitable for satellites supporting telemedicine applications, in particular for international medical communications.

TELEMEDICINE

Definition

Telemedicine is defined as medical assistance using communication methods (medical information distribution, consultation, patient information distribution, conference, home care assistance, and so on).

Classification

Telemedicine is classified into three purposes:

- Doctor to doctor: Medical consultation;
- Nurse to doctor: Medical consultation; and
- Patient to doctor: DPC (Direct Patient Care)

by communication partners. While the world's legal systems sanction the first two purposes, DPC is not generally permitted though movements are underway to sanction it world wide.

Here is described the effectiveness of communication between medical facilities using satellite, in particular of international medical communications, targeted to communications between medical practitioners.

Image Quality and Speed

Hospitals in developed countries are general hospitals including relatively independent departments for internal medicine, surgery, pediatrics, and so on. This structure provides advantage in connection to other medical institutes and implementing medical construction. To smoothly perform telemedicine, a corresponding structure should provide specialized diagnosis or advice. Therefore, telemedicine in developed countries inevitably requires high-quality images and high-speed communications to meet the demands from the users i.e., doctores.

On the other hand, hospitals in developing countries function at widely disparate standards and do not provide internal structure to support specialized residence. To integrate distributed medical professionals, there is a great expectation for telemedicine in these countries. Cooperation between hospitals could be facilitated to advantage in developing countries. I believe, for this reason, there is high feasibility of telemedicine even in the case of low image quality and low-speed communications.

Communications Between Medical Facilities

Communications between medical facilities have the following features:

- The call origination count is significantly lower than for switched lines.
- A relatively large capacity of power could be supplied, so the radio station could communicate over large distances, much further than current mobile stations.
- The communication time is relatively long because there is a large amount of information including static images, literature, and so on during connection.
- In addition, in order for international organizations such as WHO to operate networks, the following is mandatory:
- Broadcasting all over the world
- Communications for islands or out-of-the-way places where there is no switched line.

Universal Services

It was already pointed out that telemedicine is useful for remote locations. There has been a disadvantage in that medical line charges for telemedicine are not discounted, but are the same as for entertainment. In May, 1997 the U.S. Conference decided on universal communication rules that were submitted by the FCC. In these rules, communications companies and associated organizations are obliged to provide schools, medical facilities, and physically handicapped peoples in remote areas with various public communication services. The rules are expected to accelerate the spread of telemedicine.

Non-INTELSAT Satellite Communications

There is no natural boundary in the need for medical services. In a similar sense, there is no boundary to radio waves from space. Radio waves that can cross over boundaries are a useful method for saving patients as far as they are properly used. Under such a background, there has been attempted telemedicine by means of non-commercial, non-inlets satellite communications.

PEACESAT

One of the attempts of several non-INTELSAT related international satellite communications is PEACESAT(Pan-Pacific Education and Communication Experiments by Satellite). This is a style in which volunteers operate non-commercially. At the initial stage, in 1971 Prof J. Bistro, Hawaii University performed small-scaled communication experiments in Hawaii using ATS-1 satellite (VHF/130 MHz band). Then, through Japanese Hawaiian Prof Nose's donation activities, 144 MHz amateur radio equipment was modified for ATS-I global station, and over 100 terminals were installed in the Pacific area. In 1985, ATS-1 operation was canceled because of the expiration of the service life ofATS-1 satellite, however, around 1990 PEACESAT operation was restarted by new members using L band transponder of meteorological satellite GOES-3. The communication lines are intended for voice communication only, and image communication facilities are not installed. This network is extremely important in social community investigation, has high experience in educational activities, and is tested for medical applications with slow scan TV.

SPACEBRIDGE TO MOSCOW

NASA and Russia realized that telemedicine was effective for medical services in Armenia Earthquake in 1988. For one year from 1993, teleconference in medical fields was carried out using real-time two-way analog video and audio via geostationary satellite GTE Spacenet Gstar 2 (the U.S.), and WADRN Satellite (Russia). Spacebridge to Moscow was a multidisciplinary demonstration linking two diverse medical cultures during 22 clinical sessions. Case histories and patient information were provided prior to session so that clinical experts can became familiarized with the patient's case.

VITASAT

VITA (Volunteers in Technical Assistance) is a volunteer organization that provides technical training and education training overseas. At the initial stage VITA took active parts in the spread of training and education in communication technology, but is focusing attention to medical services. VITA uses UoSAT of the same type as amateur satellite in July 1991 and supports packet communications for activities against AIDS in Africa (Satellite is called Healthsat, SateLife mission:50% and VITASAT mission: 50% shared at the transponder level). VITA designs and develops satellites by their own based on technical experience of UoSAT. Unfortunately, their low-cost satellite VITASAT (43 kg producted by CTA/ex-DSI) failed and was lost in August 1995 because of rocket (LLV)failures on launch. The successive satellite launching is expected inside and outside the U.S. This plan took the Pioneer's Preference prize from the U.S. Goverment FCC in January 1993. This prize was the first given for Low Earth Orbiter (LEO) and it was significant in that this prize was given for "medical" service other than business and education.

AMINE/PARTNERS

In the PARTNERS project using Japanese Engineering Test Satellite No.5 (ETS-V) which the Ministry of

Posts and Telecommunications (MPT) and National Space Development Agency (NASDA) carried out, a group (Telemedicine Society of Japan) in Tokai University School of Medicine mainly expanded medical network in Asia and Pacific areas between October 1992 and March 1995. This network was constructed at Earth Station that was installed in six medical organizations in Japan and 20 medical organizations outside Japan.

Referred to as Network 2, this network is also called AMINE (Asia-Pacific Medical Information Network using with ETS-5). Five stations in Papua New Guinea, five stations in Thailand, five stations in Cambodia, four stations in Fuji, and one station in China were installed, opened, and operated for almost 24 hours a day, 7 days a week (see the Figure 1). In AMINE/PARTNERS, since the transmission band-width uses 16 kHz band-width FM modulation, it is limited in operation to color static images (18 pages; NTSC full color), voice, and slow packet radio (MX.25 HDLC) protocols. One advantage is that international communications and intra-national communications between developing countries is supported without the need for a dedicated and expensive hub station. The communication lines were used to carry out borderless teleconference, teleconsultation, personal computer communications, and so on.

The AMINE/PARTNERS network was linked with 26 VSATs via the south beam of ETS-S (L Band). Many clinical cases were consulted to Japanese specialist by doctors in developing country during 1992 to 1995.

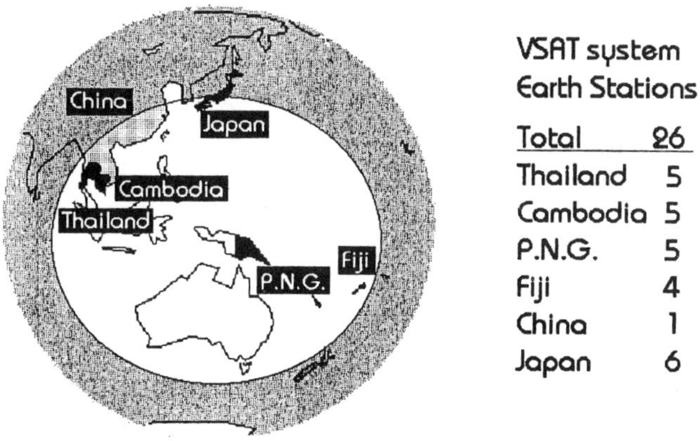

Figure 1. AMINE/PARTNERS project .

Satellie Number	1	2	3	4	5	6
Height(km)	5,004					
a (km)	11,382.14					
i (deg)	138.67					
e	0.000					
ω (deg)	90.0					
Ω (deg)	0.0		120.0		240.0	
True Anomaly(deg)	0.0	180.0	0.0	180.0	0.0	180.0

1. Orbit type
   Sun Synchronous quasi-recurring
   Easy TT&C operation
   Simple onboard EPS
2. Parameters
   Recurring Days 13 days
   Recurring Revolution 93 rev.
   Orbital motion 7 rev./day
3. Epoch Feb. 1, 1998

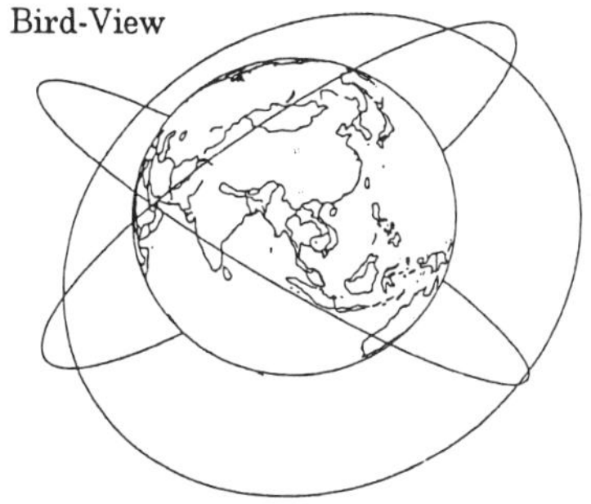

Figure 2. Satellite constellation

EMPLOYING MEDIUM EARTH ORBIT (MEO)

Communications between medical facilities are characterized by low call origination count, the range of the radio station is relatively large, and the communication time is relatively long. An example of MEO is shown in consideration of this feature.

Satellite Orbit

Six orbit elements and parameters over Sun-Synchronous Quasi-Recurrent Orbit (n=7) were shown in the Figure 2 and in the Figure 3. Specifically, six satellites to be launched, three orbit planes, and two satellites for each plane are planned, and the users targeted are those living in areas from the equator as the center to the latitude of 70 degree.

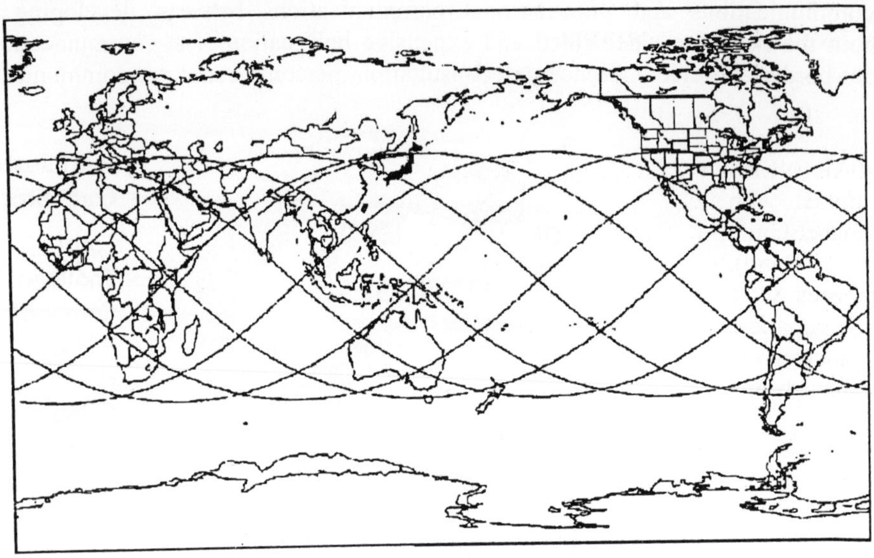

Figure 3. Orbit on the world map

Bus Specifications

The items of buses are shown in the Figure 4.

Bus Specifications

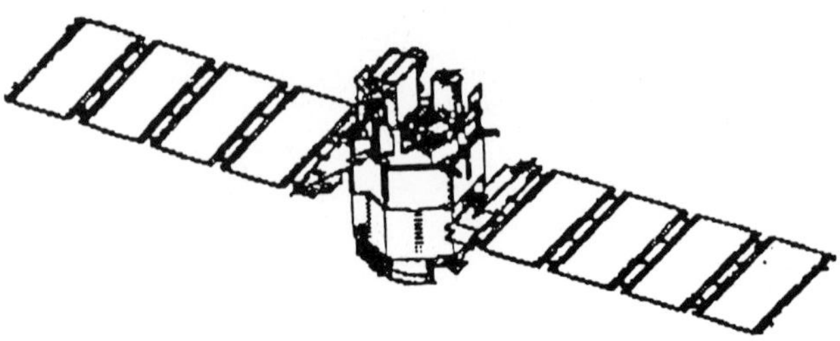

Size:	59 in.hex by 80 in.
Weight:	385,6kg (850lbs)
Electrical Power :	370 W
Attitude Control:	3-axis stabilized; zero momentum bias
Propulsion:	Eight 1-lbf hydrazine thrusters
Telemetry:	S Band (GSIDN cimpatible)
Design Life:	5 years (goal)
Reliability:	0.83 at 5 years

Figure 4. Bus Specification

Mission

Communications concept are listed in the Table 1, in the Figure 5 and in the Figure 6. Since each satellite has a different callsign/ID (software division), communications interference is prevented by spatial separation due to antenna (space division), packet timing separation(time division), and relay frequency separation(frequency divison). The link design in the Table 2. Assuming that the average visibility time per path is 40 minutes, 16 MB file data (approximately 30 JPEG images conforming with NTSC) can be transmitted at a 128 kbps line rate, in approximately 4 minutes 10 seconds. A total of 12 (twelve) channel 128 Kbps lines is available, therefore, up to 108 ground stations are available for use in a single path. In broadcasting mode, up to 640 MB can be distributed to all stations.

Table 1. Telemedicine System Specifications

Telemedicine System Specifications

Frequency: L Band, S Band
Modulation: BPSK/QPSL, SS

Digital Transponder (40 Wattes, Bandwidth:2 MHz)
HDLC Packet Communication
Broadcasting FEC 1/2 Viterbi Coding
Bit Rate Maximum 1500 kbps 1 ch. (Broadcasting)
Normal 128kbps I2ch.(HDLC)
Calling. 9.6 kbps 2 ch.(HDLC)
Call Channels Pure-ALOHA
Other Ch. Reserved

Analog Transponder (2 Wattes, Bandwidth: 200 KHz)
Two-way Voice available
Experimental Digital Communication

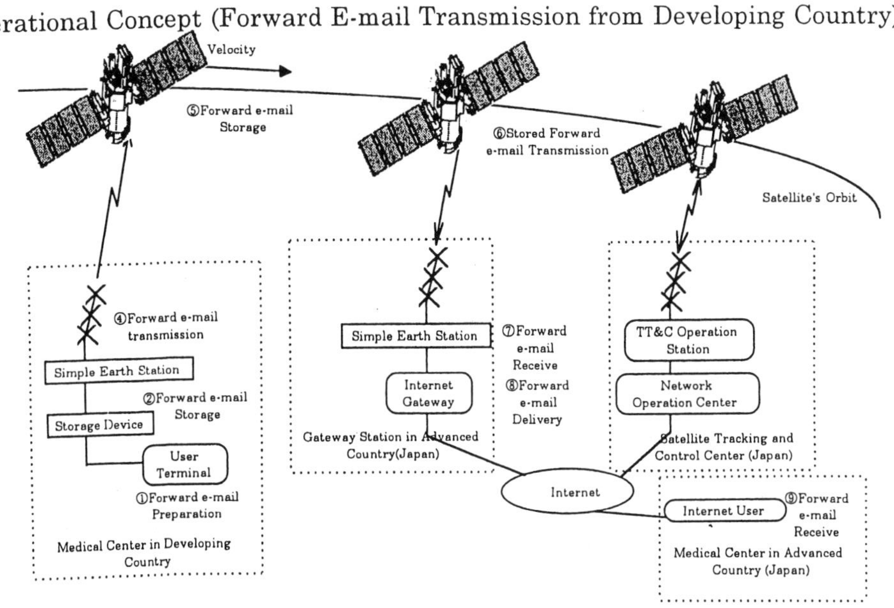

Figure. 5 Operational concept

HiCAT/MEO Concept:Tele-consultation using e-mail via communications satellite

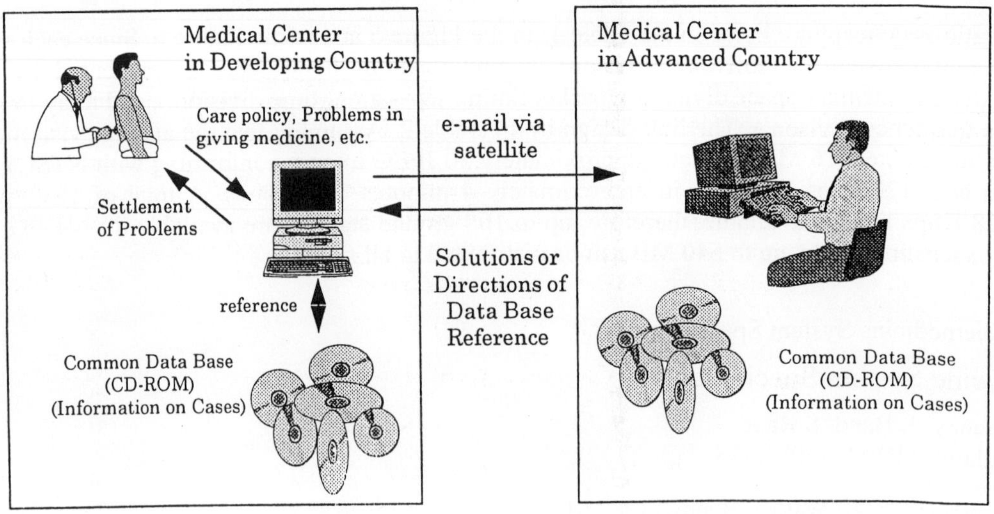

Figure 6. Concept of Tele-consultation

Table 2. Link design of digital transponder of HiCAT
HDLC packet communication at 128 kbps

HiCAT UPLINK 128 kbps			HiCAT DOWNLINK 128 kbps		
Freq.	(GHz)	2.64	Freq.	(GHz)	1.54
Ant. Dia.	(m)	1.00	Ant. Dia.	(m)	0.40
Ant. Gain.	(dB)	24.61	Ant. Gain	(dB)	11.97
Tx PWR	(dBW)	10.00	Tx PWR	(dBW)	6.00
EIRP	(dBW)	34.61	EIRP	(dBW)	17.97
Path. Loss	(dB)	179.96	Path. Loss	(dB)	175.28
Abs. Loss	(dB)	0.10	Abs. Loss	(dB)	0.10
Ant. Dia	(m)	0.40	Ant. Dia	(m)	1.00
Ant. Gain	(dB)	16.66	Ant. Gain	(dB)	19.93
Rcv. PWR	(dBW)	-128.79	Rcv. PWR	(dBW)	-137.47
Tsys	(K)	1160.00	Tsys	(K)	550.00
No	(dBW/Hz)	-197.96	No	(dBW/Hz)	-201.20
G/T	(dB Hz)	-13.99	G/T	(dB Hz)	-7.47
C/No up	(dB Hz)	69.17	C/No down	(dB Hz)	63.73
Margin	(dB)	9.57	Margin	(dB)	4.13
Data Rate	(Mbps)	0.128	Path(km) 9000 km		
BER		1.0E-06	Antenna Eff. 60%		
Eb/No	(dB)	10.53	Cable loss 2.00 (dB)		
Coding Gain(dB)		2.00			

DISCUSSION

Comparison with LEO

The number of visibility intervals and the average in-view time is shown in the Table 3 and in the Figure 7 when compared with Iridium, a LEO system, with satellites in much lower orbit. The average angle of elevation is approximately twice that of Iridium. These advantages; higher visibility and higher elevation angles can be explained to make MEO suitable for communications between medical facilities. Call origination count is less, the coverage of the radio station is relatively large, and the communication time is relatively long during line connection. The detailed advantages of MEO satellite of the sun-synchronous quasi-recurrent orbit are summarized in the Table 4.

Table 3. Visibility comparison between Iridium and HiCAT

		Iridium	HiCAT
Ulan Bator Lat.: 42.8 N (deg) Long.: 107.0 E (deg)	Contact per Day	4	6
	Total Contact Time (sec/day)	2,385.2	15,775.9
	Average Contact Time (sec)	596.3	2,629.3
	Maximum Elevation (deg)	83.2	84.3
	Minimum Elevation (deg)	5.0	5.0
	Average Elevation (deg)	16.3	33.1
	Maximum Range (km)	2,469.5	8,903.4
	Minimum Range (km)	647.0	5,027.0
	Average Range (km)	1,847.6	6,887.5
Lanzhou Lat.: 36.0 N (deg) Long.: 104.0 E (deg)	Contact per Day	4	6
	Total Contact Time (sec/day)	2,278.3	16,259.8
	Average Contact Time (sec)	569.6	2,710.0
	Maximum Elevation (deg)	79.1	87.9
	Minimum Elevation (deg)	5.0	5.0
	Average Elevation (deg)	15.8	35.2
	Maximum Range (km)	2,464.4	8,902.1
	Minimum Range (km)	662.0	5,013.5
	Average Range (km)	1,837.5	6,781.8
Kunming Lat.: 25.0 N (deg) Long.: 102.7 E (deg)	Contact per Day	4	7
	Total Contact Time (sec/day)	2,096.0	18,146.1
	Average Contact Time (sec)	524.0	2,592.3
	Maximum Elevation (deg)	67.7	89.2
	Minimum Elevation (deg)	5.0	5.0
	Average Elevation (deg)	14.8	32.6
	Maximum Range (km)	2,454.7	8,899.2
	Minimum Range (km)	694.7	5,008.2
	Average Range (km)	1,870.8	6,885.5
Bangkok Lat.: 13.8 N (deg) Long.: 100.5 E (deg)	Contact per Day	3	8
	Total Contact Time (sec/day)	1,630.3	20,085.2
	Average Contact Time (sec)	543.4	2,510.7
	Maximum Elevation (deg)	53.8	73.3
	Minimum Elevation (deg)	5.0	5.0
	Average Elevation (deg)	14.5	28.3
	Maximum Range (km)	2,444.4	8,894.9
	Minimum Range (km)	764.7	5,125.2
	Average Range (km)	1,866.2	7,100.6
Singapore Lat.: 1.2 N (deg) Long.: 103.9 E (deg)	Contact per Day	4	9
	Total Contact Time (sec/day)	1,985.6	22,092.0
	Average Contact Time (sec)	496.4	2,454.7
	Maximum Elevation (deg)	77.2	79.5
	Minimum Elevation (deg)	5.0	5.0
	Average Elevation (deg)	15.2	24.9
	Maximum Range (km)	2,434.5	8,888.4
	Minimum Range (km)	657.9	5,051.6
	Average Range (km)	1,841.5	7,320.0

Table 4. Advantages of Sun-Synchronous Quasi-Recurrent Orbit

Advantages of Sun-Synchronous Quasi-Recurrent Orbit

1. Orbit groundtracks return back on the same tracks by 13 days period.
   This makes signal acquisition and communication operations easier.
2. Local time at equator ascent/descent is constant.
   This makes broadcasting at constant local time easier.
3. Satellite design is easy because sun incident angle is stable.
   Constant sun incident angle for whole year makes designs of solar array, electric power system, and thermal control system easier.

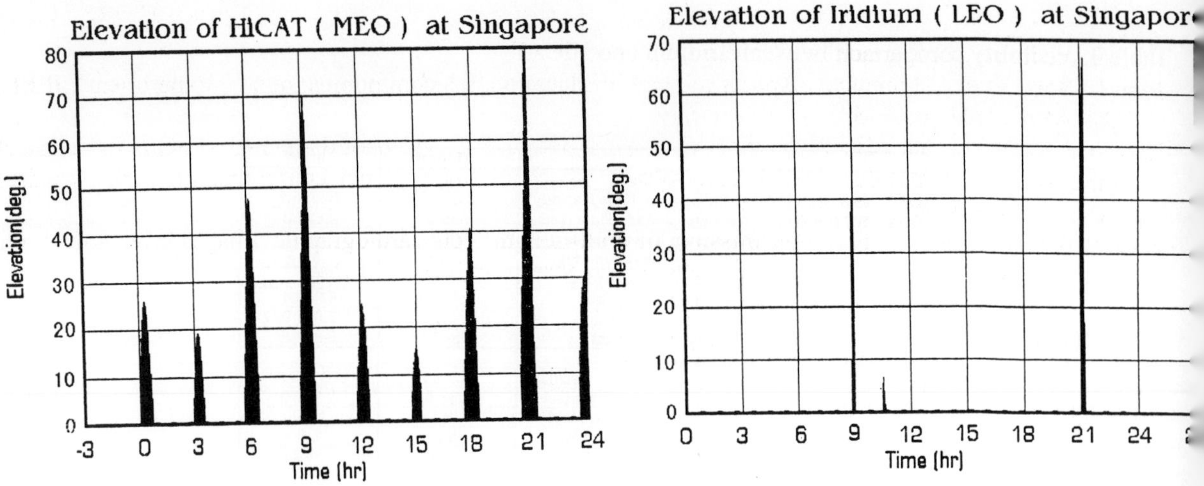

Figure 7. Comparison of elevations with Iridium at Singapore

Importance of International Cooperation

In developing countries, the fetus death rate is high, the average life is short, the interest for medical services for people is high. The governmental and administrative organizations are keen to obtain advanced medical news and therapy. In particular, since 1980, AIDS infection is a great medical problem in these countries. At that time, since there was insufficient information regarding AIDS diagnosis and therapy, the incidence of AIDS was drastically increased. Because of AIDS, development of economy of developing countries becomes slower and slower. For example, even in Thailand which made remarkable development of economy growth, 30% of total population around Chiang-Mai city of the northern part of Thai is HIV positive. In African and Asian developing countries, it is important to effectively and speedily introduce advanced medical technology(which the U.S. and Japan achieved in the past 30 years and to reduce medical expense in order to ensure further development of economy).

Likewise communications, medical services are important in the future society and must be provided in developing countries, and remote areas, islands, and less populated places. There is a variety of available medical services, including primary care, high-level therapy for serious diseases. However, it is difficult to distribute high-level medical services to various areas from the viewpoint of economic efficiency. Teleconsultation with clinical data as a store-and-forward packet communications, clinical image diagnosis with JPEG, medical educations, and broadcasting medical news distribution services using

MEO are believed to be one of the optimal applications of small satellite.

CONCLUSION

MEO is more suitable to medical communications of a large amount of data than LEO. In addition, MEO is more inexpensive than geostationary satellite is worth discussing launching of MEO through international corporation.

REFERENCE

B.E.Dunn, et al.,;Use of telepathology for routine surgical pathology, Review in a test bed in the department of Veterans Affairs, Telemedicine Journal, 3, 1;1-10,1997.

Mark Goldberg;Image data compression: Decreasing the time and cost of data transmission. The Annual Telemedicine 2000, Track B3: Chicago, IL 1996

John L. Boor, et al., "Technical Aspects of Health/Education Telecommunications Experiment", IEEE, Vol. AES-11, No.6;1975

Isao Nakajima, et al.,; How to use an INS-64 interactive videoconferencing system for clinical applications. Program and Book of Abstracts, The 3rd International conference on the medical aspects oftelemedicine;207, 1997

Ernerio T. Alboliras, et al.,:"Transmission of Full-Length Echocardiographic Images over ISDN for Diagnosing Congenital Disease", Telemedicine Journal 2 ,4; 251-258,1996

A GLOBAL VIEW OF THE MAGNETOSPHERE USING MICROSATELITES

A.D. Johnstone

Mullard Space Science Laboratory, University College London, Holmbury St Mary, Dorking Surrey RH5 6NT, UK

ABSTRACT

Our knowledge of space plasmas in general and the Earth's magnetosphere in particular depends largely on in-situ measurements from spacecraft. Such measurements transformed the science in the 1960's by revealing that the behavior of plasma in space was more complex than it was possible to deduce from ground-based observations. Since then, although the quality and coverage of the measurements has improved dramatically, and the advances in scientific understanding have been commensurate, awareness of the two main shortcomings of the method has also increased. The first, the ambiguity between temporal and spatial changes as observed from the spacecraft frame of reference has long been appreciated but the second, the impossibility of obtaining a global picture of events and the need for such a perspective has become forced on us over the past decade. There have been several technical advances which bring us to the point where it is feasible to obtain a global picture for the first time. Those advances include; the development of techniques for magnetospheric imaging; the deployment of large constellations of identical satellites for communications purposes; the miniaturization of instrumentation; and the development of smaller and smaller spacecraft. This is leading to a return to the satellite sizes of the very first space missions hut with substantially enhanced capability.

OBTAINING A GLOBAL PICTURE OF THE MAGNETOSPHERE

The traditional method for obtaining the global picture has been to use networks of ground-based observatories to provide the context for in-situ measurements made from spacecraft. In recent years the ground networks have expanded in coverage and have added instruments such as HF radar with new capabilities. The number of spacecraft in orbit has been increasing, so that there are currently more than a dozen spacecraft making magnetospheric observations. Despite the increase in numbers, the density of sampling in the magnetosphere is still too sparse to resolve many of the important questions. The new approach to obtain a global picture of the magnetosphere is to use i) remote sensing based on imaging techniques such as neutral atom imaging and He 304A imaging; and ii) the deployment of very large numbers of identical vehicles to sample the magnetosphere as densely as possible with in-situ measurements. In this report the role of microsatellites in both these new developments will he discussed.

A PROPOSED UK PROGRAMME

In the UK the Solar Terrestrial Physics community, which has been strong both in space instrumentation for magnetospheric physics and in ground-based techniques has combined its efforts to tackle the question of the global scale studies. The conclusion reached has general application even though the current plans are presented in a UK context. It is that a series of microsatellites building on existing small satellite expertise can make an important contribution to scientific studies of the masnetosphere. The proposed series includes:

i) MAGPIES – He^+ imaging at 30.4nm. The project consists of two microsatellites in geostationary transfer orbit observing the sunlight at 30.4 nm resonantly scattered from cold singly-ionised helium in the magnetosphere. This would be backed up by ground-based observations using radar to obtain the ionospheric electric field and magnetometer chains to probe for the effect of the plasmapause on magnetic pulsations.
ii) Faraday - the dayside reconnection rate. Two microsatellites in low earth polar orbit will measure the dayside reconnection rate by observing the entry of solar wind ions in combination with ionospheric observations by ground-based instrumentation on Svalbard. The project uses both in-situ and ground-based techniques in a synergistic fashion. The orbit will be targeted to maximize its daily number of overpasses of Svalbard and is an attempt to obtain a global parameter of the magnetosphere as nearly continuously as possible.
iii) SWARM - magnetosphere constellation. A large number of satellites are deployed throughout the magnetosphere to obtain as dense sampling of the magnetosphere with in-situ measurements as possible.

Since the instrumentation for Magpies and Faraday can be accommodated on small satellites then microsatellite technology is the appropriate and most cost effective way to carry out the projects. For Swarm microsatellite technology is essential. It would be prohibitively expensive to deploy large numbers of traditional magnetospheric spacecraft for Swarm so the most promising method of reducing the overall cost to that of a traditional space plasma mission such as Polar or Wind is to use microsatellite technology. Swarm is a longer term objective and one not likely to be carried out by a single nation. Our series of projects is designed to carry out science investigations which tackle questions of global coverage immediately while making the necessary technical developments which could lead to a Swarm mission at the earliest possible date.

MAGPIES (MAGnetospheric Plasma Imaging Experimental Satellites) (Johnstone et al 1997)

The cold plasma in the magnetosphere can be observed remotely by detecting sunlight resonantly scattered from singly-ionised Helium at a wavelength of 30.4 nm. Singly-ionised Helium is only a minor constituent of the cold plasma in the magnetosphere but the radiation it emits is the most effective one for imaging. The distribution of singly-charged Helium provides a good representation of the overall distribution of plasma in the magnetosphere. The He 30.4nm line can be imaged using a conventional reflecting telescope which can provide detailed images of the plasmasphere with resolution of the order of 1 degree.

Figure 1 is a simulation of what such an instrument might see from a viewpoint near dawn when there is a fully developed plasmasphere. The magnetic structure of the magnetosphere is clearly visible and the pictures should be detailed enough to follow the distortion of the magnetic field during major storm events. A particularly exciting prospect would be to watch the sequence of events when a large solar storm, such as the CME recorded in early January 1997, strikes the front surface of the magnetosphere. First of all the dayside compression of the magnetosphere and the distortion of the field lines would be seen; then the outer layers of the plasmasphere would be swept away into the outer magnetosphere and be lost in the solar wind. Such scenarios have been imagined for decades, and the magnetosphere has been sampled at many points as such events occur but it is not known what happens to the three-dimensional

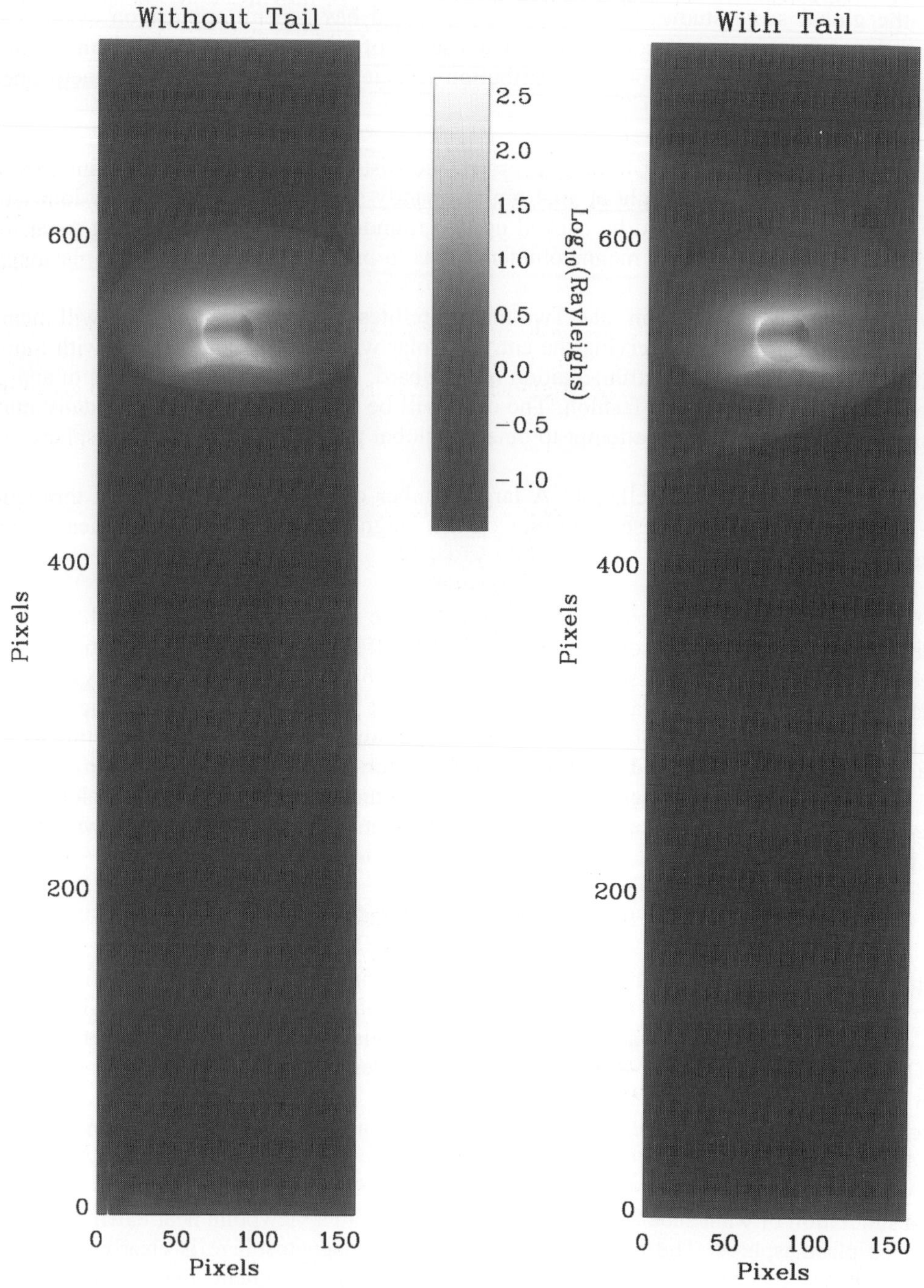

Figure 1. A simulation of a two contrasting images of the plasmasphere in the He^{+} 30.4 nm emission line. Both views are from the equator at a geocentric distance of 6.6Re and from 90^{0} solar zenith angle (at dawn). The pixel size is 0.5^{0} and the fielcd-of-view corresponds to that of the proposed imager. The frame on the left shows the plasmasphere under quiet conditions while that on the left shows a plasma tail developed as the result of a storm onset. (courtesy R.R. Meier, 1997)

structure of the magnetosphere. This project has the capability to show these events as a succession of images in a film. Not only will the images be of great scientific interest but they are also likely to be of great public and educational interest.

The scientific objectives are to study:-
- the global electric field variations during magnetic storms and substorms;
- the mechanisms by which flux tubes are filled;
- the distortions of the magnetic field during storms;

Instrumental Requirements

The size of the telescope required for any task is determined by the sensitivity it needs to detect the source of light. So the first question to ask is whether a microsatellite is big enough to carry the size of telescope needed to detect the images. The brightness of various features in the source are given in Table 1.

Table 1. Brightness of He^+ 30.4nm emissions in the magnetosphere

Plasmasphere Internal Structure:	5-150 Rayleighs
Plasmapause Location:	1-5 Rayleighs
Plasma Tails, Detached Plasma:	0.05-5 Rayleighs
Plasma Trough Filling:	0.05-5 Rayleighs

The angular resolution should be good enough to resolve the key features of the plasmapause seen at the limb at approximately 4 Re geocentric distance when the spacecraft is at apogee at 6.6 Re. 1^0 resolution corresponds to a feature size of approximately 800kms under these conditions. This is the most unfavorable aspect for the observations, since the spacecraft will often be much closer to the plasmapause, but such resolution is still good enough to enable the objectives to be met.

It would be ideal if the instrument were able to see the plasmasphere all the time, but this is not feasible from geostationary transfer orbit. Therefore the design calls for the field of view to be as wide as possible but without including the sun since direct sunlight would damage the detector. The observation is enhanced by having a stereoscopic view from two spacecraft from different angles which improves the interpretation of the images and increases the temporal and spatial coverage. Since the source is optically thin and the brightness is an integration over the whole line of sight the two views also allow the three dimensional structure to be deduced quantitatively.

The fastest time scale for variability in the observations is of the order of one minute for the times when the spacecraft is crossing a sharply defined plasmapause. At apogee 5 to 10 minutes time resolution would be adequate.

Instrument Description

The telescope consists of three optical elements; a concave mirror with a surface consisting of alternating layers of molybdenum and silicon and a radius of curvature of 300mm; a microchannel plate detector at the focal plane; and a thin filter of Aluminium and Saran over the aperture. The mirror acts as an interference filter selecting bands at wavelengths corresponding to the various successive orders of interference. The filter itself allows only one of the orders, that centred at 30.4nm, to pass. The sensitivity is a function of the size of the mirror which is determined by its radius of curvature. The field of view of the telescope is 40^0 in a meridional plane of the spacecraft coordinate system whose polar axis is the spin axis, and 16^0 in the azimuthal direction. As the spacecraft rotates, the telescope sweeps out a field of view $40^0 \times 360^0$. In order to get a more comprehensive view two telescopes are used in a back to back configuration to give a $80^0 \times 360^0$ coverage. The spin axis is directed towards the sun thus ensuring that the aperture is never directly illuminated by sunlight.

Studies of the optical structure (Sims 1997) have shown that the sensitivity allows 1 Rayleigh to be detected with good signal-to-noise in 60 seconds integration. The resolution is 0.4^0 on the telescope axis and 1.3^0 at edge of field of view. The mass of the instrument is of the order of 3-4kg per telescope making 7kg total, and its power requirements are 7 watts total. Thus a plasmaspheric imager capable of high quality observations can be accommodated in a microsatellite of 50kg. There are no significant technological risks since all the major elements, optics, detectors, electronics have flight heritage.

FARADAY (Lockwood 1997)

The primary scien6fic objectives are
- to measure the reconnection rate at the dayside magnetopause
- to determine the X line characteristics and locations
- to locate the dayside cusps
- to measure polar cap size and convection

The reconnection rate is given by (Lockwood and Smith 1992, Lockwood and Davis 1996) as

$$\varepsilon(t_0) = \frac{B_i(V_c - V_c)\dfrac{dy}{dy'}}{\left[1 + \left[\dfrac{d_0}{2}\right]\left[\dfrac{m}{2}\right]^{\frac{1}{2}} E_{ic}^{\frac{-3}{2}} \dfrac{dE_{ic}}{dt_s}\right]}$$

where

$$t_0 = t_s - d_0\left[\frac{2E_{ic}}{m}\right]^{-\frac{1}{2}}$$

and t_s, is the observation time. The ionospheric magnetic field B_i is effectively constant and can be obtained from models to within 2%. The satellite velocity, V_s normal to the Open/Closed Field Line Boundary is known from the orbit and from the Open/Closed Field Line Boundary orientation obtained by onboard cameras. The convection velocity, V_c normal to the Open/Closed Field Line Boundary is obtained from a Retarding-Potential Analyser onboard. The distance to the reconnection site, d_0 can be obtained from the observations of two spacecraft at a known separation. The ion mass, m, can be measured but may be assumed to be protons since they dominate in the cusp entry. The length mapping factor dy/dy' can be taken from a magnetic field model. The ion cut-off energy E_{ic}, is measured by an ion detector designed to give equal increments in t_0. The instrumentation consists of an ion detector for precipitating magnetospheric ions and upwelling ionospheric ions; cameras to image 630nm and 557.7nm auroral emissions; an electron precipitation detector and a cold-ion retarding-potential analyser to obtain drift velocities. Such instrumentation is easily accommodated in a microsatellite. The preferred orbit is Sun-synchronous in the noon-midnight plane at 800 km altitude. Both spacecraft should be in the same orbit with a fixed separation of 200-4000km(giving 20s to 8min separation in time). The orbit should be chosen to maximize the number of passes over the Svalbard ground observatories. The spacecraft should be aligned in the direction of the local magnetic field.

SWARM (Schwartz 1997)

This project will employ a large number (30 or more) of minimally-instrumented inexpensive satellites to study the 3-D, time-dependent magnetosphere of the Earth. Specific objectives could include:

Mapping magnetic field configurations throughout a substorm;
Mapping the shape and response of key boundary layers (bow shock, dayside magnetopause, cusp, tail current sheet) to varying solar wind input parameters and transients;
Simultaneous study of 3-D time dependence over many scales, ranging from microscales to global scales;
Studies of naturally occurring turbulence (e.g., magnetosheath, foreshock, and auroral acceleration regions) including the planarity, coherence, propagation directions and mode identification;
Testing and validating geospace and magnetospheric models.

The mission complements Cluster II, which will be launched in mid-2000 and employs 4 identical fully-instrumented spacecraft to unravel the detailed local micro-processes which mediate the transport through key regions and boundaries (bow shock, high altitude cusp, geomagnetic tail). By contrast, SWARM has large numbers of spacecraft with a minimal payload to map and follow large-scale, global processes and transients. The proposed scientific payload consists only of a de magnetometer and combined ion/electron E/q plasma analyser. The orbits need not be identical, but should include a number of highly elliptical orbits out to 15-20 Earth radii. Throughout the run of the seasons such orbits sample all of the magnetosphere including the solar wind and bow shock, magnetosheath and magnetopause, cusp/auroral regions, and near geomagnetic tail. The scientific capabilities of each individual satellite are well-tested and basic but the mission as a whole, include the satellites, provides several key areas where technological advances are required and where novel use of existing methods needs to be made.

THE TECHNICAL CHALLENGE OF CONSTELLATIONS

There is absolutely no technical difficulty in deploying a large number of spacecraft of conventional type throughout the magnetosphere. The problem is that it would be prohibitively expensive. The technical challenge is therefore to bring the overall cost within acceptable limits which will be of the order of the cost of a current single-spacecraft magnetospheric mission. Some unit costs are reduced with the number produced such as a) satellite manufacture which can be based on assembly line procedures; b) electronics packaging and instrument miniaturisation which can use techniques with a large start-up cost but which allow subsequent units to be produced cheaply. Some costs, such as launch costs, increase approximately linearly with mass. Some costs, such as communications and control, could increase nearly exponentially if the spacecraft were operated in a conventional fashion with substantial human intervention and the use of dedicated ground facilities. The obvious first step in reducing costs is to reduce the size of the spacecraft to reduce both manufacturing costs and launch costs. Hence the need to miniaturize instruments, to centralise onboard data processing, and to compact electronics.

Launch Costs

There are two ways of containing the costs of deploying the spacecraft in orbit. The First is to make use of piggy back launches with a prime customer. Such missions will only go into common operational orbits which will not always be the scientifically required orbit. The main orbits accessible this way include:-

a) low Earth polar orbit which may be sun-synchronous but targeted at the wrong local time and altitude for magnetospheric missions;
b) geostationary transfer orbit which gives a good starting point for highly elliptical orbits but which carries a high radiation risk;
c) the highly elliptical, 63^0 inclination Molniya orbit which has a number of advantages for magnetospheric missions.

The second way is to launch as many spacecraft as possible on a single launch vehicle. However the spacecraft must be dispersed after launch and so some on-board means of orbit development will be required. However there is a limit to the need to reduce the size for launch considerations. A spacecraft of 50kg is easily accommodated on most piggyback opportunities and there is a limit to the number which can be deployed into the same starting orbit. If size reduction is carried to an extreme for its own sake it could lead to an increase in overall costs.

Ground Segment Costs

Traditional methods of operating spacecraft could be completely superseded by the new global communications infrastructure being developed. Within a few years there will be new constellations of communications satellites which link directly to hand-held telephones on a worldwide scale. The user has direct access to the system, and the handset equipment is small enough to be carried on a microsatellite. The satellite itself could therefore initiate a transmission, and would be able to contact any other telephone in the world simply by issuing a number to the system itself. The hardware required is a modem and telephone at either end. Both devices are standard and mass produced in suitably miniature form. The picture which emerges is that the satellite could "phone home" every day and dump files of data in a computer using standard hardware and software without requiring human intervention. There are a number of technical questions to be resolved before this approach can be proven but the equipment is already available for communication to existing geosynchronous satellites.

This allows autonomous operation by the satellite in one significant area but if the ground segment is to be reduced autonomous operation must be carried through into all other areas where there has traditionally been human intervention such as power conditioning, stabilization, spacecraft health. The necessary hardware, sensors and control equipment, probably already exist. What is required is the software so that it will be possible to switch the spacecraft on and leave it to look after itself.

One obvious question to be asked is the data rate which can be transmitted through telephone links. Scientifically the push will always be to increase the amount of data. What is required is to have data transmitted at the highest possible rate at times of important scientific changes. Since these times cannot be predicted the spacecraft has to transmit at the highest possible rate all the time and the data selection made on the ground. With the development of large-capacity solid-state memories new strategies become possible. The data may be recorded at a high rate over a long period of time, say several days, during which time the scientific situation can be reviewed and, if appropriate, the data from the most important periods selected for transmission. The less interesting data can then be overwritten by new data. The selection could be made on the basis of other ground-based observations such as the detection of a magnetic storm, or anticipated by observation from a CME at the Sun, or it could be made on the basis of summary data transmitted from the satellite. Alternatively the selection could be made onboard by the use of autonomous event recognition software. The summary data could be used for statistical studies.

Key Developments

A partial (in the sense that these are what we have thought of but they are not likely to be all that is required) list of the development areas includes:-

a) communication by phone
b) GPS for location (already under development)
c) miniaturized instruments
d) centralised data processing hardware and software
e) autonomous operation
f) compact large capacity memory
g) orbit adjust capability
h) radiation protection for cheaper and more capable components

It is proposed to develop these capabilities in carrying out the Magpies and Faraday projects so that the deployment of SWARM could then be considered.

CONCLUSION

Microsatellites are the key technology for global, multiscale magnetospheric physics. The technical developments required are likely to find application more widely than just in magnetospheric physics.

ACKNOWLEDGEMENTS

The proposed UK small satellite programme has been developed by a group consisting of David Hall, Mark Lester, Andre Balogh, Mark Sims, Peter Cargill, Andrew Fazakerley, Mike Lockwood, Stan Cowley and the author. The work here is mainly based on the results of those stimulating discussions which it is a pleasure to acknowledge.

REFERENCES

Lockwood, M., and M.F. Smith, The variation of reconnection rate at the dayside magnetopause and cusp ion precipitation. J. Geophys. Res., 97, 14841 - 14847, (1992)

Lockwood, M., and C.J. Davis, On the longitudinal extent of magnetopause reconnection bursts, Annales Geophys., 14, 865-878, (1996)

Lockwood, M., S.W.H. Cowley, and M. Grande, Faraday - a microsatellite project to measure cusp reconnection rates, Proposal to the UK Particle Physics and Astronomy Research Council, (1997)

Johnstone, A.D., A.N. Fazakerly, D.M. Walton, M. Lester, M. Sims, S.W.H. Cowley, R.R. Meier, C. Brown, J. Seely, R. J. Moffett, D. Orr, Magnetospheric plasma imaging explorer satellites, Proposal to the UK Particle Physics and Astronomy Research Council, (1997)

Schwartz, S.J., SWARM a fleet of microsatellites to explore the magnetosphere, http://www.space-plasma.qmw.ac.uk/SWARM/ (1997)

Sims, M.J., Optical design of an EUV imaging telescope, private communication, (1997)

THE USE OF MICROSATELLITES IN MONITORING THE IONOSPHERE/PLASMASPHERE

E A Essex[1], P A Webb[1], I Horvath[1] and H A Cohen[2]

[1]*Department of Physics, La Trobe University, Bundoora, Victoria 3083, Australia*
email: e.essex@latrobe.edu.au
[2]*Department of Computer Science and Computer Engineering, La Trobe University, Bundoora, Victoria 3083, Australia*

ABSTRACT

The advent of microsatellites with their advantages of low overall cost in construction and launch and short time span between conception and launch has given a new impetus to the further study of the geosphere. By using a combination of space based and ground based receivers it is possible to undertake new and exciting experiments directed towards furthering our knowledge of the ionosphere. Combinations of high earth orbit satellites such as GPS and low earth orbit (LEO) micro satellites are providing the capability for satellite to satellite occultation experiments to reconstruct the vertical profile of the ionosphere. The topside ionosphere and plasmasphere ionisation content may also be explored with satellite to satellite experiments. This paper will describe some of the experiments proposed for future micro satellites such as those planned for the Australian FedSat1 satellite. Model studies of some of these experiments will be discussed.

INTRODUCTION

The various regions of the ionosphere have been monitored to different degrees on a long-term basis. Even instruments such as incoherent scatter radars, while providing profiles of both the topside and bottom side ionosphere, do not operate on a continuous basis and provides poor resolution for the lower ionosphere. The ground-based ionosondes provide non-homogeneous coverage of the bottom side ionosphere, as they are limited to observations from land. This is particularly true in the Southern Hemisphere where the oceans cover the majority of the Hemisphere. The topside ionosphere and the region above, the plasmasphere are not easily monitored on a long-term basis. Only a few instruments such as incoherent scatter radars, topside sounders and in situ satellite based diagnostics are able to provide details on the structure, composition and dynamics of the topside ionosphere. The properties of the plasmasphere are even less well known and techniques to explore it are very limited. The advent of the Global Positioning System (GPS) with 24 satellites in 12 hour orbits at 20,000 kilometres, provides the opportunity to monitor on a global basis the variation of the ionisation content of the ionosphere and plasmasphere. The recent deployment of space based GPS receivers have demonstrated the feasibility of using satellite to satellite experiments for monitoring the ionosphere. In this paper we discuss some of the satellite to satellite experiments proposed for FedSat1, the Australian satellite to be launched in 2001.

FEDSAT1

FedSat1, the first Australian satellite to be launched in over 30 years, will be a microsatellite of around 50 kg launch mass. It is planned to launch into a low earth orbit between 500 to 1000 km at an inclination of 60 to 70 deg. FedSat1's mission is basically a science and engineering one. The planned scientific payloads include fluxgate and induction magnetometers for geomagnetic field and current studies as well as a dual frequency GPS receiver for ionospheric and atmospheric research. The following sections detail some of the proposed experiments to be carried out using the on board GPS receiver.

IONOSPHERIC MEASUREMENTS

Observing the GPS constellation of satellites in space from a low earth orbit (LEO) satellite such as FedSat1, orbiting at 700km provides a powerful means of imaging the ionosphere and the inner magnetosphere, the plasmasphere. The provision of occulting geometry from a space based receiver enables effectively a horizontal scan through the ionosphere to be obtained (Hajj et al 1994). The lack of horizontal scans from ground based observations of GPS satellites has limited the current development of computerised ionospheric tomography. Because of the geometry, there is a lack of horizontal total electron content (TEC) information that causes a lack of information along the vertical direction in the reconstructed tomographic image. To overcome this problem most researches use some form of a priori information (Villani and Essex 1996). A review of ionospheric tomography algorithms may be found in Raymund (1994). In the planned experiment for FedSat1, this problem will be overcome by using a combination of ground based as well as space based GPS occultation data which will provide both vertical and horizontal information for the application of the tomographic reconstruction. The principal advantages of the tomographic reconstruction techniques lie in the large geographic coverage and the cost effectiveness. Incoherent scatter radars are only able to provide the information from a smaller geographical area at the cost of millions of dollars to build and operate. Recent interest in the development of space based tomography has seen the launch of the first LEO, GPSMET and its use in the radio occultation experiments. Various researchers have recently performed ionospheric reconstruction from the GPS occultation data. (Ruis et al 1997, Hajj and Romans, 1998).

PLASMASPHERE MEASUREMENTS

A further area of interest is the electron content of, as well as the irregularities in, the plasmasphere between FedSat1 and the GPS satellites orbiting above FedSat1. With a GPS antenna located on the top of the satellite, it will be possible to undertake TEC measurements vertically above FedSat1 to the GPS satellites for coincidence occurrences in the orbits. For a cone of 5 degrees above FedSat1 these coincidences would enable information to be obtained on the topside ionosphere and the plasmasphere. Figure 1 shows model calculations of the TEC content for three cases: below 500km, between 500km and 1000km and above 1000 km. The model plasmasphere used in these calculations is based on diffuse equilibrium and in combination with the International Reference Ionosphere (IRI), forms a global ionospheric-plasmasphere model (Webb and Essex, 1997). At low latitudes at sunspot maximum, there is clearly significant ionisation above 500km. For the high latitudes in the Australian region of the Southern Hemisphere, where the magnetic pole is offset from the geographic pole toward Australia, the plasmapause is located at lower geographic latitudes. Hence for coincidences at higher latitudes, it will be possible to estimate the location of the plasmapause. As at these locations the ray paths from the GPS satellites to the FedSat1 satellite will not intersect the plasmasphere and there would normally be only a negligible amount of ionisation above 500km, the TEC should approach zero (Klobuchar et al, 1994). Recent analysis of ground based GPS TEC data has shown the existence of a large build up of ionisation equatorward of the mid latitude trough in the plasmapause region. This increase in ionisation may be related to heating of the topside ionosphere by turbulent dissipation of ring current energy at the plasmapause (Horvath and Essex, 1998, Titheridge, 1976). FedSat1 may also be used to investigate this ionisation buildup, especially over the Southern Hemisphere where the ground based measurement are sparse.

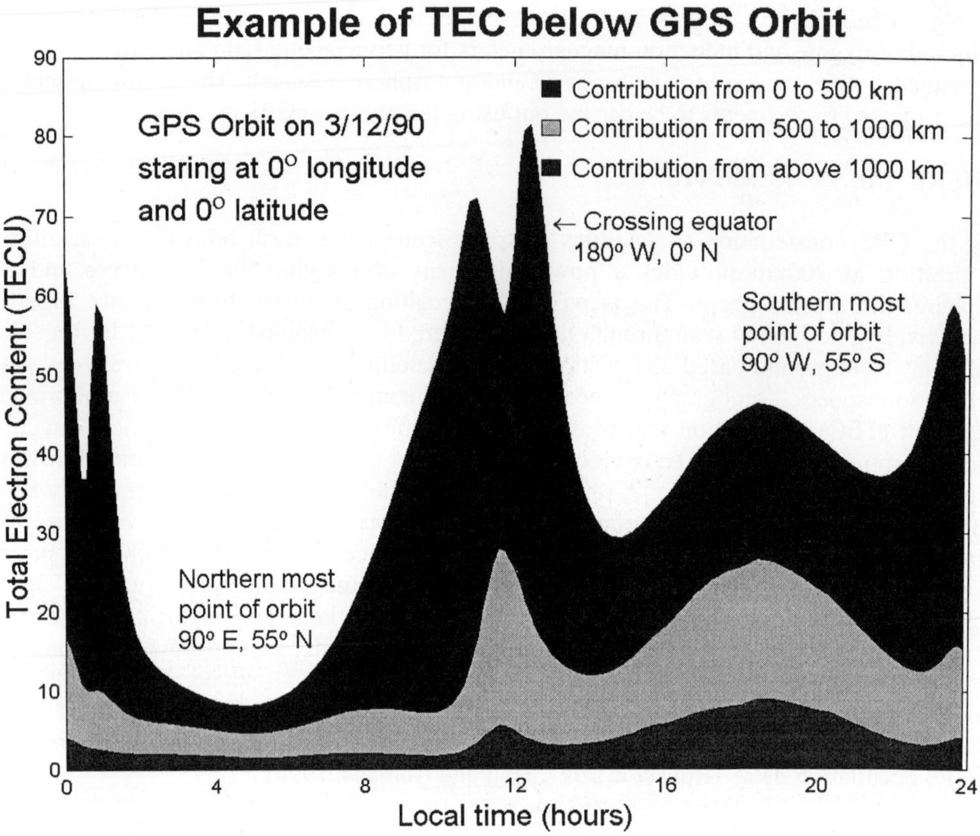

Figure 1. Electron content contributions from 0 to 500km, 500km to 1000km and from above 1000km to the total electron content for a GPS satellite orbit commencing at 0^0 longitude and 0^0 latitude.

IRREGULARITIES

Irregularities in the plasma between FedSat1 and the GPS satellites should also be detectable as scintillation in the phase path data. These irregularities may be located in either the topside ionosphere or the plasmasphere. Jacobson et al (1996) has used sensitive ground based techniques to investigate field aligned irregularities in the plasmasphere. Scintillation activity in the nighttime equatorial anomaly regions of the low latitude ionosphere is known to peak at sunspot maximum, the next maximum being around 2000. Hence the incidence of scintillation in the S band signals from GPS measured in both the occultation experiments and in the vertical direction is another area for study. There is some indication that the occurrence of well developed equatorial anomaly crests and their location is a precursor of the development of nighttime equatorial irregularities (Jayachandran et al, 1997). The combination of both ground based and satellite-based measurements would provide an excellent opportunity to further our understanding of the low latitude ionosphere.

CONCLUSIONS

LEO micro satellites such as the planned FedSat1 provide the opportunity for researchers in Space Sciences to further investigate those regions of the ionosphere for which existing equipment and techniques are either difficult or impossible to apply. In the future, constellations of microsatellites may provide a global coverage of the ionosphere, previously not possible with ground based equipment.

ACKNOWLEDGMENTS

This work is supported by the Australian Cooperative Research Centre for Satellite Systems and by an ARC grant. I.H. is the holder of an Australian Postgraduate Award and P.W. is the holder of a La Trobe University Postgraduate Award.

REFERENCES

Hajj, G. A., R. Ilbanez-Meir, E. R. Kursiniski and L. J. Romans, Imaging the Ionosphere with the Global Positioning System, *Intl. J. Imaging Sys. Tech.*, 5(2), 174-184 (1994)

Hajj, G. A. and L. J. Romans, Ionospheric Electron Density Profiles obtained with the Global Positioning System: Results from the GPS/MET Experiment, *Radio Sci.*, 3(1), 157 (1998)

Horvath, I and E. A. Essex, GPS and TOPEX/Poseidon Observations of the Southern Hemisphere Ionosphere, Submitted *J. Atmos. Solar-Terr.Phys.* (1998)

Jacobson, A. R., G. Hoogeveen, R. C. Carlos, G. Wu, B. G. Fejer and M. C. Kelley, Observations of inner plasmasphere irregularities with a satellite-beacon radio-interferometer array, *J. Geophys. Res.*, 101(A9), 19,665-19,682 (1996)

Jayachandran, P. T., P. Sri Ram, V. V. Somayajulu and P.V.S. Rama Rao, Effect of Equatorial Ionisation Anomaly on the Occurrence of Spread-F, *Ann. Geophysicae, 15*, 255-262 (1997)

Klobuchar, J. A., P. H. Doherty, G. J. Bailey and K. Davies, Limitations in Determining Absolute Total Electron Content from Dual-Frequency GPS Group Delay Measurements, *Proceedings of the International Beacon Satellite Symposium,* Ed. L. Kersley, Aberystwyth, UK, 11-15 July pp 1-4 (1994),

Raymund, T. D., Ionospheric Tomography Algorithms, *Intl. J. Imaging Sys. Tech.*, 5(2), 75-85, 1994

Ruis, A., G. Ruffini and L. Cucurull, Improving the Vertical Resolution of Ionospheric Tomography with GPS Occultations, *Geophys. Res. Lett.*, 24(18), 2291-2294 (1997)

Titheridge, J. E., Plasmapause Effects in the Top Side Ionosphere, *J. Geophys.Res.*, 81(19), 3227-3233 (1976)

Webb, P. A. and E. A. Essex, A Simple Model of the Ionosphere Plasmasphere System, *Proceedings of the Workshop on Applications of Radio Science, WARS'97,* Eds. A. Kulessa, G. James, D. Bateman and M. Tobar, Barossa Valley, Australia, 21-23 September, pp 190-195 (1997)

Villani, A. and E. A. Essex, Computerised Ionospheric Tomography Down the Eastern Coast of Australia, *Proceedings of the Ionospheric Effects Symposium,* Ed. J. M. Goodman, Alexandria, Virginia, USA, May, pp 19-24 (1996)

A MICROSATELLITE IMAGING MISSION FOR THE MULTI-PHASE-ANGLE INVESTIGATION OF VEGETATION

K. Brieß, R. Sandau, M. v. Schönermark, I. Walter

Institute of Space Sensor Technology of DLR,, Rudower Chaussee 5, D-12489 Berlin, Germany

ABSTRACT

The space remote sensing of vegetation parameters gives important information concerning environmental conditions. The combination of new technologies, scientific methods and a strict design-to-cost philosophy makes a microsatellite to an interesting research tool. Basing on scientific objectives in the field of vegetation analysis from space a technical solution for a micro satellite mission is proposed in this paper. The presentation describes all segments of the proposed space mission with their characteristics:

- the launch segment: low-cost launch by use of a piggyback opportunity,
- the space segment: use of new technologies in complement to the use of standard technologies to keep it “small“ for a piggyback launch,
- ground segment: low-cost ground station for field experiments in addition to conventional ground stations
- programme segment: mission objectives according to the capabilities of a microsatellite mission with limits in mass, power, volume, data rates and other

The presentation should point out the unsolved questions and the new chances to start with design-to-cost mission instead of a more science driven imaging mission.

INTRODUCTION

The observation of environment parameters from space becomes more and more important because of the increasing impact of antropogenic effects on our Earth. One of the most important issue of remote sensing consists in the observation of the condition and the change of vegetation in a regional or local scale. Typical parameters describing the vegetation in an image from space are vegetation indeces (e.g. NDVI), the Leaf Area Index (LAI) and the Absorbed Photosynthetically Active Radiation (APAR). They are derived from a multi-spectral image of few VIS/NIR spectral channels which are directed to nadir. They allow the coarse classification of vegetation to a certain extent. But for a classification of different types of vegetation, the estimation of the degree of ripeness, a crop yield analysis or for the analysis of vegetation stress the existing space sensor systems do not give the most useful primary information like for instance the bi-directional reflectance of the target. One way to improve essentially the primary information is a sensor system consisting of few spectral channels looking in different angles to the target.

After a proper atmospheric correction it is possible to use the measured reflectance factors of the vegetation target for an essential improved LAI retrieval. Basing on the wide-angle stereo camera WAOSS (designed for the Mars-96 mission) a dedicated sensor system can be developed within a low budget frame. This sensor system can be implemented on a microsatellite. It allows to propose a scientific microsatellite mission within a low-cost frame. This microsatellite imaging mission can meet some objectives of regional observation but it doesn`t fit very well global coverage requirements. The proposed dedicated micro satellite mission can be a supplement to the planned MISR experiment on the EOS platform.

MISSION OBJECTIVES

The mission objectives are determined by the investigation and analysis of vegetation from space by means of new methods and technologies. It should support the study of the relationship between the vegetation index, the reflectance factors, the leaf area index, the photosynthesis, the vegetation condition, the atmospheric influence under different illumination and seasonal conditions. The mission objectives are summarised in table 1.

Table 1. Mission Objectives of the Vegetation Mission.

	Veg-SAT Mission Objectives
Primary objectives	• to identify and classify surface types • to study the condition of vegetation and to detect changes of vegetation in regional or local areas • to improve the estimation of the terrestrial primary production of selected areas • to improve the retrieval of vegetation parameters by using multi-spectral and geometrical information • to investigate the biosphere-atmosphere interaction
Secondary objectives	• to study the environmental impacts of atmospheric aerosols • to investigate heterogeneous cloud fields

Information on the spatial and temporal variability of biomass and diversity has turned out to be really essential for scientific estimations concerning the climate change as well as the ecological effect of human impact. Monitoring of vegetation is required for forecasts of crop growth and yield. Thus, for instance, information on vegetation conditions allows one to identify forest damage as well as stress on drylands and to deliver early warnings against food shortage. For these tasks imaging systems on air planes or satellites are pretty convenient, because they deliver area averages which can be stored in digital form and used for comparisons at any time. The larger the area of which an inventory has to be taken the more significant is the advantage of remotely sensed data. As measurements of plant canopies, especially in the time of growth and ripening, have to be repeated very often, satellite data should be preferred.

Usually, for monitoring vegetation and deriving plant parameters spectral information is used. The red and the near infrared (NIR) spectral region are the most important ones (see Figure 1). In the red channel the reflectance of plant canopies has a minimum due to the absorption by chlorophyll. On the other hand in the NIR region the high reflectance is governed by multiple internal reflections at interfaces between hydrated cell walls and intercellular air spaces. If the vegetation is diseased, the cell wall structure is damaged and the high reflectance within this channel is reduced providing a diagnostic remote sensing technique for assessing the health of crops. The abrupt and characteristic change in the reflectance of vegetation at about 700 nm (s. Figure 1) is the basis of a number of techniques for assessing the amount of green vegetation present in a remotely sensed image. Usually it is estimated by using various kinds of vegetation indices which are defined by ratios or linear combinations of radiances in these two channels. The most common used one is the Normalised Difference Vegetation Index (NDVI) calculated by

$$NDVI = (L_{NIR}-L_{RED})/(L_{NIR}+L_{RED}) \tag{1}$$

where L is the radiance measured in the channel given by the index.

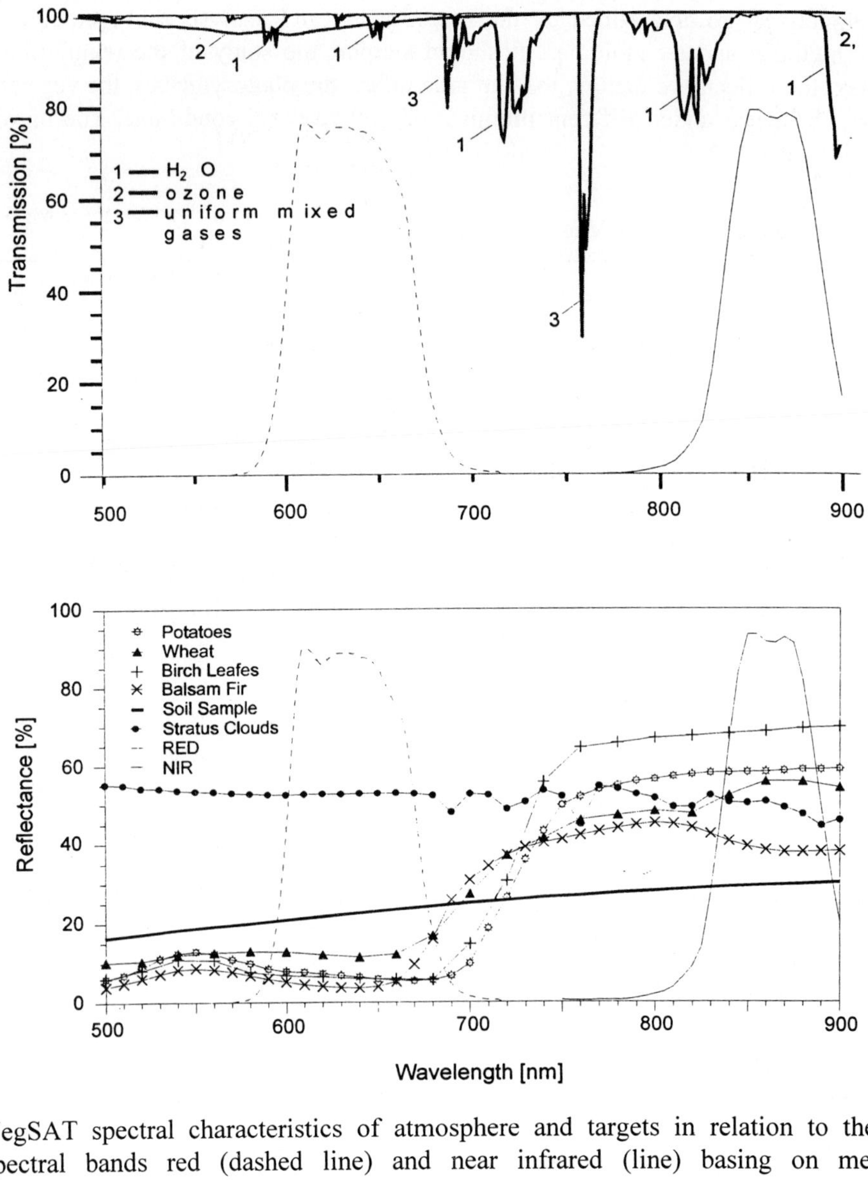

Figure 1. VegSAT spectral characteristics of atmosphere and targets in relation to the WAOSS-V spectral bands red (dashed line) and near infrared (line) basing on measured filter characteristics.

A good measure for characterising the biomass is the leaf area index (LAI). It is given by the area of leaves related to the area of surface belonging to one plant. Usually it is related to an vegetation index, for instance the NDVI. From Figure 2 it can be seen that the relationship is linear only up to a LAI of about 3 due to the chlorophyll concentration in the leaves. As the LAI depends only on structure of the plant but not on the chlorophyll content we defined an other vegetation index based only on the reflectance factor in the NIR region. As in this region the multiple scattering dominates we receive a lot of information on the plant structure. In addition in this new vegetation index ONVI (Off-Nadir Vegetation Index) we used also the geometrical information obtained by observing in different off-nadir angles. On the basis of a huge synthetic data set we calculated the coefficients of the linear regression. For only the nadir looking geometry the relationship is given in Figure 2. It can clearly be seen that the determination of the LAI using the ONVI only for the nadir is much more successful than using the NDVI. Furthermore using six different observing angles the standard deviation of the regression built up can be reduced to one third.

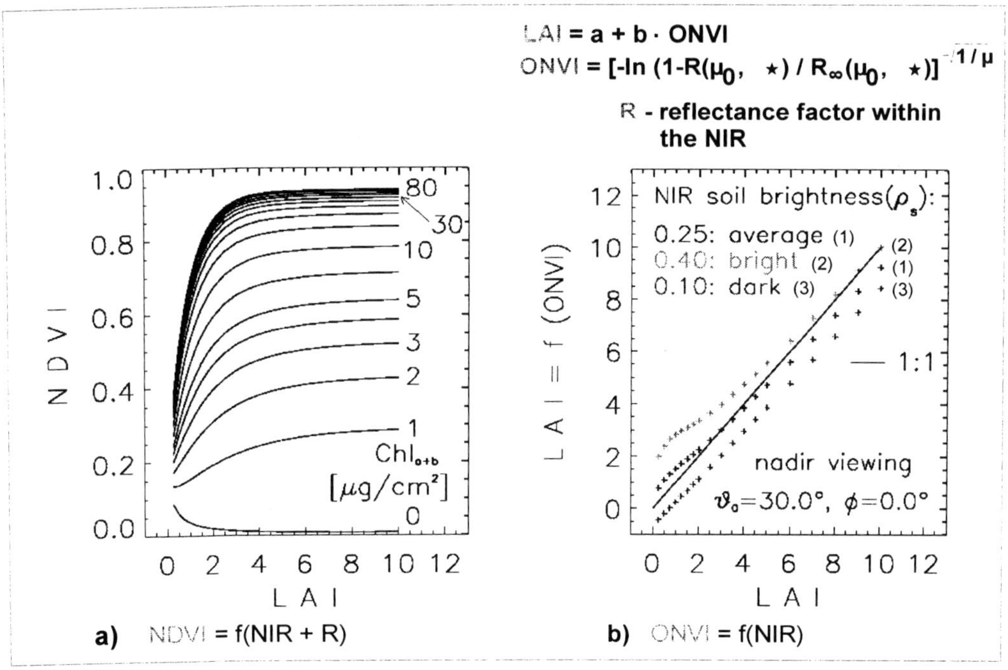

Figure 2. Determination of the Leaf Area Index (LAI).

As the Fraction of the Absorbed Photosynthetically Active Radiation (FAPAR) depends on the NDVI as well as the LAI it is also possible to give a new approach for the estimation of FAPAR and phytomass production using the nadir and off-nadir NIR-channels for calculation the LAI and the red and NIR channels for the determination of the vegetation index.

The measurement of the reflected radiation by VegSAT will provide information on reflectance factors at the nadir- and off-nadir angles. Using these reflectance factors corrected with respect to the atmospheric influence the Bi-directional Reflectance Distribution Function (BRDF) can be generated for various vegetation types (forests, grassland, sparely vegetated areas). The data will improve vegetation classification and the models estimating the natural and antropogenic induced effects on the biosphere and climate. In this connection the NIR-channel is of particular interest, because in contrast to the visible region the BRDF within the NIR is only slightly smoothed by atmospheric scattering and the polarisation of the reflected light in this region is low. On the other hand the radiance in the NIR has a high variability in dependency on the type of vegetation (s. Figure 1).

In a cloudy atmosphere the WAOSS-V of VegSAT will measure the reflectance factors of the cloud layer in the red and NIR. Using the ratio of the radiances in the NIR and red spectrum we get the possibility of differentiating between cloudy and clear areas (s. Figure 1). On the other hand the reflectance factors of the clouds allow to obtain information on the BRDF of clouds and to study the heterogeneity of the different cloud types and the variability of cloud albedo in time and space.

Finally the VegSat will provide data to determine the properties of the coarse aerosol mode over selected areas. This information can be used for the atmospheric correction as well as for climatic and environmental studies.

Table 2. Major technical and performance data of the instrument WAOSS-V.

Parameter	Data
Focal Length (in vacuum)	21.65 mm ± 0.30 mm
Field of View (FOV), across track	± 25.0°
FOV, along track	-60°, -40°, -20, 0°, +20°, +40°, +60°
Aperture number	2.8
Instantaneous Field of View	1.1 arcmin
MTF within the total FOV	50% @ 33 lp/mm >10% @ 60 lp/mm
Geometric distortion	< 0.5% (goal: <0.1%)
Vignetting	<10%
Spectral channels: @ -20°, +20°	600-670 nm
@ -60°, -40°, -20, 0°, +20°, +40°, +60°	840-890 nm
Number of optics with 3 CCDs each	3
Pixel dimension	7 μm x 7 μm
Total number of pixels per CCD	5184
Exposed pixels	2884
Radiometric resolution	8bit
Dynamic range	11bit
Absolute radiance inaccuracy	≤ 5%
Relative precision inaccuracy	≤ 1% non-linearity
Band-to-band precision (relative) inaccuracy	≤ 5%
Data compression	2...20; DCT, JPEG
Computer performance	20MIPS
Mass	8kg
Power consumption	18W

LAUNCH SYSTEM

The low-cost requirement is the design driver for the mission design. Only a piggyback launch option can meet the low-cost requirement on the launch. For this reason the microsatellite must be compatible to the auxiliary payload adapter of several launcher like Ariane (A.S.A.P.), Zenit, Zyklon, Cosmos, Delta II, PSLV and other. That means the dimensions of the spacecraft should be less than about 500x500x800 mm^3 and the total mass less than 100 kg.

SPACE SEGMENT

Payload

The payload follows the mission objectives in Table 1 and the derived mission requirements. It consists in a wide angle camera system with 3 camera heads looking nadir and off-nadir in 7 different phase angles and with 2 spectral bands.

The heritage of the proposed payload was the Wide-Angle Optoelectronical Stereo Scanner WAOSS developed by the German Institute of Space Sensor Technology of DLR (Sandau and Oertel, 1992) for the Russian Mars-96 mission.

Some characteristics of the modified vegetation sensor WAOSS-V are summarised in Table 2. The imaging principle and the different viewing angles of the of the payload for the multi-phase angle observation is depicted in Figure 3.

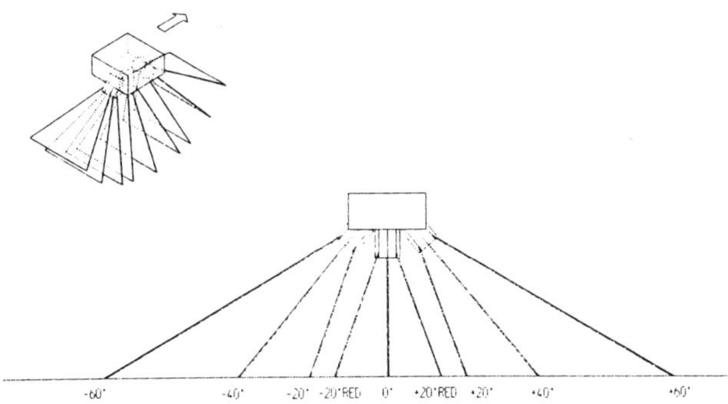

Figure 3. Multi-phase-angle imaging principle and viewing angles for the investigation of vegetation.

Spacecraft Bus

The microsatellite mission is characterised by a three axis stabilised satellite with a size of 450 mm x 450 mm x 550 mm and a mass of 50 kg. The availability of a piggy-back launch is an important design driver for the spacecraft. Table 3 gives a summary to the spacecraft configuration. The main body is a cubic box compatible to most of the launch adapters for secondary payloads. 3 sides of the spacecraft are completely covered with solar arrays and 1 side bears the instrument and antenna platform. The remaining 2 sides are occupied by solar arrays and radiators.

Spacecraft Configuration

The spacecraft represents a microsatellite design with a mass of 50 kg within a volume of 130 dm³. Due to the operation constrains one of two opposite sides of the rectangular solid carries the main payload WAOSS-V oriented to the Earth and the other the solar panel system oriented to the Sun.

To reach the required solar radiant area two side walls will be deployable from start to flight position as shown in the Figure 4. So an solar cell area of about 0.65 m² is available after deployment.

Table 3. Spacecraft volume, mass and power budgets.

SPACECRAFT SYSTEMS	**Remarks**	**Dimensions Mm × mm × mm**	**Mass kg**	**Power W (peak)**	**Power W (avg.)**	**Duty**
Payload						
WAOSS-V	Camera Heads	265 × 268 × 167	5,0	18,0	4,5	0,25
	Electronic Unit	265 × 268 × 160	4,0			
Structure						
Spacecraft structure elements	2 Plates	435 × 550 × 20	4,0			
	6 Supports	dia.22 × 160	2,0			
Eject interface	Ring Structure	dia.240×60	2,0			
AOCS						
3 Reaction Wheels	0,1-2,0Nms	70×80dia.	3,0	3,0	3,0	1,00
3 Magnetic Torquer	max.15Am2	400×18dia.	1,2	4,8	0,5	0,10
Star Sensors	Sensor Head	80×80×45	0,5	1,0	1,0	1,00
	Electronic Unit	120×80×40	0,5			
Magnetometer		60×60×120	0,4	2,0	2,0	1,00
Gyro-Assembly		120×120×120	2,0	2,0		
GPS Receiver	RPU	127×241×56	1,6	3,5	0,9	0,25
	Preamplifier	97×83×12	0,2			
	Antenna	96×102×13	0,2			
Power System and Harness						
Solar Arrays (GaAs: 0,53m2)	3 plates	400(435)×550	3,0			
2xNi-Cd battery pack, 28V	2 x 2,4A-hrs	226×160×60	4,8			
Power regulation			1,9	8,0	2,0	
Power control unit		300×300×40	1,5	14,0	5,0	
Wiring mass			1,0		0,0	
Thermal Control System						
MLI, Coating			0,5			
Radiators	2 Plates	160×380	0,5			
OBDH						
2× DPU + Mass memory	0,5Gbit	200×160×40	1,5	6,0	6,0	1,00
TT&C						
2x S-band transmitter	max. 1Mbps, 5W	150×86×35	1,0	22,0	5,5	0,25
S-band receiver	max. 4kbps	170×150×40	1,0	2,8	0,3	0,10
2x Command receiver (UHF)	150 to 400Mhz	135×165×32	1,8	0,7	0,2	0,25
2x S-band antenna		dia.63×70	0,3			
2x UHF antenna		dia.7×500	0,8			
Margin			1,9	3,0	2,5	1,00
TOTAL MASS AND POWER BUDGET			**48,0**	**90,8**	**33,3**	

The spacecraft structure mainly consists of the front and the back instrument plates (reinforced carbon fibre construction) connected by 6 cylindrical supports. The instrument units mounted between are attached to each of the instrument plates. So a rigid core compartment of the satellite can be built.

The ejected adapter is fixed on the below front of this part to spread the launch loads to the structure elements. Also on this side and on the top the heat radiators are foreseen shadowed from sun light by the small solar panels for the initialisation of S/C-positioning and emergency modes.

The arrangement of the pay-load and bus units can be taken from the Figures 4 and 5.

The solar panels are based on self-sustaining structures. The completed bonnet-shaped system will be integrated on the VegSAT after surrounding with multi-layer-insulation (not shown in the figures).

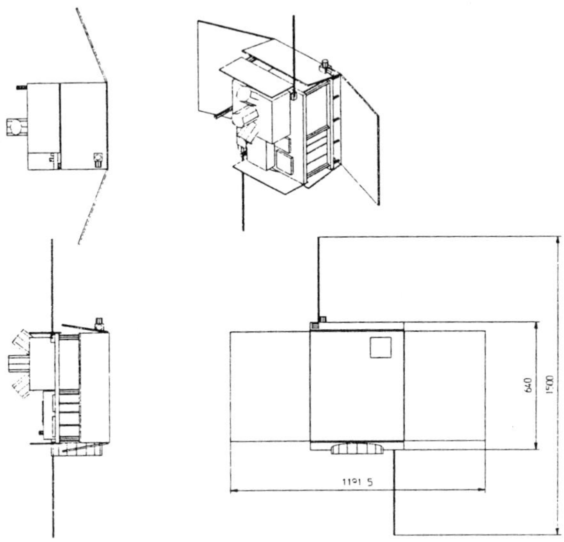

Figure 4. VegSAT- flight configuration.

GROUND SEGMENT AND COMMUNICATION ARCHITECTURE

The ground segment consists usually of three main components:

the ground station(s),
the mission control centre,
and the scientific data processing centre.

The satellite is able to store on-board 0.5Gbit. A data compression is foreseen to collect data of 2 orbits and send it down to a ground station. The mean contact time is sufficient for downlink of the data of the current orbit and the half of the mass memory data, equivalent to one additional orbit using a compression factor 3.

The main ground station should realise all bidirectional communication tasks between the mission control centre and the satellite. The scientific data processing centre should be connected with the prime TT&C ground station and the mission control centre.

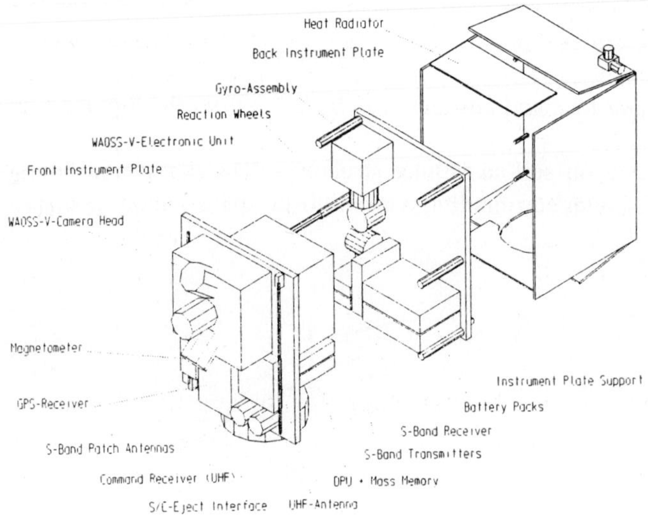

Figure 5. VegSAT- Integration scheme.

MISSION OPERATIONS

The payload has a duty time of 9:00 to 15:00 local time. The mission operations are dedicated to the payload operations. The spacecraft and the payload should be controlled by the mission control centre (MCC) via the prime TT&C ground station only. All operational needs should be collected by the MCC which will establish the mission operations plan.

The mission operations are characterised by 5 operation modes of the payload and spacecraft (Figure 6).

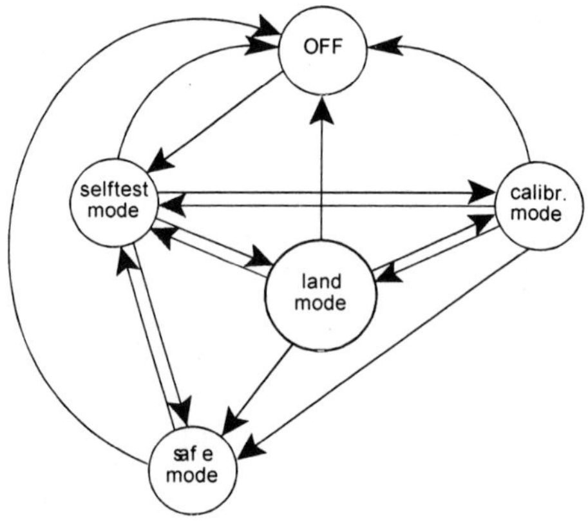

Figure 6. Reduced state graph of operational modes of the VegSAT.

The following short view over the different modes shows the feasible close relation between payload and spacecraft mode for microsatellite missions.

OFF	The payload is within the power-off mode and the spacecraft is in a low power mode.
selftest mode	The selftest mode include three main activities: the satellite check-out, the initialisation of the spacecraft and the check-out and initialisation of the payload.
calibration mode	The imager payload looks to the dark sky and to certain calibration areas by changing of the satellite attitude.
land mode	This is the main mode of the spacecraft. The satellite is nadir pointing, the payload in land mode and the telemetry in transmitting mode. Different payload regimes allow different levels of parameter control. The integration time can be controlled by command for avoiding saturation effects.
safe mode	If it is an unexpected situation or if the spacecraft detect a failure the spacecraft goes into a safe mode. That means the spacecraft turns the solar arrays into the sun direction and the telemetry is in a command reception mode. Periodically spacecraft housekeeping data will be sent.

SUMMARY

- The proposed Vegetation Imager instrument for proposed the Veg-SAT mission could supplement the planned MISR experiment on EOS-1. Especially the time coverage and the area coverage of the multi-phase data of the MISR can be improved essentially by a the Veg-SAT mission.
- A dedicated spacecraft design, driven by the scientific objectives, the orbit, the instrument and the launcher requirements leads to the result, that a microsatellite mission with the proposed imager WAOSS-V (Vogt 1997) seems to be feasible (spacecraft mass <50kg, dimensions less then 500mmx500mmx650mm). A main advantage of the microsatellite solution consists in the low launch costs as a secondary launcher payload.
- One of the peculiarities of the presented Veg-SAT mission is the planned opportunity of the direct reception of the vegetation data by users without delay. The single data user needs a steerable S-band antenna, a S-band receiver and data processing and archiving hard- and software. The hardware can be a PC in the most simple case and the software for antenna steering and data processing should be standardised in the user community.
- The received GPS-data by the satellite will be used for transmission of the ephimeris data to the user. Additional the attitude data can be transmitted. Then the user can use them for geo-referencing or geo-coding of the received data. In this case the user|日 data processing is completely independent from the payload control centre and other payload ground stations.
- The presented paper should give a feeling about the feasibility of an advanced vegetation multi-phase imaging mission by means of an microsatellite and should show, that a possible implementation can fill up some gaps in the area and /or time coverage of the other existing and planned space remote sensing systems for vegetation analysis and research.

REFERENCES

Sandau, R., Oertel, D., The Wide-Angle Optoelectronic Stereo Scanner WAOSS for the Soviet Mars 94/96 Missions, Proc. of the Third International Conf. SPACE 92, 2241-2251, Mai 29 - June 5, 1992 , Denver, Colorado, USA

Vogt, P., Retrieval of the leave area index and the absorbed, photosynthetic active radiation from bidirectional reflectance factors at canopy level, DLR-Forschungsbericht 97-25 (1997)

A STUDY ON KA BAND RAIN ATTENUATION FOR LEO SATELLITE OVER TAIWAN AREA

Yen-Hsyang Chu and Shun-Peng Shih

Institute of Space Science, National Central University, Chung-Li, Taiwan, R.O.C.

ABSTRACT

It is recognized that rain attenuation is a primary factor in the degradation of the Earth-satellite communication at the frequency of Ka band. In this paper, on the basis of 8 years (1988-1995) rainfall rate data at one-minute time resolution recorded by more than 70 tipping bucket rain gauges distributed over Taiwan island, the spatial distribution of the long-term statistics of rainfall rate is investigated. The results show that there is a pronounced latitudinal variation in the percentage of the time that the rainfall rate exceeds a specified level, indicating that more severe rain attenuation will be encountered in the southern part than that in the northern part of Taiwan. In addition, the sky noise temperature at the frequency of 19.5 GHz is also measured by using a radiometer both in the conditions of clear-air and precipitation. In the case of clear-air sky, the observed sky noise temperature, increasing exponentially with the decrease of the zenith angle in an excellent agreement with the theoretical prediction, at the elevation angles of 90°, 60°, 30°, and 15° are respectively 40K, 46K, 72K, and 128K. For the case of precipitation, the comparison between observed sky noise temperature and surface rainfall rate is made, showing a good correlation between them.

INTRODUCTION

According to the national fifteen-year space program implemented by the National Space Program Office (NSPO), a low-earth orbit (LEO) experimental satellite, ROCSAT-1, is scheduled to launch at the end of 1998. The satellite will orbit the Earth at an altitude of 600 km with an inclination of 35° and an orbital period of 97 minutes. The frequency of the downlink beacon signal of the ROCSAT-1 is 19.5 GHz. The frequency of occurrence that the beam downlink telemetry signals of collected data to local receiving stations is about 6 times per day. There are three payloads mounted on the ROCSAT-1. First one is the Ocean Color Imager (OCI) for the remote sensing of the pigment distribution in the low-latitude oceans for the oceanography research. Second payload is the Ionospheric Plasma and Electrodynamics Instrument (IPEI) used for the measurements of ionospheric parameters such as plasma density, plasma temperature, electric field strength, plasma drift velocity, and so on. Third payload is the Experimental Communication Payload (ECP) designed for the communication experiments, in which the Ka band propagation experiment is included. It is generally recognized that the electromagnetic waves at the Ka band are susceptible of the weather-related precipitation, impairing the quality of the Earth-satellite communication (Hogg and Chu, 1975). Except for the rain attenuation through the absorption and scattering effects, the absorption by oxygen molecules and water vapor, scattering and diffraction by the atmospheric refractivity irregularities, and abnormal refracting by the stratification of atmospheric structures play the important roles in the degradation of the earth-satellite communication link. In order to design an optimal earth-satellite communication link over Taiwan area, the environmental effects influencing the quality of satellite communication mentioned above should be investigated thoroughly. The plausible objectives of the Ka band propagation experiment in ECP are to study the characteristics of rain attenuation, behavior of scintillation caused by refractivity fluctuations, depolarization effect on Ka band signal due to non-spherical rain drop, development of the techniques of overcoming rain attenuation, multi-path and low-elevation propagation effects.

Because a LEO satellite continuously changes its angular position (or range with respect to a ground receiving station) during its course of the pass, it is expected that the free-space path loss, a vital factor in the performance of LEO satellite, will be varied accordingly. In addition, the effects of the rain attenuation and gaseous absorption are also the essential factors required to take into account in the determination of LEO satellite communication link. Their effects on the characteristics of the EM wave propagating in the atmospheric channel for a LEO satellite should be investigated thoroughly. The elevation angle variation in the Ka band transmission loss for a LEO satellite has been studied by Ippolito and Russell (1993). It shows that as the elevation angle varies, so does the length of the radio transmission path and the free space loss incurred on the path. After analyzing the annualized outage probability at Ka band derived from rain statistics for a single pass of a LEO satellite, Ippolito and Russell (1993) concludes that the link reliability varies drastically over the course of a LEO satellite pass and the outage probability averaged over the entire pass should be considered as an essential design criteria.

In this paper, the long-term statistics of the rainfall rate at the time resolution of one minute from 1988 - 1995 observed by using more than 70 ground-based tipping bucket rain gauges over Taiwan island are analyzed and investigated. The results indicate that a latitudinal variation in the percentage of the time that the rainfall rate exceeds a certain level is found, it is larger in the southern part of Taiwan and smaller in the northern part of Taiwan. This feature implies that the rain attenuation effect on the Ka band propagation is more severe in the southern part than in the northern part of the island. Furthermore, in order to estimate the possible impairment on LEO satellite communication link due to rain attenuation, the variation in the rain attenuation at the Ka band with the zenith angle is also studied in this article. In addition, the comparison between the rainfall rate measured by a high performance optical rain gauge and the sky noise temperature observed by a radiometer centered at 19.5 GHz is also made in this article.

STATISTICS OF RAINFALL RATE OVER TAIWAN AREA

The data employed for the statistics analysis are the rainfall rates at one-minute time resolution collected by more than 70 tipping bucket rain gauges distributed over Taiwan island from 1988 to 1995. From the cumulative distribution of the rainfall rate, the percentage of time in a year rainfall rate is exceeded can be obtained. Figure 1a shows an example of the cumulative distribution of rainfall rate recorded by the tipping bucket rain gauge located at Chung-Li station (24.9°N, 121.5°E). It indicates that the percentages of time rainfall rate exceeds 20, 50, 80, and 100 mm/hr are respectively about 0.12% (10.51 hr), 0.033% (2.88 hr), 0.006% (31.5 min), and 0.0038% (20 min). In order to examine the quality of the statistics of rainfall rate recorded by tipping bucket rain gauge, the rainfall rate observed by another kind of rain gauge, i.e., optical rain gauge, is employed for the comparison. Figure 1b presents the cumulative distribution of rainfall rate measured by optical rain gauge from November 1, 1996 to October 31, 1997. Comparing Figure 1a and 1b reveals that the general behavior of the statistics obtained by tipping bucket and optical rain gauge is very similar, irrespective of the slight difference in the curve shape and the magnitude of the percentage between them. This result implies that the rainfall rate date collected by tipping bucket rain gauge have good quality and the statistics thus obtained is acceptable.

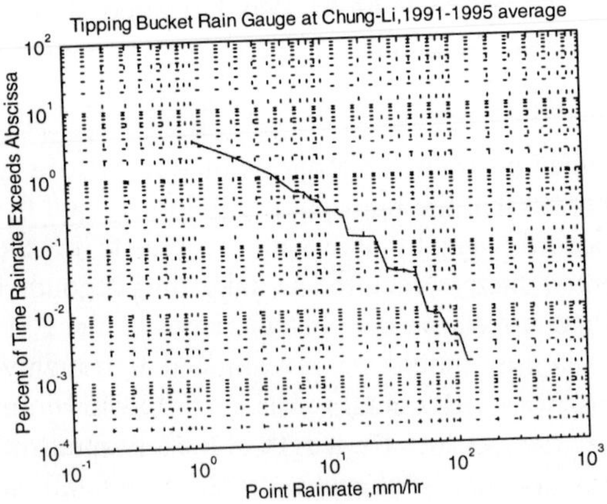

Figure 1a. An example of the cumulative distribution of rainfall rate recorded by the tipping bucket rain gauge located at Chung-Li station (24.9_N,121.5_E) from 1988 to 1995.

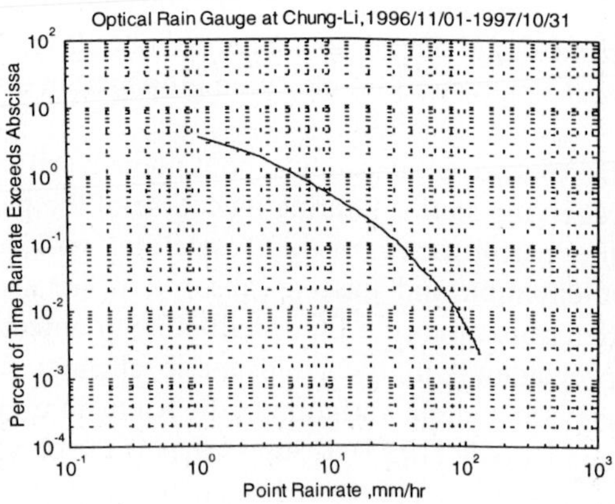

Figure 1b. Cumulative distribution of rainfall rate measured by optical rain gauge located at Chung-Li from November 1, 1996 to October 31, 1997.

The magnitude of rain attenuation A for terrestrial and earth-satellite links can be calculated from the point rainfall rate R_p in accordance with the following equation

$$A = L(R_p)\,\gamma\,(R_p) \tag{1}$$

where $\gamma(R_p)$ is the specific rain attenuation in the unit of dB/km and L(Rp) is the effective path length. The relation between γ (R_p) and R_p has been shown empirically and theoretically that it can be taken in the form of (Olsen et al., 1978)

$$\gamma\,(Rp) = aR^b \tag{2}$$

where a and b are the functions of radio wave frequency, drop-size distribution, and rain temperature. A number of investigators have examined the power-law relation as shown in equation (2) and obtained the empirical values of a and b (Harrold, 1967; Semplak and Turrin, 1969). In the case of Laws and Parsons drop-size distribution and 20°C rain temperature, the corresponding values of a and b for frequencies 20 and 30 GHz are 0.0699, 1.10 and 0.170 and 1.075, respectively (Crane, 1980). An empirical relation connecting effective path length $L(R_p)$ with the elevation angle of satellite θ and the point rainfall rate R_p has been obtained by Rogers (1980) as shown below

$$L(Rp,\ \theta) = [0.00741R_p^{0.766} + (0.232 - 0.00018R_p)\sin\theta\,]^{-1} \tag{3}$$

The total amount of rain attenuation A can thus be calculated in accordance with equation (1) ~ (3). It is noteworthy that equation (3) is obtained through the long-term statistical analysis and is inappropriate to estimating the instantaneous rain attenuation of the satellite signal. On the basis of the long-term statistics of rainfall rate recorded at Chung-Li, figure 2 presents the variations of the rain attenuation with the elevation angle of satellite for the specific point rainrate 5.5, 32.5, and 88 mm/hr, which correspond respectively to the percentage of time 1.0%, 0.1%, and 0.01% in a year rainfall rates are exceeded. As shown, a considerable variation in the rain attenuation with the elevation angle is seen, changing from 17 dB at 70° to 39 dB at 10° for the rainfall rate of 32.5 mm/r and from 31 dB at 70° to 58 dB at 10° for 88 mm/hr. In view of large variation in rain attenuation with elevation angle, the propagation margin should be large enough to achieve high link reliability in designing the communication link for a LEO satellite.

Taiwan is an island with rough terrain, two-third of the island is covered by the huge mountains and small hills. The altitude difference between the highest and lowest points in the island is about 3900 m. Because the orographic effect combined with the wind pattern associated with the monsoon greatly governs the behavior of the rainfall, it is expected that the statistical characteristics of the rain over Taiwan area will be different with the location. With employing the rainfall rate data recorded by more than 70 tipping bucket rain gauges distributed over Taiwan island, the spatial distribution of the rainfall rate can be obtained. Figure 3a, 3b and 3c present the spatial distributions of the specific rainfall rate for the percentages of the time of the year 0.01%, 0.1%, and 1.0%, respectively. Comparing figures 3a, 3b and 3c indicates that for the case of intense rainfall rate (i.e., 0.01% percentage of time) the rainfall rate in the southern part of Taiwan is greater than that in the northern part of Taiwan, while for the light rain case (i.e., 10.% percentage of time) the rainfall rate in northern part of Taiwan is larger than that in southern part of Taiwan. Because the rain attenuation is directly related to the intensity of the rain, the feature of the latitudinal variation of the rainfall rate over Taiwan area reveals that the reliability of earth-satellite communication link operated at Ka band in the northern part of Taiwan will be much higher than that in the southern part of Taiwan.

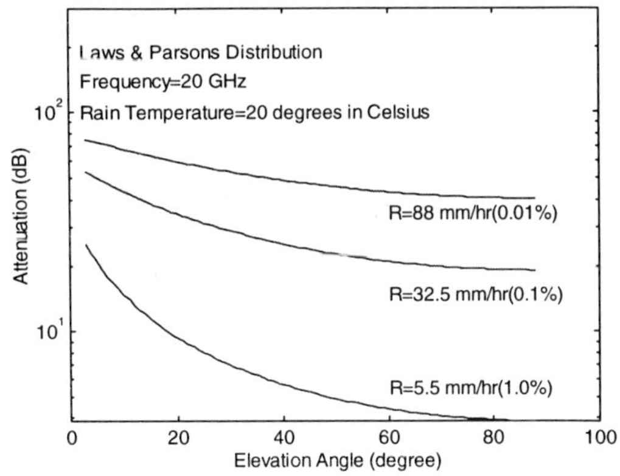

Figure 2. Variations in the estimate rain attenuation with the elevation angle of satellite for the specific point rainrate 5.5, 32.5, ad 88 mm/hr , which correspond respectively to the percentage of time 1.0%, 0.1%, and 0.01% in a year rainfall rates are exceeded. The longterm statistics of the 8 years rainfall rate used for calculation is taken from the data collected at Chung-Li.

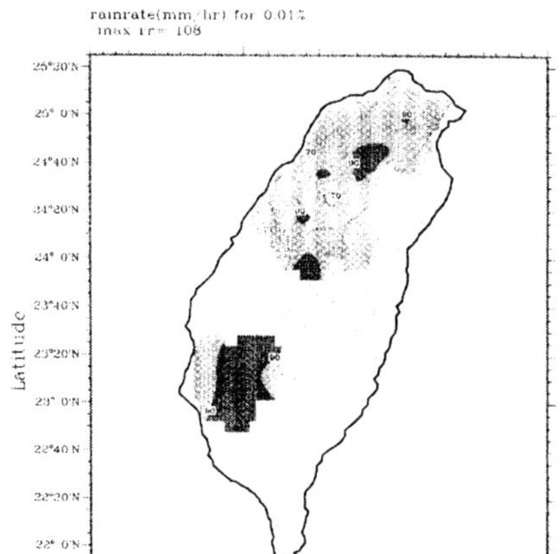

Figure 3a. Spatial distributions of the specific rainfall rate corresponding to the percentages of the time of the year 0.01% rainfall rate exceeds.

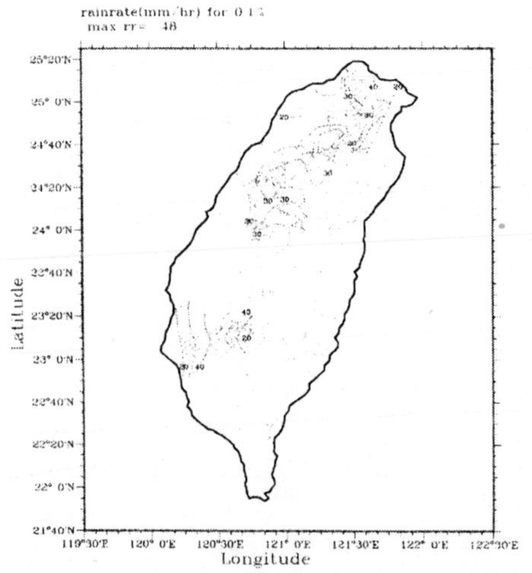

Figure 3b. Same as figure 3a, but for the percentage of time 0.1% rainfall rate exceeds.

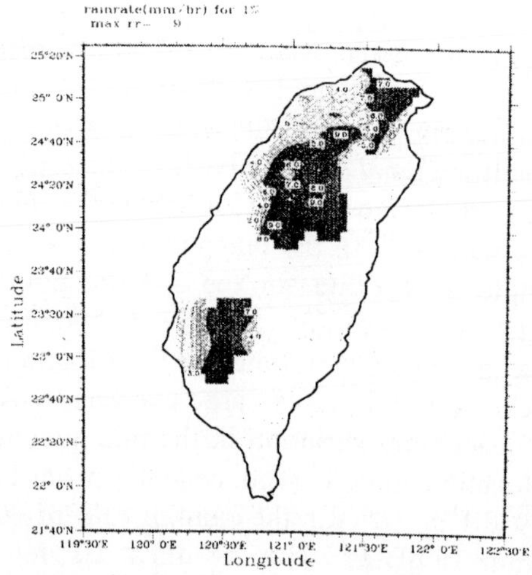

Figure 3c. Same as figure 3a, but for the percentage of time 1.0% rainfall rate exceeds.

RADIOMETER MEASUREMENTS

Radiometer is an important instrument in the investigation of the propagation effects on the microwave communication channel, especially for the rain attenuation and gaseous (water vapor and oxygen molecule) absorption effects. A well-calibrated radiometer at the central frequency of 19.5 GHz was set up on the campus of National Central University at Chung-Li in the spring of 1997 to measure the background sky noise temperature and the radiation from the precipitation particles. The characteristics of the radiometer is listed in Table 1. Figure 4 presents the variation of the measured clear-air sky noise temperature with the elevation angle of the horn antenna of the radiometer, in which asterisk points are the observed values and the solid curve is the theoretical result. To calculate the theoretical clear-air sky noise temperature for slant path, we follow the method suggested by Ippolito (1986) by assuming the height variation of the concentrations of the water vapor and oxygen molecule to be exponential and the scale height of water vapor and oxygen molecule to be 2.8 km and 8.6 km, respectively. As indicated in figure 4, the theoretical calculation in an excellent agreement with the observations and the observed sky noise temperatures at the elevation angles 90°, 75°, 60°, 45°, 30°, and 15° are respectively 40 K, 42 K, 46 K, 53 K, 72 K, and 128 K.

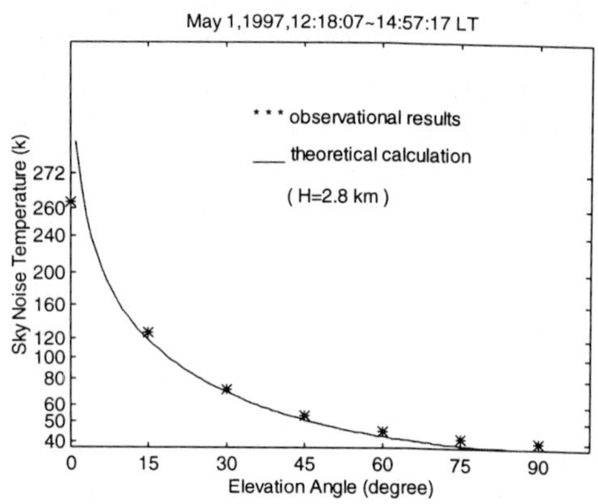

Figure 4. Comparison of the observed clear-air sky noise temperature with the theoretical calculation, where the scale heights of water vapor and oxygen molecule are adopted as 2.8 Km and 8.6Km, respectively.

Table 1.Characteristics of Microwave Radiometer.

Central Frequency		19.5 GHz
Input Bandwidth		1 GHz
IF Bandwidth		500 MHz
Half Power Beam Width of Horn Antenna		6°
Noise Figure		
	Mixer-Preamplifier	4.0 dB
	Isolator	0.3 dB
	Filter	0.2 dB
	Latching Circulator	0.3 dB
	Waveguide	0.2 dB
	In Total	5.0 dB
Integration Time		5 msec
Dynamic Range		0 ~ 350 K

The sky noise temperature generated from the rain drop is also measured in this study. The mathematical relation converting the observed sky noise temperature T_b radiated by the precipitation particles with the temperature T_a into the rain attenuation can be expressed as follows

$$A = 10\log_{10}(T_a/(T_a-T_b)) \tag{4}$$

where A is the attenuation in unit of dB, T_b is the observed sky temperature in unit of K, T_a is the apparent rain temperature and the value of 273 K is used in converting T_b into A in this study as suggested by Hogg and Chu (1975). Investigation shows that T_b increases with the increase of rainfall rate (Hogg, 1989), implying a positive correlation between A and rainfall rate. Figure 5 shows the one-minute resolution time series of rainfall rate recorded by optical rain gauge (middle panel) and the corresponding sky noise temperature observed by the radiometer (upper panel) which is pointed obliquely toward northeast at the elevation angle of 30°. It is obvious that a positive correlation between these two series is seen, irrespective of a time lag of 60 sec (lower panel) between them. After shifting the time axis to compensate the time lag and replotting the scatter diagram of rainfall rate versus sky noise temperature as presented in figure 6, we can obtain an empirical expression relating the rainfall rate R to the sky noise temperature T_b as below

$$T_b = 102.4 + 8.044R \tag{5}$$

where the assumption that the apparent rain temperature is 275 K is made. With the help of equations (4) and (5), the rain attenuation can be estimated from the observation of rainfall rate.

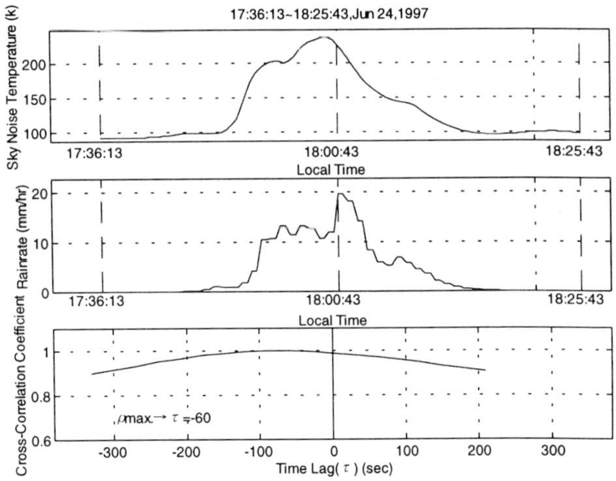

Figure 5. Time series of rainfall rate at one-minute resolution recorded by optical rain gauge (middle panel) and the corresponding sky noise temperature observe by the radiometer (upper panel) which is pointed obliquely toward northeast at the elevation angle of 30°. The cross correlation function between them is presented in the lower panel.

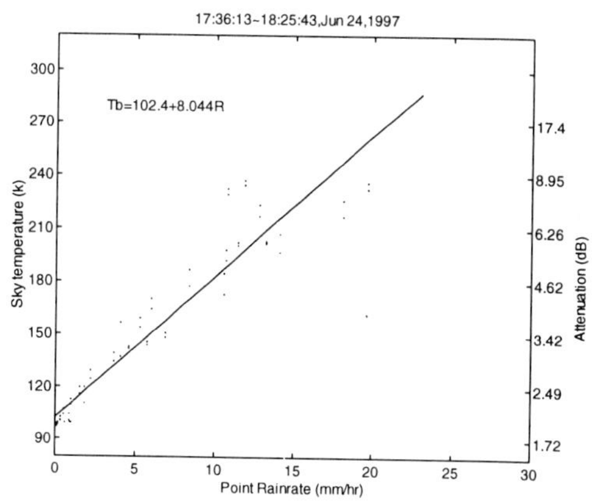

Figure 6. Scatter diagram of rainfall rate versus sky noise temperature, in which the time lag between them is compensated as shown in figure 5.

CONCLUDING REMARKS

The long-term statistical analysis of the rainfall rates at one-minute time resolution recorded by more than 70 tipping bucket rain gauges distributed over Taiwan island from 1988 to 1995 is carried out in this article. It shows that a salient latitudinal variation of the rainfall rate intensity is found, greater in the southern part and weaker in the northern part of Taiwan. Using the 19.5 GHz microwave radiometer and high performance optical rain gauge implemented on the campus of National Central University at Chung-Li, we measure the clear-air sky noise temperature at different elevation angle and the results show that the clear-air sky noise temperature varies from 38 K at the elevation angle of 90° to 138 K at 15°, in an excellent agreement with the theoretical calculation. The relation between the surface rainfall rate and the sky noise temperature from the rain drop is also analyzed and a regression line is obtained.

ACKNOWLEDGMENT

This work was supported by the National Space Program Office (NSPO) of the Republic of Ching under the grant NSC-86-NSPO-ECP-A-008-01

REFERENCE

Crane, R.K., Prediction of Attenuation by Rain, Proc. IEEE Trans. Comm., COM-28, 1717 (1980)

Harrold, T.W., Attenuation of 8.6 mm Wavelength Radiation in Rain, Proc. Inst. Elec. Eng., 114, 201 (1967)

Hogg, D.C., Rain, Radiometry, and Radar, IEEE Trans. Geosci. Remote Sensing, 27, 576 (1989)

Hogg, D.C., and T.S.Chu, The Role of Rain in Satellite Communications, Proc. IEEE, 63, 1308 (1975)

Ippolito, Jr., L.J., Radiowave Propagation in Satellite Communication, Van Nostrand Reinhold Compay, New York, NY (1986)

Ippolito, Jr., L.J., and T.A.Russell, Propagation Considerations for Emerging Satellite Communications Applications, Proc. IEEE, 81, 923 (1993)

Olsen, R.L., D.V.Rogers, and D.B.Hodge, The aR^b Relation in the Calculation of Rain Attenuation, IEEE Trans. Antenna Propagator., AP-26, 318 (1978)

D.V.Rogers, Propagation Analysis Package (PAP-2) Users Guide, COMSAT Labs. Tech. Mem. CL-9-80, January, sec.1.2.2. (1980)

Samplak, R.A., and R.H.Turrin, Some Measurements of Attenuation by Rainfall at 18.5 GHz, Bell Syst. Tech. J., 48, 1767 (1969)

THE HUMBLE SPACE TELESCOPE: THE ACCOMMODATION OF A SMALL ASTRONOMICAL TELESCOPE ON THE MINISIL BUS

A. J. Barrington-Brown, A. K. Ward, A. N. K. Wicks, L. Boland

Space Innovations Limited, The Paddock, Hambridge Road, Newbury, Berkshire, RG14 5TQ, UK

ABSTRACT

Humble Space Telescope mission is intended to provide astronomers with high quality images from a small astronomical telescope placed in Earth orbit. The success of the Hubble Space Telescope is well documented although its inherent drawback is that investigators may have to wait several years until the schedules allow for their observations. Humble is intended to provide a much more rapid response time for users requiring less demanding images than Hubble, but still wanting to overcome the severe limitations placed on astronomers by the Earth's atmosphere. The MiniSIL small satellite bus, designed and built by Space Innovations Limited (SIL), is a semi-standardised system comprising S-band uplink and X-band downlink communications at rates up to 10Msps for the downlink, attitude sensing and control to 0.1° or better, power conversion and distribution, and on-board data handling including sizeable solid state memory devices. It is inherently modular and readily adaptable to different satellite configurations. MiniSIL is an ideal platform for Humble as it combines high performance with relatively low cost. The spacecraft would use the full suite of SIL sub-systems integrated with a 25cm astronomical telescope. The detectors for the spacecraft will allow a wide range of scientific studies to be conducted, and the flexible and low-cost ground system proposed will ensure that data is with the investigator within minutes of it being transmitted by the satellite.

INTRODUCTION

There is currently a large gap between the quality of facilities available to large organisations and the smaller groups of users, such as amateur astronomers and educational institutions, who make do with a limited range of ground-based services. In order to satisfy demands for a better service, which will be of high educational and research value and also to help encourage the growing public interest in the space sector, the Humble project aims to provide a small telescope, on board a MiniSIL satellite bus, coupled with a ground station and readily accessible data distribution channels.

The project was originally conceived by Dr Michael Martin-Smith in response to a national call for proposals for Millennium Projects and is being carried out by Space Innovations Limited (SIL). The UK Department of Trade and Industry (DTI) is already partially financing the Qualification Model of the MiniSIL which is to carry the Humble payload via its Sector Challenge initiative.

Humble will provide images of the universe in the visible and ultra violet spectrum. Images in the visible spectrum taken by Humble's telescope in its low Earth orbit (LEO) will provide a regular and reliable source of high quality data for researchers and amateur astronomers, unencumbered by weather conditions or other atmospheric disadvantages. The UV spectrum images will provide a completely new opportunity for many users to investigate this aspect of the universe.

Figure 1. A graphic representation of Humble in orbit.

The service will be available to anyone with access to the Internet. Users will simply send their requests via e-mail to the Operations Control Centre where imaging schedules will be drawn up. Individuals will receive notification of the success of their application and details of when the imaging is expected to take place.

As is the case with the Hubble Space Telescope and large ground based telescopes, the services of Humble are expected to be heavily over-subscribed. A fixed percentage of observation time will be allocated to satisfy individual amateur astronomers, with another time-slice allocated to schools and colleges and the remainder allocated to higher educational institutions and researchers. Whilst the services of Humble will be available to users across the globe, preference will of course be given to applications received from countries who have contributed to the design and development of the Humble mission. At present the Humble programme is entirely a UK mission.

Humble is being designed for a three-year lifetime, although there are no components strictly limiting it to this life span. During this lifetime it will provide visible and ultraviolet images, as well as supporting an additional range of payloads which are yet to be determined. MiniSIL-P, a variant of the SIL-designed mini-satellite platform, MiniSIL, was chosen for Humble. The design incorporates a load bearing structure with a pair of standard or adapted interface rings, allowing it to be placed in a stacked configuration on launchers of the Pegasus class or larger, or launched on an Ariane 5 ASAP, thus greatly increasing launch opportunities and minimising launch costs. The baseline orbit is assumed to be a circular 800km orbit with an 80° inclination, but the overall design is capable of comfortably accommodating a wide range of orbits. The final orbit will depend on the chosen launch as Humble will not carry a propulsion system to allow for orbit manoeuvres. Attitude manoeuvres will regulate the imager's pointing direction, as it will be spacecraft fixed, as well as providing pointing capabilities during ground passes to allow for greater data rates on the downlink. Images, other payload data and house-keeping data, will be stored in a solid state mass memory and later transmitted to the ground station via the X-band downlink. The uplink will use an S-band frequency. The MiniSIL satellite bus, provides an adaptable semi-standardised platform from which to develop the spacecraft design. Most of the subsystems on the platform are semi-standard or standard off-the-shelf SIL systems and are detailed in the following sections.

THE HUMBLE MISSION

The need for more access to space telescopes is evident, however Humble proposes some more unique advantages to its users. The telescope will be space-based, this has several distinct advantages over its ground-based counterparts: Firstly, it is unencumbered by the Earth's atmosphere, which hinders certain wavelengths such as ultraviolet and also X-ray and gamma rays. It will also provide live pictures to schools during daylight hours. At the higher end of the education scale, students and others carrying out projects with limited time scales will be able to access their desired targets in shorter time scales. The Humble telescope will feature quicker request response time than the Hubble telescope as well as being suitable for targets not deemed important enough for the Hubble's overloaded schedule. It will also help support some of the work left off by the International Ultraviolet Explorer (IUE).

The telescope and satellite performance will limit the overall project capabilities, however much useful work will be carried out and the schedule is expected to be heavily subscribed to. The telescope will be used to obtain images of stars and planets, discover new asteroids or comets (well rewarded by the Benson prize), carry out 1 to 2 minute duration photometry and search for extra-solar-system planets. Some more specific projects, such as the study of putative dust clouds in the Earth Moon L4,L5 points and studies on stars, not interrupted by the Earth's rotation. The coupling of UV and visible wavelengths will allow simultaneous views across frequencies of any targeted objects, and may allow for wide spectrum bolometry and fascinating images to help raise the profile of this field.

HUMBLE SATELLITE PLATFORM

MiniSIL Satellite Structure

Humble's structural design makes use of SIL's MiniSIL, a standard mini-satellite platform suitable for applications in the 100-400 kg range. There are two types of MiniSIL, MiniSIL-P, designed for narrow fairing Pegasus class launchers, or the wider MiniSIL-L, which is suitable for wider launch vehicles allowing an increase in surface area and payload volume. Both these buses are available in spinning or 3-axis stabilised configurations. The current design is based on the former, a 3-axis stabilised MiniSIL-P. Should the additional payloads necessitate a larger volume or power supply the design may be adapted to a MiniSIL-L with relative ease.

The primary structure constitutes an aluminium alloy thrust tube, stiffened by stringers, and a pair of V-band interfaces that make MiniSIL compatible with almost all launch vehicles in use today. The use of a load bearing structure allows the satellite to be launched in a stacked configuration as a secondary passenger and is also adaptable for the Ariane 5 ASAP. This results in a significant reduction in costs whilst retaining its suitability for a wide range of launch opportunities. In the case of Humble, the thrust tube incorporates a relatively large diameter hole to mount the telescope radially, allowing it to look out the side of the spacecraft. This is the more favourable option not only for telescope design but also for power performance profiles available from the solar arrays during typical Humble pointing scenarios. In the case of a nadir pointing imaging mission a typical imager would usually be accommodated axially to point from the upper or lower closure panel which would form the nadir pointing face of the MiniSIL.

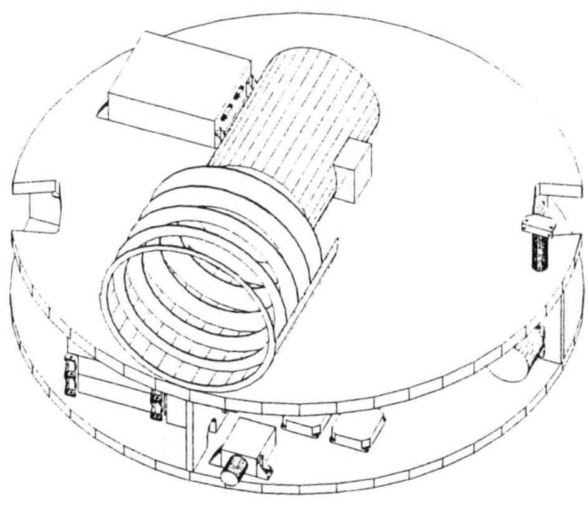

Figure 2. The MiniSIL secondary structure and sub-systems with the Humble payload.

The subsystems and payloads are accommodated on opposite sides of an Aluminium honeycomb shelf. Two more Aluminium honeycomb panels are used to enclose the volume inside the thrust tube.

This set-up allows the payload area to be a mono-volume dedicated solely to the payload and also allows the payload to be integrated separately from the platform sub-systems during the Assembly, Integration and Test (AIT) phase. The third large panel encloses the payload volume, on top of which antennas and attitude sensors may be placed.

The solar arrays are mounted on four side panels around the thrust tube circumference and on the outside surface of the top shelf. The adopted solar array configuration can be clearly seen in Figure 3. The side panels are mounted opposite the telescope aperture.

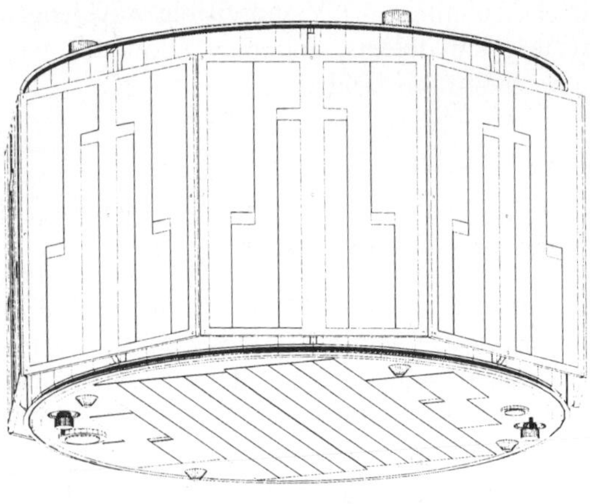

Figure 3. Humble Solar Array Configuration.

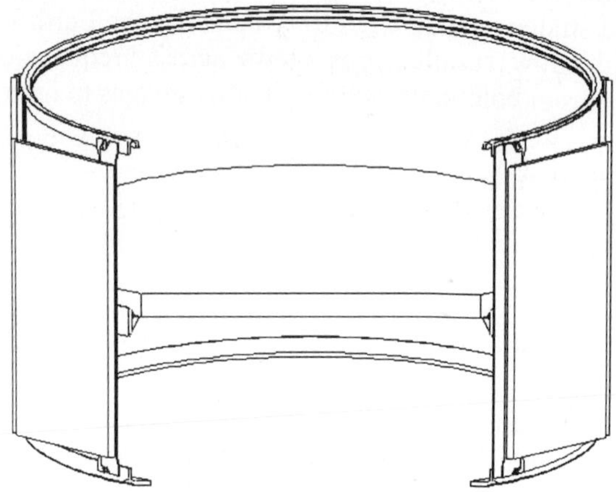

Figure 4. MiniSIL-P basic structure cut-through.

Alternatives using a MiniSIL-L, 2P or 2L may be considered. The MiniSIL-L has shear panels set further out from the thrust tube allowing sub-system accommodation outside the tube and allowing the entire volume within the thrust tube to be dedicated to payload. This version also provides an increased usable solar array area. The MiniSIL-2 series have increased spacecraft height, improving both payload volume and available array power. As already mentioned, however, MiniSIL-P provides more than enough resources to satisfy the Humble imaging payload alone and other versions will be considered only if additional payload requirements dictate such a need. Shown in Figure 5 below is the approximate mass distribution for the Humble spacecraft, including an assumed 50kg of additional payload.

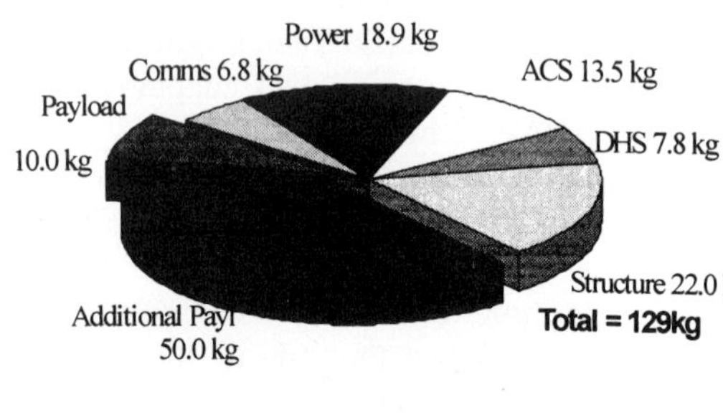

Figure 5. Mass distribution on Humble, including an estimated 50kg additional payload.

Overall Systems Design

The Humble mission utilises a standard MiniSIL satellite bus and subsystems. The various subsystems are, in general, standard SIL designs with most changes in general restricted to the Attitude Control System (ACS), due to its key role in most aspects of the satellite's mission.

The ACS is responsible for pointing the imager and providing pointing knowledge during imaging sessions, as well as pointing the X-band antennas during downlink operations and, finally, ensuring the desired pointing of the solar arrays so as to ensure a constant high power supply. The physical accommodation of the systems to be used in Humble is illustrated in Figure 6 above.

The Humble electrical system block diagram is given in Figure 7 below and a brief description of each of the systems follows.

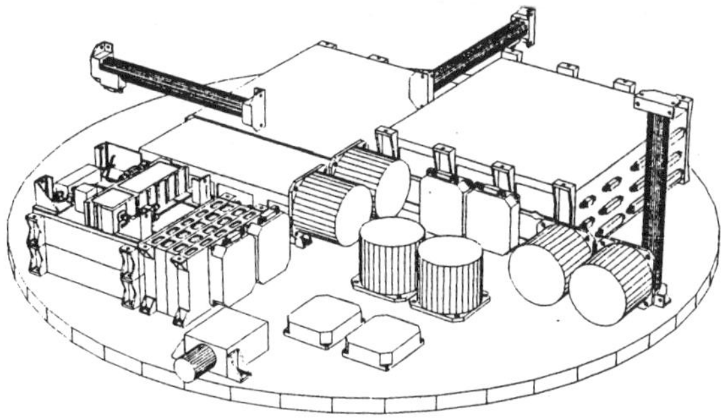

Figure 6. A schematic of the Humble systems shelf showing the various subsystems.

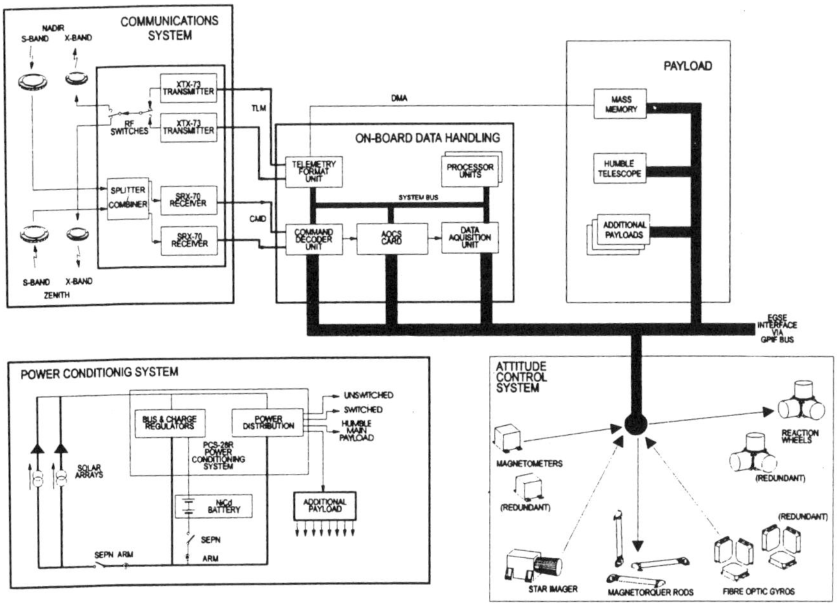

Figure 7. Electrical Systems Block Diagram.

Communications

The uplink uses S-band, whereas the downlink will make use of an X-band link set at around 8500 MHz, allowing the high data rates desirable for the transfer of images and other payload data.

The stored data will be formatted following the CCSDS packet standard recommendations by the DHS Telemetry Format Unit (TFU) which will have a Direct Memory Access (DMA) link to the on-board mass memory enabling the necessary high speed, efficient transfer of data.

The X-band downlink will then relay both the payload data and the housekeeping data via a system using one of two cold redundant XTX-73 3W transmitters and two switchable X-band patch antennas, one nadir and one zenith facing. During these ground passes the attitude control system will ensure a pointing accuracy required for downlink data rates of around 10Msps.

The S-band uplink is a BPSK coded signal modulated on a subcarrier allowing data rates up to 4kbps. The signal will be received by one of two SPA-20 S-band patch antennas, one zenith and one nadir pointing, and follow through to two hot redundant SRX-70 receivers via a combiner/splitter.

The ground segment for the link will be based on the SIL-XGS series of small ground stations. This is covered in more detail later in this paper.

Command and Data Handling

The data handling will be carried out by a DHS-S32 data handling system, at the heart of which are two high performance 32-bit SPARC-based processor units. The system will process direct and time tagged commands and react to house keeping data. The Attitude Control Electronics (SACE) will also be accommodated within the DHS-S32.

The DHS-S32 will manage all on-board housekeeping and payload management duties. In the case of the Humble mission the usual internal Mass Memory Unit of the DHS will be replaced by a larger (5Gbit) separate external Mass Memory Unit. It will process, store and format data conforming to CCSDS Packet Telemetry and Telecommand Standards.

Power System

The backbone of the power system is the PCS-28R/200 power conditioning system (PCS), which will receive power from GaAs solar arrays when sunlit and the lithium ion battery during eclipses.

The PCS will sustain and regulate the power bus to within 2% of its nominal 28V voltage. It utilises a majority voting system of three controllers operating dual hot redundant charge/discharge regulators to supply a protected output to ten switched and four unswitched loads.

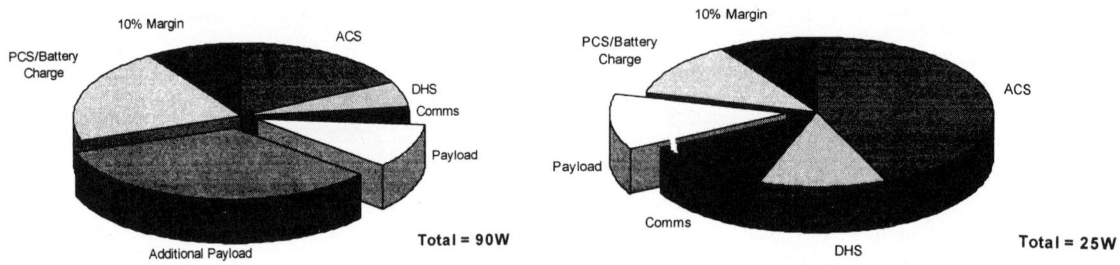

Figure 8. Left: Maximum power consumption during sunlit portion of orbit.
Right: Reduced power consumption while Humble is eclipsed.

The solar arrays, utilising strings of 37 GaAs cells, are mounted on the top side and wrapped around the rear side of the spacecraft. The top array is comprised of 13 strings and four 5-string panels form the rear side arrays, providing a combined output to the PCS of around 106W during peak power attitude at end of life. Due to the continuous attitude control employed by Humble the solar array configuration is independent from the chosen orbit.

The lithium ion batteries to be used provide a 4Ah capacity. Lithium ion technology offers the advantage of relatively fast charge rates and a high depth of discharge without detriment to the battery lifetime. Increased payload power consumption and duty cycles brought on by the additional payloads may necessitate the adoption of batteries with larger capacities and recharge rates.

The above pie charts illustrate some typical power usage scenarios for the Humble spacecraft. During eclipse payload operations are likely to be reduced or suspended for power efficiency reasons, although the present demands of the Humble payload will not preclude eclipse operations. Due to the operational pointing flexibility in the Humble mission it will be possible to provide significant flexibility in payload scheduling, including during eclipse, due to the opportunities to enter a 'recharge mode' with optimal solar array illumination if necessary. Operations during eclipse, however, will almost certainly result in the need to use a larger battery capacity, although until additional payload powers and operational scenarios can be estimated the baseline battery capacity of 4Ah is assumed sufficient for mission requirements.

Attitude Control System

Humble is a three axis stabilised design requiring continuous attitude control for its imaging, down link and recharging operations, as well as a high level of pointing accuracy and knowledge and low jitter needed for the imager. A star mapper aligned with the imager will provide accurate pointing information during imaging sequences, additional attitude information will be provided by fibre optic gyros and a magnetometer. Attitude manoeuvres will be performed by reaction wheels set in the three axes, fully backed by a further cold redundant set. Desaturation of the reaction wheels will be carried out using MTR-25 rod magnetorquers. Use of a GPS orbit determination system is being considered.

The attitude instruments are interfaced with the on-board data handling (OBDH) system via an attitude control electronics (SACE) card, accommodated within the OBDH box in place of the usual mass memory board. This is a departure from the usual configuration where the SACE is accommodated in a dedicated box and treated as a separate unit from the DHS. However, the electrical configuration will not change.

The SSM-20/2 star mapper comprises an optical head with a passively cooled 770x567 element CCD array and a space-rated SPARC processor unit, the latter being currently under development at SIL. It will be capable of a 3σ accuracy of 20 arcseconds within its 16° by 12° field of view with an update rate greater than 2Hz in mapping mode. The MFM-3L magnetometers will provide a field sensitivity of 250nT, although these magnetometers may be used in a Hi-Gain mode to allow greater sensitivity over a smaller range.

The reaction wheels will be desaturated using a combination of three MTR-25 magnetorquers aligned in each axis. Although the magnetorquers are not physically duplicated on the spacecraft, internal redundancy is provided by separate coils. Reaction wheels and fibre optic gyros will not be provided by SIL but are available in suitable form from a number of external suppliers.

PAYLOAD

The Humble telescope is the primary payload for the mission, others will be added to make full use of available resources and aid in the reduction of costs per user.

The Humble Telescope

The Humble payload consists of an imager coupled with a mass memory, used to store the large volume of data associated with images.

The imager itself is still in its early design stages and options are still being looked at; the current design is based around a 25cm (8") telescope with detectors for the visible and ultra violet spectrum. The telescope will be based on a space qualified ordinary Earth-bound astronomical telescope, and therefore be of relatively low-cost. Options include adding another low cost telescope or splitting the signal from one telescope to two detectors.

The images taken will be stored in the mass memory until the next ground pass. The solid state recorder will dedicate most of its 5Gbit for the images and other payload data, whilst allowing a small percentage for use by the satellite bus housekeeping data.

Other Payloads

Opportunities for additional payloads, to raise funds or other benefits for the project, are still open. The telescope will retain privileges of a primary user, thus additional users must be compatible with the Humble mission. The current bus design utilises the smallest MiniSIL variant, MiniSIL-P, although this may be replaced by a larger or higher power type should the need arise. However, the current payload mass of 80kg and peak power use of 60W are well within the MiniSIL-P design specifications.

Examples of expected payloads are debris sensors, radiation sensors, technology demonstration programmes and on-board experiments.

OPERATIONAL PERFORMANCE

Orbit

The orbit requirements for Humble remain flexible within the basic constraints of it being a circular Low Earth Orbit (LEO) with a high inclination; the exact details will depend on the launch opportunity selected. The baseline orbit assumed for design purposes is an 800km circular orbit inclined at 80°.

The orbit will typically allow two or three sunlit ground passes a day, each an average of around ten minutes, during which images and data from the on board experiments will be downloaded via the X-band link. Depending on operation scenarios and power usage it may also be feasible to transmit during eclipsed passes.

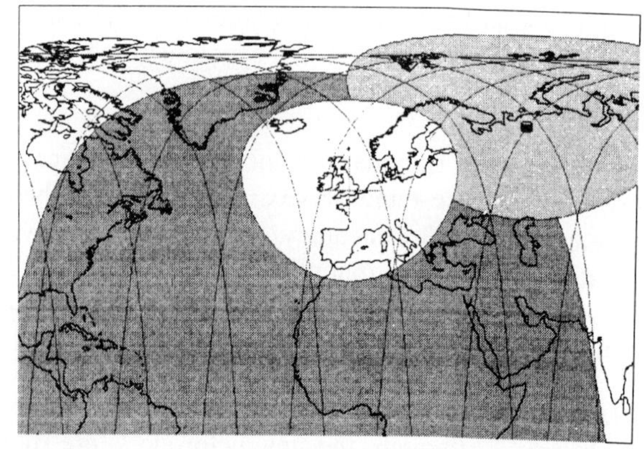

Figure 9. A likely groundtrack assuming 800km altitude and 80° inclined circular orbit. The chart shows the nightime zone, the satellite horizon and the ground mask.

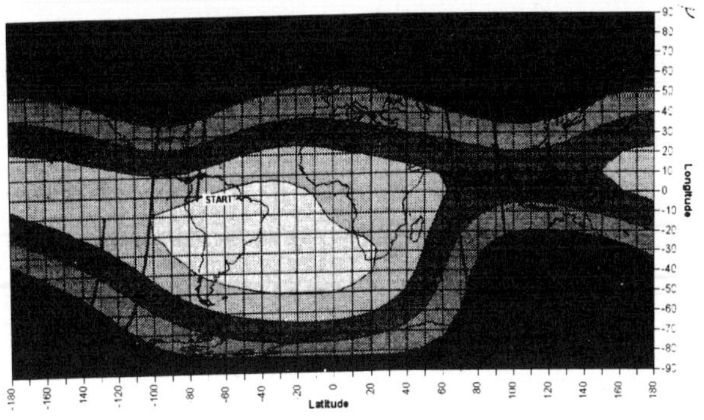

Figure 10. An orbit track together with a plot of the Earth's geomagnetic field strength as will be encountered by HUMBLE in an 800 km orbit, as generated using the SIL in-house software package SILMAG. [The model used is the International Geomagnetic Reference Field (IGRF 95)].

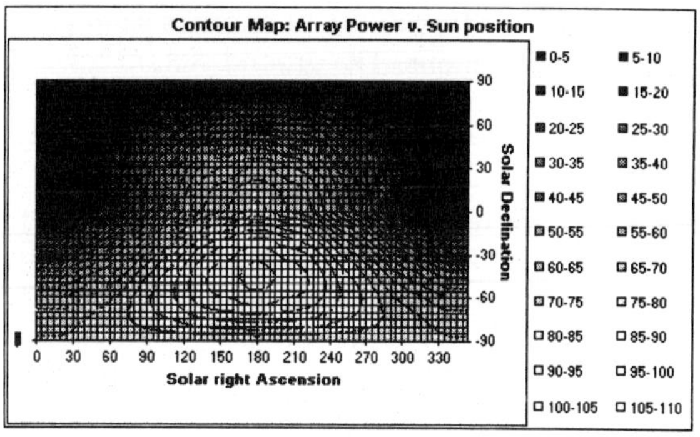

Figure 11. A plot of the power availability with respect to satellite attitude to the Sun.

Imager Operations

The imager is satellite fixed, pointing being achieved by use of the reaction wheels provided with accurate pointing knowledge by the star mapper.

The solar arrays are positioned so as to allow pointing in most directions whilst retaining some power output from the arrays. There are some attitudes relative to the sun which provide very low power, thus will be avoided during ordinary operation. The imager itself has an exclusion zone based around sun pointing angles to protect the sensitive imager. As the orientation of the imager is in an opposing direction to the rear side arrays much of the exclusion zone will overlap the low power attitude positions, hence the combined effect is limited. The regions affected by this exclusion zone will vary as the Earth revolves around the Sun, but will allow full sky coverage in the course of a year. Other constraints may be caused by the albedo from the Earth and the Moon although these will be much more temporary effects. As with other on-board equipment, imager activity may be stopped or limited during periods of eclipse.

GROUND SEGMENT

The ground station to be used for the Humble mission will be one of SIL's XGS series illustrated in Figure 12 below.

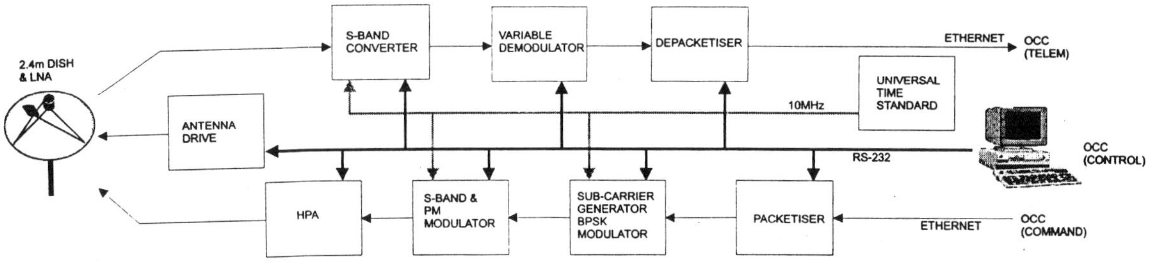

Figure 12. Ground Station Block Diagram.

Much of the heritage of the XGS series stations lies in SIL's currently available series of SGS low cost S-Band ground stations. The ground station for the Humble mission will be based in England and will be purpose built for the Humble mission, linked to an Operations Control Centre (OCC) and Science Operations Centre (SOC).

Housekeeping data will be distributed to the appropriate hosts within the OCC for real-time analysis whilst the payload data will be temporarily archived for processing and distribution after the pass. As soon as the pass is finished the Humble images will be processed in the SOC and distributed by e-mail to the persons who requested the data. Humble images will also be made available via the Internet. In a similar manner to the Humble image data, the data from the additional payloads will be temporarily archived in the OCC and distributed to the concerned parties after the pass.

CONCLUSIONS

This paper has recognised the many benefits of the Humble project and also demonstrated the feasibility of the mission. The project may be realised at low cost by means of a careful design process and use of semi-standardised systems on a standard platform and also by sharing the satellite platform with other payloads. The satellite platform described is capable of supporting a number of other payloads with considerable flexibility, although primary consideration will be given to the needs of the Humble Imager in operations.

The Humble project will provide a valuable addition to present astronomical facilities. The images, both in the visible and the ultra violet wavelengths will not only provide a valuable observing platform for scientific research but will be of immense value to students of all levels at educational institutions around the world. It will provide previously unimaginable opportunities for ordinary members of the general public to actively participate in and benefit directly from a low cost scientific space mission. Such a raised profile will help achieve an increased interest in space activities and pave the way for future similar projects.

A COMPUTER BASED SYSTEM FOR ON-GROUND TESTING OF THE ATTITUDE CONTROL SYSTEM OF SMALL SATELLITES*

V. Alexandrov (*), M. Ovchinnikov (+), D. Bugrov (*), S. Lemak (*), S. Mirer (+), A. Nevidomsky (*)

(*) Lomonosov Moscow State University, Vorobjovy Gory, Moscow 119899 RUSSIA
Fax: +7-(095)939-2090, E-mail: valex@moids.math.msu.su
(+) Keldysh Insitute of Applied Mathematics, 4 Miusskaya Sq. Moscow 125047 Russia
Fax: +7-(095)972-0737, E-mail: ovchinni@spp.keldysh.ru

ABSTRACT

The paper decribes the system realized as a software package for computer simulation, verification and quality testing of attitude control system of small satellites. The system permits to check up the quality of chosen control algorithms and to estimate the influence of faults in determination of current phase vector of satellite owning to errors of sensors, failures in control loops, influence of external dynamical perturbations.

INTRODUCTION

When a satellite is being developed it often occurs that the costs of on-ground stand tests of satellite's control system are comparable with the cost of satellite itself. The situation is especially typical for small satellites that are developed extensively last years. Many countries that just used or bought satellites manufactured in USA, France, UK, and Russia, now intend to construct their own satellites. At present, these countries are facing with necessity of performing stand tests of satellite systems before the satellite is taken into exploitation, in particular the orientation system requires a very expensive testing. Using of on-ground stands is one of possible ways to solve this problem. Alongside with mechanical stands, one of which, for example, is described in [1], the computer based stands are expediently used. It allows to simplify considerably and to make cheaper the process of development and testing of a satellite.

In the Testing System considered here the testing is based on the theory of extremum problems. Extremum situation searching is performed to get an estimate of quality of stabilization algorithms and constructive decisions used, without performing expensive stand testing. The system may operate in two modes: open mode and close-loop mode.

Open mode is used for Passive Attitude Control System (PACS) testing. Closed-loop mode is used for stabilization quality testing when the satellite is provided with an Active Attitude Control System (AACS). In open mode the strategy of the extremum situation searching is realized. This searching takes

* Presented at COSPAR Colloquium on Small Scientific Microsatellites: Microsatellites as Research Tools, 14-17 December, 1997, National Cheng Kung University, Tainan, Taiwan.

place on the set of perturbations that is direct product of the set of initial perturbations and the set of continuous in-path perturbations (such as magnetic, aerodynamic, heat-caused, etc.)

Closed-loop mode is used to test an AACS. In this mode the process of testing is realized as differential game with one player discriminated [3]. The searching of existing saddle point is performed at the preliminary stage of testing. On the final stage the optimal contrstrategy of initial and permanently acting perturbations ia realized, that permits to evaluate the quality of the AACS operating.

The software package is being designed to perform modeling and testing of wide range of space units. It has highly flexible modular structure to meet the needs of easy tuning and reconfiguration for different tested systems. The package consists of the modeling shell that performs mostly nonspecific operations, and the number of loadable modules, which perform the calculations.

The parameters of ACS are suggested to be already chosen. For such PASCs as gravity-gradient, magnetic, aerodynamic it may be performed using simplified models [2]. If the effects of external forces and torques may be approximately periodic functions of time and variables than the methods similar to Liapunov-Floquet are appropriate to optimize the ACS's parameters.

As to AACS is tested then the control system includes an attitude determination system, which processes angular position and velocity sensor's data, and control algorithms. Naturally, ACSs' configuration, control algorithms and actuators used depend on purposes of the satellite.

DESCRIPTION OF THE TESTING SYSTEM

The computer testing system is designed to verify the operation of attitude control algorithms. The scheme of one possible configuration for testing of active magnetic or gyroscopic stabilization system is shown on Fig.1.

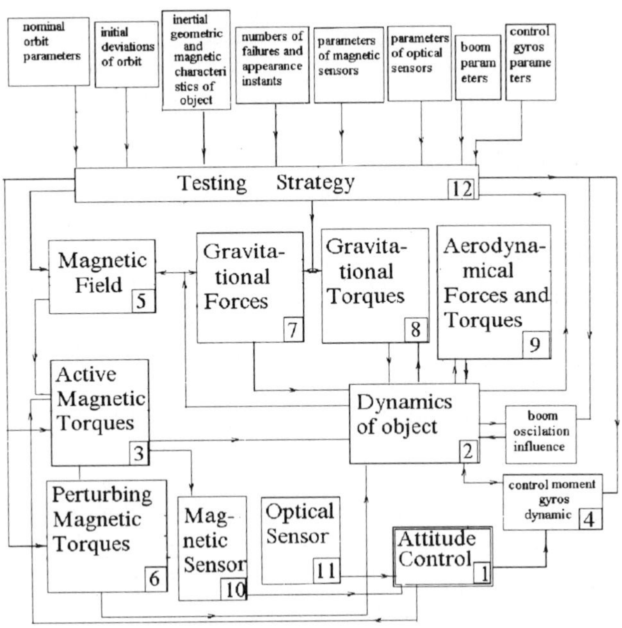

Figure 1. Scheme of Testing System.

The module 1 contains orientation algorithms, which are provided by satellite's designers as a "black box" with known inputs and outputs. The system forms an operating environment for such module by means of system dynamics module 2.

Module 2 numerically integrates the center of mass trajectory and angular dynamics under control and disturbing torques. This moment for purposes of modeling there are used dynamics equations of the system as of the set of rigid bodies including satellite courp, actuators (magnetic coils, flysheels, etc) are used. It is possible to include flexible elements in model - solar arrays, booms, etc.

The control over satellite may be performed using flysheels and/or magnetic coils. The signals for modules 3 and 4 forming control torques are sent from tested module 1. Module 3 uses also an information about magnetic field, which is formed in module 5. The magnetic field may be modeled on different levels of precision depending on the needs of the problem solved by using different number of terms in magnetic potential expansion.

The disturbing magnetic torques are formed in module 6.

Gravitational forces and torques are modeled in modules 7 and 8. Module 9 estimates aerodynamical effects due to atmosphere presence.

Instrumental errors of sensors, errors of executive mechanisms, residual and induced magnetic moments, and the number of possible defects and failures that occur during exploitation, form internal functioning media. It is supposed now that system uses two types of sensors: magnetic and solar. To imitate solar sensors the position of the Sun on the sky is calculated.

The errors of sensors have complex structure - additive, multiplicative, and random components are considered. The statistical description of random components in measured values is used both in modules 10,11 while modeling sensors, and in module 12 while searching worst deviations.

Module 12 coordinates the work of all described modules. It can change parameters in runtime, stop evaluating, and start it again with new data, etc. Input parameters for module 12 include specifications for allowable orbit deviations, mass and geometrical constructional deviations of satellite and its components, sensor information errors, atmospheric, magnetic, and gravitational deviations.

In open mode the current trajectory deviations are analyzed. For stabilization quality estimate the extremum characteristics of uniform stability are obtained with estimate on finite time interval. As shown in [3], it is an extension of known Bulgakov's problem of maximum deviation. In many cases the same approach is applicable to infinite time intervals.

For closed-loop mode the quality estimate is obtained using a differential game with one player discriminated. The discrimination means that the testing system possesses unobstructed information about the tested system behavior though does not know its algorithms. In this case as shown by N.Krasovsky [4] the saddle point exists - it is obtained at the preliminary stage of testing.

Let us introduce following values: u—vector of driving signals of active orientation v—vector of deviations, which spoils the orientation quality. It includes start point deviations, instrumental errors, different types of torques - aerodynamical, magnetic, etc. Let, $\varphi_0(u, v)$ be the orientation precision criterion. Usually the driving signals are formed as a functions depending on time and deflection estimates $u = u(t,\tilde{x})$. As to deviations vector, in case of testing the dependency may be supposed of v vector-function on deflections and on control values $v=v(t, x, u)$. Thus, the maximin problem with some discrimination on control

$$\min_{u(t,\bar{x})} \varphi_0(u,v) \to \max_{v(t,x,u)}$$

is stated. In this case corresponding to Krasovsky theorem the differential game has one saddle point

$$\max_v \min_u \varphi_0(u,v) = \min_u \max_v \varphi_0(u,v) = \varphi_0^0(u^0,v^0) \tag{1}$$

The existence of saddle point permits to introduce the term of "perfect" functioning of active orientation algorithms. Following it the testing procedure reduces to these two steps:

1st step. After the orientation method and dynamics data are offered the differential game (1) is solved that provides the saddle point ("perfect result") and optimal contr-strategy $v^0(t,x,u)$.

2nd step. The testing protedure begins: the raw sensor information is supplied to active orientation module (1) in real time and control signals are obtained from it. These signals are realized providing $\varphi_0(u,v^0)$. The value obtained can be compared with "perfect" one $\dfrac{\varphi_0(u^0,v^0)}{\varphi_0(u,v^0)} \vee 1$, which is optimal for the game.

MODULE OF TESTING STRATEGY

Let us show how the functioning of the testing module 12 performs the active magnetical stabilization system testing. The equations of small oscillations relative to center of mass point in projections on satellite's coordinate axes may be formulated as

$$\begin{aligned} A\ddot{\alpha}_1 + \omega_0(C+B-A)\dot{\alpha}_3 + 4\omega_0^2(B-C)\alpha_1 &= M_1^a + M_1^b + \Delta M_1^g, \\ B\ddot{\alpha}_2 + 3\omega_0^2(A-C)\alpha_2 &= M_2^a + M_2^b + \Delta M_2^g, \\ C\ddot{\alpha}_3 + \omega_0(B-A-C)\dot{\alpha}_1 + 4\omega_0^2(B-A)\alpha_3 &= M_3^a + M_3^b + \Delta M_3^g. \end{aligned} \tag{2}$$

Here A, B, C are inertia moments, $w_0(t)$—angular velocity of center of mass, $\vec{\alpha} = (\alpha_1, \alpha_2, \alpha_3)$—vector of small turn of implied system $Mx_1x_2x_3$ relative to orbital one $My_1y_2y_3$; denoting $\delta = \{\delta_{ij}\}$—matrix of turn of construction axes $Mz_1z_2z_3$ relative to implied system Mx;$u=(u_1,u_2,u_3)$—projections of active torque made by magnetic coils on Mz axes; (I_1'', I_2'', I_3'')—projections of own magnetic moment due to self magnetization; $\vec{T} = (T_1, T_2, T_3)$—projections of the Earth magnetic field intensity on orbital system axes My; $\Delta M_1^g, \Delta M_2^g, \Delta M_3^g$—projections on own axes of those torques, which are described statistically. They can be evaluated, say, as partial Fourier sums

$$\Delta M_i^g = d_{1i} + d_{2i}\cos\omega_0 t + d_{3i}\sin\omega_0 t + d_{4i}\cos 2\omega_0 t + d_{5i}\sin 2\omega_0 t + d_{6i}, \quad i = 1,2,3, \tag{3}$$

where d1i, d2i, d3i, d4i, d15 = const and d6i is a high frequency component of white noise type.

Projections of active magnetic torque on own axes $\vec{u}_x = \left[E - \hat{\delta}\right]u_z$, where $\hat{\delta}$ is a skew-symmetric matrix of the turn vector δ. Let us write the projections of the control torque $\vec{M}^a = \vec{u} \times \vec{T}$ as

$$M_1^a = u_2\left(T_3 - T_2\delta_1\right) + u_1\left(T_2\delta_2 + T_3\delta_3\right) - u_3\left(T_2 + T_3\delta_1\right) + \sum_{i,j}\alpha_i u_j \lambda_{ij1}(\delta),$$
$$M_2^a = u_3\left(T_1 - T_3\delta_2\right) + u_2\left(T_1\delta_1 + T_3\delta_3\right) - u_1\left(T_3 + T_1\delta_2\right) + \sum_{i,j}\alpha_i u_j \lambda_{ij2}(\delta), \tag{4}$$
$$M_3^a = u_2\left(T_3 - T - 2\delta_1\right) + u_1\left(T_2\delta_2 + T_3\delta_3\right) - u_3\left(T_2 + T_3\delta_1\right) + \sum_{i,j}\alpha_i u_j \lambda_{ij}(\delta),$$

$\lambda_{ijk}(\delta)$ denote coefficients of non-linear components of (4).

Similar equations may be written for components of magnetic deviation torque

$$M_1^b = I_2''T_3 - I_3''T_2 + \sum_{i,j}\delta_i I_j'' \lambda'_{ij1},$$
$$M_2^b = I_3''T_1 - I_1''T_3 + \sum_{i,j}\delta_i I_j'' \lambda'_{ij2}, \tag{5}$$
$$M_3^b = I_1''T_2 - I_2''T_1 + \sum_{i,j}\delta_i I_j'' \lambda'_{ij1}.$$

Assuming α_i, δ_j, u_j are small we neglect the non-linear terms in (4) and (5).

Adding to equations (2) the formation equations describing constants in (3) we have

$$\dot{d}_{1i} = 0, \quad \dot{d}_{3i} = \omega_0^2 d_{2i}, \quad \dot{d}_{5i} = 4\omega_0^2 d_{4i},$$
$$\dot{d}_{2i} = d_{3i}, \quad \dot{d}_{4i} = d_{5i}, \quad i = 1,\ldots,3. \tag{6}$$

The equations of controlled system may be written as

$$\dot{y} = A^0(t)y + B^0(t, v^2) + C^0(t)v^3 + D^0 v^4, \tag{7}$$

where

$$y = \left(\alpha_1, \alpha_2, \alpha_3, \dot{\alpha}_1, \dot{\alpha}_2, \dot{\alpha}_3, d_{1i}, d_{2i}, d_{3i}, d_{4i}, d_{5i}\right), \quad i = 1,2,3,$$
$$u = \left(u_1, u_2, u_3\right), \quad v^2 = \left(\delta_1, \delta_2, \delta_3\right), \quad v^3 = \left(I_1'', I_2'', I_3''\right), \quad v^4 = \left(d_{61}, d_{52}, d_{63}\right).$$

The deviation v^4 is modelled as a unit vector white noise.

In case when center of mass position M is known the measurements vector may be represented as $\vec{z} = \vec{T}_z^{m} - \vec{T}^*$, where $\vec{T}_z^{m}$ is a compound vector of measured magnetic field components in point M, T^*— onboard information about Earth magnetic field projections. When the position of the center of mass is known $T^* = T_y$,

$$\vec{T}_z = [E + \Delta][E + \hat{\delta}'][E + \hat{\delta}][E + \hat{\alpha}]\vec{T}_y + \vec{\xi}_0 + \vec{\xi}_1 \sin\omega_0 t + \vec{\xi}_2 \cos\omega_0 t + \vec{\xi}_3, \tag{8}$$

up to the second order of smallness, $\vec{\xi}_0 = \left(\xi_1^0, \xi_2^0, \xi_3^0\right)$ is a systematic errors (sensors zero drift); $\vec{\xi}_1, \vec{\xi}_2$— time-dependent systematic errors components; $\vec{\xi}_3$—high-frequency deviation components modelled as white noise; δ'—vector of small turn of sensitivity axes relative to M_z; Δ—matrix of relative measurements errors Diagonal elements; (Δ_{11}, Δ_{22}, Δ_{33}) are scale errors of magnetometers. Non-diagonal elements are cross-sensitivity of magnetic sensors due to non-orthogonality of sensitive axes.

Projecting (8) on M_x we have

$$\begin{aligned}
z_1 &= \Delta_{11}T_1 + \alpha_3 T_2 - \alpha_2 T_3 + \sum_i \delta_i' \kappa_{i1}^1(T_y) + \sum_i \delta_i' \kappa_{i1}^2(T_y) + \sum_i \delta_i' \kappa_{i1}^3(T_y) + \\
&\quad \sum_{i,j} \alpha_i T_j \kappa_{ij1}^4(\Delta_{ij}, \delta, \delta') + \xi_1^0 + \xi_1^1 \sin\omega_0 t + \xi_1^2 \cos\omega_0 t + \xi_1^3, \\
z_2 &= \Delta_{22}T_2 + \alpha_1 T_3 - \alpha_3 T_1 + \sum_i \delta_i' \kappa_{i2}^1(T_y) + \sum_i \delta_i' \kappa_{i2}^2(T_y) + \sum_i \delta_i' \kappa_{i2}^3(T_y) + \\
&\quad \sum_{i,j} \alpha_i T_j \kappa_{ij2}^4(\Delta_{ij}, \delta, \delta') + \xi_2^0 + \xi_2^1 \sin\omega_0 t + \xi_2^2 \cos\omega_0 t + \xi_2^3, \\
z_3 &= \Delta_{33}T_3 + \alpha_2 T_1 - \alpha_1 T_2 + \sum_i \delta_i' \kappa_{i3}^1(T_y) + \sum_i \delta_i' \kappa_{i3}^2(T_y) + \sum_i \delta_i' \kappa_{i3}^3(T_y) + \\
&\quad \sum_{i,j} \alpha_i T_j \kappa_{ij3}^4(\Delta_{ij}, \delta, \delta') + \xi_3^0 + \xi_3^1 \sin\omega_0 t + \xi_3^2 \cos\omega_0 t + \xi_3^3,
\end{aligned} \tag{9}$$

Writing again formation equation for constants in (9)

$$\begin{aligned}
&\dot{\xi}_i^0 = 0, \qquad \dot{\kappa}_i = 0, \qquad i = 1,2,3 \\
&\dot{\xi}_i^1 = \xi_i^2, \qquad \xi_1^2 = \omega_0^2 \xi_i^1,
\end{aligned}$$

and including new variables in system vector, we may write the equations of the controlled system as

$$\begin{aligned}
\dot{y}' &= A^0(t)y' + B^0(t,v^2)u + C^0(t)v^3 + D^0 v^4, \\
\dot{\tilde{y}} &= A^0(t)\tilde{y} + B^0(t,v^2)u + C^0(t)v^3 + \tilde{K}(z - H^0\tilde{y}), \\
z &= H^0(t,v^2,v^7,\Delta)y' + H^1(t)v^5 + H^2(t)v^2 + H^3(t)v^6 + v^7,
\end{aligned} \tag{10}$$

where $\tilde{y}$ is an estimate of y'; $\tilde{K}(t)$—is an amplification matrix of estimation algorithm. Putting $v^2=v^3=v^5=v^6=0$ the optimal values of $\tilde{K}(t)$ are that of Kalman filter.

The criterion of quality of control may be chosen as

$$J = M[y'(t_1)^T F_0 y'(t_1)] + M[\int_0^{t_1} y'^T N_y + u^T R_0 u dt], \tag{11}$$

where F_0, N, R_0 are nonnegative symmetric matrixes, M—mathematical expectation operator.

Corresponding to the saddle point theorem "perfect" result is

$$J^0 = \max_v \min_u J(u,v). \tag{12}$$

Having constants (v^2,v^3,v^5,v^6,Δ) known the problem of optimal stabilization (internal minimum problem) is reduced to problem of choosing of linear stochastic regulator under quadratic criterion in case of incomplete measurements. The separation theorem states that problem of obtaining of "perfect" (12) testing result for system (10) may be formulated as

$$J_0^0 = \max_v J(v), \quad \text{where}$$
$$J(v) = \min_{K,\widetilde{K}} M[y'(t_1)^T F_0 y'(t_1) + \int_0^{t_1} y'^T Ny + u^T R_0 udt], \tag{13}$$

K, $\widetilde{K}$ are feedback matrixes of optimal stabilization $u = K\widetilde{y} +^0 (t)v^3$ and estimation $\widetilde{y}$ is optimal estimate. They are derived from $K = -R_0^{-1}KL$, where L is symmetric,

$$\dot{L} + LA + A^T L + N - LBR_0^{-1}B^T L = 0, \quad L(t_1) = F_0. \tag{14}$$

Optimal estimator is formulated as follows:

$$\dot{\widetilde{y}} = A^0\widetilde{y} + B^0 u + \widetilde{K}(z - H^1 v^5 - H^2 v^2 - H^3 v^6 - H\widetilde{y}),$$
$$\widetilde{K} = PH^T R^{-1}, \tag{15}$$
$$\dot{P} = A^0 P + PA^{0T} + Q - \widetilde{K}R\widetilde{K}^T, \quad P(0) = P^0.$$

To find the "perfect" result one needs to solve finite maximum problem (13) to find the maximum value of $J(v)$ over deviations $v=(v^2,v^5,v^6,\Delta)$. Let $J_0^0 = J(v^0)$ be the solution.

On the second step of testing procedure, after modeling sensors and getting the controlling signals u from orientation algorithm, one needs to calculate the current coordinate deflections $y(t)$. Processing values of $y(t)$ statistically one gets the value of quality functional J. Comparing it with "perfect" value we get estimation of quality of functioning for given orientation algorithm.

CONCLUSION

The system described permits to check up the quality of choosen control algorithms for active and parameters for passive attitude control systems and to estimate the influence of faults in current attitude determination owing to errors of sensors, failures in control loops, and influence of external dynamic perturbations. This moment the Package is being developed and can operate with passive ACSs.

ACNOWLEDGEMENTS

This work is partially supported by the Russian Basic Research Foundition's Grant #95-01-00211 and #97-01-00910.

REFERENCES

M.Ovchinnikov, A.Dyachenko, Mathematical Modelling of Attitude Motion for Universal Small Space Platform and for on-Ground Satellite Dynamics Simulation Stand. Proceedings of the International Symposium on Space Flight Dynamics, St.Peterburg-Moscow, 22-28 May, 1994, Moscow, IKI RAS, pp.34-47

M.Ovchinnikov, The Simple Methods of a Small Satellite Orientation. Preprint # IAA-95-IAA.11.2.06 for 46th Congress IAF, 2-6 Oct., 1995, Oslo, Norwey, 7p.

V.Alexandrov, Quality Testing of the Nonstationary Motion Stabilization. Moscow, Vestnik MGU, Serie:Math.k Mech. N 3, 1997.

N.Krasovski, Dynamical System Control. Moscow, 1985

POSITIONING SYSTEM USING LOW EARTH ORBIT CONSTELLATIONS

Der-Ming Ma[1], Shen-You Zhai[2] and Huan-Ein Shen[3]

[1] *Associate Professor, Department of Aerospace Engineering, Tamkang University, Tamsui, Taiwan 25137, Republic of China.*
Tel: +886-2-2623-7600, Fax: +886-2-2620-9746, E-mail: derming@aero.tku.edu.tw.
[2] *Graduate Student, Department of Aerospace Engineering, Tamkang University, Tamsui, Taiwan 25137, Republic of China.*
[3] *Graduate Student, Department of Mechanical Engineering, Tamkang University, Tamsui, Taiwan 25137, Republic of China. Currently, in military service.*

ABSTRACT

A scheme of designing low Earth orbit satellite positioning system is proposed. An example low Earth orbit satellite constellation providing the positioning function over certain region is presented. The proposed constellation consists of 75 satellites with four different inclinations. The satellites are in the orbit of height of about 800 Km. It can provide the positioning function over the region with radius of about 3000 Km center at the chosen site with longitude of 122° E and latitude of 25°N. The construction of the constellation is based on the requirement of good values of the Dilution of Precision over the specific region at same time. Also the study of the relativistic effects of the system shows that the requirement of keeping the orbit of the low Earth orbit satellite positioning system circle is less stringent than the GPS's.

INTRODUCTION

As small satellites becoming more capable through miniaturized electronic and on-board processing, the construction of low-cost satellite constellations becomes economically feasible. With the feasibility of LightSats and the possibility of smaller, relatively inexpensive launch systems, the low-altitude satellite constellations receive increased attention. The potential of low-altitude satellite constellations' applications is significant. The positioning function is one of them.

The purpose of this paper is to present a low Earth orbit satellite constellation which provides the positioning function as the GPS does over certain region. In order to provide good accuracy of point positioning, the constellations have at least four satellites with good Dilution of Precision (DOP)[1,2] over the specified region at same time. The DOP factor is a measure for the geometry of the visible satellites. The good geometry should mirror a low DOP value. It is the reciprocal value of the volume of the geometric body, which is formed by the intersection points of the site-satellite vectors with the unit sphere centered at the reference site. In the application of GPS, the purposes of DOP are twofold. First, it is useful in planning a survey and, secondly, it may be helpful in interpreting processed baseline vectors.

In this paper the concept of DOP is further extended to be the design criterion in designing the low Earth orbit satellite constellation.

The DOP is expressed in the topocentric coordinate system with its axes along the local north, east and vertical. The position DOP is then has tow components: HDOP, the dilution of precision in the horizontal position and the VDOP, denoting the corresponding value for the vertical component. The time dilution of precision is an another important factor to good accuracy of positioning. In the paper both position and time DOPs are considered in designing the constellation. In the next two sections the concept of geometric dilution of precision (GDOP) and then the positioning system design procedure using the idea of GDOP are present.

As to the GPS, several relativistic effects those associated to the low Earth orbit satellite positioning system are non-negligible. These effects include the following: 1) gravitational field of Earth and Earth rotation; and 2) velocities of satellites. The major effects cause an increase in the satellite clock frequency as observed by a user on Earth. These relativistic effects are also discussed in the section after the designing procedure of the constellation.

GEOMETRY DILUTION OF PRECISION

Figure 1 illustrates an Earth-centered inertial coordinate system. At $t = 0$, the x-axis passes through the intersection of the equator and the Greenwich meridian, the z-axis passes through the Earth rotation axis, and the y-axis completes the right-handed coordinate system. Shown in the Figure 1 are the user position (x, y, z), and the position of satellite No. 1 (x_1, y_1, z_1). The range distance between the user and the satellite No.1 is indicated as R_1. The basic equations using four satellites are

$$\begin{aligned}
\sqrt{(x-x_1)^2+(y-y_1)^2+(z-z_1)^2}+T&=R_1\\
\sqrt{(x-x_2)^2+(y-y_2)^2+(z-z_2)^2}+T&=R_2\\
\sqrt{(x-x_3)^2+(y-y_3)^2+(z-z_3)^2}+T&=R_3\\
\sqrt{(x-x_4)^2+(y-y_4)^2+(z-z_4)^2}+T&=R_4
\end{aligned} \qquad (1)$$

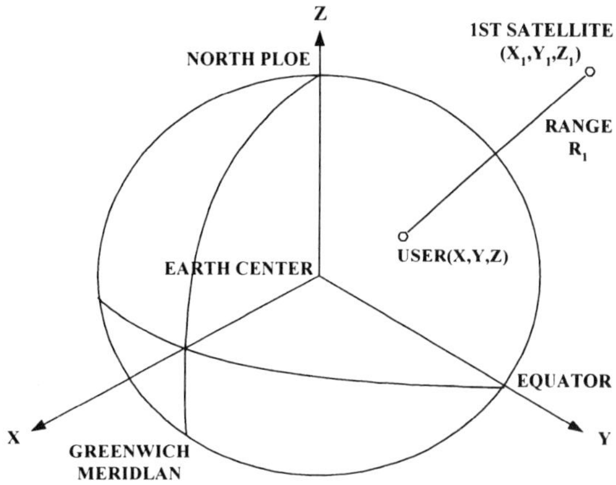

Figure 1. The relationship between the user and satellite.

where T is clock bias; x_i, y_i and z_i are the ith satellite position; and R_i is the pseudo-range measurements to the ith satellite. Note that the pseudo-ranges are the sum of the actual range plus the offset due to user clock error. For convenience, units have been selected such that the velocity of light is unity. In the equations above, the four pseudo-ranges are the measured quantities. The satellite positions are known, and the four unknowns are the user position (x, y, z), and the user clock error. The equations are nonlinear. While it is possible to solve these equations directly as they are shown. The equations can be linearized. Let x_n, y_n, z_n, and T_n be nominal values of x, y, z, and T, respectively. The Δx, Δy, Δz, and ΔT are corrections to these nominal values. R_{ni} is the nominal pseudo-range measurements from the ith satellite; ΔR_i be the difference between the actual and nominal measurements. Therefore

$$
\begin{aligned}
x &= x_n + \Delta x, \\
y &= y_n + \Delta y, \\
z &= z_n + \Delta z, \\
T &= T_n + \Delta T, \\
R_i &= R_{ni} + \Delta R_i
\end{aligned}
\tag{2}
$$

and

$$R_{ni} = \sqrt{(x_n - x_i)^2 + (y_n - y_i)^2 + (z_n - z_i)^2} + T_n. \tag{3}$$

Substituting the Eqs. (2) and (3) into Eq. (1) yields

$$\left[(x_n + \Delta x - x_i)^2 + (y_n + \Delta y - y_i)^2 + (z_n + \Delta z - z_i)^2\right]^{1/2} = R_{ni} + \Delta R_i - T_n - \Delta T \tag{4}$$

By ignoring second-order error terms, these equations can be written as

$$
\begin{aligned}
\sqrt{(x_n - x_i)^2 + (y_n - y_i)^2 + (z_n - z_i)^2} + \frac{(x_n - x_i)\Delta x + (y_n - y_i)\Delta y + (z_n - z_i)\Delta z}{\sqrt{(x_n - x_i)^2 + (y_n - y_i)^2 + (z_n - z_i)^2}} \\
= R_{ni} + \Delta R_i - T_n - \Delta T.
\end{aligned}
\tag{5}
$$

Furthermore, substituting Eq. (3) into Eq. (5), we obtain

$$\frac{x_n - x_i}{R_{ni} - T_n}\Delta x + \frac{y_n - y_i}{R_{ni} - T_n}\Delta y + \frac{z_n - z_i}{R_{ni} - T_n}\Delta z + \Delta T = \Delta R_i. \tag{6}$$

The above four equations are linearized equations that relate pseudo-range measurements to the desired user navigation information as well as the user's clock bias. The known quantities of the right-hand side of the equations are actually incremental pseudo-range measurements. They are the differences between the actual measured pseudo-range and the measurements that had been predicted by the user's computer based on the knowledge of satellite position and the user's most current estimate of his position and clock bias. The quantities to be computed, Δx, Δy, Δz, and ΔT, are corrections that the user will make to his current estimate of position and clock biases.

The coefficients of these quantities on the left-hand side are the direction cosines of the line of sight from user to the satellites as projected along the *x*-, *y*-, and *z*-axis. These linearized equations can be conveniently expressed in matrix notation and appear as

$$\begin{pmatrix} \alpha_{11} & \alpha_{12} & \alpha_{13} & 1 \\ \alpha_{21} & \alpha_{22} & \alpha_{23} & 1 \\ \alpha_{31} & \alpha_{32} & \alpha_{33} & 1 \\ \alpha_{41} & \alpha_{42} & \alpha_{43} & 1 \end{pmatrix} \begin{pmatrix} \Delta x \\ \Delta y \\ \Delta z \\ \Delta T \end{pmatrix} = \begin{pmatrix} \Delta R_1 \\ \Delta R_2 \\ \Delta R_3 \\ \Delta R_4 \end{pmatrix}. \tag{7}$$

Note that α_{ij} is the direction cosine of the angle between the range to the *i*th satellite and the *j*th coordinate. By the use of matrix notation, the above equation can be expressed as follows. Let

$$\mathbf{A} \equiv \begin{pmatrix} \alpha_{11} & \alpha_{12} & \alpha_{13} & 1 \\ \alpha_{21} & \alpha_{22} & \alpha_{23} & 1 \\ \alpha_{31} & \alpha_{32} & \alpha_{33} & 1 \\ \alpha_{41} & \alpha_{42} & \alpha_{43} & 1 \end{pmatrix}, \quad \mathbf{x} \equiv (\Delta x \quad \Delta y \quad \Delta z \quad \Delta T)^T, \text{ and } \mathbf{r} \equiv (\Delta R_1 \quad \Delta R_2 \quad \Delta R_3 \quad \Delta R_4)^T. \tag{8}$$

Then

$$\mathbf{Ax} = \mathbf{r} \quad \text{or} \quad \mathbf{x} = \mathbf{A}^{-1}\mathbf{r}. \tag{9}$$

Since this relationship is linear, it can be used to express the relationship between the errors in pseudo-range measurement and the user quantities. This relationship is therefore

$$\varepsilon_x = \mathbf{A}^{-1}\varepsilon_r \tag{10}$$

where ε_r represents the pseudo-range measurement errors and ε_x the corresponding errors in user position and clock bias.

Let now consider the covariance matrix of the expected errors in pseudo-range measurements and the covariance matrix of the user quantities. The first covariance measurement is a 4X4 array composed of the expected values of the squares and products of the errors in the pseudo-range measurements. The diagonal terms in the matrix, namely the squares of the expected errors, are the variances; i.e., the squares of the expected 1σ values of the pseudo-range measurement errors. The off-diagonal terms are the covariance between the pseudo-range measurement measurements and reflect the correlation to be expected in these measurements. The covariance matrix is given by

$$\text{cov}(\mathbf{r}) = E\{\varepsilon_r \varepsilon_r^T\} = A^{-1}\,\text{cov}(\mathbf{x})\mathbf{A}^T = [\mathbf{A}^T\,\text{cov}(\mathbf{x})\mathbf{A}]^{-1} \tag{11}$$

Now making the assumption that the pseudo-range errors are uncorrelated from satellite to satellite and equal for each satellite, the pseudo-range covariance matrix reduces to

$$\text{cov}(\mathbf{x}) = \sigma_x^2\mathbf{I}. \tag{12}$$

where $\mathbf{I}$ is the 4×4 unit vector. Thus

$$\text{cov}(\mathbf{r}) = \sigma_x^2[\mathbf{A}^T\mathbf{A}]^{-1}. \tag{13}$$

We see that the position and time error variances are functions of the diagonal elements of $[\mathbf{A}^T\mathbf{A}]^{-1}$. This leads to the concept of GDOP. GDOP is defined by

$$\text{GDOP} = \sqrt{Trace[\mathbf{A}^T\mathbf{A}]^{-1}}. \tag{14}$$

It is the amplification factor of pseudo-range error variance to the combined position and time error variance. In this paper the idea of GDOP is further extended as a criteria of designing the low Earth orbit satellite positioning system.

LEO SATELLITE POSITIONING SYSTEM

The relationship between the satellite's height, h, and the coverage area can be approximately as

$$h = R_e[\frac{1}{\sin \beta} - 1] \tag{15}$$

where R_e is the radius of Earth, and β is the angle measured at the center of the Earth of the region seen by the spacecraft (Figure 2). Note that in deriving Eq. (15) we have assumed a spherical Earth. For the satellite constellation with the height of about 800 Km, covers a region with radius of 3000 km centered at the sub-satellite point. Therefore the low Earth orbit satellite constellation provides the positioning function for the region with radius of about 3000 Km centered at the chosen site.

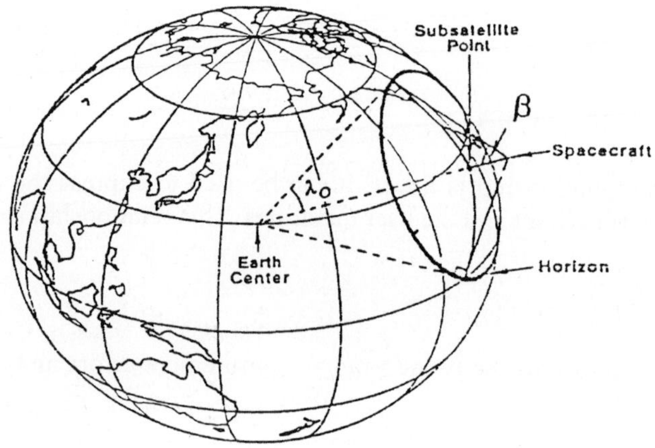

Figure 2.

To provide a positioning capability over a certain region (for example, in the study, the region is centered at the point with longitude (λ) of 122°E and latitude (ϕ) of 25°N), a scheme to orbit a sufficient number of satellites to ensure that four are always electronically visible is developed. The planned constellation will provide four satellites in good geometric position 24 hours a day. Two basic requirements are needed to be fulfilled: 1) the satellites of the constellation must be visible during one day; and 2) the four satellites visible must have good geometric relationship or with good value of GDOP. Therefore, to satisfy the requirements above the region, four sets of constellations providing the continuous coverage are required. The four satellites in good geometry position can be determined from calculation of the value of GDOP. By reviewing Eqs. (8) and (14) it is clearly that GDOP is the functions of the direction cosines of the lines of sight from the observer site to the visible satellites. The direction cosines of the line of sight from the observer site to the ith satellite in the Earth-center coordinate can be expressed as:

$$\begin{aligned} \alpha_{i1} &= \vec{i} \cdot \vec{\rho}_i, \\ \alpha_{i2} &= \vec{j} \cdot \vec{\rho}_i,. \\ \alpha_{i3} &= \vec{k} \cdot \vec{\rho}_i. \end{aligned} \tag{16}$$

Note that $\vec{i}$, $\vec{j}$, and $\vec{k}$ are the unit vector of the Earth-center coordinate system and $\vec{\rho}_i$ is the unit vector of the line of sight pointing from the observer to the ith satellite.

The constellations with continuous coverage and having repeated ground track over the region are the candidates of the system. The procedure of designing the low Earth orbit satellite positioning system is described as following. First, using the algorithm devised in Ref. 3, the constellations with continuous coverage and having repeated ground track over the chosen site are obtained. For the chosen site, the satellite constellations with the inclinations between 24° and 57° can provide the continuous coverage. Table 1 lists the orbit elements of the constellations those can provide continuous coverage and repeated ground track over the site.

Table 1. Constellations with Continuous Coverage and Repeated Ground Track.

Inclination, i,(deg.)	**Number of Sat.**	**Semi-Major Axis, a, (Km)**	**Initial Node of Ascending, Ω_0, (deg.)**	**Initial Mean Anomaly, M_0, (deg.)**
24	19	7183.85	128, 148, 168, 188, 208, 228, 248, 268, 288, 308, 328, 348, 8, 28, 48, 68, 88, 108, 128	0.00, 79.997, 159.993, 239.99, 319.987, 39.984, 119.981, 199,977, 279.974, 359.971, 79.97, 159.95, 239.96, 319.96, 39.955, 119.95, 199.95, 279.945, 359.941
27	25	7185.48	128, 302.98, 117.97, 292.56, 107.94, 282.93, 97.91, 272.90, 87.88, 262.87, 77.85, 252.84, 67.83, 242.81, 57.80, 232.73, 47.77, 222,75, 37.74, 212.72, 27.71, 202.70, 17.78, 192.67, 7.65	0.00, 70.20, 140.41, 210.61, 280.81,351.01, 61.22, 131.42, 201.62, 271.83, 342.03, 52.23, 122,43, 192.64, 262.84, 333.04, 43.25, 113.45, 183.65, 253.85, 324.06, 34.23, 104.46, 174.66, 244.87
30	23	7187.29	128, 189.84, 251.67, 313.53, 15.37, 77.22, 139.06, 200.9, 262.75, 324.59, 26.43, 88.27, 150.12, 211.96, 273.81, 335.65, 37.49, 99.33, 161.19, 223.02, 284.86, 346.70, 48.55	0.00, 214.19, 68.38, 282.58, 136.78, 350.97, 205.16, 59.36, 273.55, 127.75, 341.94, 196.13, 50.33, 264.53, 118.72, 332.91, 187.11, 41.30, 255.50, 109.69, 323.88, 178.08, 32.27
33	19	7189.28	128, 148, 168, 188, 208, 228, 248, 268, 288, 308, 328, 348, 8, 28, 48, 68, 88, 108, 128.01	0, 78, 159.98, 239.97, 319.96, 39.96, 19.95, 199.94, 279.93, 359.93, 79.92, 159.91, 239.90, 319.89, 39.89, 119.88, 199.87, 279.86, 359.85
36	22	7191.46	128, 304.23, 120.47, 296.70, 112.94, 289.17, 105.41, 281.64, 97.88, 274.11, 90.35, 266.58, 82.82, 259.05, 75.28, 251.52, 67.75, 243.99, 60.22, 236.46, 52.69, 228.93	0, 52.71, 105.43, 158.14, 210.86, 263.57, 16.29, 9, 61.72, 114.43, 167.15, 219.86, 272.58, 325.29, 18, 70.72, 123.44, 176.15, 228.87, 281.58, 334.30, 27.01
39	17	7193.81	128, 149.18, 170.36, 191.54, 212.71, 233.89, 255.07, 276.25, 297.43, 318.61, 339.79, 0.97, 22.15, 43.32, 64.50, 85.68, 106.86	0, 63.49, 126.99, 190.49, 253.98, 317.48, 20.97, 84.47, 147.96, 211.46, 274.95, 338.45, 41.95, 105.44, 168.94, 232.43, 295.93
42	18	7196.33	128, 163.41, 198.82, 234.22, 269.63, 305.04, 340.45, 15.86, 51.27, 86.68, 122.09, 157.50, 192.91, 228.32, 263.72, 299.13, 334.54, 9.95	0, 224.27, 88.55, 312.82, 177.10, 41.37, 265.65, 129.92, 354.19, 218.47, 82.74, 307.02, 171.29, 35.57, 259.84, 124.12, 348.39, 212.67
45	22	7199.02	128, 306.70, 125.39, 304.09, 122.79, 301.49, 120.18, 298.88, 117.58, 296.28, 114.97, 293.67, 112.37, 291.06, 109.76, 288.46, 107.16, 285.85, 104.55, 283.25, 101.95, 280.64	0, 18.24, 36.47, 54.71, 72.95, 91.18, 109.42, 27.66, 145.89, 164.13, 182.37, 200.61, 218.84, 237.08, 255.32, 273.56, 291.79, 310.03, 328.27, 346.50, 4.74, 22.98
48	24	7201.88	128.00, 306.79, 125.59, 304.38, 123.17, 301.96, 120.76, 299.55, 118.34, 297.14, 115.93, 294.72, 113.51, 292.31, 111.10 . 289.89, 108.68, 287.48, 106.27, 285.06 . 103.86, 282.65, 101.44, 280.23	0.00, 16.90, 33.80, 50.70, 67.60, 84.50, 101.41, 118.31, 135.21, 152.11, 169.01, 185.91, 202.81, 219.71, 236.61, 253.51, 270.42, 287.32, 304.22, 321.12, 338.02, 354.92, 11.82, 28.72
51	22	7204.90	128.00, 146.94, 165.89, 184.83, 203.78, 222.72, 241.66, 260.61, 279.55, 298.50, 317.44, 336.39, 355.33, 14.27, 33.22, 52.16, 71.11, 90.05, 108.99, 127.94, 146.88, 165.83	0, 94.78, 189.57, 284.35, 19.13, 113.91, 208.70, 303.48, 38.26, 133.04, 227.83, 322.61, 57.39, 152.17, 246.96, 341.74, 76.52, 171.30, 266.09, 0.87, 95.65, 190.44
54	29	7208.08	128.00, 197.73, 267.46, 337.20, 46.93, 116.66, 186.39, 256.12, 325.85, 35.59, 105.32, 175.05, 244.78, 314.51, 24.24, 93.98, 163.71, 233.44, 303.17, 12.90, 82.64, 152.37, 222.10, 291.83, 1.56, 71.29, 141.03, 210.76, 280.49	0.00, 103.76, 207.51, 311.27, 55.02, 158.78, 262.53, 6.29, 110.04, 213.80, 317.55, 61.31, 165.06, 268.82, 12.57, 116.33, 220.08, 323.84, 67.59, 171.35, 275.10, 18.86, 122.62, 226.37, 330.13, 73.88, 177.64, 281.39, 25.15
57	30	7211.41	128, 279.66, 71.32, 222.98, 14.64, 166.30, 317.97, 109.63, 261.29, 52.95, 204.61, 356.27, 147.93, 299.59, 91.25, 242.91, 34.57, 186.24, 337.90, 129.56, 281.22, 72.88, 224.54, 16.20, 167.86, 319.52, 111.18, 262.85, 54.51, 206.17	0, 36.75, 73.49, 110.24, 146.99, 183.73, 220.48, 257.23, 293.98, 330.72, 7.47, 44.22, 80.96, 117.71, 154.46, 191.20, 227.95, 264.70, 301.44 , 338.19, 14.94, 51.69, 88.43, 125.18, 161.93, 198.67, 235.42, 272.17, 308.91, 345.66

Second, to reduce the computation work, among the candidates the sets of four constellations with fixed inclination difference are combined and the values of GDOPs are calculated. Based on the GDOPs' values several sets with good values are selected. Several values of the fixed inclination differences are studied.

Finally, several points in the region are chosen to check the values of GDOP for different sets and the set of constellation having smallest value of GDOP is the low Earth orbit satellite positioning system. Figure 4 shows the values of GDOPs during one day of the sites. Table 2 lists the satellites of the positioning constellation for the chosen site.

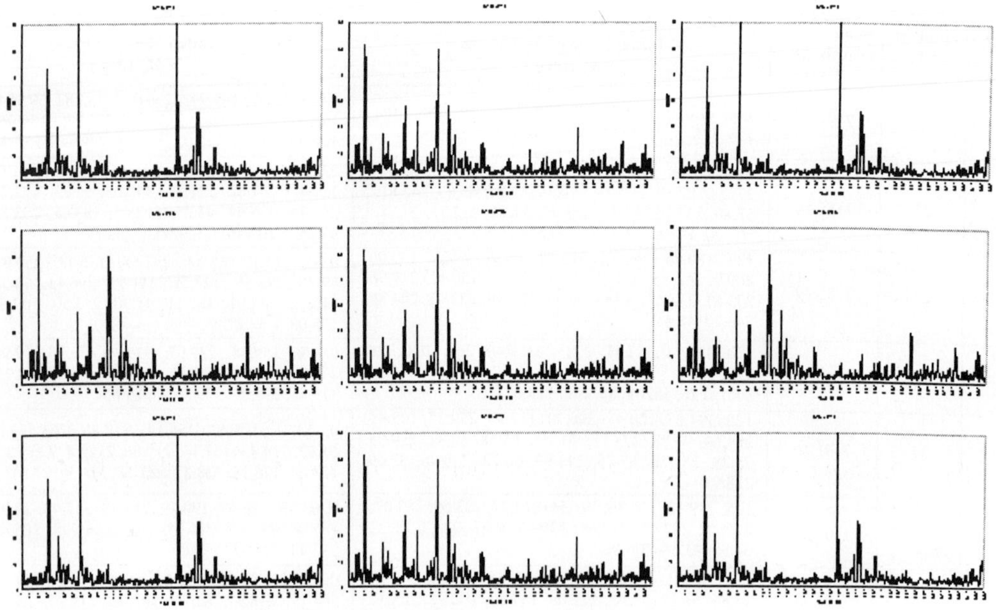

Figure 3. The GDOP value of several points.

Table 2. Low Earth Orbit Satellite Positioning System.

	Number of Sat.	Semi-Major Axis, *a*, (Km)	Inclination, *i*,(deg.)	Initial Node of Ascending, Ω_0, (deg.)	Initial Mean Anomaly, M_0, (deg.)
Set 1	27	25	7185.48	128, 302.98, 117.97, 292.56, 107.94, 282.93, 97.91, 272.90, 87.88, 262.87, 77.85, 252.84, 67.83, 242.81, 57.80, 232.73, 47.77, 222,75, 37.74, 212.72, 27.71, 202.70, 17.78, 192.67, 7.65	0.00, 70.20, 140.41, 210.61, 280.81, 351.01, 61.22, 131.42, 201.62, 271.83, 342.03, 52.23, 122,43, 192.64, 262.84, 333.04, 43.25, 113.45, 183.65, 253.85, 324.06, 34.23, 104.46, 174.66, 244.87
Set 2	30	23	7187.29	128, 189.84, 251.67, 313.53, 15.37, 77.22, 139.06, 200.9, 262.75, 324.59, 26.43, 88.27, 150.12, 211.96, 273.81, 335.65, 37.49, 99.33, 161.19, 223.02, 284.86, 346.70, 48.55	0.00, 214.19, 68.38, 282.58, 136.78, 350.97, 205.16, 59.36, 273.55, 127.75, 341.94, 196.13, 50.33, 264.53, 118.72, 332.91, 187.11, 41.30, 255.50, 109.69, 323.88, 178.08, 32.27
Set 3	33	19	7189.28	128, 148, 168, 188, 208, 228, 248, 268, 288, 308, 328, 348, 8, 28, 48, 68, 88, 108, 128.01	0, 78, 159.98, 239.97, 319.96, 39.96, 19.95, 199.94, 279.93, 359.93, 79.92, 159.91, 239.90, 319.89, 39.89, 119.88, 199.87, 279.86, 359.85
Set 4	36	22	7191.46	128, 304.23, 120.47, 296.70, 112.94, 289.17, 105.41, 281.64, 97.88, 274.11, 90.35, 266.58, 82.82, 259.05, 75.28, 251.52, 67.75, 243.99, 60.22, 236.46, 52.69, 228.93	0, 52.71, 105.43, 158.14, 210.86, 263.57, 16.29, 9, 61.72, 114.43, 167.15, 219.86, 272.58, 325.29, 18, 70.72, 123.44, 176.15, 228.87, 281.58, 334.30, 27.01

THE RELATIVISTIC EFFECTS OF LEO SATELLITE POSITIONING SYSTEM

Since the satellites have a large velocity, there is a non-negligible gravitational potential difference between that of satellites and that of the users on Earth's surface, and there is significant Earth rotation effect. Because clocks near Earth's surface experience a gravitational field may be in motion in addition to the motion associated with the Earth's rotation), they are subject to gravitational frequency shifts and time dilation. Let the gravitational potentials experience for the satellites and the receiver on the Earth's surface are ϕ_T and ϕ_R respectively, then the gravitational frequency shifts' difference is

$$f_T - f_R = (-\frac{\phi_T - \phi_R}{c^2}) f_R \tag{17}$$

or

$$\frac{f_R}{f_T} = 1 + (\frac{\phi_T - \phi_R}{c^2}) \tag{18}$$

where c is the speed of light. Let the periods of the clocks of satellite and of receiver on Earth being T_T and T_R respectively, then

$$\frac{f_R}{f_T} = \frac{1/T_R}{1/T_T} = \frac{T_T}{T_R} \tag{19}$$

Assume the velocities of satellite and of receiver on Earth are V_R and V_T respectively, then

$$T_R = \sqrt{1 - (\frac{V_R}{c})^2}\, T \,,\; T_T = \sqrt{1 - (\frac{V_T}{c})^2}\, T \tag{20}$$

In Eq. (20), T is the period of zero velocity. Substituting Eq. (20) into Eq.(19), yields

$$\frac{f_R}{f_T} = \sqrt{\frac{1 - (V_T / c)^2}{1 - (V_R / c)^2}} \cong \sqrt{1 + (\frac{V_R^2}{c^2} - \frac{V_T^2}{c^2})} \cong 1 + \frac{1}{2}(\frac{V_R^2}{c^2} - \frac{V_T^2}{c^2}) \tag{21}$$

Also the Doppler frequency shift is obtained as

$$f_R = \frac{1}{1 - \frac{\vec{e} \cdot (\vec{V}_T - \vec{V}_R)}{c}} f_T \tag{22}$$

where $\vec{e}$ is the unit vector along the line of sight.

$$\frac{f_R}{f_T} = \frac{1}{1 - \frac{\vec{e} \cdot (\vec{V}_T - \vec{V}_R)}{c}} \cong 1 + \frac{\vec{e}}{c} \cdot (\vec{V}_T - \vec{V}_R) \tag{23}$$

Combining Eqs. (17), (21) and (23), we obtain the ratio of the receiver's and of satellite's clocks frequencies

$$\frac{f_R}{f_T} = 1 + \frac{1}{c^2}(\phi_T - \phi_R) + \frac{1}{2c^2}(V_R^2 - V_T^2) + \frac{\vec{e}}{c} \cdot (\vec{V}_T - \vec{V}_R) \tag{24}$$

If the satellites of the low Earth orbit satellite positioning system are on the circular orbits, then the first two terms of the right side are

$$\frac{1}{c^2}(\phi_T - \phi_R) = \frac{1}{c^2}[(-\frac{GM}{r}) - (-\frac{GM}{r_E})]$$

$$= \frac{1}{(2.997925E9)^2}[(-\frac{3.9860015E14}{7.18385E6}) - (-\frac{3.9860015E14}{6.378137E6})] = 7.79875365^{-13} \quad (25)$$

and $\frac{1}{2c^2}(V_R^2 - V_T^2) = \frac{1}{2(2.997925E9)^2}((4.651E2)^2 - (7.4489E3)^2) = -3.07476702^{-12}$ (26)

respectively. The sum of the two terms is -2.29489166^{-12}. The amount is the factory presetting to partially accommodate the relativistic effects.

Besides the purposely setting lower in frequency, the time shift due to the deviation from the circular orbit is[4]

$$\Delta t' = 4.4428 \times 10^{-10} \frac{1}{\sqrt{m_e}} e\sqrt{a} \sin E \quad (27)$$

where E is the eccentric anomaly and

$$m_e = \mu / c^2 = 4.435028687 \times 10^{-3} m \quad (28)$$

The amount of the time shift is proportional to the square root of the semi-major axis. Figure 4 shows the amounts of the time shift for different values of eccentricities of the LEO satellite positioning system. For the eccentricity of 0.03 the amount of shift is 20ns. It is less than the amount of 23ns for the GPS satellites with the eccentricity of 0.01. In other words, the requirement of keeping the orbits of LEO satellite positioning system being circle is less stringent.

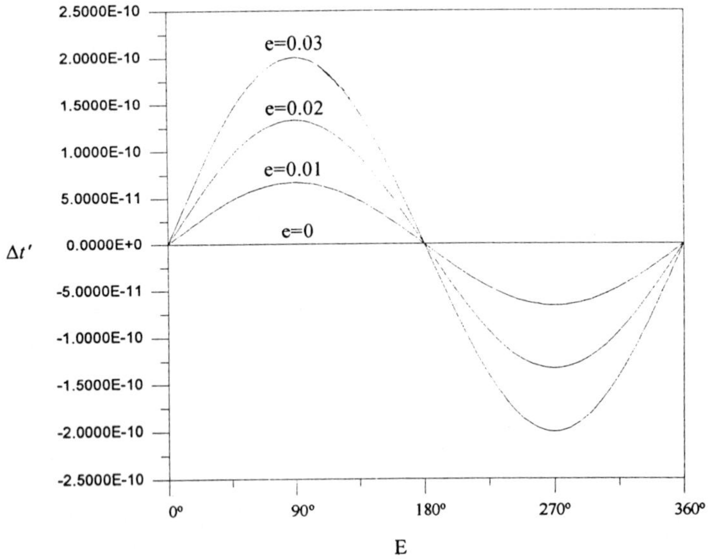

Figure 4. Time shift for different values of eccentricities of the LEO satellite.

CONCLUSION

In the paper we have proposed a scheme to orbit the satellites over the region centered at point with longitude of 122° E and latitude of 25°N. The satellite constellation can be used as the positioning system as the GPS does. The proposed constellation consists of 75 satellites with four values of inclinations. The satellites are in the orbit of height of about 800 Km. It can provide the positioning function for the region with radius of about 3000 Km center at the chosen site. After the study of the relativistic effects, we find that the requirement of the keeping the orbit circle is less stringent than the GPS's and to partially accommodate the relativistic effects the satellite clock can be purposely set lower in frequency by the factor of -2.29489166^{-12}.

ACKNOWLEDGMENT

The study was sponsored by the National Science Council of Republic of China under the contract number NSC85-2612-E-032-001.

REFERENCES

B. Hofmann-Wellenhof, H. Lichtengger and J. Collins, GPS: Theory and Practice, 2nd edition, Springer-Verlag Wien New York, 1993.

P.S. Jorgensen, "Navstar/Global Positioning System 18-Satellite Constellations", Global Positioning System, Vol. II, The Institute of Navigation, 1984, pp. 1-12.

D.-M. Ma and W.-C. Hsu, ``Exact Design of Partial Coverage Satellite Constellations over Oblate Earth'', Journal of Spacecraft and Rockets, Vol.34, No.1, 1997, pp.29-35.

Bradford W. Parkinson and James J. Spilker Jr., editors, Global Positioning System: Theory and Applications, Vol.1, AIAA, 1996, Part I, Chapter 18.

ANALYSIS THE UNSTEADY-STATE TEMPERATURE DISTRIBUTION OF MICRO-SATELLITE UNDER STABILIZATION EFFECTS

S. R. Lee[1], Z. C. Hong[2], K. T. Peng[2] and L. S. Leu[1]

[1] *Kuang-Wu Institute of Technology and Commerce, Peito, Taipei 112, ROC*
[2] *Department of Mechanical Engineering, National Central University, Chungli, Taiwan, ROC*

ABSTRACT

This paper presents the analysis of thermal control of the micro-satellite TUU-SAT1. TUU SAT-1 is a university micro-satellite program initiated at Satellite Engineering Lab. National Central University, Taiwan in 1995. We use the thermal modeling of network method to resolve the value of temperature from a nonlinear energy equation by iteration. It is also investigated the temperature distribution of the micro-satellite running in the sun-synchronous orbit. Square, the selected geometry of micro-satellite in this paper, is used to discuss the temperature of micro-satellite affected by spinning speeds, different altitudes (800km, 1000km and 2000km), material properties and radiator areas. Proper arrangements of radiator area and coatings to maintain micro-satellite in spin-stabilization or gradient-gravity stabilization under its working temperature range can be accessed in the present analysis.

INTRODUCTION

Generally speaking, it is defined the mass below 500 kg as small-satellite, the mass below 200 kg as mini-satellite, the mall below 100 kg, and the mass below 10 kg as nano-satellite. Small satellite owns, special, many advantages, including economic, reliable, high successful lunching and people training in satellite technology, to attract university's attention for working researches. Similarity, the first design of micro-satellite that the mass is about 50 kg is done by TUU-SAT1 at National Central University in Taiwan.

The satellite thermal design between cold space (4°K) environment, sun earth and the heat sources which the equipments operate inside the satellite means to control the temperature range of equipment (-20°C~40°C) [1,2]. The technology must maintain electrical components working normally and reduce the temperature change inside the satellite, even in the worst space environment. There are usually two ways to do thermal control system: passive and active [3,4], but the difference between both is not strictly defined. Generally, active thermal control system contains pumped-loop system, heater, cooler, heat-pipe, temperature transformer, etc. Passive thermal control system usually uses coating with different radiative properties and radiator plate to reach the purpose of temperature control. Passive thermal control system has the advantages of the less weight, less power needed, less error and cheaper prices. Therefore, passive thermal control becomes very useful around micro-satellite (50~100kg).

The orbit parameters of micro-satellite shown in Fig. 1 and Fig. 2 are assumed below [4,5]:

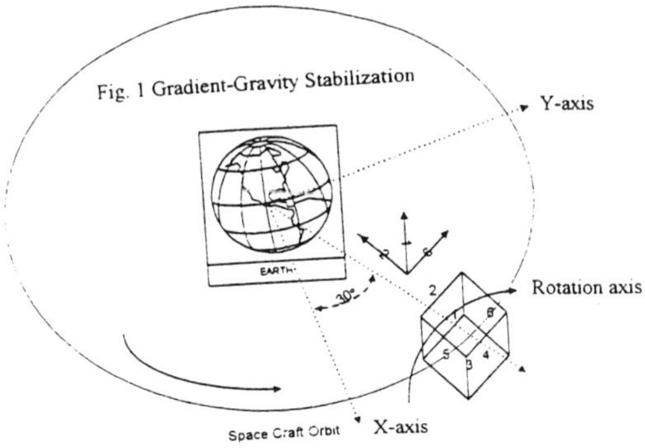

Figure 1. Gradient-Gravity Stabilization

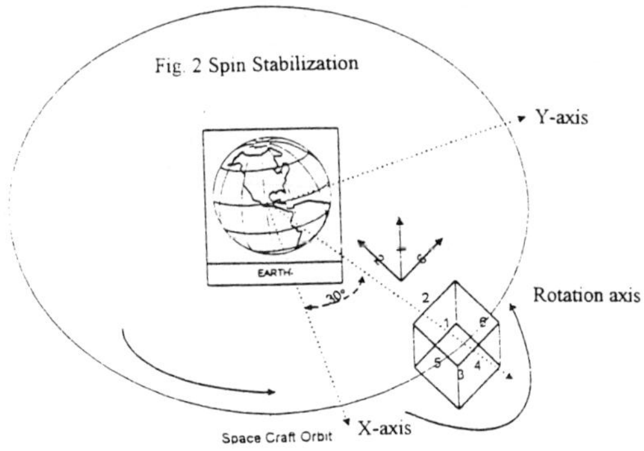

Figure 2. Spin Stabilization

Orbit type:	Sun-Synchronous
Altitude:	800, 1000, and 2000 km
Period:	15.22 rounds/day
Inclination Angle:	97.4°
Velocity:	7.6 km/s
Self-Spinning:	6mpr
Eclipse Fraction:	34.23%
Sun Angle:	Autumnal equinox (75.0°)
	Vernal equinox (75.0°)
	Summer solstice (79.37°)
	Winter solstice (73.29°)
Attitude Control:	Gradient-Gravity and Spin Stabilization

The analysis of the paper is focused on the temperature requirement of subsystem of thermal control in TUU-SAT1. Three materials of satellite structure are chosen for design to reach temperature's requirement[3,6]:

(1) Multilayer Insulation, Al-kapton filler 20 layers/inch, thermal conductivity λ=0.0004 W/m-K (in vacuum), density ρ=200 kg/m^3, specific heat C_p=168 kJ/kg-K;
(2) Honeycomb, Al H/C, thermal conductivity λ=1.7 W/m-K (in vacuum), density ρ=100 kg/m^3, specific heat C_p=950 kJ/kg-K;
(3) Radiator, thermal conductivity λ=90 W/m-K (in vacuum), density ρ=2700 kg/m^3, specific heat C_p=878 kJ/kg-K, solar absorptivity α=0.14, infrared emissivity ε=0.25;
(4) The solar cells selected with absorptivity α of 0.74 and reflectivity ε of 0.82 are adhered on outer surface excluding the top and the bottom.

THEORETICAL ANALYSIS

It is very complicated to simulate temperature variation when satellite is running in the orbit. Two simplifications can be done before solving the problem.

(1) Use mathematical model to express real physical meaning.
(2) Use numerical analysis method to solve mathematical model.

The network method [7,8] is chosen to solve numerical model. According to physical modeling of satellite, it is divided many unites. Each unit represent and individual node. Therefore, the total nodes of square pillar satellite are 38 points, which 1 to 20 nodes are inside and the other 18 nodes are outside, and the total nodes of hexagon pillar satellite are 54 points, which 1 to 28 are inside. Fig.3 is shown the nodes location for both geometries.

The heat transfer process of satellite is mostly based on radiation and conduction. Therefore, the governing equation of thermal balance in a micro-satellite between several heat source which include radiation energy of Sun solar, Albedo, and Earth infrared, the conduction between components and radiation from outer surfaces is written below for each node surface.

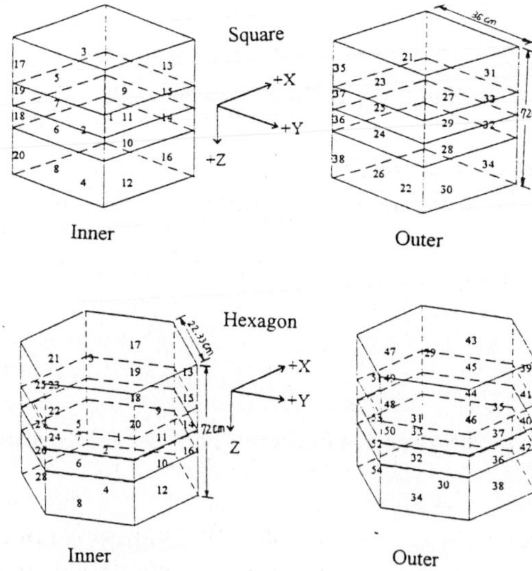

Figure 3. Nodes location for square pillar and hexagon pillar of micro-satellite

$$(\alpha_{si} S\phi_{1i} + \alpha_{si} E_r \phi_{2i} + \varepsilon_{ei} E_e \phi_{3i}) A_i + \sum_{j=1}^{N} E_{i,j}\sigma(T_j^4 - T_i^4) + P_i + \sum_{i=1}^{N} D_{i,j}(T_j - T_i) = m_i c_i \frac{\Delta T_i}{\Delta \tau} \quad (1)$$

Where $E_{i,j}=A_i\varepsilon_i B_{i,j}$ and $D_{i,j} = \frac{A_i k_{i,j}}{L_{i,j}}$

For square pillar, nodal points i=1~38; For hexagon pillar, nodal points i=1~54. The first term of left hand side of Eq.(1) which represents outer heat source affects deeply the temperature variation of each node at satellite. The relative position between satellite, sun and earth, shown in Fig. 1, will change the value of three outer heat source.

The Orbit of Satellite

The parameter of orbit beta angle β is a very important factor in the analysis of the satellite on the low earth orbit. It effects the external heat source from the sun for each node and the eclipse area of the orbit. The definition of β is the angle between orbit plane and sunlight that is varied from –90° to 90°. The mathematical form of β is described as below [9,10]:

$$\beta = \sin^{-1}(\cos\delta_s \cdot \sin RI \cdot \sin(\Omega - \Omega_s) + \sin\delta_s \cdot \cos RI) \quad (2)$$

The sun angle of the sun-synchronous orbit is a constant value that benefits by operation of CCD camera. Also we can design the velocity for regression nodes of sun-synchronous orbit as below:

$$\frac{d\Omega}{dt} = 360 \text{ deg./365.242 days} = 0.9856 \text{ deg./day} \quad (3)$$

And the orbit inclination angle, RI, can be resulted by following:

$$\frac{d\Omega}{dt} = \frac{-9.9639}{(1-e^2)^2}\left(\frac{Re}{Re+H}\right)^{3.5} \cos RI \, {}^{deg}\!/_{day} \tag{4}$$

RI varies with orbit altitudes In Eq.(3) and Eq.(4). The variation of β is very small and is assumed constant value $\beta = 15°$ in this case from Eq.(2).

The Eclipse of orbit

When the satellite runs behind the earth, sunlight will shade off by the earth. Therefore, the satellite is in the eclipse of orbit. By the view of sun, as Fig. 4, the earth receives parallel sunlight, so that the foreword circle of the earth is in light and the backward circle of it becomes a pillar shadow. Satellite runs a cycle as 360° on the orbit in Fig. 4. We define the position of the sunlight is 0°, then the period of earth from A' to D' (90°~270°) is in the shadow, and the eclipse area of the satellite orbit is between B (180°- θ_o) and C (180° + θ_o). In order to get the position of B and C, we must calculate the value of θ_o firstly.

$$\theta_0 = \cos^{-1}\frac{\sqrt{1-\left(\frac{R_e}{R_e+H}\right)^2}}{\cos\beta} \tag{5}$$

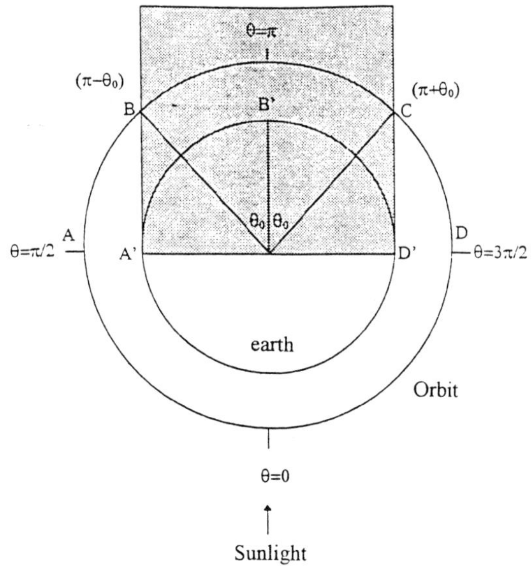

Figure 4. Satellite orbit plane as the view of sun

Where the radius of earth Re=6371 km, the altitude of orbit H=800 km . θ_o is calculated 61.63° by Eq.(5). It means that the satellite entries the eclipse at B (θ=118.37°) and exits at C (θ=241.63°). In the mean while, the eclipse fraction of the circle orbit is (241.63-118.37)/360x100%=34.23%.

External heat sources of the satellite

The first item of Eq. (2), $(\alpha_{si}s\phi_{1i}+\alpha_{si}E_r\phi_{2i}+\varepsilon_{ei}E_e\phi_{3i})A_i$, is total external heat source of the satellite. It effects the temperature of every parts of the satellite deeply, and also play an important role for the energy sources. The so-called external heat sources include radiation of sun solar $\alpha_{si}s\phi_{1i}$, albedo of earth $\alpha_{si}E_r\phi_{2i}$ and infrared radiation $\varepsilon_{ei}E_e\phi_{3i}$. All of them are varied with the relation position of the sun, earth and satellite, especially obvious if the satellite is in self-spin attitude controlling. The definitions of ϕ_{1i}, ϕ_{2i} and ϕ_{3i} are as below:

$$\phi_{1i} = \cos\alpha \tag{6}$$

Where α_i is the angle between sunlight and the normal vector of node i.

$$\phi_{2i} = \phi_{3i}\cos\Phi \tag{7}$$

Where Φ is the angle between the line of sun and earth and the line of earth and satellite.

$$\phi_{3i} = \frac{1}{\pi} \iint (\cos\theta \cos\beta + \sin\theta \sin\beta \cos\varphi) \sin\theta d\theta d\varphi \quad (8)$$

THE COMPUTATION OF TEMPERATURE

If the term of Eq.(1) $\triangle T_i/\triangle\tau$ is assumed to be zero, the equation becomes steady state. Actually, the steady-state temperature is just a starting point to the actual operation of micro-satellite. The results for both cases will be discussed later. Before solving the nonlinear Eq.(1), we must linearize Eq.(1)by Patankar's method [11].

For inner nodes,

$$T_i^{n+1,k} = T_i^{n,k} + \frac{\sum_{j=1}^{N} E_{i,j}\sigma(T_j^{(n,k)^4} - T_i^{(n,k)^4}) + D_{i,j}(T_j^{(n,k)} - T_i^{(n,k)}) + q_{T_i}^k - \frac{(mc)_i}{\Delta\tau}(T_i^{(n,k)} - T_i^{(n,k-1)})}{(\sum_{j=1}^{N} 4E_{i,j}\sigma T_i^{(n,k)^3} + D_{i,j} + \frac{(mc)_i}{\Delta\tau})} \quad (9)$$

Where $q_{Ti}=(\alpha_{si}s\phi_{1i} + \alpha_{si}E_r\phi_{2i} + \varepsilon_{ei}E_e\phi_{3i})A_i + P_i$; $T_i^{n,k-1}$ means the temperature of I node in previous time step (or location). For outer nodes,

$$T_i^{n+1,k} = T_i^{n,k} + \frac{\sum_{j=1}^{N} E_{i,j}\sigma(T_j^{(n,k)^4} - T_i^{(n,k)^4}) + D_{i,j}(T_j^{(n,k)} - T_i^{(n,k)}) + q_{T_i}^k - \frac{(mc)_i}{\Delta\tau}(T_i^{(n,k)} - T_i^{(n,k-1)})}{(4E_{i,j}\sigma T_i^{(n,k)^3} + D_{i,j} + \frac{(mc)_i}{\Delta\tau})} \quad (10)$$

In Eq.(9) and Eq.(10) the time step ($\triangle\tau$) chosen is 15.767 seconds, because the satellite takes 94.6 minutes for a cycle. As in Fig. 4 shown, the starting point of computation is steady-state temperature in location of $\theta=0$, then processing the calculation by iteration at next position, until the tolerance $|T_i^{n+1} - T_i^{n,k}| \le 10^{-6}$ is reached at each node. Also average temperature $\bar{T}$ defined is to explain micro-satellite temperature in inner surface, $\bar{T} = \frac{\sum_{i=1}^{N} T_i}{N}$.

RESULTS AND DISCUSSIONS

According previous numerical method of computational simulation, three parameters of angle coefficient ϕ_{1i},, ϕ_{2i} and ϕ_{3i} are input to the left handside of Eq.(1) to obtain three external heat sources for each node in micro-satellite orbit. The results studied are presented with factors influencing temperature variation listed below for square and hexagon pillar micro-satellite under stabilization effect.

The variation of total external heat source

The computational value of total external heat energy for gradient-gravity and spin stabilization for different geometries are shown from Fig. 5 to Fig. 8 in 800 km altitude and 0.5 m^2 radiator area. It is obvious to know from figures above that spin stabilization receives higher slightly energy than gradient-gravity in the same orbit parameters, but the energy distribution in spin stabilization changes more uniform except in eclipse period. Also, comparing the effect of geometry effect, hexagon pillar is better than square pillar if standing the point of distribution of total external energy.

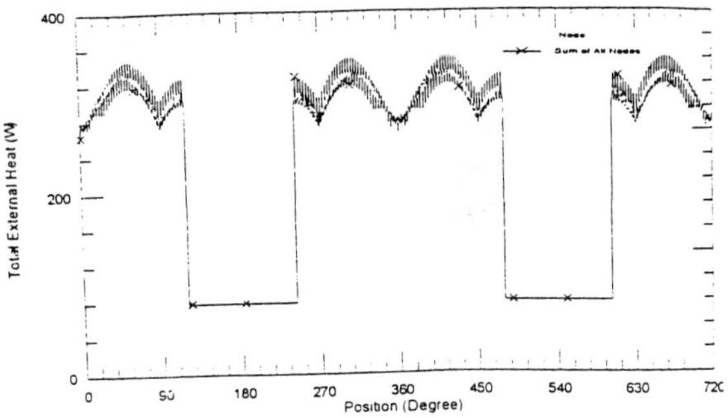

Figure 5. The total external heat of square pillar in gradient-gravity stabilization

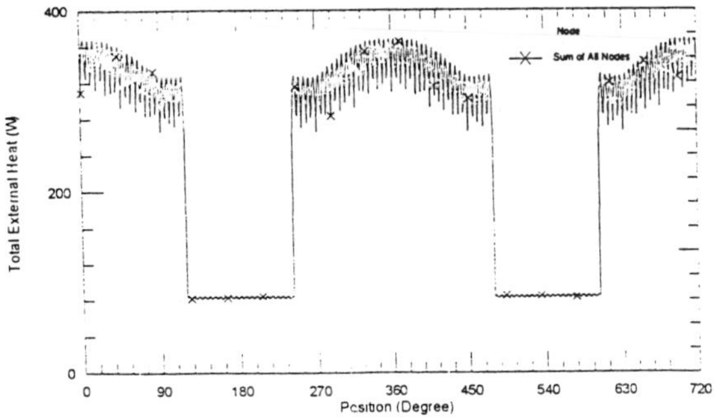

Figure 6. The total external heat of square pillar in spin stabilization

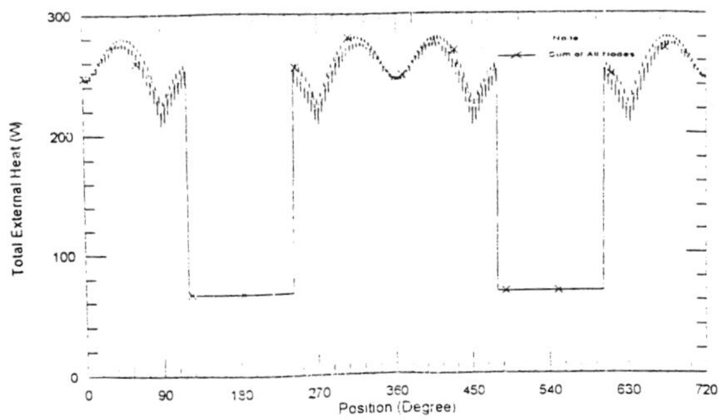

Figure 7. The total external heat of hexagon pillar in gradient-gravity stabilization

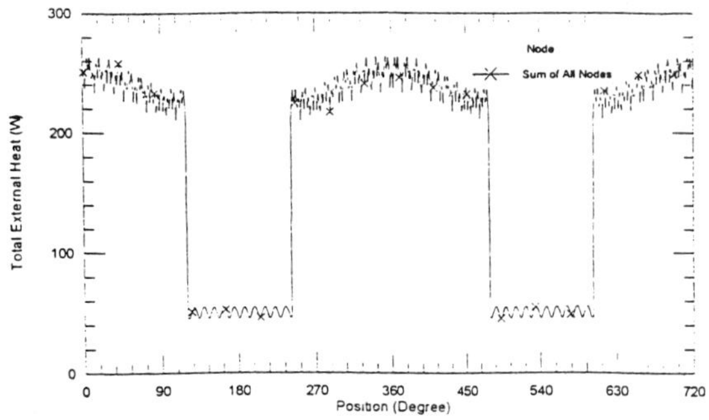

Figure 8. The total external heat of hexagon pillar in spin stabilization

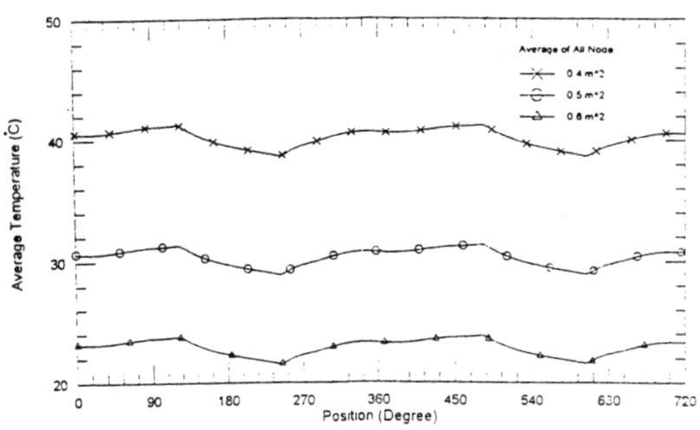

Figure 9. The average temperature of inner nodes for different radiator area of square pillar in gradient-gravity stabilization

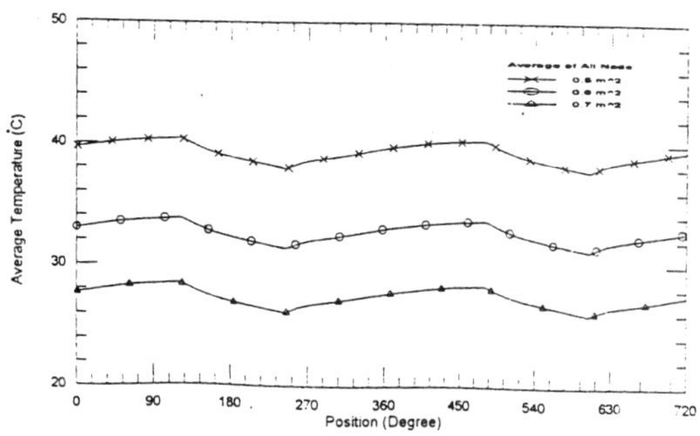

Figure 10. The average temperature of inner nodes for different radiator area of square pillar in spin stabilization

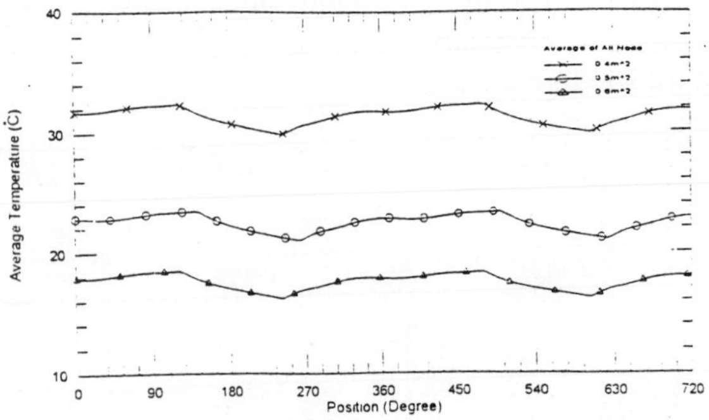

Figure 11. The average temperature of inner nodes for different radiator area of hexagon pillar in gradient-gravity stabilization

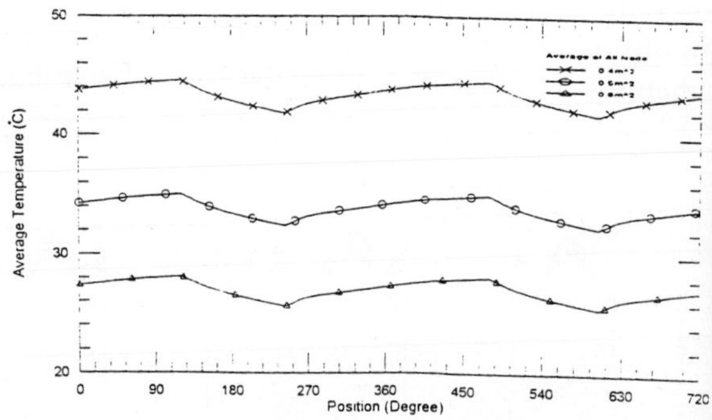

Figure 12. The average temperature of inner nodes for different radiator area of hexagon pillar in spin stabilization

The Effect of Radiator Area

The thermal control design is based on the requirement of temperature range (-20 ° C~40 ° C) in TUU-SAT1. Therefore, proper arrangement of radiator area is required to reach design's goal. The effects of radiator areas in temperature are shown from Fig. 8 to Fig. 12 in 800 km altitude. The results obtained are observed if radiator area increase $0.1m^2$ in the middle surface, the temperature decreases about 8 ° C~9 ° C. Also in the same geometry, the inner average temperature of spin stabilization is higher than gradient-gravity stabilization.

Altitude's Effect

In the analysis of temperature effect by changing altitudes, the parameter s of radiator area 0.6m, and self-spinning 6mpr are chosen to discuss inner average temperature of different altitudes, 800, 1000 and 2000 km, under stabilization effect. The simulated results are presented in Fig. 13~Fig. 16. As apparently, T_i is decreasing by following the satellite altitude increasing. The difference of inner average temperature from 800 km to 2000 km of satellite is small and about 1 °C~3 °C.

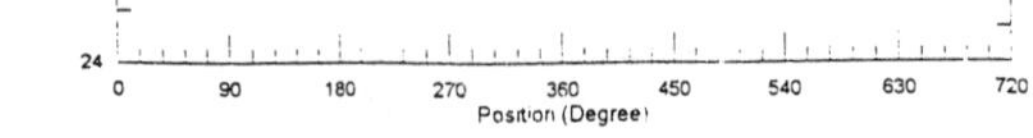

Figure 13. The average temperature of inner nodes for different altitudes of square pillar satellite in gradient-gravity stabilization

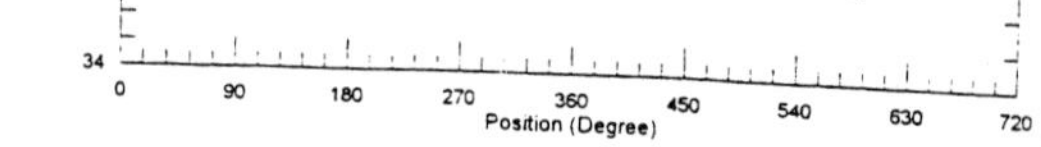

Figure 14. The average temperature of inner nodes for different altitudes of square pillar satellite in spin stabilization

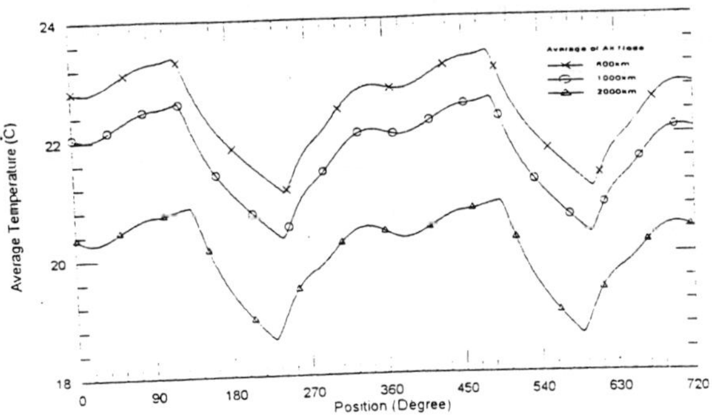

Figure 15. The average temperature of inner nodes for different altitudes of hexagon pillar satellite in gradient-gravity stabilization

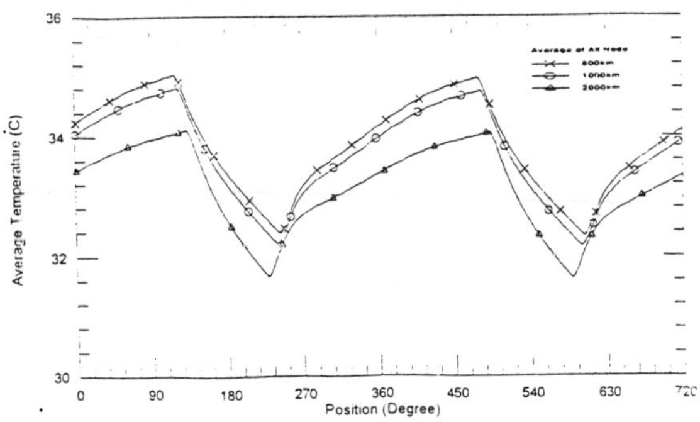

Figure 16. The average temperature of inner nodes for different altitudes of hexagon pillar satellite in spin stabilization

CONSLUSION

The temperature distribution of micro-satellite orbiting can be properly designed and arranged by using radiator area, coating properties and structural materials to obtain allowable temperature working range. However, the design must be compromised with the requirements of other subsystem such as power system which needs more surface area receiving the energy from the sun to keep electrical components working.

Based on the design requirement of temperature range (-20 °C ~ 40 °C) in TUU-SAT1 thermal subsystem, the results are summarized as the table following.

stabilization	Gradient-Gravity		Spin	
Geometry	Square	Hexagon	Square	Hexagon
Radiator	$0.6m^2$ above	$0.4m^2$ above	$0.7m^2$ above	$0.5m^2$ above
Altitude	800 km ~2000 km			

ACKNOWLEDGEMENT

The authors would like to thank the paper viewers for their comments. The support of National Science Council for this research under the contracts, NSC-86-2612-E-008-001 and NSC-86-2612-E-008-002, is gratefully acknowledged.

NOMENCLATURE

A_i Surface area of node i
$B_{i,j}$ Absorptivity from node i to node j
C_I Specific heat of node i
D_j Conduction coefficient between node i and j
T_i Temperature of node i
α_{Si} Absorptivity of outer surface
m_i ---Mass of node i
P_i ---Inner power rate of node i
q_{Ti} ---Total heat source
S ---Solar intensity
E_e ---Radiation density of albedo
$E_{i,j}$ ---Radiation coefficient between nodei and j

β Angle between sun and orbit plane

ε_{ei}---Emissivity of node i

E_r Average radiation density of Earth reflection from sun

ϕ_1---Radiation shape factor of sun solar in node i

ϕ_2---Albedo shape factor of Earth in node i

$E_{i,s}$---Radiation coefficient from i node to Space

ϕ_{3i}---Shape factor if infrared radiation

Φ ---Angle between Earth, Sun and satellite

$L_{i,j}$---Conduction distance between node i and j

ρ ---Reflectivity

σ ---Stephan-Boltzman constant coefficient

k_j ---Conductivity coefficient between node i and j

$\triangle\tau$---Time step

REFERENCE

J. Gulpen and J.P.Bouchez, "OTS: Two Years of Thermal Control Experience in Orbit", Vol.19, No.1, AIAA 80-1500R.

Gilmore, “Satellite Thermal Control Handbook”, The Aerospace Corporation Press. EI SEGUNDO, 1994.

G. R. Min, “Thermal Control Technology”, China Aerospace Pub., 1991.(in Chinese)

J .R. Wertz and W. J. Larson, “Space Mission Analysis and Design”, Kluwer Academic Publisher, 1991.

P. Forescue and J. P. W. Stark, “Spacecraft System Engineering”, John Wiley and Sons Inc., 1991.

T. Appourchanx and G. Cislaghi, “Passive Thermal Control of a Multilayer Filter for Space-Based Solar Observation”, Optical Engineering, Vol.31, No.8, 1992, pp.1715~1722.

T. F. Pan, “Network Analysis of Steady Temperature of Satellite”, China Space Science Technology, 1987, No.6, pp.37~44.(in Chinese)

J. H. Chin, T. D. Panczak and L. Fried, “Spacecraft Thermal Modeling”, International J. for Numerical Method in Eng., Vol.35, 1992, pp.641~653.

Chobotov, “Orbital Mechanics”, AIAA Inc., 1991, pp.3~99.

Brown, “Spacecraft Mission Design”, AIAA Inc., 1992, pp.81~94.

S. V. Patankar, "Numerical Heat Transfer and Fluid Flow", Hemishphere Publishing, New York, 1972.

ROBUST PERFORMANCE DESIGN TO SATELLITE ATTITUDE CONTROL

Ciann-Dong Yang* (Professor to whom all correspondence should be sent) , Yun-Ping Sun[1](Ph. D. candidate), Chia-Yuan Chang[2] (Senior program engineer, program engineering management, local vehicle team)

[1] *Institute of Aeronautics and Astronautics, National Cheng Kung University, Tainan, Taiwan, 701, Republic of China, E-mail:p4883109@sparc6.cc.ncku.edu.tw*

[2] *Ford Lio Ho Motor Company Ltd., Chung Li 320, Taiwan, Republic of China*

ABSTRACT

A satellite has an astronomical survey mission requiring accurate pointing of a scientific sensor package. This package must be maintained in the quietest possible environment, which entails isolating it from the vibrations and electrical noise of the main service body. The pointing requirement arises when it is necessary to point the unit in another direction, it can be met by dynamics with a transient settling time of 20 seconds and an overshoot of no more then 15%. We model the satellite as two masses connected by a spring with torque constant *k* and a viscous-damper with damping constant *b*. Physical analysis leads us to assume that the parameters *k* and *b* vary as a result of large temperature fluctuation. The attitude pointing requirement must be satisfied for any parameter values in a prespecified range. We will synthesize the robust controller to achieve the specified performance requirement in the presence of parameter uncertainties.

INTRODUCTON

Engineers encounter the robust control problems with respect to structured mixed (real and complex) uncertainty in many practical applications. Such concept of mixed uncertainty provides control engineers with a more proper description and less conservative design to the general uncertain systems. Unfortunately the computation of structure singular value with respect to mixed perturbation sets is very complex and known to be NP (nonpolynomial) hard [1]. As a result, the mixed μ-synthesis becomes a more challenging problem.

There are two representative approaches, *D,G-K* iteration [2] and μ-*K* iteration [3], to mixed μ-synthesis. Although a few successful designs have been presened, the practical use of *D,G-K* iteration is obstructed by the need of fitting purely complex scalings with high order all-pass transfer functions as well as the need of fitting both in phase and magnitude. By comparison with *D,G-K* iteration, the advantage of μ-*K* iteration is that one need fitting in magnitude only ; however the disadvantage is that one need to compute not only the mixed but also the corresponding complex μ upper bounds at each iteration. Therefore, the μ-*K* iteration is still quite involved and computationally demanding.

A new approach, called H_2-based loop-shaping iteration, has been proposed and successfully applied to the H_∞-synthesis [4], the complex μ-synthesis [5] and the mixed μ-synthesis [6]. This H_2-based mixed μ-synthesis approach is totally different from the mentioned H_∞-based mixed μ-synthesis approaches (D,G-K iteration and μ-K iteration). The mixed μ controller is synthesized by performing a sequence of weighted H_2 optimizations, not H_∞ optimizations; therefore this approach brings about a lower computational demand.

REVIEWS ON μ

Consider a generalized augmented plant P partitioned as

$$P = \begin{bmatrix} P_{11} & P_{12} \\ P_{21} & P_{22} \end{bmatrix} \tag{1}$$

To test for robust performance, we first pull out the uncertain perturbations and rearrange the uncertain system into $N\Delta$-structure [7]. N is the lower linear fractional transformation of augmented plant P closed by controller K, and M is the upper linear fractional transformation of N closed by uncertainty Δ,

$$N = F_l(P,K) = P_{11} + P_{12}K(I - P_{22}K)^{-1}P_{21} \tag{2}$$

$$M = F_u(N,\Delta) = N_{22} + N_{21}\Delta(I - N_{11}\Delta)^{-1}N_{12} \tag{3}$$

The uncertainty Δ is a member of the bounded subset $\mathbf{B}\Delta$:

$$\mathbf{B}\Delta = \left\{\Delta \in \Delta_s \,\middle|\, \overline{\sigma}(\Delta) < 1\right\} \tag{4}$$

where $\overline{\sigma}$ denotes the largest singular value and Δ_s represents the structure perturbation set

$$\begin{aligned}\Delta_s = \{&\mathrm{diag}(\delta_1^r I_{r_1}, \cdots, \delta_{m_r}^r I_{r_{m_r}}, \delta_1^c I_{r_{m_r+1}}, \cdots, \delta_{m_c}^c I_{r_{m_r+m_c}}, \Delta_1, \cdots, \Delta_n) \\ &\big|\ \delta_i^r \in \mathbf{R}, \delta_i^c \in \mathbf{C}, \Delta_j \in C^{r_{m_r+m_c+j} \times r_{m_r+m_c+j}}\}\end{aligned} \tag{5}$$

The structured singular value μ of N with respect to Δ is defined as

$$\mu_\Delta(N) \equiv \frac{1}{\min\{\overline{\sigma}(\Delta) : \Delta \in \mathbf{B}\Delta, \det(I - N\Delta) = 0\}} \tag{6}$$

Give an H_∞ norm performance specification

$$\left\|F_u(N,\Delta)\right\|_\infty < 1 \tag{7}$$

Rearrange the uncertain system into $N\Delta$-structure by introducing a fictitious uncertainty block Δ_P, which itself is a full complex matrix stemming from the above H_∞ norm performance specification. Then the closed loop system is performance robust, i.e., $\left\|F_u(N,\Delta)\right\|_\infty < 1$ for all $\Delta \in \mathbf{B}\Delta$ iff

$$\left\|\mu_{\tilde{\Delta}}(N)\right\|_\infty \le 1 \tag{8}$$

where the enlarged perturbation set $\tilde{\Delta}$ is

$$\tilde{\Delta} = \left\{ \mathrm{diag}(\Delta, \Delta_P) \,\middle|\, \Delta \in \mathbf{B}\Delta, \Delta_P \in C^{k \times k}, \bar{\sigma}(\Delta_P) < 1 \right\} \quad (9)$$

ALGORITHM OF H_2-BASED LOOP-SHAPING ITERATION

Corollary 1. Suppose there exists a scalar frequency-dependent weight $\hat{W}(s)$ such that the optimal H_2 controller K^* obtained from

$$K^* = \arg\inf_K \left\| \hat{W} F_l(P, K) \right\|_2 \quad (10)$$

satisfies

$$\mu_{\tilde{\Delta}}\left(F_l(P, K^*)(j\omega) \right) = \left\| \hat{W} F_l(P, K^*) \right\|_2 \quad (11)$$

and $\left\| \mu_{\tilde{\Delta}}\left(F_l(P, K^*)(j\omega) \right) \right\|_\infty < 1$ (12)

then K^* is a H_2-based μ controller with robust performance ability.

The procedure of H_2-based loop-shaping iteration for mixed μ-synthesis is summarized as follows:

Initialization: $i = 0$

(i) Set $\hat{W}_0(s) = 1$
(ii) Compute $K_0 = \arg\inf_K \left\| \hat{W}_0 F_l(P, K) \right\|_2$ using H_2 optimization technique.
(iii) Compute $\gamma_0(w) = \mu_{\tilde{\Delta}}\left(\hat{W}_0 F_l(P, K_0)(j\omega) \right)$.
(iv) Compute $\lambda_0 = \left\| \hat{W}_0 F_l(P, K_0) \right\|_2$.

Recursive formula: $i = 1, 2, 3\ldots$
(i) Set $\deg(\hat{W}_i) = n_w$, where $\hat{W}_i(s)$ is a scalar, minimum-phase transfer function.
(ii) Fit $\gamma_{i-1}(\omega)/\lambda_{i-1}$ by $\left| \hat{W}_i(j\omega) \right|$.
(iii) Compute $K_i = \arg\inf_K \left\| \hat{W}_i F_l(P, K) \right\|_2$.
(iv) Compute $\gamma_i(\omega) = \mu_{\tilde{\Delta}}\left(\hat{W}_i F_l(P, K_i)(j\omega) \right)$.
(v) Compute $\lambda_i = \left\| \hat{W}_i F_l(P, K_i) \right\|_2$.
(vi) Repeat the recursive formula until no further reduction in $\left\| \mu_{\tilde{\Delta}}\left(F_l(P, K_i)(j\omega) \right) \right\|_\infty$ can be achieved.

Practically the structured singular value $\mu_{\tilde{\Delta}}\left(F_l(P, K_i)(j\omega) \right)$ may become flat and $\left\| \mu_{\tilde{\Delta}}\left(F_l(P, K_i)(j\omega) \right) \right\|_\infty$ approximates to a constant for a wide frequency range after a few iterations. The extent of flatness depends heavily on the accuracy of calculating $\mu_{\tilde{\Delta}}\left(\hat{W}_{i-1} F_l(P, K_{i-1})(j\omega) \right)$ and curve fitting $\gamma_{i-1}(\omega)/\lambda_{i-1}$ by $\left| \hat{W}_i(j\omega) \right|$.

Remark. The H_2-based loop-shaping iteration depends heavily on good curve fitting in recursive formula (i) and (ii), preferably by a transfer function $\hat{W}_i(s)$ of lower order. The main reason for preferring a low-order fitting is that it reduces the order of H_2 problem, which usually improves the numerical properties of the corresponding H_2 optimization and also yields a controller of lower order. However this way will probably cause a slow iteration convergence.

NONCOLLOCATED SATELLITE ATTITUDE CONTROL

Consider a satellite, which is suggested by the need to control the pointing direction, or attitude, in orbit about the earth [8]. Physical analysis leads us to assume that there are parameter variations as a result of temperature fluctuation. Moreover, in order to measure the attitude angle of the vehicle, the sensor and actuator are noncollocated. This makes the system much harder to control. The objective of this problem is to maintain an acceptable performance in the presence of parameter variations. The satellite structure is modeled as a simple spring-mass-damping system, where the spring constant k and damping constant b are assumed to vary in certain ranges. However, this simplified system has crucial features of many more complex space structures.

The equations of motion are derived as

$$J_1\ddot{\theta}_1 + b(\dot{\theta}_1 - \dot{\theta}_2) + k(\theta_1 - \theta_2) = T_c \tag{13a}$$

$$J_2\ddot{\theta}_2 + b(\dot{\theta}_2 - \dot{\theta}_1) + k(\theta_2 - \theta_1) = 0 \tag{13b}$$

$$0.09 \le k \le 0.4 \tag{14}$$

$$0.0036 \le b \le 0.04 \tag{15}$$

where T_c is the control torque and θ_2 is the output to be controlled. With inertias $J_1 = 1$ and $J_2 = 0.1$, the transfer function from T_c to θ_2 is

$$G(s) = \frac{10bs + 10k}{s^2(s^2 + 11bs + 11k)} \tag{16}$$

The pointing requirement of satellite arises when it is necessary to point the unit from one direction to another direction. Explicitly this can be specified by the dynamics of step response with settling time t_s less than 20 seconds and overshoot M_p no more than 15%. According to the performance specifications in time-domain, we select the performance weighting $W_p(s)$ by shaping complementary sensitivity function

$$W_p(s) = \frac{0.85(s^2 + 0.696s + 0.36)}{(s + 0.696)(s + 0.0001)(0.0001s + 1)} \tag{17}$$

After performing H_2-based loop-shaping iteration 8 times, we obtain a mixed μ controller with 14 states. The magnitude plot of $\mu_{\tilde{\Delta}}(F_l(P, K_i)(j\omega))$ is shown in Fig. 1. It can be easily recognized that the final curve is approximately flat and is actually less than 1, indicating that the robust performance requirement is certainly satisfied. Finally as shown in Fig. 2, the performance robustness achieved by the resulting mixed μ controller can be further verified by the step responses of 121 perturbed plants, where the parameters k and b are selected as:

$$k = \{0.245 + 0.155\delta_k \mid \delta_k \in [\pm 1,\ \pm 0.8,\ \pm 0.6,\ \pm 0.4,\ \pm 0.2,\ 0]\} \tag{18}$$

$$b = \{0.0218 + 0.0182\delta_b \mid \delta_b \in [\pm 1,\ \pm 0.8,\ \pm 0.6,\ \pm 0.4,\ \pm 0.2,\ 0]\} \tag{19}$$

REMARKS ON THE SATELLITE EXAMPLE.

In all approaches to μ-synthesis (including mixed μ-synthesis), a resulting high order controller is almost inevitable. By applying model reduction methods it is possible to reduce the controller order to a lower level with acceptable degradation in control performance. However, using controller approximation with fewer states implies an increase in μ. Still there are open problems with regards to numeric in connection with actual implementation in μ-synthesis.

The robust performance (RP) condition eq.(8) is identical to a robust stability (RS) condition with an additional perturbation block Δ_P. In other words, the RP condition is more severe than the RS condition. So there is no doubt of robust stability in this illustrative satellite case, where the robust performance condition is satisfied.

CONCLUSION

In this paper we extend the H_2-based loop-shaping method to mixed μ-synthesis. A noncollocated satellite's attitude control design is formulated to the mixed μ-synthesis problem. The resulting controller with robust performance property from μ criterion is obtained by a sequence of weighted H_2 optimizations. Also the simulation results show that the time-domain performance specifications on settling time and overshoot are completely satisfied under real parameter variations. This successful design stresses the superior ability of the H_2-based loop-shaping method for mixed μ-synthesis.

REFERENCE

Braatz, R. P., Young, P. M., Doyle, J. C., and Morari, M., "Computational Complexity of μ Calculation," *IEEE Transactions on Automatic Control*, Vol. 39, No. 5, 1994, pp. 1000-1002.

Young, P. M., "Controller Design with Mixed Uncertainties," *Proceedings of the American Control Conference*, Baltimore, USA, 1994, pp. 2333-2337.

Steen TØffer-Clausen, "Mixed μ Synthesis — μ-K Iteration," *System Identification and Robust Control — A Case Study Approach*, Springer, London, 1996, pp. 81-86.

Yang, C.-D., and Chang, C.-Y., "Synthesis of Optimal H_∞ Controllers via H_2-Based Loop-Shaping Design," *Journal of Guidance, Control, and Dynamics*, Vol. 19, No. 4, 1996, pp. 878-885.

Yang, C.-D., and Chang, C.-Y. "μ-Synthesis Using Linear Quadratic Gaussian Controllers," *Journal of Guidance, Control, and Dynamics*, Vol. 19, No. 4, 1996, pp. 886-892.

Yang, C.-D., Chang, C.-Y. and Sun, Y.-P. "Mixed μ-Synthesis via H_2-Based Loop-Shaping Iteration," *Journal of Guidance, Control, and Dynamics*, Vol. 21, No. 1, 1998, pp. 182-184.

Skogestad, S., and Postlethwaite, I. "Robust Stability and Performance Analysis," *Multivariable Feedback Control — Analysis and Design*, John Wiley & Sons, 1996, pp. 291-348.

Franklin, G. F., Powell, J. D., and Emami-Naeini, A., "Design of a Satellite's Attitude Control," *Feedback Control of Dynamic Systems*, 3rd ed., Addison-Wesley, 1994, pp. 669-684.

Table 1. Iteration results of satellite's attitude control.

Iteration No. i	$\left\|\mu_{\tilde{\Delta}}\left(F_l(P,K_i)(j\omega)\right)\right\|_\infty$	$\left\|\hat{W}_i F_l(P,K_i)\right\|_2$
0	2.5646	62.6941
1	1.6297	0.9103
2	1.3212	0.9279
3	1.1756	0.9271
4	1.0853	0.9181
5	1.1109	0.8975
6	1.0311	0.8923
7	1.1201	0.8863
8	0.9715	0.8820

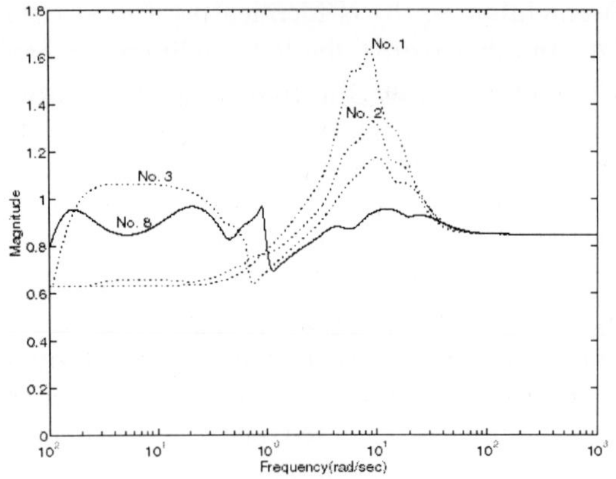

Figure 1. Results of H_2-based loop-shaping iteration on the satellite's attitude control. Shown are upper bounds of $\mu_{\tilde{\Delta}}\left(F_l(P,K_i)(j\omega)\right)$ for iteration No. 1, 2, 3, and 8.

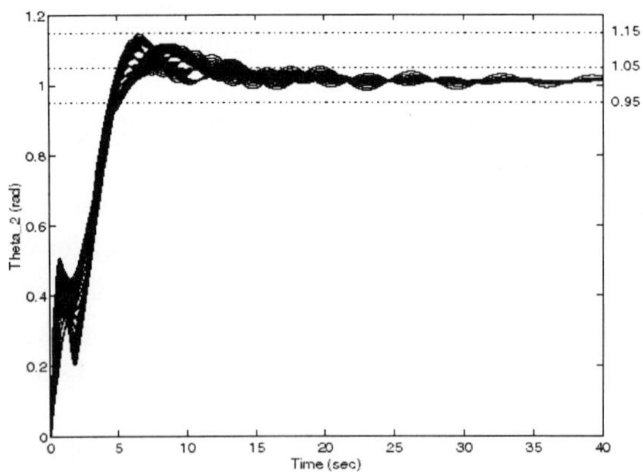

Figure 2. Step responses of 121 perturbed satellite plants with mixed μ controller, where the parameters $k = \{0.245 + 0.155\delta_k \mid \delta_k \in [\pm 1,\ \pm 0.8,\ \pm 0.6,\ \pm 0.4,\ \pm 0.2,\ 0]\}$, and $b = \{0.0218 + 0.0182\delta_b \mid \delta_b \in [\pm 1,\ \pm 0.8,\ \pm 0.6,\ \pm 0.4,\ \pm 0.2,\ 0]\}$.

AN EUV SPECTROMETER FOR MONITORING THE AURORA

Roberto Stalio[1], A. Lyle Broadfoot [2]

[1]*CARSO and University of Trieste* , [2]*LPL West, University of Arizona, Tucson*

SCIENCE OBJECTIVES

Aurora is a spectrometer optimized for measurements of the EUV flux of the aurora and the Earth plasmasphere in the emission lines of O^+, He and He^+ . It is designed to be small and to operate from a microsatellite. It is basically an optical bench consisting of two grazing incidence spectrographs (channels). The first channel, EUV-1, operates in the wavelength band from 15 to 45 nm while the second channel, EUV-2, is sensitive from 55 to 85 nm. Aurora spectral resolution allows to obtain integrated fluxes of the aurora and/or airglow in the lines of He at 58.4 nm and of O^+ at 83.4 nm with the EUV-2 spectrometer. The EUV-1 channel provides integrated fluxes of the He^+ Lyman-α line at 30.4 nm. The detector system, intensified CCD, and the instrument electronics are derived from our experience with UVSTAR [(1) (2)]. A prototype of *Aurora* has flown attached to UVSTAR on the Discovery Shuttle flight no. STS-85.

Aurora is aimed at monitoring the EUV most intense line fluxes emitted by the Earth upper atmosphere. The goal is to complement the numerous studies of the aurora and the plasmasphere made from ground based observatories and from satellites: measurements of the aurora in the EUV He and O^+ lines allows a more detailed study of the energy balance in the upper atmosphere of the Earth. This energy is built by both the absorption of UV sunlight and by the auroral energy generated by particle precipitation and Joule heating. It is especially during magnetic substorms, which is when the aurora is best visible at optical wavelengths, that 1) the balance between auroral and solar radiation input and 2) the relative densities of the auroral energy (confined to the auroral ovals) and the solar radiation energy (distributed over the sunlit hemisphere) are less understood. Another of the key questions of the auroral physics is the composition and the origin of the precipitating particles, as well as their distribution in the polar upper atmospheric regions. Finally, the long term monitoring characteristics of the program contributes to the understanding of the physics and dynamics of the aurora during the various phases: quietness, growth, expansion, etc. The EUV-1 spectrometer, which measures the He^+ line at 30.4 nm, provides the link with the solar radiation flux through measurement of the scattering properties of the EUV solar photons with the Earth plasmasphere particles and, at the same time, the mapping of the He^+ distribution.

In summary the simultaneous measurements of the Earth upper polar atmosphere fluxes at the EUV wavelengths of neutral, ionized helium and ionized oxygen are the main characteristics of the program which has as the main objectives:

- to study of the energy balance in the aurora and characteristics of the particle fluxes;
- to produce a temporal mapping of the plasma density and distribution even during periods of

sudden and intense variations;

- to correlate of the aurora phenomenon with the plasmasphere ion flux and the solar activity cycles;
- to study the global dynamics of both aurora and plasmasphere in order to discriminate between different theories of plasma density variations during strong geomagnetic activity.

TECHNICAL DESCRIPTION

Aurora is a grazing incidence spectrometer of low mass and reduced size. Its size has been studied to fit into a microsatellite of the class up to 50 kg, including the scientific payload. The microsatellite, a cube of 40 cm, is built by a newly formed Italian company, Megsat. The satellite power is generated by GaAs solar arrays both body mounted and deployable. The total power is about 35 W while the continuum power is about 25 W. The satellite orbit is circular with an altitude of 750-800 km and an inclination between 70 and 100 degrees. The attitude, LVLH, has three axis stabilization by means of wheels. Data transmission occurs once a day at a rate of 64 Kbit/s; the ground station is located in Italy.

Aurora spectrometers integrates over the O^+, He, He^+ EUV fluxes every 0.1 sec to maintain the nominal field of view of 3.5× 6 degrees. Each acquisition provides three values of monochromatic flux; each observing period lasts 5 minutes and allows to cover an area of about 3.5× 26 degrees. Observations will be made when the satellite is near the celestial poles.

The necessary sensitivity is obtained with the following technical choices for the basic instrument:

◊ 2 grazing incidence spectrometers,
◊ 2 intensified diode detectors,
◊ stable electronics.

Further increase in sensitivity could be obtained with the following improvements:

◊ anticoincidence system,
◊ chopping system,
◊ internal and external baffling,
◊ extra-stable electronics,,
◊ microchannel plates with low potassium content,
◊ holographic gratings. .

In figure 1 we display the opto-electronic instrument and the accommodation of its components: the two grazing incidence spectrometers, the detector, the slits are easily seen. The EUV-1 spectrometer covers the wavelength band 15-45 nm; the EUV-2 spectrometer the band 55-85 nm. The EUV-1 grating has 1065 lines/mm, an area of 25× 10 mm and a curvature radius of 230 mm; the EUV-2 grating has 950 lines/mm, the same area as EUV-1 and a curvature radius of 250 mm. The gratings are gold coated. The spectrograph slits are closed

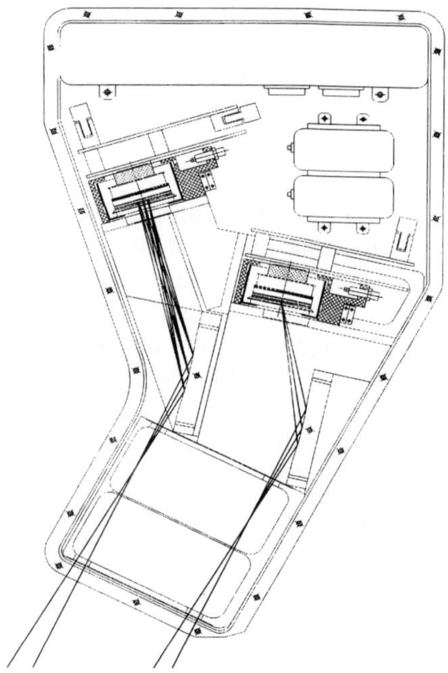

Figure 1. Optomechanical scheme of *Aurora* .

during integration in the laboratory and before launch. During these phases the spectrographs are filled with ambient pressure nitrogen. The slits will be open in flight to collect the radiation through an aperture of $0.3\times\ 5$ mm.

The EUV radiation incident on the gratings is dispersed and focused onto the detectors. The detectors are intensified CCD of the type used for UVSTAR [1][2]. The intensification stage is a " are" MCP of 2.5 cm diameter. Each intensified image is formed onto an anode covered by phosphor and is transferred by means of an optical fiber bundle to the CCD for the read-out operation. In practice, due to the dominant emission strength of the spectral lines in study over any other signal, the EUV-1 detector measures He^+ and the sky background while the EUV-2 detector measures O^+, He and the sky background. Since the spectrometers have very limited imaging capabilities, we basically have a photometric system: each 0.1 sec the EUV-1 channel measures the He^+ signal from an angular area of about 0.006 sr while the EUV-2 channel measures the O^+ and He emission.

The improvements described above could increase the instrument sensitivity by an estimated factor from 10 to 50. Another improvement which could be implemented is to open three slits of $3.5\times\ 6$ degrees onto the EUV-1 channel to obtain simultaneous mapping of three regions using the same grating and the same intensified CCD. In this way *Aurora* would have the monochromatic EUV-1 channel collecting data from three different plasmasphere regions and the EUV-2 channel collecting date at two wavelengths from a single region.

Aurora electronics is based on a 24 MIPS Analog Device fixed point DSP system. The system has 4 modules: CPU board (figure 2), interface and housekeeping board, detector board, voltage converters. The main characteristics of the CPU board are:

- memory: 32 Kbyte RAM data, 32 Kbyte RAM program, 128 Kbyte EEPROM;
- 12 bit AD converter for science data acquisition and housekeeping monitoring;
- 2 synchronous serial doors with up to 12 Mbaud velocity;
- DAC gain control of the intensifiers.

During operations the satellite central processor acts as a master: it turns on/off Aurora and communicates with the instrument electronics via synchronous serial line. In figure 2 we present the functional scheme of the CPU board.

CONCLUSIONS

Aurora will fly on Megsat-1 at the end of 1999.

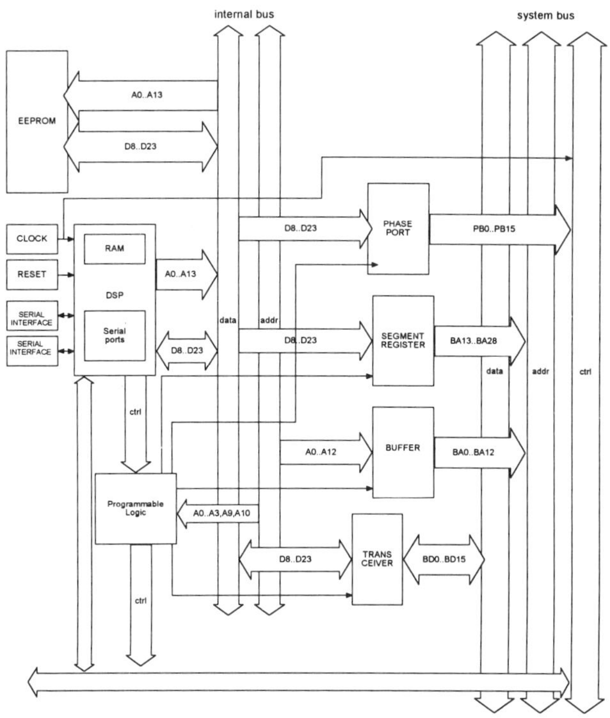

Figure 2. CPU board functional scheme.

REFERENCES

UVSTAR: an imaging spectrograph with telescope for the shuttle Hitchhiker-M platform, B.R. Sandel, A.L. Broadfoot, R. Stalio, 1993, in "Optical Engineering", 32, 3009

Preliminary scientific results from UVSTAR: the FUV spectra of BD+28° 4211 and NGC 246, R. Stalio, V. Bratina, J. Holberg, 1998 in " IUE Final Archive Symposium", ESA-SP, in press

A STUDENT SATELLITE PROJECT

K. C. Hsieh[1] and C. A. Lewicki[2]

[1]*Department of Physics, University of Arizona, Tucson, AZ 85721, USA*
[2]*Department of Aerospace & Mechanical Engineering, University of Arizona, Tucson, AZ 85721, USA*

ABSTRACT

Student satellites are not new. Many such projects are represented at this colloquium. We report on the Student Satellite Project at the University of Arizona that is unique in its science-driven initiative and broad-based participation. In addition to describing the satellite, we shall emphasize the importance of interdisciplinary collaboration in science, the cross-fertilization between science and engineering, the leadership role of the students, and the partnership among the university, the community and industry. The progress of the project will be described in terms of its procedures and structures. Presently, it is too early to set a launch date for this particular student satellite.

INTRODUCTION

A satellite carrying out its mission in orbit is truly a marvelous work of our minds and hands. It harnesses the energy required to keep it functional for a purpose. It has sensors connected to an onboard computer to collect and process the data it was designed to gather. It delivers the gathered data to its operators on Earth. It obeys the commands sent from Earth to perform specific tasks required of its mission. Yes, an operating satellite is truly a marvelously complex system. The marvel of a microsatellite is perhaps best described by a Chinese saying about sparrows, "Small though it is, it is complete with all its organs." Such a small and yet so complete a complex system and our innate fascination with space invite us to consider seriously a microsatellite as a useful tool for education.

Student satellites are not new, but how each one comes about from inception to operational makes it unique. Here is a story of a beginning and the beginning of a story of a student satellite project, the Student Satellite Project at the University of Arizona, which is just one year old since its inception on 7 November 1996. We shall describe the project's present status, the process that brought it about, its features and the favorable conditions surrounding it, and finally, what we expect to benefit from it.

WHAT DO WE HAVE?

On 22 November 1997, the Student Satellite Project at the University of Arizona -- SSP for short -- passed its Conceptual Design Review conducted by the Evaluation & Selection Panel (ESP). The satellite, yet to be named and built, will be one suitable for inserting into a low Earth orbit by the Hitchhiker Ejection System (HES) on NASA's Space Shuttle. The HES is chosen from the outset for its frequent and predictable schedule and orbit, and its well-defined design envelope, in comparison to some of the "piggy-back" rides.

Figure 1 shows the dimensions of the HES. Our satellite will be a circular cylinder 42 cm in diameter and 52 cm in length with a mass of 68 kg, all of which befits the HES. Carrying out the designated scientific objectives of the mission, i. e. a) Detection and observation of lightning, b) Detection and observation of sprites and c) UVBRI stellar photometry, a refractive optical system shall use the central axis of the cylindrical satellite as its primary axis. A block diagram of the science instrument is shown in Figure 2. This single optics system will provide for all of the science observations, as well as serve as the system for the laser communications experiment.

Viewing the lightning and sprites is best done by looking at Earth's limb, at the proposed orbit of 400 km in altitude. Performing stellar photometry, on the other hand, requires the instrument to look the night sky further above the limb. If laser communication with ground is to be tested, the instrument must then be nadir pointing. To accommodate the three different pointing requirements, the three-axis stabilized satellite will be steered by two pairs of student-designed reaction wheels with a pointing accuracy of approximately 0.5°. The satellite will determine its attitude through a suite of sensors including a coarse sun sensor, magnetometer, micromechanical gyros, and a crude star tracker utilizing the science instrument. A state diagram of the GNC system is illustrated in Figure 3.

Processing the data and controlling the satellite requires an onboard micro-processor that will be able to address the many activities being carried out on the satellite, each with their own priority. Of particular challenge is the bursts of high data rate that the laser experiment and science instrument will produce. Trade studies are currently underway to determine what approach (centralized or distributed) will provide the best solution to this challenge.

Our telemetry system is fortunate enough to have a head start on the SSP, as the ground station was already under construction at the time of the inception of SSP. Members of the University of Arizona Students for the Exploration and Development of Space, and the Ham Radio Club had begun a satellite ground station project in support of another student satellite: SEDSAT (Maier and Wu, 1998). SSP is modeling as much of its approach to its ground station as possible after the SEDSAT approach, to limit duplicate development. A picture of the ground station under construction is shown in Figure 4. The satellite portion of the telemetry system is currently planning to utilize wavelengths in four Amateur bands (2m, 70cm, 23cm, 13cm). The lower frequency bands will provide a packet store and forward service to the amateur radio community, while the higher frequencies will be used for command uplink, and data downlink. In particular, it is hoped that the S-Band (13cm) link along with advanced modulation schemes can provide data rates in upwards of 1Mbps!

To perform all the above described tasks required of the mission objectives, the circumference of the satellite will be covered with a thin film solar cell that is currently under development by Global Solar, a subsidiary of Tucson Electric Power. Commercial Ni-CD D-Cells are the current plan for

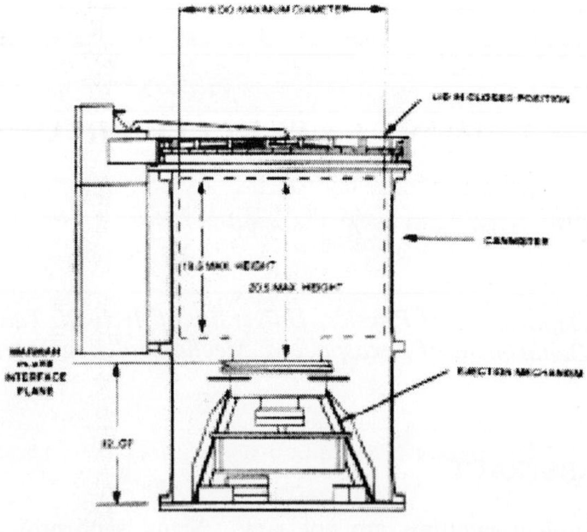

Figure 1. Hitchhiker ejection system

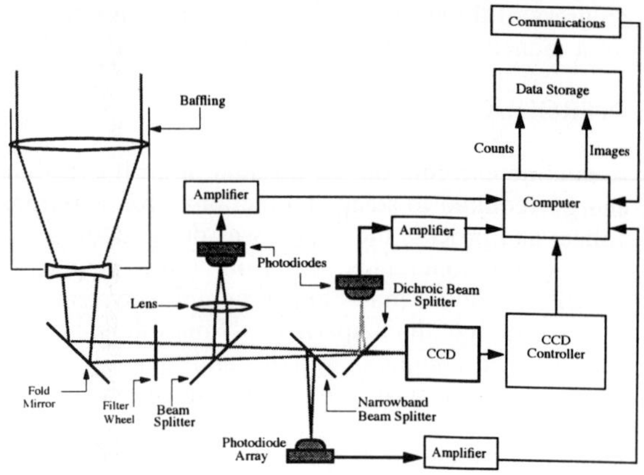

Figure 2. Science instrument block diagram.

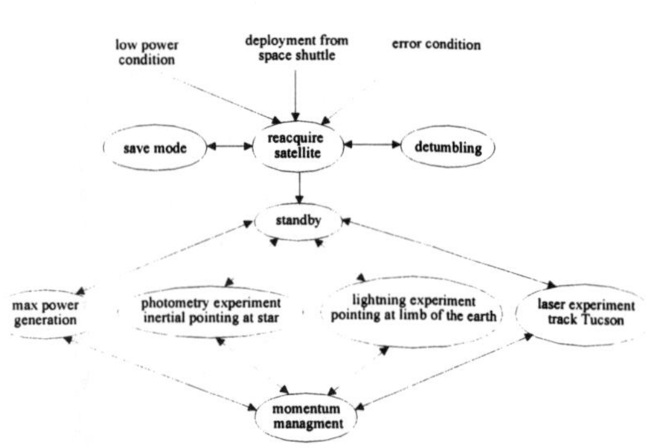

Figure 3. State diagram of GNC system.

primary on-board power storage. Additional battery capacity may be added in order to support high power draw when the laser downlink experiment is operating. This separate power block would be trickle-charged between tests of the laser experiment. The photovoltaic efficiency of the $CuInSe_2$ (CIS) cells is estimated at 8-10%, although this efficiency is comparable with other cells due to its increased performance over a wider temperature range. With the system described, a continuous power capability of between 20 and 35 watts is estimated.

Figure 4. SSP Project Lead Chris Lewicki, assembling satellite groundstation, on a cloudy day.

With the conceptual design described above, SSP will begin its Preliminary Design Phase in January 1998. The Preliminary Design Review (PDR) is tentatively scheduled for April 1998. Meanwhile, the Hitchhiker Launch Manifest Request (Form 1628) will be submitted to NASA, requesting for a launch sometime in 2000. The sequence of tasks accomplished and events yet to come is illustrated in Figure 5.

HOW DID WE GET TO WHERE WE ARE?

The idea of a satellite designed, fabricated and operated by students at the University of Arizona emerged in an informal conversation between some faculty members of the Department of Aerospace & Mechanical Engineering and K. C. Hsieh, after his seminar on 7 November 1996. What followed was a series of informal meetings involving a handful of interested faculty members from Aerospace & Mechanical Engineering, Astronomy, Atmospheric Science, Electrical & Computer Engineering, Optical Science, Physics, and Planetary Science. Although the initiative came from the Department of Physics, all participants recognized from the very beginning that this project has to be interdisciplinary in nature and in practice. The use of electronic mailing across campus enabled the rapid communication among participants, so that no paper documents were exchanged and meetings were kept infrequent.

At first, the faculty members in discussion were oblivious of the burgeoning activities of the Students for the Exploration & Development in Space (SEDS) on campus. The president of that national organization for 1996 was an Aerospace & Mechanical Engineering student in his senior year, C. A. Lewicki. It was through S. A. Brew, coordinator of the Arizona Space Grant Consortium, a state-wide NASA funded program, the connection between the faculty in planning and C. A. Lewicki was made; and student participation began in the planning phase of SSP. With the support of Arizona Space Grant, J. Rademacher of NASA's Jet Propulsion Lab, a graduate of the Arizona State University and the first student leader of the ASUSat, was brought in to describe his experience with ASUSat and programs at JPL that may be of assistance to SSP.

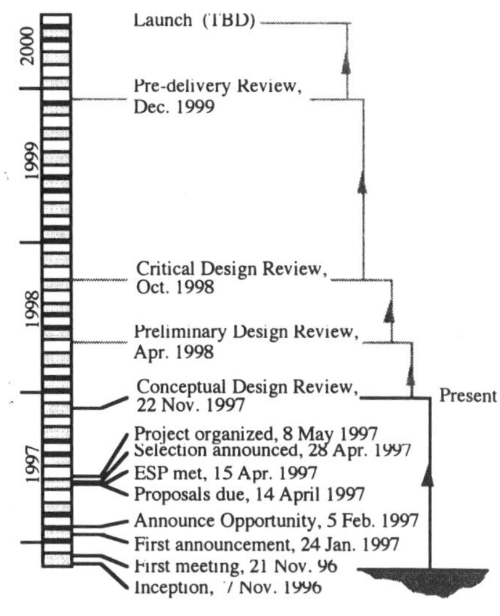

Figure 5. Milestones of SSP at the University of Arizona.

After three meetings, the basic ideas of SSP appeared sound and worth pursuing. By then, a SSP website prepared by A. Levine, Communications Director of the Physics Department, and her student assistant was ready to become accessible on campus. On 24 January 1997, SSP made its first announcement on the website. On 5 February 1997, the Announcement of Opportunity followed. Students began to register on the website indicating the areas of their respective interests and intentions. By the end of February 1997, over 100 students indicated their intent to proposed. (See Table 1 for the distribution of students in academic standings and disciplines. Students with double majors appears as 0.5 in each of the majors)

Table 1. Tally of student distribution

	Fr.	So.	Jr.	Sr.	Grad	Totals
Aerospace Engineering		1	1	4	1	7
Agr. Engineering			1		1	2
Astronomy		0.5	2	1	2.5	6
Atmospheric Physics					2.5	2.5
Chemistry			1	1		2
Chemical Engineering		2	1			3
Civil Engineering			1			1
Computer Engineering		2	5.5			7.5
Computer Sciences	1.5	2	3.5	4.5		11.5
Electrical Engineering		4	6	1	1.5	12.5
Engineering Math			1.5			1.5
Engineering Physics	1		1.5	1		3.5
English				0.5		0.5
Geosciences			0.5			0.5
Hydrology				1	1	2
MIS				0.5	1.5	2
Math	0.5	0.5	3	2		6
Mechanical Engineering		4	4	3	3	14
Optical Engineering		2		1	0.5	3.5
Optical Sciences					1	1
Physics	1	1.5	6	2.5	3	14
Planetary Sciences		0.5	1		0.5	2
Pre-Med			0.5			0.5
Systems Engineering				1		1
	4	20	40	24	19	107
17 students withdrew during semester						

Working together, H. S. Tharp of Electrical & Computer Engineering, E. D. Fasse of Aerospace & Mechanical Engineering, E. K. Hege of Astronomy, and C. A. Lewicki assigned the students from 22 departments spanning from sophomore to graduate students into 17 teams -- 5 for payload instrumentation, 12 spread among the five subsystems, mechanical structure & analysis, data & command handling, guidance, navigation & control, tracking, telemetry & command, and power generation & distribution. The teams were instructed to organize themselves, seek their own mentors, and decide on their own approaches to their proposals. The proposals were due on 14 April 1997. In anticipation of the 17 proposals, the Evaluation & Selection Panel (ESP) was formed. We are extremely fortunate to have the highly qualified experts in a variety of fields willing to serve on the ESP (see Table 2). ESP has a significant representation from the well-established space industries.

All proposals were submitted on time and most of them in electronic form. The Chair of ESP, R. D. Lorenz, distributed the copies to the ESP at their first meeting the next day. The fact that some members had to drive 200 km to the meeting did not result in any absence. At the end of the two-hour meeting, each member went home with a stack of proposals to review. By mid-night two days later, reports from all ESP members were received by R. D. Lorenz in electronic form for collation and distribution to all ESP members. After much discussion over the electronic mail, the summary of the selection process was announced by ESP member, D. R. Huffman, to the students gathered for the first time on 28 April 1997. The result of the selection set the objectives for the first SSP mission and the goal for the Conceptual Design Review, mentioned at the beginning of this presentation, and set the stage for the organization of SSP, as shown in Figure 6. The Team Mentors who helped in the preparation for the Conceptual Design review are listed in Table 3.

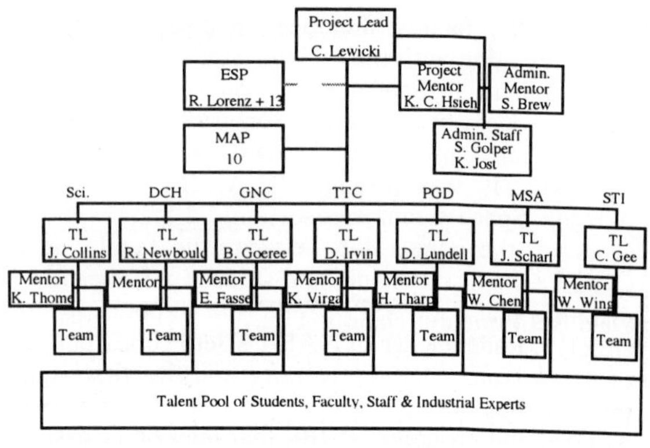

Figure 6. Organization chart of the Student Satellite Project at the University of Arizona as of end of 1997. The structure will evolve as SSP enters into different phases with different requirements.

Table 2. The Evaluation & Selection Panel (ESP).

[LPL = Lunar & Planetary Lab; AME = Aerospace & Mechanical Engineering; ECE = Electrical & Computer Engineering; MMT = Multi-Mirror Telescope. All these are working units of the University of Arizona.]

Name	Current Position	Expertise
Ralph Lorenz (Chair)	Res. Assoc. LPL	Nine years of experience on small satellites at Univ. of Surrey, Univ. of Kent, & ESA.
Jill Bechtold	Assoc. Prof., Astronomy	Active in space-based astronomy. Member of NASA selection panels.
Bob Brown	Prof., LPL	Planetary surface science. Mars missions.
Chuan F. (Tony) Chen	Prof., AME	Fluid dynamics & complexity
Don Huffman	Regents Prof., Physics	Optical properties of matter. Nobel nominee for his work on C_{60}.
Don Hunten	Regents Prof., LPL	Senior member of space physics & planetary aeronomy communities. Member of National Academy of Sciences and various national panels.
Matthew Jones	Asst. Prof., AME	Radiative heat transfer
Ron Jost	Chief Engineer, IRIDIUM Project, Motorola	System engineering, integration and testing. DoD & NASA programs since 1969.
Bill Kerwin	Prof. Emeritus, ECE	Analog & digital circuits
Pitu Mirchandani	Head, System & Industrial Eng.	Complex systems
James Palmer	Assoc. Res. Prof., Optical Sci.	Spacecraft solar-cell engineering.
Ed Pierce	Adj. Assoc. Prof., ECE; Senior Engineer, Hughes	Microwave specialist. 34 yrs. with Hughes from design to project management.
John Reagan	Head, ECE	Space radiometry & remote sensing.
Rich Van Riper	Chief Engineering Fellow, Honeywell Satellite Systems	Attitude control & guidance, data handling systems. 35 yrs. in space engineering.
Richard Schotland	Prof., Atmospheric Sciences	Space-borne LIDAR experiments.
Bobby Ulich	Vice-Pres., Kaman Aerospace Res. Prof., Astronomy Chief Engineer, MMT	Electro-optics development. First space-borne adoptive optics.

Table 3. Team mentors

Team	Mentor/Title	Department
Science	Kurt Thome, Asst. Prof.	Optical Sciences
Data & Command Handling	Vacant	
Tracking, Telemetry & Command	Kathleen Virga, Asst. Prof.	Electrical & Computer Eng.
Guidance & Navigation Control	Ernie Fasse, Asst. Prof.	Aerospace & Mechanical Eng.
Power Generation & Distribution	Hal Tharp, Assoc. Prof.	Electrical & Computer Eng.
Mechanical Structure & Analysis	Wayne Chen, Asst. Prof.	Aerospace & Mechanical Eng.
Strategic Technology Initiative	Bill Wing, Prof.	Physics and Optical Sciences

The organization of SSP will evolve as different needs arise. For example, at the Conceptual Design Review, the ESP rightly pointed out the need for a System Team to ensure the interface requirements between the subsystems are clearly defined and carefully monitored. This recommendation will be heeded as SSP moves into its Preliminary Design Phase. Obviously, still later, other teams will be formed to meet the new needs; e. g., an Integration Team to perform the task of integration and testing prior to delivery. The evolution will continue as the tasks progress.

The SSP could not get to its present state relying on enthusiasm and voluntarism alone. Expenses for student support (especially in the summer), student travel to the Annual Conference of the American Institute of Aeronautics and Astronautics and Utah State University Small Satellite conference in the neighboring state of Utah, communications, and some computer and clerical supplies had to be paid. Again, SSP has been fortunate to have received a total of $52,703, including the first prize awards at the Student Showcase on campus, as listed in Table 4.

Table 4. Funding received as of November 1997

Source of Funding	Amount	Remarks
Department of Physics	$5,000.	Summer student stipends; office supplies; and partial support for student travel to AIAA in Utah.
Prizes from Student Showcase	$500.	SSP won first place in engineering undergraduate and first place in engineering graduate.
Department of Atmospheric Sciences	$5,000.	Summer student stipends
WAESO (Western Alliance to Expand Student Opportunities)	$3,386.	Two Undergraduate Grants $1100 Materials & Supplies $286 Admin. Fee $2000 Student Stipends
Space Grant	$38,187.	$6785 5 undergraduate students $25402 1 graduate fellowship $6000 Cash and 486 computer $68 Joel Rademacher's airfare to informational meeting at UA $1000 Partial travel to AIAA in Utah

In addition to the generously contributed funds, SSP also received contributions in kind. These include the use of services and spaces in different departments. For example, the Department of Physics provided the crucial service of establishing a SSP website that made publicity and out-reach possible within one month of SSP's inception and the use of a lab space as the SSP Headquarters. The Aerospace & Mechanical Engineering Department also provided the use of two teaching labs for the MSA and the GNC teams, and the use of the Department's machine shop facilities.

WHAT IS SO SPECIAL?

The SSP at the University of Arizona is indeed very special in what has already been described above and in the additional features to be described below.

Student-run. After the announcement of the result of evaluation and selection on 28 April 1997, the original 17 proposing teams re-grouped into 7 teams, as shown in Figure 6. Each team elected its own Team Leader. Except the DCH Team, all teams found their respective Team Mentors. With or without its Team Mentor, each team met weekly to carry out self-assigned tasks towards its defined goal. The Team Leaders, fully aware of its own team's progress and problems, met weekly with the Project Coordinator, C. A. Lewicki., to ensure the smooth progress of the project by mutual assistance and coordination. All meeting agenda and minutes were distributed through the electronic mailing network and archived electronically in a central location accessible to all. The timely completion of the Conceptual Design Review is a proof of the strength of the students, because most of the work in preparation for the Review was done during the summer months of 1997, when most of the mentors and advisors, including the Project Mentor, were away from the campus!

Abundant talents. The smooth progress of SSP up to now has to be attributed to the highly motivated students coming from all disciplines that are necessary for the complexity of the project. These students demonstrated their ability to obtain necessary technical knowledge and to work together in harmony. In addition, we have right on campus all the experts in space science and technology. A group of experts on and off campus offered themselves to meet SSP's needs on a day-to-day and issue-by-issue basis form the Mission Advisory Pool (MAP), as shown in Table 5. A wide variety of expertise of the members of MAP can be seen in Table 5. Not to be forgotten is the pool of talents that are not yet involved, but potentially available. This is represented as the base of Figure 6.

Table 5. Mission Advisory Pool (MAP)

Name	Current Position	Expertise
Steve Bell	Senior Staff Eng., LPL	Space-borne electronics
Bill Bickel	Prof., Physics	Particle accelerator for detector calibration and radiation effects on space-borne instruments
Lyle Broadfoot	Senior Res. Scientist, LPL	Space-borne EUV imagers. PI on numerous space missions.
Matt Cheselka	Res. Specialist, Steward Obs.	
Elliot Chu	Asst. Prof., Physics	Particle detection & digital electronics
Charles Curtis	Res. Assoc. Prof., Physics	Space-borne particle detectors. Co-I on numerous space missions.
Roger Davies	Assoc. Prof., Atmospheric Sci.	Cloud physics from space
Eustace Dereniak	Prof. Optical Sci.	IR detector & imagers.
C. Y. Fan	Prof. Emeritus, Physics	Space physics since 1950's
Uwe Fink	Prof., LPL	Stellar photometry
Sy Goodman	Prof., Managem. & Info. Systems	Management
Larry Head	Asst. Prof., Sys. & Industrial. Eng.	Complex systems.
Keith Hege	Assoc. Astronomer, Steward Obs.	Opto-electronics
Ben Herman	Head, Atmospheric Science	Remote sensing via GPS
Jeff Jacobs	Assoc. Prof., AME	Fluid dynamics
Philip Krider	Prof., Atmospheric Science	Lightning & sprites
Larry Lebofsky	Sr. Res. Sci., LPL	Solar system small bodies
Alfred McEwen	Assoc. Res. Sci., LPL	Remote sensing surfaces
Kumar Ramahalli	Prof., AME	Space-borne new materials
Bill Sandel	Sr. Res. Sci., LPL	Space-borne EUV imaging
K. R. Sridhar	Assoc. Prof., AME	Spacecraft engineering
Tom Vincent	Prof., AME	Control & guidance.

Broadly based interest & support. Aside from the initial support received from the University in the form of faculty and staff volunteers and contributions of funds, services and facilities, SSP has already received the attention it needs from the industry and local community. Already mentioned is the dedicated service of experts from industries such as, Motorola, Honeywell, Hughes, and Kaman to the ESP. Not mentioned in the listing of members in the MAP (Table 5) are six staff engineers from SatCON, covering mechanical components, as well as electronics, power, guidance and control aspects of a spacecraft. Other industrial partners have also offered advisory help. Talks of possible material support coming from industries are already in the planning. Another important sector has also shown interest in supporting SSP. This is the local community of Tucson. On 4 June 1997, K. C. Hsieh gave a presentation to the Greater Tucson Economic Council, a body of community leaders and financial magnates interested in improving the economic lot of the City of Tucson and its surroundings. The reaction was enthusiastic. Ways to attract industrial and financial support for SSP are being planned under the leadership of D. Finley and R. Walkup, both of the Greater Tucson Economic Council.

WHAT CAN WE EXPECT FROM SSP?

The overwhelming response from the students, faculty, staff, industries, and community to SSP has its base on at least five potential benefits SSP could offer. These are:

1. A hands-on experience through teamwork in the design, fabrication, testing, and operation of a complex system with a well-defined objective
2. A needed channel for many students to gain self-confidence and employable skills
3. An example of intercollegiate, inter-departmental, and interdisciplinary collaboration
4. An avenue to enhance beneficial interactions among university, industry and community
5. A test-bed for innovative ideas in a wide variety of areas

Each of these benefits is delineated below.

1. A hands-on experience through teamwork in the design, fabrication, testing, and operation of a complex system with a well-defined objective

There is no substitute for experience. Bad experience may teach us caution and good experience courage; but no experience teaches us nothing.

Complexity and specialization are inseparable. As individual and societal goals become harder to achieve, more complex systems and devices are sought. The more complex the system or device becomes, the more specialized skills and techniques in a wide variety of fields are needed. The whole process of making a complex system work, from conceptual design to fully operational, demands closely coordinated large-scale teamwork. Such teamwork requires people skills as much as technical skills. The development of a complex system also demands time -- a duration longer than the average dwell time of its participants. Such a disparity in time-scale brings on the challenge for patience and dedication to the continuity of a larger self. A modern university is to provide its students the opportunity to face these challenges and to develop these skills.

Test-oriented courses, result-driven research projects and departmentally-defined programs are necessary building blocks of a university. They alone, however, do not provide the kind of learning just described. Talents already nested in these building blocks, however, can indeed be organized to provide a "hands-on experience through teamwork in the design, fabrication, testing, and operation of a complex system with a well-defined objective " to the students to gain the experience and the varied skills that are more and more demanded of them upon graduation.

2. A necessary channel for many students to gain self-confidence & employable skills

Realistic self-confidence enables further endeavors, and it grows out of excelling in one's performance along the way. Not all students can excel in classroom-oriented and test-driven learning. Students who excel in test-oriented learning are not necessarily profitably employable. If the university can provide a training ground for professional athletes, who do not always excel in regular classes, there should be room for programs that will enable a larger number of students, who may not necessarily excel in regular classes, to become competent workers in a variety of fields.

As a state university, we serve a large number of students with a wide range of interests and abilities. We must try to provide additional modes of learning to attract and retain a larger cross section of students to become future productive citizens.

3. An example of intercollegiate, inter-departmental, and interdisciplinary collaboration

The knowledge and skills required for SSP thread through departments and colleges, within our university. The breadth of experience to be offered by SSP should surpass all existing interdisciplinary programs. What is learned in the process of making SSP a success can be beneficial to other interdisciplinary programs addressing other complex systems. An example of a more practical, but far more complex system, could be the design, construction and operation of a model water-supply system that could adequately, safely and economically serve a community living in a delicate environment.

While much of the problem-solving along the way utilizes analytic tools and approaches, the entire project offers a holistic view and integrating process of working together towards a common goal. The SSP and similar kinds of projects or programs re-affirm the concept of "UNI" in our "university".

4. An avenue to enhance beneficial interactions among university, industry and community

A holistic view of the university must include the community. From the community, the university draws its life-giving resources in the form of money, goods, employees, and most of all, its students. To the community, the university returns life-enriching truth, beauty and goodness in the form of profound and innovative ideas, useful technology and services, and most of all, responsible and productive graduates.

The University of Arizona is a state institution located in a city that is equipped and working for economic development through "hi-tech" industries, especially in optical- and aerospace-based science and technology. The SSP will intensify the interactions among industry, the community and the university. For example, mutual benefits are expected when the Mechanical Structure & Analysis (MSA) Team of SSP interacts with the Composite Airframe Program at Pima Community College. Another example could be SSP's use of test facilities and expertise at Hughes as SSP trains skilled potential employees. The selection of the instrument and scientific objectives for SSP's first satellite stands as an example of the electro-optics based science and technology linking the University of Arizona with the local industries.

5. A test-bed for innovative ideas in a wide variety of areas

Generating and testing new ideas are intrinsic to the mission of a university. Unlike most of the research projects -- each conducted by a faculty and professional staff plus one or two graduate students -- SSP is conducted by undergraduate and graduate students. In addition to new ideas generated or stimulated by students, rigid rules and regulations can also be made more rational and tolerant to allow testing new ideas under much less external pressure. This is in obvious contrast to the way NASA runs its missions. SSP can truly serve as a test-bed for ideas in a wide variety of disciplines, in a manner NASA could not afford or dare to pursue.

The test-bed is also available to participating industries. As an example, Tucson Electric Power Co.'s newly developed thin-film solar battery is being considered by SSP for its first satellite. SSP offers other industrial partners similar opportunities to test their ideas or components.

CONCLUSION

The Student Satellite Project at the University of Arizona (SSP) has been underway for a year. It is an interdisciplinary project in engineering and science for students. Students participating in SSP are to design, fabricate, test, and deliver a small Earth-orbiting satellite to NASA for a launch by the Hitchhiker Ejection System on the Space Shuttle, and to carry out post-launch operations, data analysis and publications. The duration of the project is estimated at 2 - 3 years from conceptual design to launch and 1 - 2 year from launch to publications. The scientific objectives of the first attempt are to monitor global distribution of lightning, observe sprites and to perform stellar photometry. In addition, a laser communication experiment using the same optical system will be attempted. In the process of accomplishing these goals, SSP will offer predictable benefits to the students, faculty and staff at the University and concerned parties in industry and the local community. This is another example of the use of microsatellites.

ACKNOWLEDGMENT

The work reported here belongs to all the people whose names are mentioned in the text and the tables and the students who are too numerous to mention. Special thanks go to E. H. Levy, D. L. Stein, and B. Herman for their generous support and to S. Golper for his able assistance. One of us (K. C. H.) thanks the organizing committee of this Colloquium for the invitation and the National Cheng Kung University for the hospitality.

REFERENCES

1. National Aeronautics and Space Administration, Goddard Space Flight Center, Hitchhiker Ejection System Specifications
2. Maier, M. W. and S.T. Wu, SEDSAT-1 Lessons Learned, this volume.
3. Mende, S. B., Sentman, D. D., and Wescott, E. M., Lightning Between Earth and Space, *Scientific American*, August 1997
4. University of Arizona Student Satellite Project Web Site, http://www.physics.arizona.edu/ssp/

MUNIN: A STUDENT NANOSATELLITE FOR SPACE WEATHER INFORMATION

Olle Norberg[1], W. Puccio[1], J. Olsen[1], S. Barabash[1], L. Andersson[1], J. D. Winningham[2], U. Jonsson[3],and M. Eriksson[3]

[1]*Swedish Institute of Space Physics, Box 812, SE-981 28 Kiruna, Sweden*
[2]*Southwest Research Institute, Drawer 28510, San Antonio, Texas 78284, USA*
[3]*Luleå University of Technology, SE-971 87 Luleå, Sweden*

ABSTRACT

The Munin satellite (Figure 1) is set to become the first of a new type of monitoring spacecraft. Using modern technology, this very small satellite (mass ≈ 5 kg) will have all the necessary functions needed to support its specific scientific mission: monitoring of the auroral activity on both the northern and southern hemispheres. With Munin we can now usher in a fleet of monitor spacecraft to cost-effectively provide global monitoring, and at the same time have a large degree of student involvement on different levels.

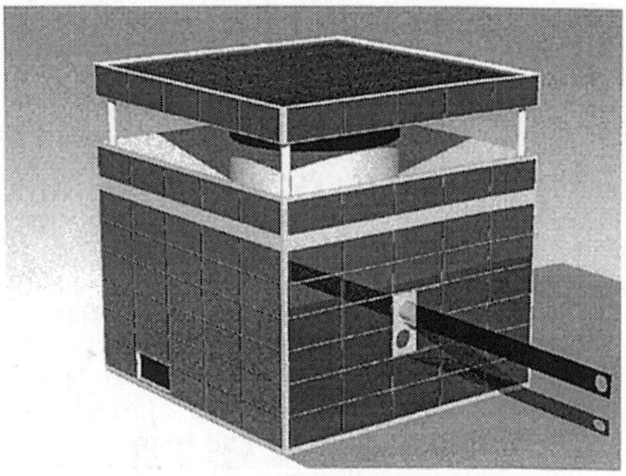
Figure 1. The Munin nanosatellite.

Munin has three scientific instruments for monitoring of auroral activity:

- MEDUSA, a combined electron and ion spectrometer with continuos coverage of all pitch angles. Covers the energy range 10 eV - 18 keV.
- DINA, measures high energy ions and neutral particles at pitch angles 0° and 90°. Covers the energy range 30 - 1200 keV.
- HiSCC, a high sensitivity CCD camera for visible and infrared wavelengths. The camera has a field-of-view of 50°, and a resolution of 340 x 240 pixels.

The payload has a total mass of 1.6 kg, and consumes 2.4 W, if power permits it will be operated continuously.

The satellite is cubic with solar cells on all six sides. The solar cells provides Munin with a power of 6.0 W. The average power consumption is 4.7 W, peak consumption (when the transmitter is on) is 9.1 W. A

Li-Ion battery will provide the peak power and also keep the satellite powered during eclipses. The structure is made from aluminum and has a mass of 1.9 kg. The total mass of the Munin spacecraft, including the separation system, is 6.0 kg. A separation system holding the spacecraft secured to the launcher using three hooks has been designed. Munin communicates on the UHF band, the up-link frequency is 449.95 MHz and down-link frequency is 400.55 MHz. The down-link is designed for a bit rate of either 9600 or 19200 bps. The modem is implemented in software in a digital signal processor (DSP), which also handles command handling and telemetry formatting. Another DSP takes care of instrument operation and data compression, The attitude will be controlled by a magnet aligning Munin with the local magnetic field. Soft-magnetic hysteresis rods will dissipate oscillation energy.

Munin will use an existing ground station located at the Swedish Institute of Space Physics, Kiruna, Sweden. The ground station was built in 1994 and has been successfully used for the two Swedish satellites Freja and Astrid.

PROJECT OBJECTIVES

Auroral Research and Forecasting

The Munin satellite is set to become the first of a new type of monitoring spacecraft. Using modern technology, this very small satellite (mass $\approx$ 5 kg) will have all the necessary functions needed to support a specific scientific mission such as the one we will describe in this document. With Munin we can usher in a fleet of monitor spacecraft to cost-effectively provide global monitoring, and at the same time have a large degree of student involvement on different levels.

Today auroral research is a wide field covering topics such as the acceleration mechanisms of precipitating particles, the physics behind the triggering of substorms, and forecasting of magnetic storms to mitigate the damage that “space weather” can cause in e.g. power distribution lines. These research areas all need data about the conditions in space. Both the solar wind conditions and the resulting disturbances on the Earth such as magnetic and electric field fluctuations, particle populations, and auroral activity are needed in order to deduce the physical mechanisms behind these phenomena.

The northern and southern auroral ovals are always present, day and night. From the Earth they can be seen by the naked eye at latitudes between 60 and 80 degrees north and south. The aurora is the visible manifestation of the particle precipitation along the Earth`s magnetic field lines at high latitudes. Since the aurora can be viewed only during relatively cloud-free nights, it is not an optimal source of information about the result of magnetospheric activity. The best tool to gather this information was (very early in the history of space flight) found to be satellites in polar orbits. There have been a number of satellites of this type over the years, with new discoveries about auroral physics made by every one.

Up until now, the data from such satellites have first been analyzed by the Principal Investigator of the instrument in question, then the results of the investigation have been published in scientific papers. With Munin, we propose to use a new paradigm of data dissemination, which we think will enhance the output from auroral research. We will make the data from the instruments on Munin available in real time to everyone interested, by using the Internet and WWW-based services. Munin will provide measurements of the electron and ion particle distributions above the auroral ovals, the fluxes of energetic particles (ions and neutral particles), as well as images of the aurora taken by a CCD camera in visible wavelengths. As the data comes down to a ground station located in Kiruna, Sweden, the data will be processed such that it can be used by scientists all over the world for auroral research. The satellite will store data from the pass over the southern auroral oval (over Antarctica) in its memory which will then be downloaded when it is over Kiruna, whereas the data from the northern oval will be down-linked in near real time.

The fact that the data is "fresh" will enhance the possibilities of using it for forecasting of auroral activity. The shape of the ovals, and the offset of their centers relative to the geographic poles, means, for example, that it is possible to use measurements over Scandinavia to predict the auroral activity in north America or Siberia at a later time. We intend to use the Munin measurements to present the present status of the auroral ovals, and let users of the data interpolate the position and size of the auroral oval a few hours into the future, in order to forecast auroral activity.

Public and Educational Outreach

The Munin team has striven to implement this project in order to promote the public interest in science, technology, space research, and international cooperation. During the last decades the public interest in these questions has declined, and we are of the opinion that a small, focused, project like Munin can catch the interest of the public. We intend to pursue this objective by visibility in newspapers, magazines, and by the Munin WWW-server, **http://munin.irf.se.**

The complexity of today`s space projects makes it extremely difficult for engineering students to grasp how the complete system works, or how it was designed. This results in more and more engineers becoming experts in just a small part of a project, with very limited insight in the overall system functions. With Munin, we intend to give students a possibility to be not only engaged in the mission, but also to gain a genuine knowledge about all parts of a space mission.

The Swedish Institute of Space Physics is engaged in two undergraduate space engineering programs, a 3-year program given by Umeå University, and a 4.5 year program given by Luleå University of Technology. Students from these programs are already contributing to the project by doing their degree projects on parts of the satellite. The intention is also to have students involved in the operation of the satellite, for example in the form of lab-courses involving hands-on satellite operations.

Orbit Considerations

In order to fulfill the scientific objectives of the Munin project, the satellite should be injected into a polar orbit with an inclination of at least 63 degrees and an altitude between 400 and 2000 km. A circular orbit at approximately 1000 - 1400 km and an orbit plane at around noon-midnight local time is optimal, since this orbit takes Munin both through the active night-side of the auroral ovals, and the polar cusps.

PAYLOAD DESCRIPTION

Miniaturized Electrostatic DUal-tophat Spherical Analyzer (MEDUSA)

The primary instrument on Munin is MEDUSA (Miniaturized Electrostatic DUal-tophat Spherical Analyzer), a combined electron and ion spectrometer. The instrument sensor is provided by the Southwest Research Institute, San Antonio, Texas. Electrons and ions with energies up to 18 keV/q will be measured simultaneously, with a maximum time resolution of 16 energy sweeps per second for electrons, and 8 seconds for ions. Particles are measured in 16 sectors in the plane of acceptance, which is aligned with the Earth's magnetic field. See Table 1 for a summary of the MEDUSA instrument characteristics.

Electrons are responsible for the major part of the energy input creating the aurora, and therefore the most important species to measure. The precipitating electrons energize (heat) ionospheric ions in the auroral region, these heated ions are then accelerated along the magnetic field lines and leave the heating region. This creates an outflow of atmospheric ions, primarily hydrogen and oxygen. With Munin we intend to measure both the electrons responsible for the aurora and ion heating, and the heated and accelerated ions. During certain circumstances, precipitation of ions also occur in the auroral region.

The MEDUSA sensor will be mounted on Munin in such a way that each of the 16 sectors in the acceptance plane always looks at particles arriving with a certain angle to the magnetic field lines, the "pitch angle". This is possible due to the simple, yet very efficient, attitude control chosen for Munin. A permanent magnet will align one of Munin`s axes along the local magnetic field line, this axis will be in the aperture plane.

Particles enter the spectrometer aperture at any angle in the 10° wide plane of incidence (see Figure 2), electrons and ions are then deflected into their respective spectrometer unit by a spherical electrostatic analyzer (deflection plates). The particles hit a microchannel plate (MCP) after being filtered in energy in the electrostatic analyzer, the hits are counted by preamplifiers connected to registers, and the number of hits per sample interval are then further processed by the data processing unit (DPU).

The MEDUSA sensor represents a completely new development in terms of highly compact sensor design. MEDUSA will be used on the Swedish microsatellite Astrid-2.

Table 1. The MEDUSA Instrument Characteristics

Instrument Characteristics	
Sensor mass	0.60 kg
Sensor power	1.0 W
Energy range (for both electrons and ions)	10 eV/q – 18 keV/q
Energy resolution ($\Delta E/E$)	25%
Acceptance angles	360°× 10°
Geometric factor per sector	4×10^{-4} cm^2 sr

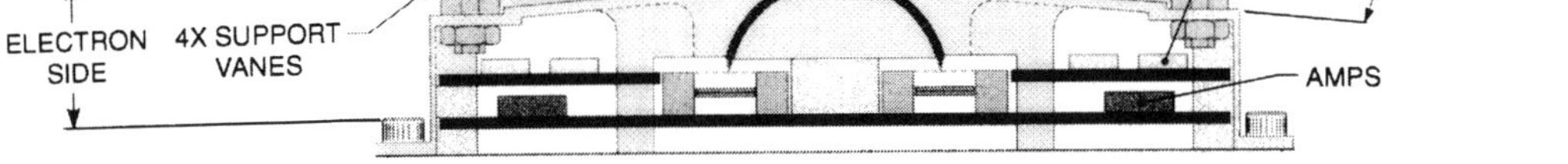

Figure 2. Cross Section of the MEDUSA Sensor Unit.

Detector of Ions and Neutral Atoms (DINA)

The main scientific objective of the instrument is to extend the measurements of the precipitating and mirroring ions to the higher energy range 30 - 1200 keV, complementing the MEDUSA experiment.

Apart from the main scientific objective, the DINA instrument will also be able to measure energetic neutral atoms (ENAs). ENAs are produced via the exchange of charge between singly charged magnetospheric ions and atoms of the upper atmosphere/exosphere. ENAs can be found in almost any space environment and are suitable agents to remotely probe ion populations of plasmas at a distance, so-called ENA imaging. In magnetospheric physics, general attention has so far concentrated on the ENAs generated within the high altitude ring current, because they can be used in global magnetospheric imaging from high altitude (> 20000 km) spacecraft (*Williams et al.*, 1992). However, in the auroral region where the ring current / radiation belt particles plunge into the dense upper atmosphere/exosphere, the charge exchange process is much more effective and ENAs emissions are much more intense. As was shown theoretically by *Roelof* (1997), ENAs from low altitudes are emitted from a very thin area near the exobase and the ENA generation region is, essentially, two-dimensional. The ENA camera PIPPI on-board the microsatellite Astrid (Norberg, 1995) has performed ENA measurements from a 1000 km polar orbit and demonstrated the potential of low altitude ENA imaging (*Barabash et al.*, 1997; *Brandt et al.*, 1997). DINA will continue studies of the low-altitude ENAs concentrating on measurements close to the generation region. The results from DINA, looking at the ring current from below at low latitudes and close to the ring current at high latitudes, will undoubtedly help the interpretation of the observations to be performed by IMAGE, the NASA mission to image the ring current from above.

Another interesting topic to be addressed is ENA albedo. The O^+ ions precipitating onto the upper atmosphere during major geomagnetic storms transform their energy and momentum into atmospheric heating and result in escaping fluxes of fast neutral atoms via charge exchange and elastic scattering (*Ishimoto et al.*, 1992). The total neutral flux escaping due to these processes could be significant for the atmospheric evolution during the life-time of the Earth (*Torr et al.*, 1974). However, no direct measurements have been performed to prove the correctness of the developed models. The DINA simultaneous measurements of the precipitating ions and ENA flux would provide the necessary inputs to evaluate the existing models of mass and energy transport in the ring current-atmosphere interaction.

DINA will also investigate the low-altitude equatorial ring current formed by stripped ENAs from the main radiation belt (see *Voss et al.,* 1995, and references therein) by measuring both the trapped ions and parent ENAs. During geomagnetic storms the ENA flux produced in the low-altitude ring current could be sufficient for detection, although for quiet conditions the flux will not be sufficient (*Bishop*, 1996).

In summary, the scientific objectives of DINA are:

1. Measurements of the ion flux in the energy range 30 - 1200 keV.
2. Measurements of the ENA flux from the exobase in the energy range 30 - 400 keV.
3. Measurements of the outflowing ENA flux (ENA albedo) from the precipitation region in the energy range 20 - 280 keV.
5. Mass resolving measurement of particles in the energy range 100 - 1200 keV.
6. Studies of the low-altitude ring current.

The instrument consists of two sensors with an aperture opening of 10° × 38° each (see Table 2). The sensor DINA-0 will detect particles with 0 degree pitch-angle, and DINA-90 will detect particles with 90 degree pitch-angle. Over the northern hemisphere DINA-0 provides measurements of the precipitating ions. While the spacecraft moves over the polar cap, DINA-90 will make measurements of ENA flux from the exobase in one local time sector. Over the southern hemisphere DINA-90 will measure the exobase ENA flux from a different local time sector due to the expected very slow spin of the spacecraft along the magnetically aligned axis. DINA-0 will be pointing toward the Earth over the southern hemisphere, and is aimed to detect outflowing ENA in the precipitation region. Information about the input ion flux can be obtained from the DINA-90 measurements because the energetic ion distributions can be considered approximately isotropic (*Lyons*, 1987). The instrument performs alternative measurements of ions and energetic atoms by turning on and off the high voltage on the electrostatic deflection system. Electrons with energies below 400 keV are always swept away by permanently magnetized broom magnets (the simulated transmittance for 400 keV electrons is 14%). Mass identification is performed by $\Delta E/E$ detectors. For the energy range 30 - 100 keV, the front detector (2μm Si detector) will provide the integral flux of all masses. For energies above 100 keV, coincidence and anti-coincidence logic provide measurements of hydrogen and helium (coincidence between the front and back detectors) and particles with $A > 4$, mainly oxygen (anti-coincidence between the front and back detectors).

Table 2. The DINA Instrument Characteristics

	Instrument Characteristics
Mass	0.9 kg
Power	0.5 W
Energy range (protons or H atoms)	30 - 1200 keV
Masses to resolve	A = 1, 4, A > 4 for E > 100 keV
Particles to measure	ions, neutrals
Aperture per sensor	$10^\circ \times 38^\circ$
Geometric factor (per detector)	3.1×10^{-3} cm^2 sr
Deflector cut-off energy	400 keV
Broom magnets transmittance	≈ 2% at for 100 keV electrons ≈ 14% for 400 keV electrons

High Sensitivity CCD Camera (HiSCC)

One of the best global space weather measurement tools is a camera imaging the auroral oval. A camera onboard a satellite supports both onboard instruments and ground facilities. Over the northern hemisphere HiSCC will take images of the night side auroral oval. This will give information about the activity level of the magnetosphere. HiSCC will support MEDUSA with identification of the spacecraft's footprint in the auroral oval, and any intensifications in the oval. Over the southern hemisphere HiSCC will operate as a star imager, enabling attitude determination based on star positions in each image. Attitude knowledge is crucial for interpretation of the measurements made by the MEDUSA and DINA instruments, as well as a major interest for the satellite operation group since the spacecraft behavior in space can be studied. This means that HiSCC can be used to verify the performance of the spacecraft attitude control system.

The camera will be the main instrument to provide the mission with pictures to the public. HiSCC data will help to educate the public about what space weather is, and the easy access to the data on the Web will increase the public interest in understanding the science of space plasmas in general and specifically the aurora.

The camera field of view is 50° with a highest spatial resolution of 3 × 3 km (nadir looking from an altitude of 1000 km). When the satellite is at 60° latitude the ground coverage ranges from 65° latitude and over the pole. At 75° the camera is almost looking in the nadir direction, covering 10° in both latitude and longitude. In order to avoid blurry pictures the exposure time will be below one second. Approximately 25 pictures will be taken during each pass over the northern auroral oval. The data will be compressed by the DPU before being down-linked to the ground station.

The CCD camera is a modified Quickcam from Connectix Corporation. It is made for connecting to a PC parallel port; for Munin this has been modified to a special interface to the DPU. A new housing and lens system assembly has been designed. We have removed the built-in infrared filter to increase the light sensitivity. The CCD chip used in the camera is a Texas TC-255 with a quantum efficiency of up to 60% in the wavelength range 500-800 nm. It can produce pictures with a resolution of 320× 240 pixels, in 64 shades of gray. The camera can operate with different exposure times, the shortest time being 1/1000 second. See Table 3 for a summary of the HiSCC instrument characteristics.

Table 3. The HiSCC Instrument Characteristics

	Instrument Characteristics
Mass	0.10 kg
Power	0.3 W
Camera Optics	focusing lens, f1.9, field-of-view 50°
Spectral passband	450 - 850 nm
Detector	TC-255 CCD chip, 50,000 e^- full well capacity
Resolution	0.2°× 0.2°, 320 × 240 pixels / 3.2 × 2.4 mm
Quantum Efficiency	20 - 60 %
Exposure time	minimum 10 ms, operationally 0.5 - 1 s

SATELLITE SYSTEM DESCRIPTION

Satellite Structure

The combined electron and ion spectrometer MEDUSA has a requirement of a 360° field-of-view in a plane, the satellite was thus designed as a natural extension of the MEDUSA instrument. The satellite can be considered to be an experiment with its own power and telemetry resources. The use of magnetic attitude stabilization means that Munin will rotate twice per orbit. All faces of the satellite will thus need to be covered with solar cells to maximize the energy input. To satisfy the power requirement of approximately 6 W it was decided to design Munin as a cube with the sides 210 mm long. The inner structure consists of the bottom platform, four support struts, the battery platform, the MEDUSA platform, and the top platform (Figure 3). The four side panels are

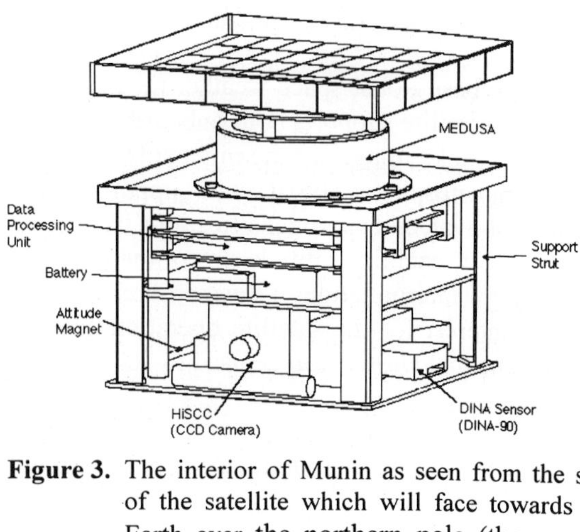

Figure 3. The interior of Munin as seen from the side of the satellite which will face towards the Earth over the northern pole (the camera side).

screwed together into a shell which can be integrated on Munin after all the units inside the satellite have been assembled and tested. There are aperture openings in some of these panels for the HiSCC and DINA instruments, as well as a mounting hole for the radio antenna. The structure is made of anodized aluminum.

Separation System

The low mass and the small size of the Munin satellite gives us the opportunity to design a very simple yet reliable separation system, see Figure 4. Munin will be clamped down to the launcher interface plate at three points. The satellite is equipped with three steel wires fitted with cylindrical end blocks, these blocks are held down by grappling hooks. Clamping will be provided for by a spring-tensioned Kevlar line. Three helical springs will be used for pushing Munin away from the launcher with a speed of about 0.6 m/s. A small launcher activated pyro-guillotine will be used to initiate the separation by cutting the Kevlar line. The mass of the separation system is 350 grams. The separation mechanism will be attached to the launcher with three bolts. The separation system will include a separation switch, to indicate the separation of Munin in the launcher telemetry. Inside Munin there is also a separation switch which is in the off position when Munin is located on the separation system. At separation this switch will power-on the satellite.

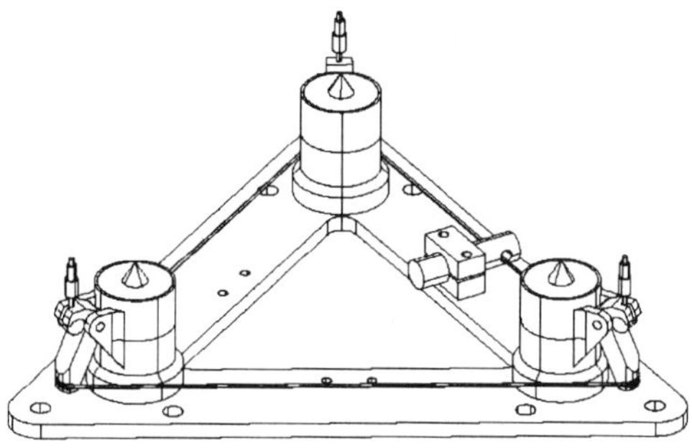

Figure 4. The Munin Separation System. With compressed springs the total height of system is 40 mm.

Power System

Each side of the satellite is covered with solar cells. We are using standard monocrystalline silicon cells, with an efficiency of 15%. These cells are cut to a size of 20 × 40 mm and are glued directly onto the satellite body with a 0.5 mm cover glass on top of each cell. Each cell interconnection is done with two leads for safety reasons. Each side of Munin will have one string with 40 cells serially connected, which will give about 6W power, and 15-20V. This power will be sufficient to run the payload instruments continuously, and the radio transmitter for 10-15 minutes per orbit. See Table 4 for the power budget.

Munin will use a Lithium-Ion battery. These batteries can store more energy per unit mass and volume than Nickel-Cadmium, Nickel-Metal-Hydride or any other commonly used battery types. Lithium-Ion batteries do not have any memory effect like Nickel-Cadmium, and a very low self-discharge rate. The battery used will have three cells stacked in series to achieve a maximum bus voltage of 12.6 Volt and a nominal voltage of 10.8 Volt, with a capacity of 4050 mAh resulting in an energy density of 44 Wh with a mass of only 560g (including mounting parts). The depth of discharge (DOD) will typically be around 9%, this will ensure the battery survives the huge number of charging and discharging cycles it will undergo during the life-time of the satellite. The practical charging of a Lithium-Ion battery is separated into two phases, first the battery should be charged with a constant current and when the voltage over the cell reaches a certain point, constant voltage charging should be used instead.

The temperature range for discharging the batteries should be kept between -10 and +50 degrees Celsius and charging between 0 and +45 degrees Celsius. A computer simulation of the thermal control, as well as a solar simulation test, was performed as a M.Sc. thesis work during 1997. This work found that the battery needs a 1. 2 W heater mounted inside the battery pack to keep the battery above the minimum level.

Table 4. Power Budget

Consumer	Power (mW)
MEDUSA (ion & electron analyzer)	1000
DINA (high energy particle detector) with electronics	500
HiSCC (CCD camera)	300
DPU (data processing unit)	600
Charging electronics and DC/DC converters	50
Radio transceiver (transmitting / receive only)	6500 / 180
Modem	200
TOTAL (transmitting / receive only)	9150 / 2830

Radio Transceiver and Modem

The communication will be done with a modified commercial Tekk-1000 radio and a dedicated 320C50 digital signal processor (DSP) acting as a modem to digitally generate and demodulate the frequency shift keying (FSK) signals carrying the data to and from the Munin satellite (see Table 5). This will enable the DSP to control the bandwidth and speed of communication with the software implemented filters and demodulation techniques. The software based modem was designed as a M.Sc. thesis work.

The spacecraft will transmit on 400.550 MHz with a maximum output power of 2.2 W; this will be sufficient for a bit rate of 9600 bps based on our experience with the Freja and Astrid satellites. During favorable conditions also 19200 bps should be possible. Munin will receive on 449.950 MHz using a narrow band 600 bps FSK since only a small amount of data needs to be uploaded to the spacecraft. The up-link will mainly be used for commands to the DPU for different operating modes and schedules, including a possibility to upload patches of the existing onboard software.

The modem DSP also has a 12-bit AD converter used for monitoring of satellite status, such as temperatures, battery and sun panel voltages and currents etc. There are also standard functions such as a real-time clock, watchdog etc. which are mostly realized in Actel PLD (programmable logic devices).

The antenna used both to transmit and receive consists of a one quarter wavelength (18 cm) whip located on one of the solar panel sides, pointing along the local magnetic field line due to the magnetic attitude stabilization scheme. The transceiver will run on a regulated 8 Volt supply requiring 20 mA in receive mode (default) and 0.8 A while transmitting with an output power of 2.2 W to the antenna. Doppler shift occurring due to the satellite motion relative to the ground will be corrected for by the ground station equipment.

Table 5. Specifications for the Tekk-1000 Transceiver.

Specifications for the Tekk-1000 Transceiver	
Receiver Frequency	449.950 MHz
Transmitter Frequency	400.550 Mhz
Operating Temperature	-30 to +60C

Operating Voltage	7.5 to 12.0 V
Receiver noise floor	-118 dBm
Dimensions	85 × 52 × 21 mm
Mass	200 g

Data Processing Unit

The DPU (data processing unit) handles the payload functions and data compression. The DPU used on Munin is the engineering model of the DPU for the MEDUSA instrument on Astrid-2. The heart of the DPU is a Texas Instruments TMS320C50 digital signal processor, which has 20 kb of internal memory and runs at a clock frequency of 40 MHz. We have equipped it with an additional 16kb of external RAM (used as program memory), 64k word EPROM (for flight software), and 32 k word EEPROM (for various data and software patches). The PROMS are only used (and powered) during the boot process and when patching of the onboard software is needed. For onboard data storage, 2 Mb of RAM are used for instrument data compression and buffering (FIFO). The data compression software in the DPU is based on a NASA-developed algorithm (CCSDS, 1994). We have successfully used this data compression scheme on previous missions we have been involved with.

Attitude Control System (ACS)

To enable continuos coverage of all pitch angles for the particle instrument MEDUSA, and to ensure correct pointing of the two DINA apertures, the spacecraft needs to be aligned with one axis along the local magnetic field. This requirement is realized by the use of a passive magnetic stabilization system. Due to the fact that Munin is a small satellite with a mass less than 5 kg, the use of this particular stabilization technique is ideal. Magnets have been used many times before, e.g. in the AMSAT microsatellites. Magnets are simple and cost-, mass- and volume effective. Soft-magnetic hysteresis rods are used for dissipation of oscillation energy,

Mass budget

The goal has been to keep the mass around 5 kg, this will be achieved by using light weight materials (mostly aluminum) for structural parts and a high degree of integration between the payload instruments and the satellite itself. By using programmable logic devices (PLD) in the electronics we will achieve highly compact and low mass systems. To handle the high degree of integration most printed circuit boards will have six layers. Table 6 gives the mass break down for the different parts of Munin.

Table 6. Mass budget

Part	**Mass (g)**
Satellite structure	1950
MEDUSA (ion & electron analyzer)	588
DINA (high energy particle detector) with electronics	900
HiSCC (CCD camera)	100
DPU (data processing unit)	500
Battery (including holder)	560
Charging electronics and DC/DC converters	110
Solar cells (including cover glasses)	300
Radio transceiver	200

Modem	120
Antenna + cable	50
Separation system	500
Magnet (attitude stabilization)	25
Soft-magnetic hysteresis rods	100
TOTAL	**6003**

On-board Data Handling

The telemetry consists of data packets of fixed size (512 bytes). Every packet will start with four synchronization bytes followed by four housekeeping bytes. The remainder of the packets will contain data from the experiments, keeping the "non-science" data at a low level of only about 2% of the total telemetry. Only instrument data from the latest orbit will be sent down in the telemetry, i.e. after each telemetry session the on-board memory will be cleared and new data will be collected.

In order to transmit as much relevant data as possible within the allocated bit rates, the data to be transmitted has to be reduced in some way. One solution is to use thresholds for measured parameters, and trigger data recording only when the threshold is exceeded. Our solution on Munin is to use data reduction by temporal integration, combined with loss less data compression. A data compression algorithm, Universal Source Encoding for Space (CCSDS, 1994), has been implemented and tested in the DPU. This algorithm will be used for all instrument data. There is also a mode in which only uncompressed satellite housekeeping data is sent down in the telemetry.

Commands up-linked from the ground station, decoded by the modem, and considered valid by the DPU are executed either immediately or at a prescribed time. If the command is "time-tagged" it will be placed in a command queue and executed when its time-tag corresponds to the satellite clock.

Ground Station

Munin will use an existing ground station located at the Swedish Institute of Space Physics, Kiruna, Sweden. The ground station was built in 1994 and has been successfully used for the two Swedish satellites Freja (for student lab courses on satellite control) and Astrid (as a secondary ground station). The hardware used in the ground station is mainly off-the-shelf equipment, except for the modem. The ground station consists of three computers, one dedicated to Doppler shift control and satellite tracking, one for the actual satellite control, and one for Web services and presentation of received data.

Operations

The ground station will be designed to work unsupervised. The Munin team has gathered knowledge in constructing this type of automatic ground station during the Freja and Astrid missions. Since the satellite operations and safety functions to a large degree are implemented in on-board autonomy, the ground station will during normal operations only receive telemetry data. If something unexpected occurs in the ground station, e.g. there is no telemetry from the satellite during a scheduled pass, the ground station will automatically alert the satellite operator on duty by the use of a mobile pager. This scheme was successfully used during the Astrid mission in 1995, and will also be used for the Astrid-2 mission.

MEDUSA and DINA data from the passes over the northern and southern auroral region will be included in the telemetry data. HiSCC images will be mainly taken over the northern auroral region, but some images will be taken over the southern hemisphere (where HiSCC looks out into space) for use in attitude determination, by comparing star positions in the image with star catalogues.

PROJECT MANAGEMENT

The Munin project is managed by young scientists and engineers at the Swedish Institute of Space Physics. Two Swedish universities and an U.S. research establishment are also involved in the project.

Public Relations

The scientific purpose of the Munin satellite is to collect data on the auroral activity. Such data are vital to prediction of space weather, but the information must be made public in order to be useful outside the Munin project team. To enable quick publication, we will use the World Wide Web (WWW), where everybody can retrieve the information they need. The idea is to use real-time uploading, from the down-link-computer to Internet, so that the information will be accessible as soon as it is downloaded from the satellite. The data will be free for everybody to use, regardless of institution, company or nationality, and will be available as soon as it is received at the ground station.

We are planning to make the received data available in several ways:

- as simple graphs and plots, in a format that illustrates the auroral activity to the layman.
- as highly processed images (for scientific use, education or illustration)
- as uncompressed scientific data, for further processing by the end user

Schedule and Milestones

The Munin project was conceived in September 1996, and work on the project has been on-going since then despite the fact that initially there was no guarantee that the satellite would be launched. The reason for this is that we see this project as an important contribution to science and the education of space engineers, and we have always been optimistic about the possibilities of finding a piggyback launch for such a small spacecraft as Munin. Munin is scheduled for a launch during 1999 as a secondary payload on a NASA mission using a Delta-II launch vehicle.

SUMMARY

Munin will function as a test bed for possible technologies and methods to be used in the miniaturization of spacecraft and instrumentation. Within the next decade there will be a need for fleets of small, autonomous spacecraft for various purposes. One objective for such a fleet will be to resolve the spatial-temporal ambiguity in satellite measurements. Another will be to form a fleet of spacecraft for global measurements of plasma parameters, as input for global modeling. Whatever the objective is, it will be necessary to keep the recurrent costs low when producing large number of identical spacecraft, here the use of commercial, off-the-shelf, equipment is essential. The miniaturization of spacecraft and instruments is necessary to enable deployment of large number of satellites. One issue to address in the future is the possibility to de-orbit satellites after they have ceased to function. Small, simple satellites are also ideal for use in educational environments, in which a Munin-class satellite can be designed, built, launched, and operated during the time of a undergraduate or graduate student career.

ACKNOWLEDGMENTS

Munin is financed by grants from the Swedish National Space Board. The project is managed by the Swedish Institute of Space Physics in Kiruna, Sweden, where the spacecraft and the instruments DINA and HiSCC are built. The MEDUSA instrument is provided by the Southwest Research Institute, San Antonio, Texas, with funds from the U.S. Office of Naval Research grant N00014-98-1-0175. Space Engineering students at Umeå University, Sweden, and students at Luleå University of Technology, Sweden, are involved in all phases of the project. The DINA instrument will be developed, and the data will be processed and analyzed, in close cooperation with our Co-Investigators Dr. K. C. Hsieh and Dr. C. Curtis of University of Arizona, Tucson, Arizon, USA, and Dr. E. C. Roelof, JHU / APL, Laurel, Maryland, USA.

REFERENCES

Barabash, S., P. C:son Brandt, O. Norberg, R. Lundin, E. C. Roelof, C. Chase, B. Mauk, Energetic neutral atom imaging by the Astrid microsatellite, Adv. Space Res., Vol. 20, pp. 1055-1060, 1997.

Bishop, J., Multiple charge exchange and ionization collisions within the ring current-geocorona-plasmasphere system: Generation of a secondary ring current on inner L shells, J. Geophys. Res., 101, pp. 17,325-17,336, 1996.

CCSDS, Source Coding for Data Compression, Consultative Committee for Space Data Systems, CCSDS 111.0-W-2, Washington, D.C., March 1994.

Brandt, P. C:son, S. Barabash, O. Norberg, R. Lundin, E.C. Roelof, C.J. Chase, B.H. Mauk, and M. Thomsen, ENA imaging from the Swedish micro satellite Astrid during the magnetic storm of 8 February, 1995, Adv. Space Res., Vol. 20, No. 4/5, pp. 1061-1066, 1997.

Ishimoto, M., G. J. Romick, and C.-I. Meng, Energy distribution of energetic O^+ precipitation into the atmosphere, J. Geophys. Res., 97, pp. 8619-8629, 1992.

Lyons, L. R., Processes associated with the plasma sheet boundary layer, Physica Scripta, 8, pp. 103-110, 1987.

Norberg, O., S. Barabash, I. Sandahl, R. Lundin, H. Lauche, H. Koskinen, P. C:son Brandt, E. Roelof, L. Andersson, U. Eklund, H. Borg, J. Gimholt, K. Lundin, J. Rynö, and S. Olsen, The Microsatellite Astrid, Proceedings of the 12th ESA Symp. on European Rocket and Balloon Programmes and Related Research, Lillehammer, Norway, 29 May - 1 June, 1995.

Roelof, E. C., ENA Emissions from Nearly Mirroring Magnetospheric Ions Interacting with the Exobase, Adv. Space Res., Vol. 20, No. 3, pp. 361-366, 1997

Torr, M. R., J. C. G. Walker, and D. G. Torr, Escape of fast oxygen from the atmosphere during geomagnetic storm, J. Geophys. Res., 79, pp. 5267-5271, 1974.

Williams, D. J., E. C. Roelof, and D. G. Mitchell, Global magnetospheric imaging, Rev. Geophys., 30, 183-208, 1992.

Voss, H. D., J. Mobilia, H. L. Collin, and W. L. Imhof, Satellite observations and instrumentation for measuring energetic neutral atoms, Optical Engineering, 32, pp. 3083-3089, 1993.

SEDSAT-1 LESSONS LEARNED

Mark W. Maier, Shi-Tsan Wu

University of Alabama in Huntsville, 301 Sparkman Dr., Huntsville, AL 35801

ABSTRACT

A number of universities have initiated small scientific satellite projects in which students take the primary roles of investigators, managers, designers, and fabricators. The University of Alabama in Huntsville's SEDSAT-1 project is among them. SEDSAT-1 is manifested on the July, 1998 Delta II launch as a secondary payload (the JPL DS-1 mission is the primary). Although not yet launched, the SEDSAT-1 project has yielded numerous lessons learned for small satellite programs. Some of these, such as the critical importance of tightly defined and controlled mission scope, are well known. But others are more specific to university student satellites, such as the tradeoffs in cost and educational value of proto-flight versus qualification unit approaches. This paper discusses the primary lessons learned and methods suggested for mitigating the identified risks and problems. Included are the outlines of project templates, both in system architecture and organizational structure, to help guide future projects.

INTRODUCTION

SEDSAT-1 is a 35 kg satellite roughly 30 cm on a side. It will be flown as a Delta secondary payload in July, 1998, although it is also compatible with Space Shuttle Hitchhiker deployment. SEDSAT will be a secondary payload on the JPL Deep Space One mission. On orbit SEDSAT-1 will perform analog relay and digital store-and-forward on the amateur radio bands, test the use of a unique panoramic imager, will collect multispectral images of the earth, will conduct Internet integration experiments, and will experiment with a variety of small satellite technologies.

An overall view of SEDSAT-1 is shown in Figure 1. The a schematic of the internals is shown in Figure . The external view shows the solar cell covered exterior, and the mixture of fixed and deployed antennas. It also shows the protruding PAL optical system. The internal view shows the experimenting plate (a removable optical bench), and the stack of electronics boards.

SEDSAT-1 HISTORY

The SEDSAT-1 project began informally about 1990 with the interest of two University of Alabama in Huntsville (UAH) students in building an orbital payload. The students, Mr. Dennis Wingo and Ms Cheryl Bankston, were particularly motivated by a desire to participate directly in a flight program and by their desire to have direct student involvement in space work. After a period of discussions with NASA Marshall Space Flight Center (MSFC) personnel plan jelled around a tether endmass for a small expendable deployer (SEDS) mission. SEDS (not to be confused with the Students for the Exploration and Development of Space also acroynmed SEDS) is a system that uses a 20 to 40 km very lightweight Spectra tether to deploy small payloads. At the time the SEDS system had been flown successfully once, the SEDS-1 mission, as a secondary payload on a Delta II. That experiment had demonstrated the operation of the SEDS system, and further missions were planned to further develop tether deployment technologies.

MSFC SEDS and the UAH student efforts joined with a plan that the UAH group would provide an endmass compatible with the SEDS deployer. The mission plan was to use the tether to deploy the endmass into a raised orbit by a "slingshot" maneuver. The endmass would collect information on tether dynamics (acceleration and attitude) during deployment and transmit the recorded data back to the ground after the maneuver. Once the endmass had completed the basic mission plan it would remain in orbit for a nominal three years. The students could add additional instrumentation to the endmass (compatible with the secondary weight and volume constraints) which would come into play after the primary mission. The initial plan was that the students would independently raise the money to build the endmass and the SEDS project would fly it at no cost. The endmass became known as SEDSAT-1.

An initial plan made SEDSAT-1 part of the SEDS-2 mission, but this proved to be an impossible schedule. The agreement was replanned for the SEDS-3 mission. During the run-up to SEDS-3 the launch plan was changed from a secondary on a Delta II to a hitchhiker on the Space Shuttle. There were several reasons for this change, one being the desire of NASA to qualify the SEDS system for use on the shuttle.

In Spring of 1995 continuing development problems with SEDSAT-1 led to a new agreement between MSFC and UAH. MSFC began to directly support several newly recruited graduate students and direct faculty supervision, the author among the latter. Between Spring 1995 and Spring 1996 development continued, moving toward a Shuttle flight in summer of 1997. A number of problems remained, especially severe delays in getting the computer hardware operational. Because the selected operating system could not be run on any but custom hardware, software development was similarly delayed. Programmatic problems soon intervened when the TSS-1R tether broke in March of 1996. The combined SEDS/SEDSAT program went through Space Shuttle safety review in May of 1996. This review was highly contentious, and resulted in extensive requirements changes to the SEDS system and minor modifications to SEDSAT. As a result MSFC decided to terminate the SEDS-3 mission.

In an effort to find some productive employment for the SEDSAT hardware that had already been fabricated UAH began negotiation with potential NASA sponsors. This resulted in a new contract awarded in December of 1996 to fly SEDSAT-1 as a free-flying secondary Delta II payload in July of 1998 on the JPL DS-1 mission. The revised SEDSAT-1 program delivered an initial configuration to environmental development testing in August of 1997 [UAH97]. That testing was largely successful, although it surfaced thermal problems in antenna deployer system and the motherboard. Additional delays occurred throughout the second half of 1997 with ongoing image processor development and a vendor failure to deliver flight qualified separation switches. SEDSAT-1 is now scheduled for delivery at the end of 1997 with final environmental testing to commence in January of 1998.

LESSONS LEARNED

Flying even a simple satellite is a complex undertaking. Thus, encountering a range of problems is hardly surprising. But a number of problems, both managerial and technical, have returned again and again to bedevil the project. It is likely that these problems will be revisited by other student satellite projects as they are fundamental to projects of this nature. Since there are an increasing number of such projects in various stages of planning around the world, it seems appropriate to review them, and to suggest methods for their mitigation. The core problems have been:

- Reconciling cost, schedule, and performance on a strictly schedule limited timeline
- Excessive dependency chains on the critical path resulting is long delays in resolving key risks. Especially a problem where custom hardware is involved.
- The need to make commitments in the absence of resources to backup those commitments.
- The need for diverse expertise and mentoring without employing a diverse set of professional experts.

Cost/Schedule/Performance Reconciliation

Being launched as a secondary payload forces a project to adopt the schedule of the primary payload. The rocket won wait if the secondary payload is not ready. For Delta II secondaries there is an established 24 month integration schedule [Boeing]. With the schedule fixed cost and/or performance will be at risk. SEDSAT faced this problem continually when technical problems arose that impacted schedule. The choice had to be made to either find and spend additional money or give up on performance objectives. In theory, a student microsatellite project can "bail-out" troubled elements with professional talent. In practice, the money is not likely to be available on short notice.

A major problem was the long dependency path in the development of the command and communications electronics. We needed to build a custom board, port an operating system, and write application code, all before we could begin to test communications. Board integration took far longer than expected. In this case we were saved only by the slip of the larger program from Shuttle to Delta II launch. Another problem was the development of the cameras. Space problems required custom construction of structural elements putting the cameras together. See Figure for how they ultimately came out. We had originally planned to rely on student labor to do all machining and optical fabrication. With experience this proved impossible. Ultimately we had to find funding to pay for professional machining of the critical parts, although from there students were able to do all the integration. Several parts of the satellite were simply too complex, or were too flight-critical to rely non-professional machine ship labor. The optics was an example of the former, the -Z faceplate (Figure) is an example of the latter.

Long Critical Paths

A first lesson in project management is to look at the critical path. That is, look at the longest path of dependencies from where you are to launch. We learned to be very wary of any long path that required success to meet launch schedule. Slips are inevitable, and there has to be room to make up or adapt for lost ground. Again, or main problem was the long dependency path of electronics and software development. Each stage of dependent on the one before to initiate. No operating system integration was possible without hardware completion. No application code could be developed without OS integration. The system architecture must support concurrent and incremental development.

Mismatch of Commitments and Resources

The easiest way to understand this situation is to consider the Delta II integration schedule. At 24 months before launch the team will be asked to make firm commitments to launch. They will be expected to have full drawings packages and be in development. Once that date is past the launch conveyor doesn't stop moving. A basic question should be, does the team have in place the resources to meet the launch commitment? At our Delta II kickoff we could do that. At our previous Shuttle commitment we could not. We were committed to building a specific system, but the resources (mainly in money) were largely hypothesized. If the needed resources (with reserve) are not in place, serious consideration should be given to back up plans. Will parts be cut? Is there a "source of last resort" for funds? Can the project be abandoned part way in if serious problems arise? These are hard questions, but we found ourselves facing them repeatedly until we had a core sponsor.

Part-Time Experts

On a major satellite project a variety of experts in different fields will be committed to the program. The fields include structural analysis, thermal analysis, radiation performance, communications, software engineering, optics, board level electronics, manufacturing, quality assurance, safety analysis, and many others. A small satellite project will likely need all the same experts, but not for nearly the level of effort. If it did, it might as well not be a small satellite project! How will these experts be brought into the program if they are needed only a limited time?

Student microsatellites tend to rely on volunteer experts with only one or two nearly full-time experts (the faculty advisor and perhaps one professional staffer). The problem is that the expert may need to put in very significant amount of time, although not over an extended period. For example, a Delta II secondary payload must deliver a finite element dynamic model of the payload to integration with the launch vehicle level dynamic model. Advanced mechanical engineering students have often learned to use finite element modeling tools. But how many will be able to make a determination of how complicated the model needs to be to be acceptable for flight? To interpret the results of the model? To know what approximations and complexity reductions will be acceptable in the light of results from a full model? This kind of expertise is learned through experience and not through the classroom.

In SEDSAT`s case we were very fortunate to obtain the services of an industry expert very near retirement. His company made him available mainly because one of the projects students was also an intern at the company. In the absence of this happy circumstance it would have been necessary to spend quite a bit of money to hire somebody comparable on a consulting basis. It would have been very difficult to get the help solely through volunteer experts as it really required 20-40 hours a week involvement for a few weeks. In other cases we had disappointing support from outside experts. Many of them were genuinely interested, but could not devote large blocks of time when it was needed. Small amounts of time over long periods are not effective. The same issue exists for all staff on the project. Students who can contribute a few hours a week have a very difficult time contributing effectively. It is important to have a core staff who work nearly full-time on the project for periods of at least six months at a time.

MITIGATION ON MICROSATELLITE PROJECTS

In the light of experience, all of these problems do have mitigating strategies. A better programmatic plan and system architecture could have alleviated our major problems, and may lead to an easier experience for other projects.

There's No Such Thing as a Free Launch

Student satellites will typically be launched as secondary payloads under an apparently free launch arrangement. The launch will be free in the sense that the satellite developing organization carries no cost for the launch on its books. Since the payload is carried as a secondary, it appears that the payload is a no cost addition to the primary launch. Unfortunately, this is an illusion. There are very substantial costs associated with a secondary launch. These costs are genuine budgetary items for the sponsor. Even where the sponsor is also the primary payload sponsor, as when NASA sponsors a secondary payload on a launch carrying a NASA primary payload. The important thing for the payload team to remember is that, even if they aren't paying anything for the launch, somebody is, and the budget amount may be big. That means, once the integration train gets rolling, that few excuses of expense will be accepted.

Thus, a student satellite must be treated as a conventional project, albeit a relatively simple and low cost project. It needs a full time staff, and a substantial cash budget for travel and manufacturing. To obtain full-time staff it may be necessary to employ students or to use a mixture of professional staff and students. This last approach is particularly recommended as it provides the best opportunities for combining training with work.

Decoupled Architecture, Spiral Process

Unless the project has a sponsor who is willing to "spend what it takes," the system needs to be arranged flexibly. This means the system architecture should be as decoupled as possible, and the system should be amenable to spiral, incremental development.

A good example of such a choice is using the three stage--breadboard, brassboard, flight--sequence for electronics and using a commercial operating system. This is most easily illustrated by example. Suppose one selected the VME bus as a backplane, the i386EX[1] processor, and the VxWorks[2] real-time operating system. In a decoupled system architecture it should be possible to assemble a hardware breadboard, begin brassboard development, and begin software development all concurrently. This set of choices would largely meet the test. The i386EX processor is available as an off-the-shelf development board (including schematics), VxWorks can be run in a complete development environment on a standard PC. Thus, once the architecture is determined, one group can wire wrap a i386EX VME board using off-the-shelf chipsets, another group can begin software design without a working board, and design can commence on a flyable board using the off-the-shelf electronic schematics. In contrast, on SEDSAT-1 we were unable to perform significant software development until after the core electronics had been designed and debugged, and a difficult operating system integration had been accomplished.

Similarly, in the recommended program plan the structure, core electronics, and instruments are decoupled during early stages. Each can be developed to a deliverable and the team replaced or refreshed as members graduate. As in the classic spiral process, the emphasis is on risk reduction at each step.

Find A Core Sponsor

A core sponsor is one who commits enough funding to stably maintain a development team and to fabricate key hardware elements. There need not be a single core sponsor throughout the project. The level of commitment is much lower in the early stages than in the later. The support level should be enough to build a minimally acceptable satellite with acceptable risk. A major contributor to mission creep during SEDSAT-1 development was the lack of a core sponsor. In the absence of a core sponsor the team will frequently face near term deadlines to get more money or shutdown. There will be pressure to accept additional missions in return for immediate funding. This can be a destructive spiral as the system

[1] An Intel processor containing a 386 core with extensions for embedded control.

[2] A product of Wind River Systems (www.wrs.com).

becomes more complex and the additional funding does not cover the incremental cost of the new functionality. With a core sponsor we have found it much easier to bring in additional support. This additional support has included both direct SEDSAT funding a related project support for team personnel when they were not needed full-time on SEDSAT.

Beware Too Much Student Enthusiasm

Almost certainly, a major objective of a student-centered satellite program will be education. Education should not mean having a group of students floundering around in their own inexperience. It should mean hands-on learning. If major management responsibility is given to students it needs to be remembered that they will not have significant lifecycle development experience. They will not have an experiential basis for making cost or effort estimates. This has to be guarded against since launch sponsors will be making million dollar (U.S.) commitments and will probably not be understanding of slips.

WHEN IN DOUBT, USE COTS

A major source of delay on the project has been custom electronics development. The SCOS operating system does not run on any off-the-shelf processor boards, much less a flight qualifiable board. It is supposed to run on a particular IBM 80186 PC card, but even experts on SCOS have been unable to boot SCOS on our examples of these boards. As a result we spent an extended period developing wire-wrap and printed circuit prototypes of our eventual flight boards without being able to develop software.

In practice we would now argue that no electronic hardware or software components should be chosen unless they meet the following criteria:

- An integrated hardware/software system usable for development can be commercially purchased. Preferably, the selected system should be in wide use.
- If the hardware is not available commercially in flyable configuration, the complete schematics should be available, preferably in electronic format for a selected design tool.

Unfortunately, choosing a truly off-the-shelf solution will produce many limitations. The requirements of the flyable system are likely to be unique enough that custom final electronics will be needed. This is acceptable, as long as development of other system parts is not dependent on a long chain custom developments. Cameras should be considered similarly. Some custom work is probably inevitable, but at the least a breadboard system should be buildable from off-the-shelf cameras.

Ask For Enough Money

Mission creep is a significant problem, but is far worse on projects of this type if the new missions come with inadequate funding. On several occasions new sponsors added roles to the mission, but the funding provided was inadequate to cover even the marginal costs of the new role. A more appropriate pricing strategy would be to require any new roles to fund both their marginal cost and a pro-rated fraction of the overall effort. This admonition applies to both core sponsors and later arrivals. The project should proceed out of each phase only as a core sponsor has been found willing to commit a realistic amount to complete. Later sponsors should not be accepted unless they are willing to contribute more than their marginal cost of inclusion. Much self-discipline is required on student projects to turn down interesting, but inadequately funded, proposals.

Professional Fabrication Of Core Parts

While volunteer labor and donated hardware may be critical to making an educational satellite affordable, there are several areas where we have found accepting donations arguably more expensive than paying prevailing commercial rates. We have developed some criteria for accepting labor donations. In general, the donations worth accepting are those in which an experienced professional is committed at least half time to the project as if it were a normal professional assignment. In this fashion we have received invaluable assistance on structural analysis and in machining major structural components. On the other hand, our acceptance of a minimal time labor donation of an inexperienced person for printed circuit board layout disastrously effected the schedule for our image processor. The money spent during the wasted time exceeded the commercial cost of the work by a factor of four, and nearly caused a major element of the system to be dropped.

PROJECT TEMPLATES

As a result of these experiences we can offer some templates for building a university payload to be launched a secondary payload on a Delta II rocket. Our advice divides along the physical system architecture, the computer system architecture, and the management plan.

Physical System Architecture

The Delta Secondary Payload Planners Guide [Boeing] gives precise constraints on the physical configuration. In general, the satellite can be up to 80 lbs (36 kg) with a volume of one to two cubic ft. (35 to 70 kcm^3). A particularly suitable configuration is a rectangular solid roughly 60 by 30 by 25 cm. If the height is kept to about 25 cm there is little risk of exceeding the center of gravity height specification, and the satellite will be completely shielded from any third stage plume by the available plume shield. Heights up to about 35 cm can be accommodated, but the outboard edge will be exposed to third stage plume heating and the payload attach adapter will have to be integrated with the structure (contributing many ground handling problems to SEDSAT-1).

Within these limits there are several particular suitable arrangements. One possibility is to make the satellite cubical with symmetric axes of inertia and the solar panels hard-mounted to the satellite`s body. This arrangement is suitable for either a very low cost unstabilized satellite, or a very maneuverable satellite using three axis reaction wheel control. If the satellite fills out the dimensions discussed above it would suitable for either a gravity gradient boom nadir pointing satellite or a momentum biased satellite with fold-out solar panels.

SEDSAT-1 is built by bolting together honeycomb milled aluminum panels with hard-mounted solar panels. In retrospect there are better alternatives. One low cost alternative would be bolt-together construction of flat panels cut with triangular holes. This would be simple to manufacture and should allow construction of a full mechanical prototype. The solar panels would be attached to the partially open frame. A better, although more expensive, alternative would be to machine a frame out of a single aluminium block and add all panels and instruments as bolt-ons or bolt-ins.

Electronics System Architecture

The electronics system (computer and communications) is a critical component of most satellites. It is also an area where very high technology components can be used, if warranted by the application of interest. There are two likely options. One is to use a low technology solution based on proven components. This is appropriate where the instrument is the focus of the mission and very high processing capabilities cannot enhanced capabilities. The other option is to push very hard on computer technology to exploit the tremendous MIPS to weight or power ratio available with modern VLSI.

The first option is exemplified by an 80186 processor solution similar to that used on SEDSAT. The UAH design has been licensed to Space Electronics, inc., and a 486 variant is forthcoming. A related option is to use commercial VME or compact PCI processors in rugged or military quality levels. The problem with this solution is that commercial boards are not generally optimized for low power consumption, and the radiation tolerance levels may not be suitable for other than short missions in low orbits. Rugged or military boards may be quite suitable in all respects, except single event upset. This has to be looked at carefully, and the option of watchdog shutdown for protection evaluated.

Going the other direction is to choose a computer system architecture providing very high computation throughput, which would then be used to enable unique applications. Some examples could include software based signal analysis, sophisticated communications handling, or real-time frame rate image processing. The ideal processing architecture would have very high MIPS to power consumption ratio, have off-the-shelf boards, be available in extended temperature ranges (and even radiation hard versions), and run widely available operating systems. Naturally, no such ideal solution exists. Some interesting options are the StrongARM processor (400 MIPS/watt), one of the very low power 486 variants, or a DSP designed for battery powered applications. However, many of these are only available in die with very poor radiation properties.

Communications is much more challenging since few space-rated transponders exist, and those that do are extremely expensive. One options is to build a custom system, but this will certainly require a full-time team of experienced engineers. A better alternative may to adapt newly appearing commercial designs, such as mil-rated amateur radio equipment.

Development Plan

Based on our experience, we recommend a three part plan. The three parts are concept formulation, breadboard demonstration, and launch program. A flight operations phase would follow. Each phase proceeds to the next on a specific set of risk-based success criteria.

The goal of the concept formulation stage is to produce packages of fundable scientific and educational objectives and compatible system architectures. The elements produced will be a structural design that accommodates launch vehicles and instruments packages of interest, a design of a core electronics and power system, designs for two or more instrument packages, and further management plans. The exit criteria are that high confidence cost and development plans exist for all elements in the breadboard phase, and moderate confidence estimates exist for the full development phase. The financial requirements for the phase are to keep a small team together at a high enough work fraction to build the packages within a year.

The second stage builds electronic, software, and structural breadboards sufficient to control the risk of a fixed schedule final development effort. A full set of production and integration drawings should also be completed during phase two. The object is to be able to work to a fixed time schedule beginning with launch vehicle integration kickoff.

The third phase, the launch program, can be matched to the integration requirements of the launch vehicle. For the Delta II the normal launch integration period is 24 months. Once launched, flight operations can begin.

CONCLUSIONS

Student microsatellites can carry out significant missions, and provide exceptional hands-on education as well. The experience of SEDSAT-1 both shows what can be accomplished, even by a relatively inexperience team, and provides important lessons in what not to do. Perhaps the most important lesson is that these projects, while much lower cost than any conventional satellite program, are still far from free. These projects need to be planned and executed with appropriate attention to systems engineering, but adapted for the mixed educational and scientific goals.

There is a need, and an opportunity, for greater standardization in system architectures. Although quite a few satellites have been launched as low-cost secondaries, they have shared relatively few elements beyond vehicle interfaces. Some common approaches to structure, electronics, and software could open new opportunities for microsatellites of this class.

REFERENCES

[UAH97] Pictures and other materials on SEDSAT-1 are archived at http://www.seds.org/sedsat and http://146.229.5.181/

[Boeing] Delta II Secondary Payload Planners Guide available from the Boeing Corporation, McDonnell Astronautics Division.

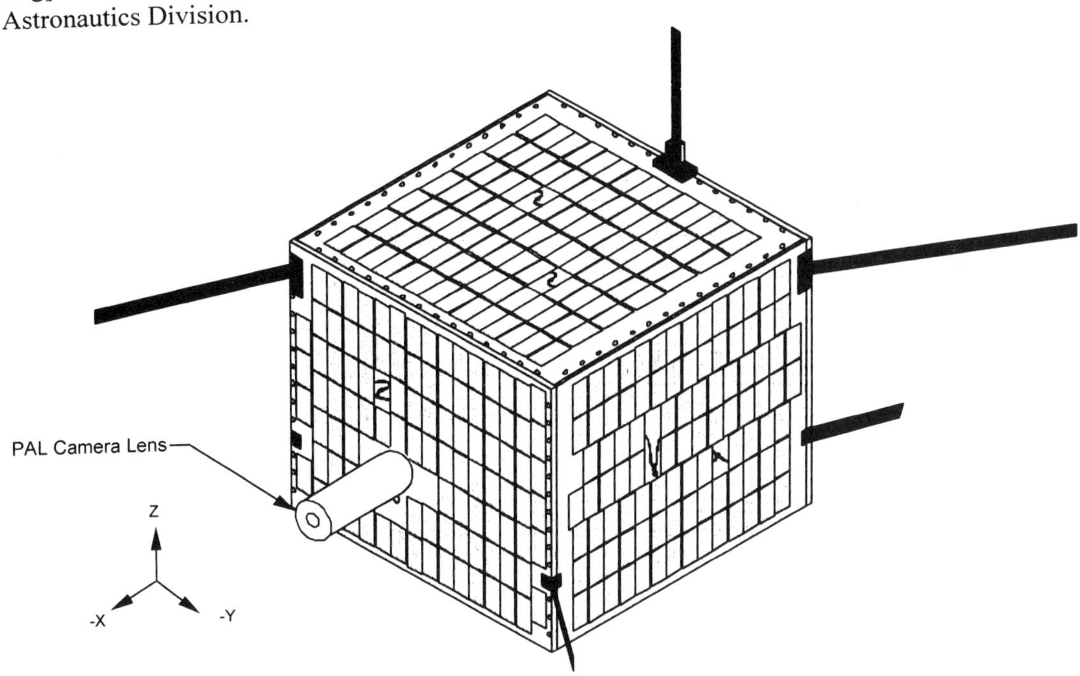

Figure 1. SEDSAT-1 External View. This view shows the external solar panel covering, the deployed antennas, and the PAL optics.

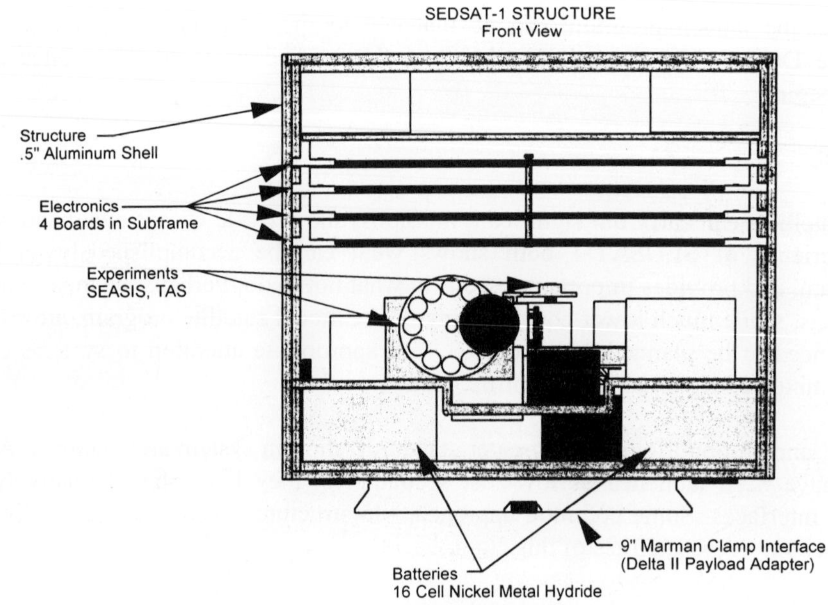

Figure 2. SEDSAT-1 Internal/Cutaway view.

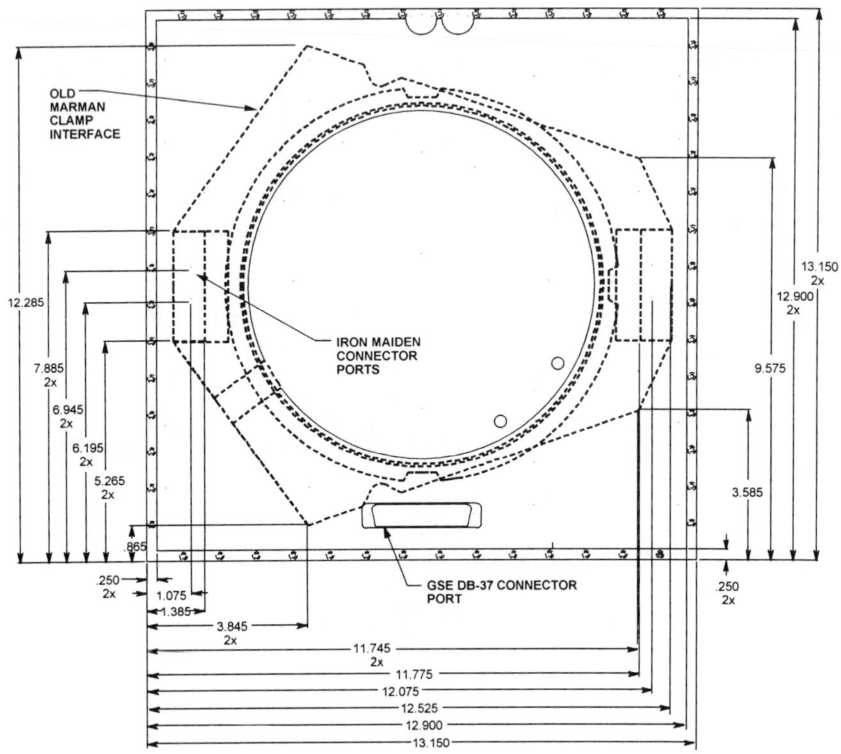

Figure 3. SEDSAT-1 Underside (-Z) view. This view shows the integrated payload attach adapter and the original GSE port.

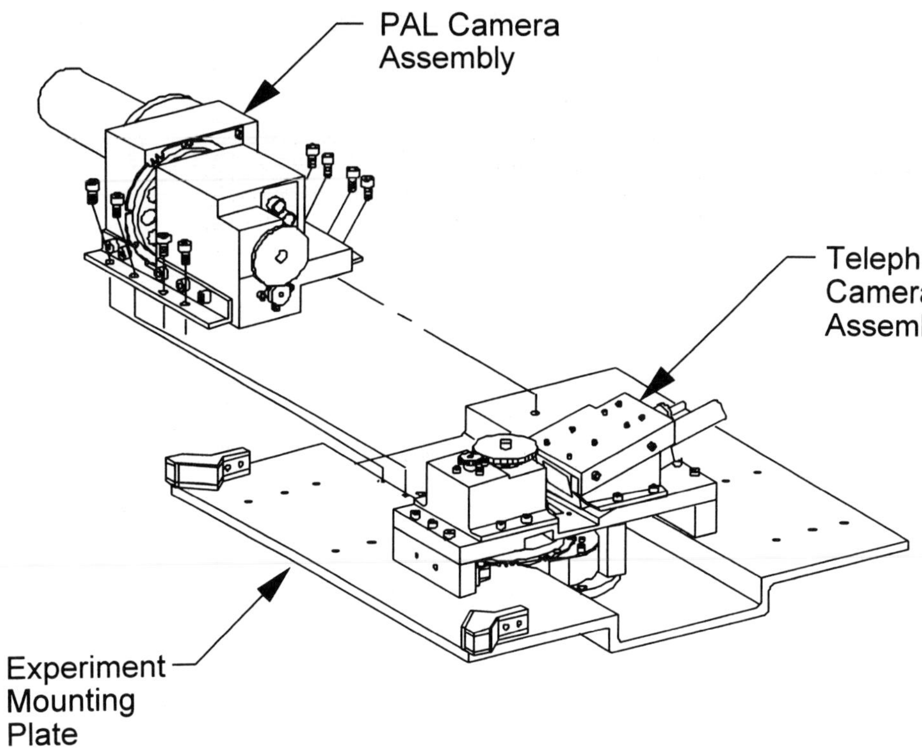

Figure 4. SEDSAT-1 optical assembly; complex parts required professional machining.

THE STUDENT EXPLORER DEMONSTRATION INITIATIVE PROJECT

John R. Sevier and Paul J. Coleman, Jr.

Universities Space Research Association, 10227 Wincopin Circle, Suite 212, Columbia, Md. 21044-3498

ABSTRACT

The Student Explorer Demonstration Initiative (STEDI) is a NASA funded, USRA managed pilot program consisting of three university-led projects. The program is designed to assess the effectiveness of small, low-cost space flight missions. In response to an Announcement of Opportunity (AO) for three STEDI space-flight missions, issued by USRA, 66 proposals were received. The AO called for complete missions which were to operate for up to one year in polar orbit and accomplish significant research or technology-development objectives. Total cost including spacecraft/instruments design and development, test and integration, and operation for one year in orbit could not exceed $4.16 million. Projects had to be ready for launch within two years from go-ahead. In addition, proposers had to demonstrate that there would be significant and meaningful student involvement over the course of the program. From the 66 proposals that were received, the three projects were selected for development. Herein, these three projects will be described, and some conclusions regarding the effectiveness of the pilot program will be presented.

THE MISSIONS

TERRIERS.

Scheduled to be launched into a 550 km circular, sun-synchronous orbit in August 1998, TERRIERS (Tomographic Experiment using Radiative Recombinative Ionospheric EUV and Radio Sources) will conduct a global upper-atmospheric study. Using a combination of ground-based and space instruments, the satellite will survey the upper atmosphere using a technique called tomography, measuring ultraviolet light emissions, to construct an image of Earth's ionosphere. Although the ionosphere has been studied in great detail with various ground and space-based instruments, currently there is no means to obtain these types of global images. Such measurements are crucial to the advancement of the understanding of many upper atmosphere phenomena. TERRIERS' primary goal is to, for the first time, demonstrate meridianal 2-D (latitude-altitude) and global 3-D imaging of the ionospheric electron density and thermospheric photo-emission profiles using EUV emissions and tomographic techniques. Secondary goals of TERRIERS focus on the study of several ionospheric and thermospheric phenomena through the use of this novel combination of techniques and observations.

The TERRIERS satellite weighs 270 lb. and its solar cells generate an average of 16 watts. The project is a collaboration between the Center for Space Physics at Boston University, AeroAstro, the University of Illinois at Urbana-Champaign (UIUC), the Naval Research Laboratory, MIT's Haystack Observatory, and the Phillips Laboratory. In addition, TERRIERS will fly experiments from a high school science class in Cleveland Heights, Ohio. For more information on TERRIERS, visit their web site at http://net.bu.edu /terriers/terriers.html.

SNOE.

Scheduled for launch in February, 1998, SNOE (Student Nitric Oxide Explorer) will investigate how variations in the solar soft x-radiation produce changes in the density of nitric oxide in the lower thermosphere and how auroral activity produces increased nitric oxide in the polar region. Nitric oxide is an important minor constituent of the upper atmosphere that exhibits strong solar-terrestrial coupling. Nitric oxide directly affects the composition of the ionosphere and the thermal structure of the thermosphere, and it may be transported downward into the mesosphere and stratosphere where it can react with ozone. Nitric oxide reacts chemically with ozone to form nitrogen dioxide which in turn reacts with atomic oxygen to reform nitric oxide. This is a catalytic cycle which destroys ozone while leaving the odd-nitrogen intact. Any nitric oxide that is transported downward from the lower thermosphere into the mesosphere and stratosphere may participate in the catalytic destruction of ozone. An opportune time for downward transport to take place is during polar night when photodissociation of nitric oxide does not occur.

The SNOE spacecraft and its instrument complement are designed, built, and operated entirely at the University of Colorado at Boulder Laboratory for Atmospheric and Space Physics (CU/LASP). The principal partner in the Colorado team is Ball Aerospace. The spacecraft, weighing 254 lb., is a compact hexagonal structure, 36" high and 39" across its widest dimension. It is launched into a circular orbit, 580 km altitude, at 97.75 degrees inclination for sun synchronous precession. It spins at 5 rpm with the spin axis normal to the orbit plane. It carries three instruments: an ultraviolet spectrometer to measure nitric oxide altitude profiles, a two-channel auroral photometer to measure auroral emissions beneath the spacecraft, and a five-channel solar soft X-ray photometer. In addition to the scientific instruments, SNOE will also fly a special Global Positioning System (GPS) technology investigation built by the Jet Propulsion Laboratory. Figure 1 is a computer-aided design (CAD) diagram of the SNOE structural system showing the launch vehicle adapter and the solar panels. For more information on SNOE as the mission progresses, visit their web site at http://lasp.colorado.edu.

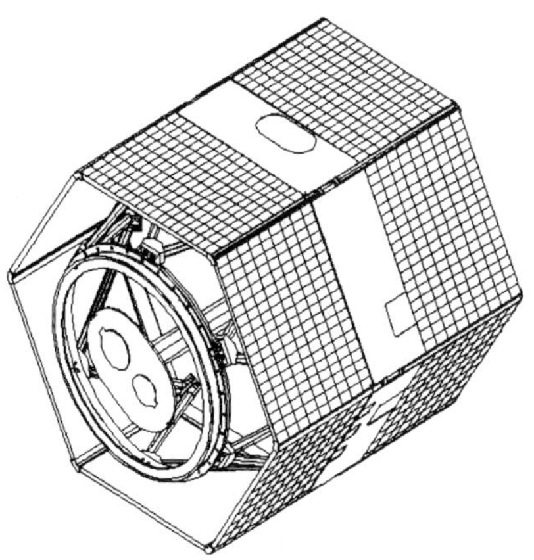

Figure 1.

CATSAT.

A team led by the University of New Hampshire is developing CATSAT (Cooperative Astrophysics and Technology Satellite). Using gamma-ray and X-ray sensors, CATSAT's objective is to estimate the distance and polarization of gamma-ray bursters, an important astronomical problem. The New

Hampshire group is partnered with Weber State University in Utah (spacecraft development and testing) and the University of Leicester in the U.K.(soft X-ray instrument development).

The CATSAT satellite weighs 265 lb. and will be launched into a sun synchronous terminator (dawn/dusk) orbit with an approximate altitude of 550 km and an inclination of 97 degrees. The altitude is a compromise between the minimum backgrounds and particle dose rates from the SAA at lower altitudes and a greater than 1 year useful orbit life. The scientific payload requires that the spacecraft be nadir pointed with an attitude positioning accuracy of ±5 degrees. A terminator orbit offers continuous illumination of the solar panels and enhances power management efficiency. It also allows a nearly 2π steradian field of view for the soft X-ray detectors free of Earth occultations and contaminating background from the Sun.

The scientific instrument portion of CATSAT includes four detector systems having mutually supportive and overlapping capabilities to provide the first direct measurement of the distance to gamma-ray bursters. The three upward looking systems measure the direct burst spectrum over the energy range 500 eV to 10 MeV in three overlapping bands. They are the soft x-ray spectrometer (SXR), the hard x-ray spectrometer (HXR), and the directional gamma-ray spectrometer (DGS). The X-ray albedo polarimeter (XAP) is a set of nine nadir looking collimated detectors that collectively provide full coverage of the visible Earth's surface and are sensitive to the 20 keV - 200 keV energy range.

The CATSAT project was initiated more than a year after the first two missions and is not scheduled to launch until late 1999 or early 2000. For more information on CATSAT, visit their web site at http://www.catsat.sr.unh.edu/catsat.html.

THE STEDI CHALLENGES

Over the course of the STEDI program there have been a number of challenges to face. First it was to prove that there was a great deal of interest in STEDI class missions on the part of the university research community. Then it was to show that the overwhelming response to the announcement represented truly excellent science and technology investigations. Finally, it was to prove that the best of those responses could actually be carried out for slightly more that $4 million, a fraction of what had traditionally been considered as "small" missions. Meeting the low-cost challenge proved to be the most demanding and required careful attention on the part of the STEDI teams. From the outset, cost had to be a major consideration, both in design trade-offs as well as the project management structure. Some of the main considerations which drive costs are mentioned below.

First, some of the design decisions which offer lessons for low-cost projects such as STEDI are:

- Choose limited, focused scientific objectives
- Collect the minimum amount of data necessary to achieve objectives
- Use instruments that require minimum of new development
- Use a simple spinning satellite
- Use sub-systems with a high heritage, but use modern computer hardware
- Take a integrated approach to subsystem and instrument electronics
- Use a basically single-string design but with selective redundancy where it is absolutely needed

Similarly, the management/oversight function can add considerable cost to a program. Ways to avoid this is on STEDI lead to the following recommendations for low-cost programs:

- Small leadership team
- Streamlined project management
- Minimum oversight/interference from sponsoring agency (USRA and NASA)
- Maintain adequate budget reserves
- Close adherence to budget and schedule
- Strict configuration control
- Minimum of formal reviews
- Widely and easily accessible "paperless" documentation for:
    - Configuration control
    - Change control
    - Design reviews
    - Monthly reports
    - Anomaly reporting and tracking

CONCLUSIONS

Two of the STEDI missions are scheduled to be launched in 1998 and are expected to be operational by the time this colloquium proceedings volume is published. As with any new concept, the development of the STEDI program to its present stage of completion (February, 1998) has been a process of learning what works and what doesn't. It has not been without its technical problems; however, the high quality of the professionals on the program and the enthusiasm of bright students eager to learn from them lead us to conclude that STEDI-class missions are ideally suited for development in a university environment. The hardware development time of 2-3 years corresponds to the typical graduate student's (master's degree) tenure. The magnitude and complexity of the tasks are about right for students to handle (properly supervised by experienced professionals), and the university environment is uniquely suited to low-cost constraints. Most importantly, the STEDI program has been an invaluable hands-on educational experience for both undergraduates and graduates alike.

The major problem encountered in our STEDI experience is one over which we had no control – the lack of timely launch vehicle assignments. The original concept of missions being ready to launch within two years of go-ahead drove the formulation of staffing, budgets, and schedules. Then, USRA (and NASA) began to encounter delays in being manifested on the Pegasus XL, which at the beginning of the STEDI program was a new and untried launch vehicle. As a result of these manifesting problems, SNOE has been delayed by 11 months, TERRIERS by 16 months, and CATSAT by 18 months.

Extending the development time is resulting in a cost impact of keeping the university teams together while awaiting a launch slot In addition, key students graduating and moving on to the workforce necessitates training replacements. Although it gives more students the opportunity to participate in the program, it nevertheless disrupts the continuity and deprives students who have worked on the project for two years from seeing the payoff of their labors. In spite of these delays, we are certain that the original concept of a two year development is a good figure and that development costs could be constrained to the original target if the launches had not been delayed.

At this writing, the SNOE launch is only a few days away; however, NASA seems encouraged enough by the STEDI concept that they have issued the first Announcement of Opportunity for the University Explorer (UNEX) program, which is based directly on the STEDI model and which is described in the NASA's submission to the Congress as the follow-on to STEDI. The UNEX program will be funded by NASA's Office of Space Science, which funded the STEDI program.

More recently, NASA's Earth Science enterprise has annnounced their intention to create a new, STEDI-like category of Earth System Science Pathfinder (ESSP) missions directed at the academic community. According to a NASA spokesman, these missions, called Earth Science University Pathfinder, have as their key objective to conduct low cost research while getting more undergraduate and graduate-level students directly involved in spacecraft engineering and Earth Science. They are expected to have a cost ceiling of $5 million ..........Q.E.D.

AUTHOR INDEX

LIST OF PARTICIPANTS

No	Name	Tel / Fax / E-mail	Address
1	Azlin, MD. Said	Tel: 60-5-367-6901 Fax: 60-5-367-7440 E-mail: ceazlin@kop.usm.my	School of Civil Engineering University Sains Malaysia Ipoh, 31750 Tronoh, Malaysia
2	Baltuck, Miriam	Tel: 61-2-6281-8501 Fax: 61-2-6281-8509 E-mail: NASA-rep.Australia@CBR.FOR.CSIRO.AU	NASA PSC 277, APO AP USA
3	Barabash, Stas	Tel: 46-980-79122 Fax: 46-980-79050 E-mail: stas@irf.se	Swedish Institute of Space Physics, Box 812, Kiruna 98128, Sweden
4	Barrington-Brown, James	Tel: 44-1635-46254 Fax: 44-1635-38785 E-mail: brown@sil.com	Space Innovations Limited The Paddock, Hambridge Road Newbury, RG 14 5TQ, UK
5	Byun, Yong-Ik	Tel: 886-3-422-8013 Fax: 886-3-426-2304 E-mail: byun@tien.phy.ncu.edu.tw	National Central University Institute of Astronomy Chung-Li, 32054, Taiwan
6	Chang, C.F.	Tel: 886-2-737-7502 Fax: 886-2-737-7668	National Science Council Taipei, Taiwan
7	Chen, A.J.	Tel: 886-3-425-7770 Fax: 886-3-425-5535 E-mail: ajchen@csrsr.ncu.edu.tw	Center for Space and Remote Sensing Research, National Central University Chung-Li, Taiwan
8	Chen, C.T.	Tel: 886-3-422-7151 ext. 7672 E-mail: ctchen@csr.px.csrsr.nsu.edu.tw	Center for Space and Remote Sensing Research, National Central University Chung-Li, Taiwan
9	Chen, Chang-Hui	Tel: 886-3-426-2302 Fax: 886-3-426-2304 E-mail: rosie@joule.phy.ncu.edu.tw	National Central University Institute of Astronomy Chung-Li, Taiwan
10	Chen, Chien-Jen	Tel: 886-3-579-5873 Fax: 886-3-578-1226 E-mail: cjchen@pidc.gov.tw	Precision Instrument Development Center National Science Council, Hsinchu, Taiwan
11	Chen, L. C.	Tel: 886-3-425-7232 Fax: 886-3-425-5535 E-mail: lcchen@csrsr.ncu.edu.tw	Center for Space and Remote Sensing Research, National Central University Chung-Li, Taiwan
12	Chen, N. Y.	Tel: 886-3-425-7232 Fax: 886-3-425-5535	Center for Space and Remote Sensing Research, National Central University Chung-Li, Taiwan
13	Chen, Wen-Shyang	Tel_1: 886-7-731-0606 ext. 300 Tel_2: 886-7-395-7992 E-mail: wschen@ee05.ccit.edu.tw	Cheng-Shiu College of Technology & Commerce, Department of Electronic Engineering, Kaohsiung 833, Taiwan
14	Chen, Yen-Hsun	Tel: 886-6-652-5633	National Central University Chung-Li, Taiwan
15	Chen, Yung-Chang	Tel: 886-3-573-1153 Fax: 886-3-571-5971 E-mail: ycchen@ee.nthu.edu.tw	National Tsing Hua University Dept. of Electrical Engineering Hsin-Chu, Taiwan

No	Name	Tel / Fax / E-mail	Address
16	Cheng, Yi-Chang	Tel: 886-3-380-9991 ext. 143 Fax: 886-3-380-1407 E-mail: yccheng@cc04.ccit.edu.tw	Department of Electrical Engineering Chung Cheng Institute of Technology P.O. 92247-30 Tahsi, Taoyuan
17	Cheng, Yu-Chih	Tel: 886-6-239-2811 ext. 270 Fax: 886-6-239-1915 E-mail: ot29@nspo.gov.tw	Aerospace Science and Technology Research Center of NCKU, 198, Hsin Shen Street, Kuei Jen, Tainan, Taiwan
18	Chern, Jeng-Shing	Tel: 886-2-673-9638 ext. 355058 Fax: 886-3-471-1304 E-mail: jschern@ms19.hinet.net	Chung Shan Institute of Science and Technology, 6F No.13, Yang Ming 10th St., Taoyuan, Taiwan 330, ROC
19	Chiu, H.H.	Tel: 886-6-275-7575 ext. 63695 Fax: 886-6-238-9940 E-mail: hhchiu@htind1.iaa.ncku.edu.tw	Institute of Aeronautics and Astronautics, National Cheng Kung University, Tainan, Taiwan
20	Choi, Soon Dal	Tel: 82-42-869-8611 Fax: 82-42-861-0523 E-mail: sdchoi@satrec.kaist.ac.kr	Satellite Technology Research Center, KAIST, 373-1 Kusung-dong, Yusung-Gu, Taejon, Korea
21	Chou, Chung-Ping	Tel: 886-6-239-2811 ext. 273 Fax: 886-6-239-1915 E-mail: ot30@nspo.gov.tw	Aerospace Science and Technology Research Center of NCKU, 198, Hsin Shen Street, Kuei Jen, Tainan, Taiwan
22	Chu, David Fei-Hon	Tel: 886-3-578-4208 ext. 9172 Fax: 886-3-577-4015 E-mail: sb01@nspo.gov.tw	National Space Program Office (NSPO), 8F, 9 Prosperity 1st Road, Science-Based Industrial Park, Hsin-Chu City, Taiwan
23	Cohen, Harvey	Tel: 61-3-9479-2596 Fax: 61-3-9479-3060 E-mail: H.Cohen@latrobe.edu.au	La Trobe University, Computer Science, La Trobe Univ. Bundoora Melbourne, Australia
24	De Groot, Wim	Tel: 216-977-7485 Fax: 216-433-8311 E-mail: Wim.deGroot@lerc.nasa.gov	NYMA, Inc., MS 301-3, 21000 Brook Parv. RD. Cleveland 44135, USA
25	Dostal, Pavel	Tel: 420-204-62-0232 Fax: 429-204-64-9314 E-mail: dostal@asu.cas.cz	Astronomical Institute, Au Czech Republic, FRICOVA 1, Ondrejov 25165 Czech Republic
26	Essex, Elizabeth	Tel: 613-9479-1485 Fax: 613-9479-1552 E-mail: e.essex@latrobe.edu.au	La Trobe University, School of Physics, La Trobe, Bundoora, Victoria 3083 Australia
27	Flaherty, Mary Katherine	Tel: 410-792-6000 ext. 3892 Fax: 410-792-5883 E-mail: Kate.flaherty@jhuapl.edu	Johns Hopkins University, Applied Physics Laboratory, 4-326 Johns Hopkins Rd., Laurel, MD 20723, USA
28	Friis-Christensen, Eigil	Tel: 45-3532-5707 Fax: 45-3536-2475 E-mail: Eigil.Friis@dsri.dk	Danish Space Research Institute Juliane Maries Vej 30, DK-2100, Copenhagen, Denmark
29	Hall, David	Tel: 44-171-215-0787 Fax: 44-171-215-0804 E-mail: David_Hall@pparc.ac.uk	British National Space Centre 151 Buckingham Palace Road London, SW1W 9SS, UK
30	Hausler, Bernd	Tel: 49-0-89-6004-2138 Fax: 49-0-89-6004-2138 E-mail: Bernd.Haeusler@Kommsrv.RZ.Unibw-Muenchen.de	Universitat Der Bundeswehr Werner-Heisenberg-Weg 39 Neubiberg, 85577, Germany

No	Name	Tel / Fax / E-mail	Address
31	Hayashi, Tomonao	Tel: 81-474-78-0370 Fax: 81-474-78-0379 E-mail: p04hayat@cc.it-chiba.ac.jp	Chiba Institute of Technology 2-17-1, Tsudanuma, Narashino Chiba 275, Japan
32	Ho, C.H.	Tel: 886-6-275-7575 ext. 63671 Fax: 886-6-238-9940 E-mail: csho@mail.iaa.ncku.edu.tw	Institute of Aeronautics and Astronautics, National Cheng Kung University, Tainan,Taiwan
33	Ho, Chung-Ru	Tel: 886-2-462-2192 ext. 6339 Fax: 886-2-462-0912 E-mail: chungru@sun4.oce.ntou.edu.tw	National Taiwan Ocean University, Department of Oceanography, 2 Pei Ning Road, Keelong 202, Taiwan
34	Hong, Zuu-Chang	Tel: 886-2-392-9289 Fax: 886-2-392-0016 E-mail: pmid0319@unet.net.tw	Department of Mechanical Engineering, National Central University, Taiwan
35	Hsiao, Fei Bin	Tel: 886-6-275-7575 ext. 63667 Fax: 886-6-238-9940 E-mail: fbhsiao@mail.ncku.edu.tw	Institute of Aeronautics and Astronautics, National Cheng Kung University, Tainan, Taiwan
36	Hsiao, Shu-San	Tel: 886-3-426-7341 Fax: 886-3-425-4501 E-mail: sshsian@cc.ncu.edu.tw	National Central University Department of Mechanical Engineering Chung-Li, Taiwan
37	Hsieh, C.	Tel: 886-6-275-7575 ext. 63670 Fax: 886-6-238-9940 E-mail: fc44hsie@mail.iaa.ncku.edu.tw	Institute of Aeronautics and Astronautics, National Cheng Kung University, Tainan, Taiwan
38	Hsieh, Ke Chiang	Tel: 1-520-621-6772 Fax: 1-520-621-4721 E-mail: hsieh@physics.arizona.edu	Dept. of Physics University of Arizona Tucson, 85721 USA
39	Hsieh, S.J.	Tel: 886-6-275-7575 ext. 63650 Fax: 886-6-238-9940 E-mail: sjhsieh@mail.ncku.edu.tw	Institute of Aeronautics and Astronautics National Cheng Kung University Tainan, Taiwan
40	Hsieh, Wen H.	Tel: 886-5-242-8170 Fax: 886-5-272-0589 E-mail: imewhh@ccunix.ccu.edu.tw	Dept. of Mechanical Engineering National Chung Cheng University Chia Yih, Taiwan
41	Hsu , Roy C.	Tel: 886-3-5784208 ext. 1271 Fax: 886-3-5784034 E-mail: sp15@nspo.gov.tw	National Space Program Office (NSPO) 8F, 9 Prosperity 1st Road, Science-Based Industrial Park, Hsin-Chu City, Taiwan
42	Hu, Wen-Hu	Tel: 886-3-426-7379 Fax: 886-3-425-5057 E-mail: s3352099@cc.ncu.edu.tw	National Central University Mechanical Engineering Department No. 38 Chung-Li, Taiwan
43	Hum, Robert	Tel: 613-990-0795 Fax: 613-952-0970 E-mail: Robert.Hum@space.gc.ca	Canadian Space Agency PO Box 7275, Vanier, KIL 8E3 Canada
44	Hwang , Chi Hung	Tel: 886-3-577-9911 ext. 521 Fax: 886-3-577-3947 E-mail: chhwang@pidc.gov.tw	Precision Instrument Development Center National Science Council Hsinchu, Taiwan
45	Ip, Wing	Tel: 886-3-422-7151 ext. 5000 E-mail: wingip@jupiter.ss.ncu.edu.tw	National Central University, Dean of Science College Chung-li, Taiwan

No	Name	Tel / Fax / E-mail	Address
46	Jan, Ying-Wen	Tel: 886-3-578-4208 ext. 9203 Fax: 886-3-577-4015 E-mail: sb26@nspo.gov.tw	8F, 9 Prosperity 1st Road Science-Based Industrial Park Hsin-Chu City, Taiwan
47	Johnstone, Alan	Tel: 44-1483-27-4111 Fax: 44-1483-27-8312 E-mail: adj@mssla3.mss1.ucl.ac.uk	UCL Mullard Space Science Laboratory Holmbury St. Mary, Dorking, Surrey, RH56NT, UK
48	Kang, Che-Shing	Tel: 886-3-578-4208 ext. 9141 Fax: 886-3-577-4015 E-mail: sb17@nspo.gov.tw	8F, 9 Prosperity 1st Road Science-Based Industrial Park Hsin-Chu City, Taiwan
49	Kao, Wei-Wen	Tel: 886-2-737-6497 Fax: 886-2-737-6460 E-mail: wwkao@mail.ntnst.edu.tw	National Taiwan University of Science and Technology, 43 Keelung Rd., Sec. 4, Taipei, Taiwan
50	Kim, Byung Kyo	Tel: 82-42-860-2480 Fax: 82-42-860-2007 E-mail: bkkim@viva.kari.re.kr	Korea Aerospace Research Institute 52 Eoeundong, Yusung-ku Taejon, 305-333, Korea
51	Kim, Ee Eul	Tel: 82-42-869-8639 Fax: 82-42-861-0064 E-mail: eek@krsc.kaist.ac.kr	Satellite Technology Research Center, KAIST, 373-1 Kusung-Dong, Yusung-gu Taejon, Korea
52	Kuo, F.S.	Tel: 886-3-422-7151 ext. 5767 Fax: 886-3-422-4394 E-mail: fskuo@jupiter.ss.ncu.edu.tw	Institute of Space Science National Central University, Chung-Li, Taiwan
53	Kuo, Nan-Jung	Tel: 886-2-462-2192 ext. 6339 Fax: 886-2-462-0912 E-mail: kuonj@sun4.oce.ntou.edu.tw	National Taiwan Ocean University Department of Oceanography 2 Pei Ning Road, Keelong 202, Taiwan
54	Lai, W.H.	Tel: 886-6-275-7575 ext. 63694 Fax: 886-6-238-9940 E-mail: whlai@mail.iaa.ncku.edu.tw	Institute of Aeronautics and Astronautics National Cheng Kung University Tainan, Taiwan
55	Lee, Chin-Chun	Tel: 886-3-380-0301	Department of Electrical Engineering Chung Cheng Institute of Technology Tahsi, Taoyuan, 335, Taiwan
56	Lee, Dong-Hun	Tel: 82-331-280-2449 Fax: 82-331-281-4964 E-mail: dhlee@nms.kyunghee.ac.kr	Kyung Hee University, Dept. of Astronomy and Space Science, Yongin, Kyunggi 449701, Korea
57	Lee, Jun-Ji,	Tel: 886-3-578-4208 ext. 1062 Fax: 886-3-577-9058 E-mail: jjlee@nspo.gov.tw	National Space Program Office (NSPO), 8F, 9 Prosperity 1st Road, SBIP Hsin Chu, Taiwan
58	Lee, Lou C.	Tel: 886-6-275-7575 ext. 65001 Fax: 886-6-237-8377 E-mail: loulee@plasma.et.ncku.edu.tw	Dean of College of Science National Cheng Kung University Tainan, Taiwan
59	Lee, W.I.	Tel: 886-3-571-2121 ext. 6151 Fax: 886-3-572-5230	National Chiao-Tung University 1001 Ta Hsueh Road, Hsin Chu, Taiwan
60	Levy, Eugene	Tel: 520-621-4090 Fax: 520-621-8389 E-mail: ehl@u.arizona.edu	University of Arizona Gould-Simpson 1025 Tucson, AZ 85721-0077 USA

No	Name	Tel / Fax / E-mail	Address
61	Li, Hsein-Wei	Tel: 886-2-462-2192 ext. 6339 Fax: 886-2-462-0912 E-mail: lihw@sun4.oce.ntou.edu.tw	National Taiwan Ocean University Department of Oceanography, 2 Pei Ning Road, Keelong 202, Taiwan
62	Li, Huan Hsin	Tel: 886-3-422-7151 ext. 5969 Fax: 886-3-494-1927 E-mail: huanhsin@ms5.hinet.net	Graduate Institute of Astronomy National Central University Chung-Li, Taiwan
63	Lin, Chin-E	Tel: 886-6-275-7575 ext. 63693 Fax: 886-6-238-9940 E-mail: chinlin@ctsun.iaa.ncku.edu.tw	Institute of Aeronautics and Astronautics, National Cheng Kung University, Tainan, Taiwan
64	Lin, Chia-Chieh	Tel: 886-2-782-1864 ext. 140 Fax: 886-2-782-1864 ext. 140	China Institute of Technology and Commerce, Taipei, Taiwan
65	Lin, Jian-Yang	Tel: 886-5-534-2601 ext. 4319 Fax: 886-5-531-2063 E-mail: linjy@mailsv.yuntech.edu.tw	National Yunlin University of Science and Technology, Dept. of Electronic, Engineering, 123 University Rd., Sec. 3, Toulin, Taiwan 640
66	Lin, Wei-Song	Tel: 886-2-363-5251 ext. 413 Fax: 886-2-363-8247 E-mail: weisong@cc.ee.ntu.edu.tw	Department of Electrical Engineering National Taiwan University No. 1, Sec. 4, Rooserelt Rd., Taipei, Taiwan
67	Lin, Y.Y.	Tel: 886-6-275-7575 ext. 63673 Fax: 886-6-238-9940 E-mail: fc16yylin@mail.iaa.ncku.edu.tw	Institute of Aeronautics and Astronautics, National Cheng Kung University, Tainan, Taiwan
68	Lindqvist, Per-Arne	Tel: 46-879-07696 Fax: 46-824-5431 E-mail: lindqvist@plasma.kth.se	Div. of Plasma Physics, Alfven Laboratory, Royal Institute of Technology SE-100 44 Stockholm, Sweden
69	Liu, C.H.	Tel: 886-3-425-4822 Fax: 886-3-425-4842 E-mail: t341426@twncu865.ncu.edu.tw	President National Central University, Chung-Li, Taiwan
70	Lundin, Rickard	Tel: 46-980-79063 Fax: 46-980-79091 E-mail: richard@irf.se	Swedish Institute of Space Physics Bok 812, Kiruna, S-98128 Sweden
71	Ma, Der-Ming	Tel: 886-2-623-7600 Fax: 886-2-620-9746 E-mail: derming@aero.tku.edu.tw	Dept. of Aerospace Engineering Tamkang University, 151 Ying-Chuan Road, Tamsui, Taiwan
72	Maier, Mark W.	Tel: 205-890-6642 Fax: 205-890-6803 E-mail: maier@ece.uah.edu	Elec. And Computer Engineering University of alabama in Huntsville Huntsville, AL 35899, USA
73	Manouchehri, Khosrow	Tel: 98-21-811-2845 Fax: 98-21-85-8566 E-mail: ops-reps@www.dci.co.ir	Training Center & SAT COM Dept. (TCI), TCI Bldg. #2, 8th Floor, Room #825, Dr. Sharlati Ave., Tehran, Iran
74	McKenna-Lawlor, Susan	Tel: 353-1-628-6788 Fax: 353-1-628-6470 E-mail: STIL@May.IE	Space Technology Ireland, Ltd. National University of Ireland, Maynooth, Co. Kildare, Ireland
75	Miau, J.J.	Tel: 886-6-275-7575 ext. 63688 Fax: 886-6-238-9940 E-mail: ff251212@mail.iaa.ncku.edu.tw	Institute of Aeronautics and Astronautics, National Cheng Kung University, Tainan, Taiwan

No	Name	Tel / Fax / E-mail	Address
76	Min, Kyoungwook	Tel: 82-42-869-2525 Fax: 82-42-869-5525 E-mail: kwmin@space.kaist.ac.kr	Korea Advanced Institute of Science and Tech., 373-1 Kusong-dong, Yusong-gu Taezon 305-701, Korea
77	Nakajima, Atsushi	Tel: 81-422-47-5911 Fax: 81-422-49-8813 E-mail: nakajima@nal.go.jp	National Aerospace Laboratory 7-44-1 Jindaiji-Higashi-Machi, Chofu-shi Tokyo 182, Japan
78	Nakajima, Isao	Tel: 81-463-91-3130 Fax: 81-463-95-5337 E-mail: is2hb@is.icc.u-tokai.ac.ip	Tokai Univ. School of Medicine Boseidai Isehara, Kanagawa 259-11 Japan
79	Norberg, Olle	Tel: 46-980-79119 Fax: 46-980-79050 E-mail: Olle.Norberg@irf.se	Swedish Institution of Space Physics Box 812, Kiruna, SE-98128 Sweden
80	Okamoto, Hiroyuki	Tel: 81-466-28-8041 Fax: 81-466-28-8045 E-mail: hlokmt@mars.jplanet.or.jp	Astro Research, Fujisawa Central Bldg, 4F, 1-1-4 Kugenuma, Tachibana, Fujisawa 251, Japan
81	Olthof, Henk	Tel: 31-71-565-4350 Fax: 31-71-565-4693 E-mail: holthof@estec.esa.nl	ESA/ESTEC, O. Box 299 Noordwijk 2200 AG The Netherlands
82	Ou, S.H.	Tel: 886-6-275-7575 ext. 62001 Fax: 886-6-275-4228	Dean of Engineering College National Cheng Kung University, Tainan, Taiwan
83	Ousley, Gilbert	Tel: 410-290-6083 Fax: 410-290-6028 E-mail: Gousley@BDM.COM	BDM International 9055 Guilford Rd, Columbia, Maryland 21046 USA
84	Ovchinnikov, Michael	Tel: 7-095-972-3714 Fax: 7-095-972-0737 E-mail: ovchinni@spp.keldysh.ru	Keldysh Institute of Applied Mathematics, 4 Miusskaya Square Moscow 125047, Russia
85	Paik, Hong-Yul	Tel: 82-42-860-2410 Fax: 82-42-860-2007 E-mail: phy@viva.kari.re.kr	Satellite Application Dept. Korea Aerospace Research Institute O. Box 113, Yu-Sung Taejon, 305-600 Korea
86	Peresty, Radek	Tel: 420-204-620167 Fax: 420-204-649314 E-mail: peresty@asu.cas.cz	Astronomical Institute of Avcr Fricova 1 Ondreoov 25165 Czech Republic
87	Phipps, Andy	Tel: 44-1483-269278 Fax: 44-1483-259503 E-mail: A.Phipps@ee.surrey.ac.uk	Surrey Satellite Technology Ltd., Centre for Satellite Engineering Research, University of Surrey, Guildford, Surrey, GU 2 5XH, U.K.
88	Reynolds, Edward L.	Tel: 301-953-5101 Fax: 301-953-6635 E-mail: edwardreynolds@jhuap1.edu	Johns Hopkins Applied Physics Laboratory, 11100 Johns Hopkins Road, Laurel Laurel, MD 20723, USA
89	Roeser, Hans Peter	Tel: 49-30-6705-5500 Fax: 49-30-6705-5502 E-mail: Hans-Peter.Roeser@dlr.de	Deutsches Zentrum fur Luft- und Raumfahrt e.V, Rudower Chaussee 5, Berlin 12489 Germany
90	Safavi, Mostafa	Tel: 9821-870-4200 Fax: 9821-882-3902	Amir-Kabir University of Technology, Electrical Engineering Dept. #424 Hafez Avenue, 15875 Tehran, Iran

No	Name	Tel / Fax / E-mail	Address
91	Sandau, Rainer	Tel: 49-30-6705-5530 Fax: 49-30-6705-5532 E-mail: rainer.sandau@dlr.de	Deutsches Zentrum fur Luft- und Raumfahrt e.V. (DLR), Rudower Chaussee 5 Berlin D-12489, Germany
92	Sato, Kenichiro	Tel: 81-46741-6138 Fax: 81-46741-6951 E-mail: killy@grape.kama.melco.co.jp	Advanced Space Systems Group Space, Systems Dept., Kamakura Works, Mitsubishi Electric Gorp., Japan
93	Sheen, Ye-Shun	Tel: 886-7-344-6601 Fax: 886-7-224-4450 E-mail: sen1212@ms5.hinet.net	International Business Group, Chunghwa Telecom Co., Ltd., Kaohsiung, Taiwan
94	Shieh , Ching J.	Tel: 886-3-578-4208 ext. 1121 Fax: 886-3-578-4034 E-mail: cshieh@nspo.gov.tw	National Space Program Office (NSPO), 8576 Boatbill Cil., Fountain Valley, CA 92708, USA
95	Shih, Shun-Peng	Tel: 886-3-427-8798 Fax: 886-3-422-4394 E-mail: spshih@jupiter.ss.ncu.edu.tw	National Central University, Institute of Space Science, Chung-Li, Box 9-97, Taiwan
96	Shyu, Jia-Ming	Tel: 886-3-578-4208 ext. 8121 Fax: 886-3-578-4210 E-mail: shyu@nspo.gov.tw	National Space Program Office (NSPO), 8F, 9 Prosperity Road, Science-Based Industrial, Park Hsin-Chu City, Taiwan
97	Su, Ching-Lun	Tel: 886-3-422-7151 ext. 5779 Fax: 886-3-422-4394 E-mail: clsu@jupiter.ss.ncu.edu.tw	National Central University Institute of Space Science Chung-Li, Box. 9-97, Taiwan
98	Su, Szu-Lin	Tel: 886-6-275-7575 ext. 62367 Fax: 886-6-234-5482 E-mail: ssl@eembox.ee.ncku.edu.tw	Department of Electronical Engineering, National Cheng Kung University, Tainan, Taiwan
99	Sugiki, Mitsuteru	Tel: 81-466-28-8041 Fax: 81-466-28-8045 E-mail: sugiki@mars.jplanet.or.jp	Astro Research Co., Fujisawa Central Bldg, 4F, 1-1-4 Kugenuma, Tachiban, Fujisawa 251, Japan
100	Sung, David	Tel: 886-2-782-1862 ext. 142 Fax: 886-2-788-0423	China Institute of Technology and Commerce Taipei, Taiwan
101	Tekinalp, Ozan	Fax: 90-312-210-1272 Emial: tekinalp@rorqual.cc.metu.edu.tr	Middle East Technical University Aeronautical Engineering Department Inonu Bulvari, Ankara 06531, Turkey
102	Ting, Lee H.	Tel: 886-3-520-5692 Fax: 886-3-577-4015 E-mail: ting@nspo.gov.tw	National Space Program Office (NSPO), 8F, 9 Prosperity 1st Road, Science-Based Industrial Park, Hsin-Chu City, Taiwan
103	Tulunay, Yurdanur	Fax: 90-312-210-1272 E-mail: ytulunay@rorqual.cc.metu.edu.tr	Middle East Technical University Aeronautical Engineering Dept. 06531 Ankara, Turkey
104	Wang , Ann Lee	Tel: 886-3-578-4208 ext. 9413 Fax: 886-3-577-0134 E-mail: pd05@nspo.gov.tw	National Space Program Office (NSPO) 8F, 9 Prosperity 1st Road, Science-Based Industrial Park, Hsin-Chu City, Taiwan
105	Wang, Hung-Chih	Tel: 886-3-577-8273	National Space Program Office (NSPO) 8F, 9 Prosperity 1st Road, SBIP, Hsin Chu, Taiwan

No	Name	Tel / Fax / E-mail	Address
106	Weng, C.I.	Tel: 886-6-275-7575 ext. 50001 Fax: 886-6-274-3981	President National Cheng Kung University Tainan, Taiwan
107	Wu, Jene Chun-Hsien	Tel: 886-3-578-4208 ext. 9205 Fax: 886-3-577-4015 E-mail: itso@nspo.gov.tw	National Space Program Office (NSPO) 8F, 9 Prosperity 1st Road, Science-Based Industrial Park, Hsin-Chu City, Taiwan
108	Wu, Jin	Tel: 886-2-356-5556 ext. 59 Fax: 886-2-397-6920	DePARTMENT OF Hydraulic Egineeing National Cheng Kung University. Tainan, Taiwan
109	Wu, Shi-Tsan	Tel: 205-890-6413 Fax: 205-890-6382 E-mail: wus@cspar.uah.edu	The University of Alabama in Huntsville Engineering Building Room 157-F Huntsville, AL 35899, USA
110	Yang, Ciann-Dong	Tel: 886-6-275-7575 ext. 63669 Fax: 886-6-238-9940 E-mail: cdyang@mail.iaa.ncku.edu.tw	Institute of Aeronautics and Astronautics National Cheng Kung University Tainan, Taiwan
111	Yeh, Fu-Rong	Tel: 886-2-567-0118 Fax: 886-2-523-2793 E-mail: yehfurong@acer.net	Eagle EG&G Aerospace Taiwan No. 7, Lane 144, Chi Lin Road, Taipei, Taiwan
112	Yeh, Huey C.	Tel: 886-3-422-7151 ext. 7645 Fax: 886-3-425-5535 E-mail$_1$: yeh@jupiter.ss.ncu.edu.tw E-mail$_2$: yeh@csrsddc.csrsr.ncu.edu.tw	National Central University Institute of Space Science Chung-Li, Taiwan
113	Yong, Kay	Tel: 886-6-275-7575 ext. 63652 Fax: 886-6-238-9940 E-mail: kayyong@mail.iaa.ncku.edu.tw	Institute of Aeronautics and Astronautics National Cheng Kung University Tainan, Taiwan
114	Yuan, H.F.	Tel: 886-6-238-9127 Fax: 886-6-238-9940 E-mail: yahn@mail.ncku.edu.tw	Institute of Aeronautics and Astronautics National Cheng Kung University Tainan, Taiwan
115	Yusuke, Matsumura	Tel: 81-298-522405 Fax: 81-298-522247 E-mail: matsumura.yusuke@nasda.go.jp	National Space Development Agency of Japan, Tsukuba Space Center 2-1-1 Sengen, Tsukuba, Ibaraki 305, Japan